ENCYCLOPEDIA OF MATHEMATICS AND ITS APPLICATIONS

EDITED BY G.-C. ROTA

Volume 76

Geometry of Sporadic Groups I
Petersen and tilde geometries

ENCYCLOPEDIA OF MATHEMATICS AND ITS APPLICATIONS

4 W. Miller, Jr. *Symmetry and separation of variables*
6 H. Minc *Permanents*
11 W. B. Jones and W. J. Thron *Continued fractions*
12 N. F. G. Martin and J. W. England *Mathematical theory of entropy*
18 H. O. Fattorini *The Cauchy problem*
19 G. G. Lorentz, K. Jetter and S. D. Riemenschneider *Birkhoff interpolation*
21 W. T. Tutte *Graph theory*
22 J. R. Bastida *Field extensions and Galois theory*
23 J. R. Cannon *The one-dimensional heat equation*
25 A. Salomaa *Computation and automata*
26 N. White (ed.) *Theory of matroids*
27 N. H. Bingham, C. M. Goldie and J. L. Teugels *Regular variation*
28 P. P. Petrushev and V. A. Popov *Rational approximation of real functions*
29 N. White (ed.) *Combinatorial geometrics*
30 M. Pohst and H. Zassenhaus *Algorithmic algebraic number theory*
31 J. Aczel and J. Dhombres *Functional equations containing several variables*
32 M. Kuczma, B. Chozewski and R. Ger *Iterative functional equations*
33 R. V. Ambartzumian *Factorization calculus and geometric probability*
34 G. Gripenberg, S.-O. Londen and O. Staffans *Volterra integral and functional equations*
35 G. Gasper and M. Rahman *Basic hypergeometric series*
36 E. Torgersen *Comparison of statistical experiments*
37 A. Neumaier *Intervals methods for systems of equations*
38 N. Korneichuk *Exact constants in approximation theory*
39 R. A. Brualdi and H. J. Ryser *Combinatorial matrix theory*
40 N. White (ed.) *Matroid applications*
41 S. Sakai *Operator algebras in dynamical systems*
42 W. Hodges *Model theory*
43 H. Stahl and V. Totik *General orthogonal polynomials*
44 R. Schneider *Convex bodies*
45 G. Da Prato and J. Zabczyk *Stochastic equations in infinite dimensions*
46 A. Bjorner, M. Las Vergnas, B. Sturmfels, N. White and G. Ziegler *Oriented matroids*
47 E. A. Edgar and L. Sucheston *Stopping times and directed processes*
48 C. Sims *Computation with finitely presented groups*
49 T. Palmer *Banach algebras and the general theory of *-algebras*
50 F. Borceux *Handbook of categorical algebra I*
51 F. Borceux *Handbook of categorical algebra II*
52 F. Borceux *Handbook of categorical algebra III*
54 A. Katok and B. Hassleblatt *Introduction to the modern theory of dynamical systems*
55 V. N. Sachkov *Combinatorial methods in discrete mathematics*
56 V. N. Sachkov *Probabilistic methods in discrete mathematics*
57 P. M. Cohn *Skew Fields*
58 Richard J. Gardner *Geometric tomography*
59 George A. Baker, Jr. and Peter Graves-Morris *Padé approximants*
60 Jan Krajicek *Bounded arithmetic, propositional logic, and complex theory*
61 H. Gromer *Geometric applications of Fourier series and spherical harmonics*
62 H. O. Fattorini *Infinite dimensional optimization and control theory*
63 A. C. Thompson *Minkowski geometry*
64 R. B. Bapat and T. E. S. Raghavan *Nonnegative matrices and applications*
65 K. Engel *Sperner theory*
66 D. Cvetkovic, P. Rowlinson and S. Simic *Eigenspaces of graphs*
67 F. Bergeron, G. Labelle and P. Leroux *Combinatorial species and tree-like structures*
68 R. Goodman and N. Wallach *Representations of the classical groups*
70 A. Pietsch and J. Wenzel *Orthonormal systems and Banach space geomery*

ENCYCLOPEDIA OF MATHEMATICS AND ITS APPLICATIONS

Geometry of Sporadic Groups I
Petersen and tilde geometries

A. A. IVANOV
Imperial College, London

CAMBRIDGE UNIVERSITY PRESS
Cambridge, New York, Melbourne, Madrid, Cape Town, Singapore, São Paulo

Cambridge University Press
The Edinburgh Building, Cambridge CB2 8RU, UK

Published in the United States of America by Cambridge University Press, New York

www.cambridge.org
Information on this title: www.cambridge.org/9780521413626

First published 1999
This digitally printed version 2008

A catalogue record for this publication is available from the British Library

Library of Congress Cataloguing in Publication data
Ivanov, A. A.
Geometry of sporadic groups 1, Petersen and tilde geometries / A.A. Ivanov.
p. cm.
Includes bibliographical references and index.
Contents: v. 1. Petersen and tilde geometries
ISBN 0 521 41362 1 (v. 1 : hb)
1. Sporadic groups (Mathematics). I. Title.
QA177.I93 1999
512′.2–dc21 98-45455 CIP

ISBN 978-0-521-41362-6 hardback
ISBN 978-0-521-06283-1 paperback

Contents

Preface

Sporadic simple groups are the most fascinating objects in modern algebra. The discovery of these groups and especially of the Monster is considered to be one of the most important contributions of the classification of finite simple groups to mathematics. Some of the sporadic simple groups were originally realized as automorphism groups of certain combinatorial–geometrical structures like Steiner systems, distance-regular graphs, Fischer spaces *etc.*, but it was the epoch-making paper [Bue79] by F. Buekenhout which brought an axiomatic foundation for these and related structures under the name "diagram geometries". Buildings of finite groups of Lie type form a special class of diagram geometries known as Tits geometries. This gives a hope that diagram geometries might serve as a background for a uniform treatment of all finite simple groups.

If G is a finite group of Lie type in characteristic p, then its Tits geometry $\mathcal{G}(G)$ can be constructed as the coset geometry with respect to the maximal parabolic subgroups which are maximal overgroups of the normalizer in G of a Sylow p-subgroup (this normalizer is known as the Borel subgroup). Thus $\mathcal{G}(G)$ can be defined in abstract group-theoretical terms. Similar abstract construction applied to sporadic simple groups led to maximal [RSm80] and minimal [RSt84] parabolic geometries, most naturally associated with the sporadic simple groups. Notice that besides the parabolic geometries there are a number of other nice diagram geometries associated with sporadic groups.

Tits geometries are characterized by the property that all their rank 2 residues are generalized polygons. Geometries of sporadic groups besides the generalized polygons involve c-geometries (which are geometries of vertices and edges of complete graphs), the geometry of the Petersen graph, tilde geometry (a triple cover of the generalized quadrangle of order (2,2)) and a few other rank 2 residues.

In the mid 80's the classification project of finite Tits geometries attracted a lot of interest, motivated particularly by the revision program of the classification of finite simple groups (see [Tim84]). It was natural to extend this project to geometries of sporadic groups and to try to characterize such geometries by their diagrams. For two classes of diagrams, namely

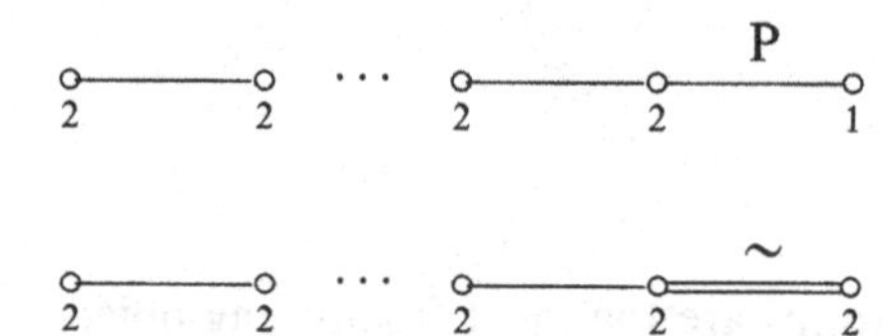

and

the complete classification under the flag-transitivity assumption was achieved by S.V. Shpectorov and the author of the present volume [ISh94b]. Geometries with the above diagrams are called, respectively, Petersen and tilde geometries. A complete self-contained exposition of the classification of flag-transitive Petersen and tilde geometries is the main goal of the two volume monograph of which the present is the first volume.

To provide the reader with an idea what sporadic group geometries look like we present the axioms for the smallest case.

A Petersen geometry of rank 3 is a 3-partite graph $\mathcal{G}$ with the partition

$$\mathcal{G} = \mathcal{G}^1 \cup \mathcal{G}^2 \cup \mathcal{G}^3$$

which possesses the following properties. For a vertex $x \in \mathcal{G}$ let $\mathrm{res}(x)$ denote the subgraph in $\mathcal{G}$ induced on the set of vertices adjacent to x. For $x_i \in \mathcal{G}^i$, $1 \leq i \leq 3$, the following hold:

$\mathrm{res}(x_1)$ is the incidence graph of vertices and edges of the Petersen graph

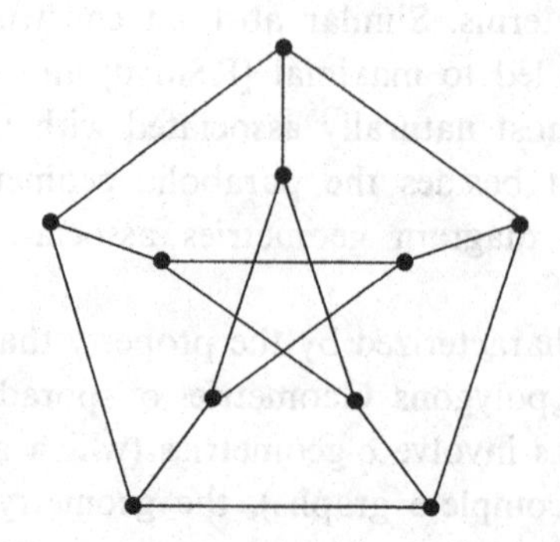

res(x_2) is the complete bipartite graph $K_{3,3}$

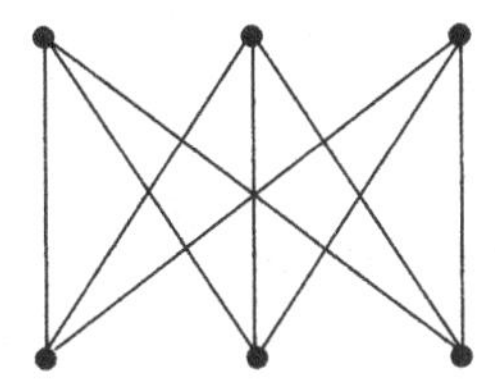

res(x_3) is the incidence graph of seven points and seven lines of the Fano plane

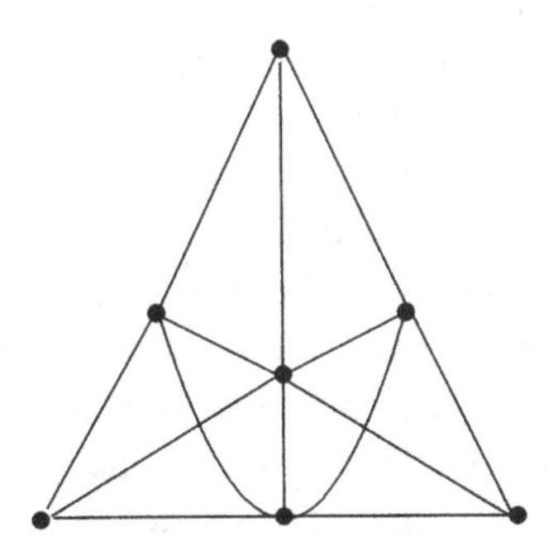

The geometry $\mathcal{G}$ as above is flag-transitive if its automorphism group acts transitively on the set of maximal complete subgraphs (such a subgraph contains one vertex from each part).

It was shown by S.V. Shpectorov in [Sh85] that there exist exactly two flag-transitive Petersen geometries of rank 3. Their automorphism groups are isomorphic respectively to the automorphism group $\mathrm{Aut}\, Mat_{22}$ of the Mathieu group Mat_{22} and to a non-split extension of $\mathrm{Aut}\, Mat_{22}$ by a subgroup of order 3. This was the first step in the classification project of Petersen and tilde geometries.

Our strategy of classification, first implemented in [Sh85], is based on analysis of amalgams of maximal parabolic subgroups and calculation of the universal covers and consists of two principal and rather independent steps.

Step 1. To describe all known pairs $(\mathcal{G}, G)$ where $\mathcal{G}$ is a Petersen or tilde geometry and G is a flag-transitive automorphism group of $\mathcal{G}$, calculate the universal cover of $\mathcal{G}$ and determine its flag-transitive quotients.

Step 2. To show that the amalgam of maximal parabolic subgroups, corresponding to a flag-transitive action on a Petersen or tilde geometry

$\mathcal{H}$, is isomorphic to such an amalgam corresponding to a pair $(\mathcal{G}, G)$ described in Step 1. By a standard principle this means that $\mathcal{H}$ is a flag-transitive quotient of the universal cover of $\mathcal{G}$.

The main goal of the present volume is to realize Step 1. The local analysis of amalgams needed for Step 2 will be given in the second volume. Here we also discuss various applications and implications of the classification of flag-transitive Petersen and tilde geometries.

In Chapter 1 we start with a review of the main notions and principles concerning the diagram geometries and their flag-transitive automorphism groups. Then we formulate and discuss the results of the classification project for flag-transitive Petersen and tilde geometries. In Chapter 2 we prove the existence and uniqueness of the (binary) Golay code and the Steiner system $S(5,8,24)$. Our approach is a mixture of the approach of Conway (in [Con71]) who constructs the Golay code as the quadratic residue code over $GF(23)$ and the approach in Lüneburg (in [Lün69]) who treats the Steiner system $S(5,8,24)$ as an extension of the projective plane of order 4. The approach provides us with a strong background to define the Mathieu groups and to study their subgroup structure. In Chapter 3 we define and study geometries of the Mathieu groups. We refer to computer calculations performed independently in [Hei91] and [ISh89a] to claim the simple connectedness of the tilde geometry $\mathcal{G}(Mat_{24})$. The simple connectedness proofs for the Petersen geometries $\mathcal{G}(Mat_{22})$ and $\mathcal{G}(Mat_{23})$ which we present here are basically the original ones from [Sh85] and [ISh90a]. In Chapter 4 we follow [Con69] and [KKM91] to establish the existence and uniqueness of the Leech lattice. This approach immediately gives the order and basic properties of the automorphism group of the Leech lattice. We present a detailed study of the action of Co_1 on $\bar{\Lambda}_4$ and of an orbital graph associated with this action. This graph is the collinearity graph of the tilde geometry $\mathcal{G}(Co_1)$. We present the simple connectedness proofs for $\mathcal{G}(Co_2)$ and $\mathcal{G}(Co_1)$, originally given in [Sh92] and [Iv92a], respectively. At the end of Chapter 4 we discuss geometries of certain subgroups in the Conway group Co_1. In Chapter 5 we prove the simple connectedness of the tilde geometry $\mathcal{G}(M)$ of the Monster. We start with an amalgam $\mathcal{M}$ similar to the amalgam of maximal parabolics associated with the action of the Monster on its tilde geometry and consider a faithful completion G of $\mathcal{M}$. We define a number of subgroups in G associated with certain subgeometries in $\mathcal{G}(M)$. Applying the simple connectedness of these subgeometries originally established in [Iv92c], [Iv94] and [Iv95] we identify in G the subgroups $3 \cdot M(24)$ and $2 \cdot BM$. By considering the subgeometry in $\mathcal{G}(M)$ formed by the fixed points of an element of order 7, we construct the tilde geometry $\mathcal{G}(He)$ of the Held group. We define a graph Γ on the set of Baby Monster subgeometries in $\mathcal{G}(M)$ (called the second Monster graph) and study its local properties. We apply the triangulability of Γ proved in [ASeg92] to establish the simple connectedness of $\mathcal{G}(M)$. In Chapter 6 we follow [ISh93a] to construct an infinite family of tilde geometries associated with some non-split extensions of symplectic groups over $GF(2)$. In the last section of Chapter 6 we follow [ISh90a] to prove the non-existence of tilde analogues of the exceptional C_3-geometry $\mathcal{G}(Alt_7)$. In Chapter 7 we construct the Petersen geometries associated with the non-split extensions

$$3 \cdot \mathrm{Aut}\, Mat_{22}, \quad 3^{23} \cdot Co_2, \quad 3^{4371} \cdot BM$$

and prove their 2-simple connectedness following [Sh92] and [ISh93b]. In Chapter 8 we discuss the identification proof of Y_{555} with the Bimonster. In this proof the simple connectedness of $\mathcal{G}(M)$ plays an essential rôle. In Chapter 9 we consider locally projective graphs and show how the classification of the flag-transitive Petersen geometries implies description of a class of locally projective graphs of girth 5. Originally this reduction was proved in [Iv88], [Iv90] (see also a survey [Iv93a]). In this volume we do not treat the fourth Janko group J_4 and its Petersen geometry $\mathcal{G}(J_4)$, and refer the reader to [IMe93] where the group and its geometry are constructed and characterized starting with very basic principles.

I would like to thank S.V. Shpectorov for the fruitful cooperation on the classification project for flag-transitive Petersen and tilde geometries which led to its completion. I am grateful to B. Baumeister, S. Hobart, G. Glauberman, C.E. Praeger, C. Wiedorn who read various parts of preliminary versions of the volume and suggested a number of corrections. I am glad to acknowledge that many suborbit diagrams presented in the volume have been computed by D.V. Pasechnik.

1
Introduction

In this introductory chapter we collect basic definitions, formulate main results and discuss some of the motivations and consequences. In Section 1.1 we start with an informal review of classical geometries in order to motivate the general notion of geometry as introduced by J. Tits in the 50's. In Section 1.2 we discuss morphisms of geometries and two of their most important special cases, coverings and automorphisms. Our main interest is in flag-transitive geometries. By a standard principle a flag-transitive geometry $\mathcal{G}$ can be uniquely reconstructed from its flag-transitive automorphism group G and the embedding in G of the amalgam $\mathcal{A}$ (defined in Section 1.3) of maximal parabolic subgroups corresponding to the action of G on $\mathcal{G}$. In Section 1.4 we formulate a condition under which an abstract group G and a subamalgam $\mathcal{A}$ in G lead to a geometry. In Section 1.5 we formulate the most fundamental principle in the area which relates the universal cover of a flag-transitive geometry $\mathcal{G}$ and the universal completion of the amalgam of maximal parabolic subgroups corresponding to a flag-transitive action on $\mathcal{G}$. In Section 1.6 we discuss parabolic geometries of finite groups of Lie type. These geometries belong to the class of so-called Tits geometries characterized by the property that all rank 2 residues are classical generalized polygons. We formulate the local characterization of Tits geometries which shows a special rôle of C_3-geometries. We also formulate a very useful description of flag-transitive automorphism groups of classical Tits geometries due to G. Seitz. A very important non-classical Tits geometry, known as the Alt_7-geometry, is discussed in Section 1.7. In Section 1.8 we apply the characterization of Tits geometries to $C_n(2)$-geometries which play a very special rôle in our exposition. In Section 1.9 we mimic the construction of $C_n(2)$-geometries of symplectic groups to produce a rank 5 tilde geometry of the Monster group. In Section 1.10 the classification result for flag-transitive Petersen and tilde geometries is stated, which shows in particular that the Monster is strongly characterized as a flag-transitive automorphism group of a rank 5 tilde geometry. In Section 1.11 we introduce and discuss a very important notion of natural representations of geometries. Section 1.12 contains a brief historical essay about the classification of flag-transitive Petersen and tilde geometries. In Section 1.13 we present some implications of the classification including the identification of Y-groups. In the final section of the chapter we fix our terminology and notation concerning groups, graphs and geometries. The terminology and notation are mostly standard and we start using them in the earlier sections of the chapter without explanations.

1.1 Basic definitions

We start this chapter with a brief and informal review of the geometries of classical groups in order to motivate the general definition of geometries.

Let G be a finite classical group (assuming the projective version). The group G itself and its geometry can be defined in terms of the natural module which is an n-dimensional vector space $V = V_n(q)$ over the Galois field $GF(q)$ of order q. Here q is a power of a prime number p called the *characteristic* of the field. There is a sesquilinear form Ψ on V which is either trivial (identically equal to zero) or non-singular and the elements of G are projective transformations of V which preserve Ψ up to multiplication by scalars. If Ψ is trivial then G is just a projective linear group associated with V. If Ψ is non-singular, it is symplectic, unitary or orthogonal and G is the symplectic, unitary or orthogonal group of a suitable type determined by n, q and the type of Ψ. We have introduced the trivial form in the case of linear groups in order to treat all classical groups uniformly.

For a subspace W of V we can consider the restriction of Ψ to W. The subspaces on which Ψ restricts trivially play a very special rôle and they are called *totally singular subspaces* of V with respect to Ψ. Clearly every subspace of a totally singular subspace is also totally singular and in the case of linear groups all subspaces are totally singular. If Ψ is a non-singular form then by the Witt theorem all maximal totally singular subspaces have the same dimension known as the *Witt index* of Ψ.

The geometry $\mathcal{G} = \mathcal{G}(G)$ of a classical group G is the set of all proper totally singular subspaces in the natural module V with respect to the invariant form Ψ together with a symmetrical binary incidence relation $*$ under which two subspaces are incident if and only if one of them contains the other one. In the case of a linear group we obtain the projective geometry associated with the natural module and in the remaining cases we obtain various polar spaces.

By the definition every element of a classical group geometry is incident to itself which means that the relation $*$ is reflexive. One can consider $\mathcal{G}$ as a graph on the set of elements whose edges are pairs of incident elements. Since two subspaces of the same dimension are incident if and only if they coincide, one can see (ignoring the loops) that the graph is multipartite. Two vertices are contained in the same part if and only if they have the same dimension as subspaces of V. It is natural to define the type of an element to be the dimension of the corresponding subspace. The Witt theorem and its trivial analogue for the case of linear groups imply that every maximal set of pairwise incident elements of $\mathcal{G}$ (a maximal clique in graph-theoretical terms) contains exactly one element of each type. This construction suggests the definition of geometry as introduced by J. Tits in the 1950s.

Geometries form a special class of *incidence systems*. An incidence system is a quadruple $(\mathcal{G}, *, t, I)$ where $\mathcal{G}$ is the set of elements, $*$ is a binary reflexive symmetric incidence relation on $\mathcal{G}$ and t is a type function which prescribes for every element from $\mathcal{G}$ its type which is an element from the set I of possible types; two different elements of the same type are never incident. We will usually refer to an incidence system $(\mathcal{G}, *, t, I)$ simply by writing $\mathcal{G}$, assuming that $*$, t and I are clear from the context. The number of types in an incidence system (that is the size of I) is called the *rank*. Unless stated otherwise, we will always assume that $I = \{1, 2, ..., n\}$ for an incidence system of rank n and write $\mathcal{G}^i$ for the set of elements of type i in $\mathcal{G}$, that is for $t^{-1}(i)$.

An incidence system $\mathcal{G}$ of rank n can be considered (ignoring loops) as an n-partite graph with parts $\mathcal{G}^1, ..., \mathcal{G}^n$. An incidence system is *connected* if it is connected as a graph.

A set Φ of pairwise incident elements in an incidence system is called a *flag*. In this case $|\Phi|$ and $t(\Phi)$ are the *rank* and the *type* of Φ, respectively. If $\mathcal{G}$ is an incidence system of rank n over the set I of types then $n - |\Phi|$ and $I \setminus t(\Phi)$ are the *corank* and the *cotype* of Φ, respectively. Let Φ be a flag in an incidence system $\mathcal{G}$. The *residual incidence system* $\mathrm{res}_{\mathcal{G}}(\Phi)$ of Φ in $\mathcal{G}$ (or simply *residue*) is the quadruple $(\mathcal{G}_\Phi, *_\Phi, t_\Phi, I_\Phi)$ where

$$\mathcal{G}_\Phi = \{x \mid x \in \mathcal{G},\ x * y \text{ for every } y \in \Phi\} \setminus \Phi,$$

$I_\Phi = I \setminus t(\Phi)$, $*_\Phi$ is the restriction of $*$ to $\mathcal{G}_\Phi$ and t_Φ is the restriction of t to $\mathcal{G}_\Phi$ The notion of residue corresponds to that of link, more common in topology. For a flag consisting of a single element x its residue will be denoted by $\mathrm{res}_{\mathcal{G}}(x)$ rather than by $\mathrm{res}_{\mathcal{G}}(\{x\})$. It is easy to see that one can construct an arbitrary residue inductively, producing at each step the residue of a single element.

Definition 1.1.1 *A geometry is an incidence system* $(\mathcal{G}, *, t, I)$ *for which the following two conditions hold:*

(i) *every maximal flag contains exactly one element of each type;*
(ii) *for every* $i, j \in t(\mathcal{G})$ *the graph on* $\mathcal{G}^i \cup \mathcal{G}^j$ *in which two elements are adjacent if they are incident in* $\mathcal{G}$ *is connected, and a similar condition holds for every residue in* $\mathcal{G}$ *of rank at least* 2.

The graph on the set of elements of a geometry $\mathcal{G}$ in which two distinct elements are adjacent if they are incident in $\mathcal{G}$ is called the *incidence graph* of $\mathcal{G}$. The incidence graphs of geometries of rank n are characterized as n-partite graphs with the following properties: (i) every maximal clique contains exactly one vertex from each part; (ii) the subgraph induced by any two parts is connected and a similar connectivity condition holds for every residue of rank at least 2. It is easy to see that a residue of a geometry is again a geometry.

Let $(\mathcal{G}_1, *_1, t_1, I_1)$ and $(\mathcal{G}_2, *_2, t_2, I_2)$ be two geometries whose sets of elements and types are disjoint. The *direct sum* of $\mathcal{G}_1$ and $\mathcal{G}_2$ is a geometry whose element set is $\mathcal{G}_1 \cup \mathcal{G}_2$, whose set of types is $I_1 \cup I_2$, whose incidence relation and type function coincide respectively with $*_i$ and t_i when restricted to $\mathcal{G}_i$ for $i = 1$ and 2 and where every element from $\mathcal{G}_1$ is incident to every element from $\mathcal{G}_2$.

The above definitions of residue and direct sum have the following motivation in the context of geometries of classical groups. Let G be a classical group with a natural module V and the invariant form Ψ. Let $\mathcal{G} = \mathcal{G}(G)$ be the geometry of G as defined above. Let W be an element of $\mathcal{G}$ that is a totally singular subspace of V with respect to Ψ. It is easy to see that $\mathrm{res}_{\mathcal{G}}(W)$ is the direct sum of two geometries $\mathrm{res}_{\mathcal{G}}^{-}(W)$ and $\mathrm{res}_{\mathcal{G}}^{+}(W)$, where the former is the projective geometry of all proper subspaces of W and the latter is formed by the totally singular subspaces containing W and can be described as follows. Let

$$W^{\perp} = \{v \mid v \in V, \Psi(v, w) = 0 \text{ for every } w \in W\}$$

be the orthogonal complement of W. Then $W \leq W^{\perp}$ and Ψ induces on $U = W^{\perp}/W$ a non-singular form Ψ'. The elements of $\mathrm{res}_{\mathcal{G}}^{+}(W)$ are the subspaces of U totally singular with respect to Ψ' with the incidence relation given by inclusion. So $\text{res}^+_{\mathcal{G}}(W)$ is the geometry of the classical group having U as natural module and Ψ' as invariant form. Certainly $\text{res}^-_{\mathcal{G}}(W)$ or $\text{res}^+_{\mathcal{G}}(W)$ or both can be empty and one can easily figure out when this happens. In any case the observation is that the class of direct sums of geometries of classical groups is closed under taking residues.

By introducing geometries of classical groups we started considering the totally isotropic subspaces of their natural modules as abstract elements preserving from their origin in the vector space the incidence relation and type function. It turns out that in most cases the vector space can be uniquely reconstructed from the geometry and moreover the geometry itself to a certain extent is characterized by its local properties, namely by the structure of residues. The theory and classification of geometries can be developed quite deeply without making any assumption on their automorphism groups. But our primary interest is in so-called flag-transitive geometries to be introduced in the next section.

1.2 Morphisms of geometries

Let $\mathcal{H}$ and $\mathcal{G}$ be geometries (or more generally incidence systems). A *morphism* of geometries is a mapping $\varphi : \mathcal{H} \to \mathcal{G}$ of the element set of $\mathcal{H}$ into the element set of $\mathcal{G}$ which maps incident pairs of elements onto incident pairs and preserves the type function. A bijective morphism is called an *isomorphism.*

A surjective morphism $\varphi : \mathcal{H} \to \mathcal{G}$ is said to be a *covering* of $\mathcal{G}$ if for every non-empty flag Φ of $\mathcal{H}$ the restriction of φ to the residue $\text{res}_{\mathcal{H}}(\Phi)$ is an isomorphism onto $\text{res}_{\mathcal{G}}(\varphi(\Phi))$. In this case $\mathcal{H}$ is a *cover* of $\mathcal{G}$ and $\mathcal{G}$ is a *quotient* of $\mathcal{H}$. If every covering of $\mathcal{G}$ is an isomorphism then $\mathcal{G}$ is said to be *simply connected.* Clearly a morphism is a covering if its restriction to the residue of every element (considered as a flag of rank 1) is an isomorphism. If $\psi : \widetilde{\mathcal{G}} \to \mathcal{G}$ is a covering and $\widetilde{\mathcal{G}}$ is simply connected, then ψ is the *universal covering* and $\widetilde{\mathcal{G}}$ is the *universal cover* of $\mathcal{G}$. The universal cover of a geometry exists and it is uniquely determined up to isomorphism. If $\varphi : \mathcal{H} \to \mathcal{G}$ is any covering then there exists a covering $\chi : \widetilde{\mathcal{G}} \to \mathcal{H}$ such that ψ is the composition of χ and φ.

A morphism $\varphi : \mathcal{H} \to \mathcal{G}$ of arbitrary incidence systems is called an *s-covering* if it is an isomorphism when restricted to every residue of rank at least s. This means that if Φ is a flag whose cotype is less than or equal to s, then the restriction of φ to $\text{res}_{\mathcal{H}}(\Phi)$ is an isomorphism. An incidence system, every s-cover of which is an isomorphism, is said to be *s-simply connected.* The universal s-cover of a geometry exists in the class of incidence systems and it might or might not be a geometry. In the present work we will mainly use the notion of s-covers either to deal with concrete morphisms of geometries or to establish s-simple connectedness. For these purposes we can stay within the class of geometries. It must be clear that in the case $s = n - 1$ "s-covering" and "covering" mean the same thing.

An isomorphism of a geometry onto itself is called an *automorphism*. By the definition an isomorphism preserves the types. Sometimes we will need a more general type of automorphisms which permute types. We will refer to them as *diagram automorphisms*.

The set of all automorphisms of a geometry $\mathcal{G}$ obviously forms a group called the *automorphism group* of $\mathcal{G}$ and denoted by $\mathrm{Aut}\,\mathcal{G}$. An automorphism group G of $\mathcal{G}$ (that is a subgroup of $\mathrm{Aut}\,\mathcal{G}$) is said to be *flag-transitive* if any two flags Φ_1 and Φ_2 in $\mathcal{G}$ of the same type (that is with $t(\Phi_1) = t(\Phi_2)$) are in the same G-orbit. Clearly an automorphism group is flag-transitive if and only if it acts transitively on the set of maximal flags in $\mathcal{G}$. A geometry $\mathcal{G}$ possessing a flag-transitive automorphism group is called *flag-transitive*.

A flag-transitive geometry can be described in terms of certain subgroups and their cosets in a flag-transitive automorphism group in the following way. Let $\mathcal{G}$ be a geometry of rank n and G be a flag-transitive automorphism group of $\mathcal{G}$. Let $\Phi = \{x_1, x_2, ..., x_n\}$ be a maximal flag in $\mathcal{G}$ where x_i is of type i. Let $G_i = G(x_i)$ be the stabilizer of x_i in G. The subgroups G_1, G_2, ..., G_n are called the *maximal parabolic subgroups* or just *maximal parabolics* associated with the action of G on $\mathcal{G}$. When talking about n maximal parabolic subgroups associated with an action on a rank n geometry we will always assume that the elements which they stabilize form a maximal flag. By the flag-transitivity assumption G acts transitively on the set $\mathcal{G}^i$ of elements of type i in $\mathcal{G}$. So there is a canonical way to identify $\mathcal{G}^i$ with the set of right cosets of G_i in G by associating with $y \in \mathcal{G}^i$ the coset $G_i h$ such that $x_i^h = y$. This coset consists of all the elements of G which map x_i onto y (assuming that action is on the right). Now with y as above let z be an element of type j which corresponds to the coset $G_j k$. By the flag-transitivity assumption y and z are incident if and only if there is an element g in G which maps the pair (x_i, x_j) onto the pair (y, z). It is obvious that g must be in the intersection $G_i h \cap G_j k$ and each element from the intersection can be taken for g. Thus y and z are incident if and only if the cosets $G_i h$ and $G_j k$ have a non-empty intersection. Notice that if the intersection is non-empty, it is a right coset of $G_i \cap G_j$. In this way we arrive at the following.

Proposition 1.2.1 *Let $\mathcal{G}$ be a geometry of rank n over the set $I = \{1, 2, ..., n\}$ of types and G be a flag-transitive automorphism group of $\mathcal{G}$. Let $\Phi = \{x_1, x_2, ..., x_n\}$ be a maximal flag in $\mathcal{G}$ and $G_i = G(x_i)$ be the stabilizer of x_i in G. Let $\mathcal{G}(G)$ be the incidence system whose elements of type i are the right cosets of G_i in G and in which two elements are incident if and only if the intersection of the corresponding cosets is non-empty. Then $\mathcal{G}(G)$ is a geometry and the mapping*

$$\eta : y \mapsto G_i h$$

(where $y \in \mathcal{G}^i$ and $x_i^h = y$) establishes an isomorphism of $\mathcal{G}$ onto $\mathcal{G}(G)$. □

1.3 Amalgams

Discussions in the previous section and particularly (1.2.1) lead to the following.

Definition 1.3.1 *A (finite)* amalgam $\mathcal{A}$ *of rank n is a finite set H such that for every $1 \leq i \leq n$ there are a subset H_i in H and a binary operation $*_i$ defined on H_i such that the following conditions hold:*

(i) *$(H_i, *_i)$ is a group for $1 \leq i \leq n$;*
(ii) *$H = \bigcup_{i=1}^n H_i$;*
(iii) *$\bigcap_{i=1}^n H_i \neq \emptyset$;*
(iv) *if $x, y \in H_i \cap H_j$ for $1 \leq i < j \leq n$ then $x *_i y = x *_j y$.*

We will usually write $\mathcal{A} = \{H_i \mid 1 \leq i \leq n\}$ for the amalgam $\mathcal{A}$ as in the above definition. Whenever x and y are in the same H_i their product $x *_i y$ is defined and it is independent of the choice of i. We will normally denote this product simply by xy. Since $B := \bigcap_{i=1}^n H_i$ is non-empty, one can easily see that B contains the identity element of $(H_i, *_i)$ for every $1 \leq i \leq n$. Moreover, all these identity elements must be equal. The reader may notice that a more common definition of amalgams in terms of morphisms is essentially equivalent to the above one.

If $(G, *)$ is a group, $H_1, ..., H_n$ are subgroups of G and $*_1, ..., *_n$ are the restrictions of $*$ to these subgroups, then $\mathcal{A} = \{H_i \mid 1 \leq i \leq n\}$ is an amalgam. This is the most important example of an amalgam, but at the same time it is not very difficult to construct an example of an amalgam which is not isomorphic to a family of subgroups of a group. The amalgam $\mathcal{A}$ as above is said to be isomorphic to an amalgam $\mathcal{A}' = \{H_i' \mid 1 \leq i \leq n\}$ if there is a bijection of H onto H' which induces an isomorphism of $(H_i, *_i)$ onto $(H_i', *_i')$ for every $1 \leq i \leq n$.

Definition 1.3.2 *A group G is said to be a* completion *of an amalgam $\mathscr{A} = \{H_i \mid 1 \leq i \leq n\}$ if there is a mapping φ of H into G such that*

(i) *G is generated by the image of φ,*

(ii) *for every $1 \leq i \leq n$ the restriction of φ to H_i is a group homomorphism with respect to $*_i$ and the group operation in G.*

If φ is injective then the completion G is said to be faithful.

Thus an amalgam $\mathscr{A}$ is isomorphic to a family of subgroups of a group if and only if $\mathscr{A}$ possesses a faithful completion. If G is a faithful completion of $\mathscr{A}$ then we will usually identify $\mathscr{A}$ and its image in G.

There is a completion $U(\mathscr{A})$ of $\mathscr{A}$ known as the *universal completion*, of which any completion is a homomorphic image. The group $U(\mathscr{A})$ has the following definition in terms of generators and relations: the generators are all the elements of H; the relations are all the equalities of the form $xyz^{-1} = 1$ where x and y are (possibly equal) elements contained in H_i for some i and $z = x *_i y$. It is easy to see that $U(\mathscr{A})$ is a completion of $\mathscr{A}$ with respect to the mapping ψ which sends every $x \in H$ onto the corresponding generator of $U(\mathscr{A})$. Moreover, if G is an arbitrary completion of $\mathscr{A}$ with respect to a mapping φ then there is a unique homomorphism $\chi : U(\mathscr{A}) \to G$ such that φ is the composition of ψ and χ. Finally, $\mathscr{A}$ possesses a faithful completion if and only if $U(\mathscr{A})$ is a faithful completion.

Let G, $\mathscr{G}$ and the G_i be as in (1.2.1). The amalgam $\mathscr{A} = \{G_i \mid 1 \leq i \leq n\}$ is called *the amalgam of maximal parabolic subgroups* in G associated with the flag Φ. The geometry $\mathscr{G}(G)$ should be denoted by $\mathscr{G}(G, \mathscr{A})$ since its structure is determined not only by G by also by the amalgam $\mathscr{A}$ and by the embedding of $\mathscr{A}$ in G. We can reformulate (1.2.1) as follows.

Proposition 1.3.3 *Let G be a flag-transitive automorphism group of a geometry $\mathscr{G}$ of rank n and $\mathscr{A} = \{G_i \mid 1 \leq i \leq n\}$ be the amalgam of maximal parabolic subgroups associated with a maximal flag. Let $\mathscr{G}(G, \mathscr{A})$ be the incidence system whose elements of type i are the right cosets of G_i in G and in which two elements are incident if and only if the intersection of the corresponding cosets is non-empty. Then $\mathscr{G}$ and $\mathscr{G}(G, \mathscr{A})$ are isomorphic.* □

Notice that by the above proposition the residues of $\mathscr{G}$ are uniquely determined by the amalgam $\mathscr{A}$. That is, $\mathrm{res}_{\mathscr{G}}(x_i)$ is isomorphic to $\mathscr{G}(G_i, \mathscr{A}_i)$ where $\mathscr{A}_i = \{G_i \cap G_j \mid 1 \leq j \leq n,\ j \neq i\}$.

For a subset $J \subseteq I = \{1, 2, ..., n\}$ let $G_J = \bigcap_{i \in J} G_i$ be the elementwise stabilizer in G of the flag $\{x_i \mid i \in J\}$. The subgroup G_J is a *parabolic subgroup* of rank r where $r = |I| - |J|$. For $i, j \in I$ we write G_{ij} instead of $G_{\{i,j\}}$. The parabolic subgroups of rank $n-1$ are the maximal parabolics. The parabolic subgroups of rank 1 are known as *minimal parabolics* and the subgroup $B = G_I$ is called the *Borel subgroup*. We will usually write P_i to denote the minimal parabolic $G_{I\setminus\{i\}}$ and P_{ij} to denote the rank 2 parabolic $G_{I\setminus\{i,j\}}$.

1.4 Geometrical amalgams

In view of (1.3.3) the following question naturally arises.

Q. Let G be a group, $G_1, G_2, ..., G_n$ be subgroups of G and $\mathscr{A} = \{G_i \mid 1 \leq i \leq n\}$ be the amalgam formed by these subgroups. Under what circumstances is the incidence system $\mathscr{G} = \mathscr{G}(G, \mathscr{A})$ a geometry and the natural action of G on $\mathscr{G}$ flag-transitive?

Below we discuss the answer to this question as given in [Ti74].

The set $\Phi = \{G_1, G_2, ..., G_n\}$ is a flag in $\mathscr{G}$ since each G_i contains the identity element and Φ is a maximal flag since for $1 \leq i \leq n$ and $g \in G$ either $G_i g = G_i$ or $G_i g \cap G_i = \emptyset$. A set $\Psi = \{G_{i_1}h_1, ..., G_{i_m}h_m\}$ is a flag in $\mathscr{G}$ if and only if $G_{i_j}h_j \cap G_{i_k}h_k \neq \emptyset$ for all j, k with $1 \leq j, k \leq m$ (which implies particularly that $i_j \neq i_k$). We say that the flag Ψ is *standard* if the intersection $\bigcap_{j=1}^{m} G_{i_j}h_j$ is non-empty and contains an element h, say. In this case $\Psi = \{G_{i_1}, ..., G_{i_m}\}^h$, which means that Ψ is the image under h of a subflag in Φ. This shows that every standard flag is contained in a standard maximal flag and G acts transitively on the set of standard flags of each type. Clearly G cannot map a standard flag onto a non-standard one. Thus the necessary and sufficient condition for flag-transitivity of the natural action of G on $\mathscr{G}$ is absence of non-standard flags.

The proof of the following result uses elementary group theory only (compare Sections 10.1.3 and 10.1.4 in [Pasi94]).

Lemma 1.4.1 *The incidence system $\mathscr{G}(G, \mathscr{A})$ does not contain non-standard flags if and only if the following equivalent conditions hold:*

(i) *if J, K, L are subsets of I and g, h, f are elements in G such that the cosets $G_J g$, $G_K h$, $G_L f$ have pairwise non-empty intersection, then $G_J g \cap G_K h \cap G_L f \neq \emptyset$;*

(ii) *for $i, j \in I$ and $J \subseteq I \setminus \{i, j\}$ if $g \in G_J$ and $G_i \cap G_j g \neq \emptyset$ then $G_J \cap G_i \cap G_j g \neq \emptyset$.* □

One may notice that, in general, existence of non-standard flags in $\mathscr{G}(G,\mathscr{A})$ depends not only on the structure of $\mathscr{A}$ but also on the structure of G.

The connectivity condition in (1.1.1 (ii)) is also easy to express in terms of parabolic subgroups. By the standard principle the graph on $\mathscr{G}^i \cup \mathscr{G}^j$ is connected if and only if G is generated by the subgroups G_i and G_j. This gives the following.

Lemma 1.4.2 *The incidence system $\mathscr{G}(G,\mathscr{A})$ satisfies the condition (ii) in* (1.1.1) *if and only if for every 2-element subset $\{i,j\} \subseteq I$ the subgroups G_i and G_j generate G.* □

Finally let K be the kernel of the action of G on $\mathscr{G}(G,\mathscr{A})$. It is straightforward that K is the largest subgroup in the Borel subgroup $B = \bigcap_{i=1}^n G_i$, which is normal in G_i for all i with $1 \le i \le n$ (equivalently, normal in G). In particular the action of G on $\mathscr{G}(G,\mathscr{A})$ is faithful if and only if the Borel subgroup contains no non-identity subgroup normal in G.

1.5 Universal completions and covers

The fact that the structure of residues in $\mathscr{G}(G,\mathscr{A})$ is determined solely by $\mathscr{A}$ plays a crucial rôle in the description of the coverings of $\mathscr{G}(G,\mathscr{A})$.

Let $\mathscr{G}$ be a geometry, G be a flag-transitive automorphism group of $\mathscr{G}$ and $\mathscr{A} = \{G_i \mid 1 \le i \le n\}$ be the amalgam of maximal parabolic subgroups associated with the action of G on $\mathscr{G}$. Then on the one hand $\mathscr{G} \cong \mathscr{G}(G,\mathscr{A})$ and on the other hand G is a faithful completion of $\mathscr{A}$. Let G' be another faithful completion of $\mathscr{A}$ and let

$$\varphi : G' \to G$$

be an $\mathscr{A}$-homomorphism, *i.e.* a homomorphism of G' onto G whose restriction to $\mathscr{A}$ is the identity mapping. As usual we identify $\mathscr{A}$ with its images in G' and G. The following result is straightforward.

Lemma 1.5.1 *In the above terms the mapping of $\mathscr{G}(G',\mathscr{A})$ onto $\mathscr{G}(G,\mathscr{A})$ induced by φ is a covering of geometries.* □

In the above construction we could take G' to be the universal completion $U(\mathscr{A})$ of $\mathscr{A}$. The following result of fundamental importance was proved independently in [Pasi85], [Ti86] and an unpublished manuscript by S.V. Shpectorov.

Proposition 1.5.2 *Let $\mathcal{G}$ be a geometry, G be a flag-transitive automorphism group of $\mathcal{G}$ and $\mathcal{A}$ be the amalgam of maximal parabolic subgroups associated with the action of G on $\mathcal{G}$. Then $\mathcal{G}(U(\mathcal{A}), \mathcal{A})$ is the universal cover of $\mathcal{G} \cong \mathcal{G}(G, \mathcal{A})$.* □

By the above proposition a flag-transitive geometry $\mathcal{G}$ is simply connected if and only if a flag-transitive automorphism group G of $\mathcal{G}$ is the universal completion of the amalgam of maximal parabolic subgroups associated with the action of G on $\mathcal{G}$.

We also present a condition for 2-simple connectedness of a geometry.

Proposition 1.5.3 *Let $\mathcal{G}$ be a geometry, G be a flag-transitive automorphism group of $\mathcal{G}$ and $\mathcal{B} = \{P_{ij} \mid 1 \leq i < j \leq n\}$ be the amalgam of rank 2 parabolics associated with the action of G on $\mathcal{G}$. Then $\mathcal{G}$ is 2-simply connected (as an incidence system) if and only if G is the universal completion of $\mathcal{B}$.* □

1.6 Tits geometries

In view of (1.3.3) and Section 1.4 a flag-transitive geometry can be constructed starting with a group G and an amalgam $\mathcal{A}$ of which G is a faithful completion. In these terms the classical geometries possess the following very natural description.

Let G be a classical group defined over a field of characteristic p. Let S be a Sylow p-subgroup of G and $B = N_G(S)$. Let $G_1, ..., G_n$ be those maximal subgroups of G which contain B and

$$\mathcal{A} = \{G_i \mid 1 \leq i \leq n\}$$

be the corresponding amalgam in G. Then the classical geometry $\mathcal{G}(G)$ defined in terms of totally singular subspaces in the natural module of G is isomorphic to $\mathcal{G}(G, \mathcal{A})$. This observation shows that the natural module is not needed for defining $\mathcal{G}(G)$ and enables one to associate geometries with exceptional groups of Lie type as well. We believe that this was the main motivation of J. Tits for introducing the notion of geometries. The geometry $\mathcal{G}(G, \mathcal{A})$ will be called the *parabolic geometry* of G.

Let us discuss residues of $\mathcal{G}(G)$. Similarly to the case of geometries of classical groups, the class of direct sums of parabolic geometries of Lie type groups is closed under taking residues. Let us consider the smallest non-trivial residues, the residues of rank 2. We know that the residue $\mathscr{H}_{ij}$ of type $\{i, j\}$ is isomorphic to

$$\mathscr{G}(P_{ij}, \{P_i, P_j\}).$$

As mentioned, $\mathscr{H}_{ij}$ is either the direct sum of two geometries of Lie type groups of rank 1 or the parabolic geometry of a Lie type group of rank 2. In the former case $\mathscr{H}_{ij}$ is a complete bipartite graph while in the latter case it is a classical generalized m-gon for $m \geq 3$.

Definition 1.6.1 *A generalized* m-gon *of order* (s, t) *is a rank 2 geometry* Σ *in which the elements of one type are called points, the elements of the other type are called lines, such that*

(i) *every line is incident to* $s+1$ *points, every point is incident to* $t+1$ *lines,*

(ii) *the incidence graph of* Σ *has diameter* m *and its girth (the length of the shortest cycle) is* $2m$.

If Σ is a generalized m-gon, then the geometry in which the rôles of points and lines are interchanged is a generalized m-gon dual to Σ. Sometimes we do not distinguish generalized m-gons from their duals, and identify them both with their incidence graphs.

If G is a Lie type group of rank 2 whose Weyl group is isomorphic to the dihedral group D_{2m} of order $2m$, then the parabolic geometry of G is a generalized m-gon. The generalized m-gons arising in this way are called *classical.*

Notice that a complete bipartite graph is a generalized 2-gon (also called a generalized *digon*). A generalized 3-gon (a *generalized triangle*) is the same as a projective plane. In this case $s = t$ is the order of the plane.

Let Σ be a generalized m-gon of order (s, t). If $s = t = 1$ then Σ is the ordinary m-gon, which clearly exists for every m. If $s > 1$ and $t > 1$ then Σ is said to be *thick.* The class of thick generalized m-gons (also called *generalized polygons*) is rather restricted, as follows from the Feit – Higman theorem [FH64].

Theorem 1.6.2 *Finite thick generalized* m*-gons exist if and only if* $m \in \{2, 3, 4, 6, 8\}$. □

For every $m \in \{3, 4, 6, 8\}$ there exists a classical generalized m-gon. Since generalized triangles are just projective planes, there are many non-classical ones. Similarly there are many non-classical generalized 4-gons (also called *generalized quadrangles.*) So far no non-classical examples of generalized 6-gons (*hexagons*) or 8-gons (*octagons*) have been constructed.

Spherical Coxeter diagrams :

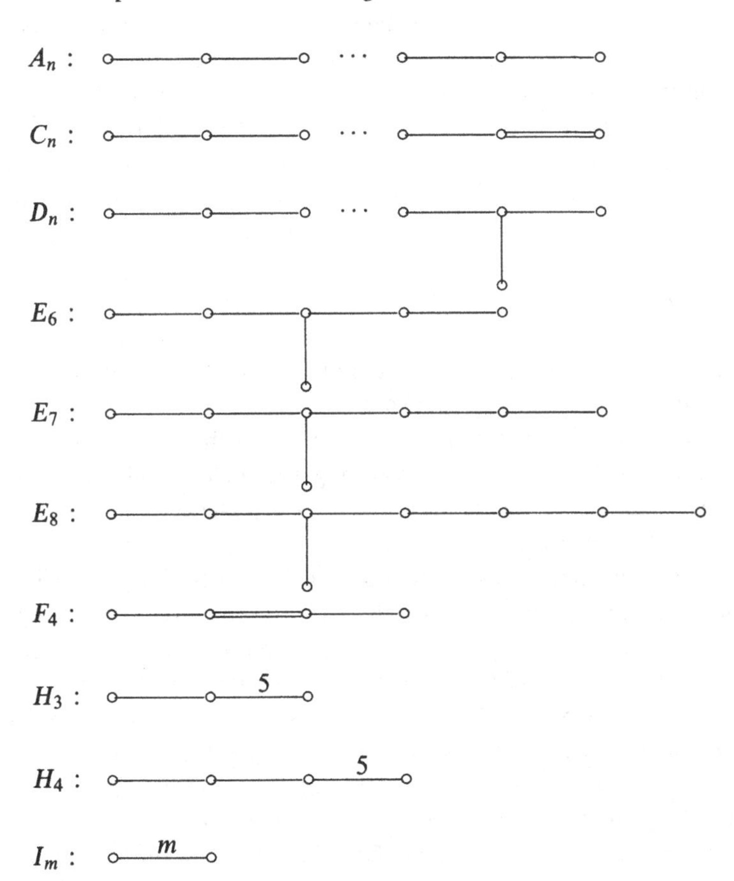

Let Δ be the *diagram* of $\mathcal{G}(G)$, which is a graph on the set I of types with nodes i and j being joined by an edge with multiplicity $(m_{ij}-2)$ (or just by a simple edge labelled by m_{ij}) if the rank 2 residues in $\mathcal{G}(G)$ of type $\{i, j\}$ are generalized m_{ij}-gons. This means particularly that there is no edge between i and j if the residues of type $\{i, j\}$ are generalized digons. Then Δ is the Dynkin diagram of the Lie algebra associated with G and also the Coxeter diagram of the Weyl group W of G. This means that W has the following presentation in terms of generators and relations:

$$W = \langle e_i,\ i \in I \mid e_i^2 = 1,\ (e_i e_j)^{m_{ij}} = 1 \rangle.$$

Since G is finite, W is a finite Coxeter group, *i.e.* the diagram Δ is spherical. Recall that a Coxeter group is a Weyl group of a finite group of Lie type if and only if its diagram is spherical and each m_{ij} is from the set $\{2, 3, 4, 6\}$.

In a similar way one can associate with an arbitrary (flag-transitive) geometry $\mathscr{G}$ the *diagram* $\Delta(\mathscr{G})$ whose nodes are the types of $\mathscr{G}$ and the edge joining i and j symbolizes the residues of type $\{i, j\}$ in $\mathscr{G}$. Under the node i it is common to put the index q_i such that the number of maximal flags containing a flag of type $I \setminus \{i\}$ is $q_i + 1$. Sometimes above the node we write the corresponding type but usually the types on the diagram are assumed to increase rightward from 1 up to the rank of the geometry. A geometry all of whose rank 2 residues are generalized polygons is called a *Tits geometry*.

Throughout the book all geometries are assumed to be *locally finite* which means that all the indices q_i are finite. If all the indices are greater than 1 then the geometry is said to be *thick*.

If $\mathscr{H}$ is the residue in $\mathscr{G}$ of an element of type i then the diagram $\Delta(\mathscr{H})$ can be obtained from $\Delta(\mathscr{G})$ by omitting the node i and all the edges incident to this node. Notice that if $\mathscr{G}' \to \mathscr{G}$ is a 2-covering of geometries then $\mathscr{G}'$ and $\mathscr{G}$ have the same diagram.

It turns out that many properties of $\mathscr{G}(G)$ can be deduced from its diagram and in many cases the diagram of $\mathscr{G}(G)$ (including the indices) specifies $\mathscr{G}(G)$ up to isomorphism. Without going into details, this important and beautiful topic can be summarized as follows.

Existence in G of the Weyl group W as a section imposes on $\mathscr{G}$ an additional structure known as a *building*. The buildings of spherical type (*i.e.* with underlying geometries having spherical diagrams) were classified in [Ti74] by showing that they are exactly the parabolic geometries of finite groups of Lie type. Later in [Ti82] it was shown that under certain additional conditions the structure of a building can be deduced directly from the condition that all rank 2 residues are generalized polygons. That is the following result was established.

Theorem 1.6.3 *Let $\mathscr{G}$ be a Tits geometry of rank $n \geq 2$. Then $\mathscr{G}$ is covered by a building if and only if every rank 3 residue in $\mathscr{G}$ having diagram*

C_3 : o——o====o *or* H_3 : o——o——o (with label 5 on the second edge)

is covered by a building. □

We should emphasize again that in view of the main result of [Ti74] the buildings of spherical type are exactly the parabolic geometries of finite groups of Lie type.

We formulate another important result from [Ti82].

Theorem 1.6.4 *Every building of rank at least 3 is 2-simply connected.* □

If G is a finite group of Lie type then its parabolic geometry $\mathcal{G}(G)$ is a building and hence it is 2-simply connected by the above theorem. In view of (1.5.3) this means that G is the universal completion of the amalgam of rank 2 parabolic subgroups associated with the action of G on $\mathcal{G}(G)$. This reflects the fact that G is defined by its Steinberg presentation. In fact every Steinberg generator is contained in one of the minimal parabolic subgroups associated with a given maximal flag and for every Steinberg relation the generators involved in the relation are contained in a parabolic subgroup of rank at most 2. Thus the Steinberg presentation is in fact a presentation for the universal completion of the amalgam of rank 2 parabolics.

The last important topic we are going to discuss in this section is the flag-transitive automorphism groups of parabolic geometries of groups of Lie type. Let G be a Lie type group in characteristic p, $\mathcal{G} = \mathcal{G}(G)$ be the parabolic geometry of G, B be the Borel subgroup and $U = O_p(B)$. An automorphism group H of $\mathcal{G}$ is said to be *classical* if it contains the normal closure U^G of U in G. In this case if G is non-abelian, then H contains the commutator subgroup of G. The following fundamental result [Sei73] (see Section 9.4.5 in [Pasi94] for the corrected version) shows that up to a few exceptions the flag-transitive automorphism groups of classical geometries are classical.

Theorem 1.6.5 *Let $\mathcal{G}$ be the parabolic geometry of a finite group of Lie type of rank at least 2 and H be a flag-transitive automorphism group of $\mathcal{G}$. Then either H is classical or one of the following holds:*

(i) *$\mathcal{G}$ is the projective plane over $GF(2)$ and $H \cong Frob_7^3$;*
(ii) *$\mathcal{G}$ is the projective plane over $GF(8)$ and $H \cong Frob_{73}^9$;*
(iii) *$\mathcal{G}$ is the 3-dimensional projective $GF(2)$-space and $H \cong Alt_7$;*
(iv) *$\mathcal{G}$ is the generalized quadrangle of order $(2,2)$ associated with $Sp_4(2)$ and $H \cong Alt_6$;*
(v) *$\mathcal{G}$ is the generalized quadrangle of order $(3,3)$ associated with $Sp_4(3)$ and H is one of $2^4 : Alt_5$, $2^4 : Sym_5$ and $2^4 : Frob_5^4$;*

(vi) *$\mathcal{G}$ is the generalized quadrangle of order* $(3,9)$ *associated with* $U_4(3)$ *and H is one of* $L_3(4).2_2$; $L_4(3).2_3$ *and* $L_3(4).2^2$;

(vii) *$\mathcal{G}$ is the generalized hexagon of order* $(2,2)$ *associated with* $G_2(2)$ *and* $H \cong G_2(2)' \cong U_3(3)$;

(viii) *$\mathcal{G}$ is the generalized octagon of order* $(2,4)$ *associated with* ${}^2F_4(2)$ *and* $H \cong {}^2F_4(2)'$ *(the Tits group).* □

1.7 Alt_7-geometry

Let us discuss the exceptional rank 3 residues from (1.6.3). By (1.6.2) there do not exist any thick generalized 5-gons, so as long as we are interested in thick locally finite Tits geometries we should not worry about the H_3-residues. On the other hand there exists a thick flag-transitive C_3-geometry which is not covered by a building. This geometry was discovered and published independently in [A84] and [Neu84] and can be described as follows.

Let Ω be a set of size 7 and $G \cong Alt_7$ be the alternating group of Ω. Let π be a projective plane of order 2 having Ω as set of points. This means that π is a collection of seven 3-element subsets of Ω such that any two of the subsets have exactly one element in common. Let $G_1 \cong Alt_6$ be the stabilizer in G of an element $\alpha \in \Omega$. Let G_2 be the stabilizer in G of a line of π containing α, so that $G_2 \cong (Sym_3 \times Sym_4)^e$ where the superscript indicates that we take the index 2 subgroup of even permutations. Finally let G_3 be the stabilizer of π in G, so that $G_3 \cong L_3(2)$ is the automorphism group of π.

Let $\mathcal{A} = \{G_1, G_2, G_3\}$ and $\mathcal{G} = \mathcal{G}(G, \mathcal{A})$. Then $\mathcal{G}$ is a Tits geometry with the following diagram:

$$C_3(2): \quad \underset{2}{\circ}\!\!-\!\!-\!\!-\!\!\underset{2}{\circ}\!\!=\!\!=\!\!=\!\!\underset{2}{\circ}$$

If $\{x_1, x_2, x_3\}$ is a maximal flag in $\mathcal{G}$ where x_i is of type i then $\mathrm{res}_{\mathcal{G}}(x_3)$ is canonically isomorphic to π, $\mathrm{res}_{\mathcal{G}}(x_2)$ is the complete bipartite graph $K_{3,3}$ and $\mathrm{res}_{\mathcal{G}}(x_1)$ is the (unique) generalized quadrangle of order (2,2) associated with $Sp_4(2)$ on which $G_1 \cong Alt_6$ acts flag-transitively (1.6.5 (iv)). Notice that $G_i \cap G_j \cong Sym_4$ for $1 \le i < j \le 3$ and $B \cong D_8$.

The C_3-geometry $\mathcal{G}(Alt_7)$ was characterized in [A84] by the following result (see also [Tim84], p. 237).

Theorem 1.7.1 *Let $\mathcal{G}$ be a flag-transitive C_3-geometry such that the residue of an element of type 1 is a classical generalized quadrangle and the residue of an element of type 3 is a Desarguesian (classical) projective plane.*

Then $\mathcal{G}$ is either a classical Tits geometry (a building) or isomorphic to $\mathcal{G}(Alt_7)$. □

It is implicit in the above theorem that $\mathcal{G}(Alt_7)$ is simply connected.

In [Yos96] with use of the classification of finite simple groups it was shown that $\mathcal{G}(Alt_7)$ is the only non-classical flag-transitive C_3-geometry. This result together with some earlier work on Tits geometries implies that $\mathcal{G}(Alt_7)$ is the only thick non-classical Tits geometry with spherical diagram of rank at least 3.

1.8 Symplectic geometries over $GF(2)$

In this section we apply the results on classical geometries discussed in the preceding sections to symplectic geometries over the field $GF(2)$ of two elements. These geometries will play an important rôle in our subsequent exposition.

Let V be a vector space of dimension $2n$, $n \geq 1$, over $GF(2)$ and let Ψ be a non-singular symplectic form on V. If $\{v_1^1, ..., v_n^1, v_1^2, ..., v_n^2\}$ is a basis of V then up to equivalence Ψ can be chosen to be

$$\Psi(v_i^k, v_j^l) = \delta_{i,j}\delta_{k,3-l}$$

(here and elsewhere $\delta_{i,j}$ is the standard Kronecker symbol). A subspace U of V is *totally singular* with respect to Ψ if $\Psi(u, w) = 0$ for all $u, w \in U$. Since Ψ is symplectic, every 1-dimensional subspace is totally singular. All maximal totally singular subspaces have dimension n and $V_n = \langle v_1^1, ..., v_n^1 \rangle$ is one of them.

Let $\mathcal{G}$ be the set of all non-zero totally singular subspaces in V with respect to Ψ, $*$ be the incidence relation on $\mathcal{G}$ with respect to which two subspaces are incident if one of them contains the other one (we say that $*$ is *defined by inclusion*), t be the mapping from $\mathcal{G}$ into $I = \{1, 2, ..., n\}$ which prescribes for a subspace its dimension. Then $\mathcal{G} = (\mathcal{G}, *, t, I)$ is a geometry. Let V_i, $1 \leq i \leq n$, be subspaces in V_n such that $\dim V_i = i$ and V_i is contained in V_j whenever $i \leq j$. Then $\Phi = \{V_1, ..., V_n\}$ is a maximal flag in $\mathcal{G}$.

For $1 \leq i < j \leq n$ put $\Phi_{ij} = \Phi \setminus \{V_i, V_j\}$. Then $\mathrm{res}_{\mathcal{G}}(\Phi_{ij}) = \mathcal{H}_i \cup \mathcal{H}_j$, where $\mathcal{H}_l$ is the set of l-dimensional totally singular subspaces in V incident to every subspace from Φ_{ij}. If i is less than $k := j-1$ then every subspace $U \in \mathcal{H}_j$ contains V_k while every $W \in \mathcal{H}_i$ is contained in V_k, which means that $U \leq W$ and hence $\mathrm{res}_{\mathcal{G}}(\Phi_{ij})$ is a generalized digon. If $i = j-1$ and $j < n$ then $\mathcal{H}_i$ and $\mathcal{H}_j$ correspond to all 1- and 2-dimensional subspaces in the 3-dimensional $GF(2)$-space V_{j+1}/V_{j-2} and

$\mathrm{res}_{\mathcal{G}}(\Phi_{ij})$ is the projective plane of order 2. Finally, if $i = n-1$ and $j = n$ then $\mathcal{H}_i$ and $\mathcal{H}_j$ correspond to 1- and 2-dimensional subspaces in the 4-dimensional space $V_{n-2}^{\perp}/V_{n-2}$ that are totally singular with respect to the symplectic form on this space induced by Ψ. In this case $\mathrm{res}_{\mathcal{G}}(\Phi_{ij})$ is the generalized quadrangle of order $(2,2)$ associated with $Sp_4(2)$. Since all indices in $\mathcal{G}$ are equal to 2, we observe that $\mathcal{G}$ is a C_n-geometry with the following diagram, where as usual the types increase rightward from 1 to n:

$$C_n(2): \quad \underset{2}{\circ}\!\!-\!\!\!-\!\!\!-\!\!\underset{2}{\circ}\ \cdots\ \underset{2}{\circ}\!\!-\!\!\!-\!\!\!-\!\!\underset{2}{\circ}\!\!=\!\!=\!\!=\!\!\underset{2}{\circ}$$

Let H be the group of those linear transformations of V which preserve Ψ. Then $H \cong Sp_{2n}(2)$, the action of H on $\mathcal{G}$ is flag-transitive and H is the automorphism group of $\mathcal{G}$. The flag-transitivity follows from the transitivity of H on the set of maximal totally singular subspaces and from the fact that the stabilizer in H of such a subspace induces on the subspace its full automorphism group $L_n(2)$. The geometry constructed above will be denoted by $\mathcal{G}(Sp_{2n}(2))$.

Let G be any flag-transitive automorphism group of $\mathcal{G} = \mathcal{G}(Sp_{2n}(2))$. Since H is simple for $n \geq 3$ we conclude from (1.6.5) that either $G = H$ or $n = 2$ and $G \cong Alt_6$.

Let B, P_i and P_{ij} be the Borel subgroup, minimal and rank 2 parabolics in G associated with Φ (here $1 \leq i < j \leq n$). Then B is a Sylow 2-subgroup in G and $P_{ij} = \langle P_i, P_j \rangle$. If $G = H$ then P_{ij} acting on $\mathrm{res}_{\mathcal{G}}(\Phi_{ij})$ induces the automorphism group of the residue isomorphic to $Sym_3 \times Sym_3$, $L_3(2)$ or $Sp_4(2)$ depending on i and j. The kernel Q_{ij} of this action is contained in B which is a 2-group. This shows that $Q_{ij} = O_2(P_{ij})$. Thus G and its parabolic subgroups of rank 1 and 2 satisfy the following conditions:

(WP1) G is a group generated by its subgroups P_i, $1 \leq i \leq n$, where $n \geq 2$;

(WP2) for $1 \leq i < j \leq n$ the intersection $B := P_i \cap P_j$ is a 2-group, which is independent of the particular choice of i and j;

(WP3) $P_i/O_2(P_i) \cong Sym_3$ for $1 \leq i \leq n$;

(WP4) if $P_{ij} = \langle P_i, P_j \rangle$ for $1 \leq i < j \leq n$ and $Q_{ij} = O_2(P_{ij})$ then B is a Sylow 2-subgroup of P_{ij} and

$$(*) \quad P_{ij}/Q_{ij} \cong \begin{cases} Sym_3 \times Sym_3 & \text{if } j - i > 1, \\ L_3(2) & \text{if } i = j-1 \text{ and } j < n, \\ Sp_4(2) \text{ or } Alt_6 & \text{if } i = n-1,\ j = n; \end{cases}$$

(WP5) if $N \leq B$ and N is normal in P_i for all $1 \leq i \leq n$, then $N = 1$ (the identity subgroup).

By (1.6.4) the geometry $\mathscr{G}$ is 2-simply connected and by (1.5.3) this implies that G is the universal completion (and in fact the only completion) of the amalgam $\mathscr{C} = \{P_{ij} \mid 1 \le i < j \le n\}$ of rank 2 parabolics. But in fact the conditions (WP1)–(WP5) show that $(G; \{P_i \mid 1 \le i \le n\})$ is a so-called *weak parabolic system* and by Theorem 3.2 in [Tim84] we have the following.

Theorem 1.8.1 *Let G be a group, P_i, $1 \le i \le n$, $n \ge 2$, be subgroups in G and suppose that conditions (WP1)–(WP5) hold. Let $G_i = \langle P_k \mid 1 \le k \le n,\ k \ne i \rangle$ and $\mathscr{A} = \{G_i \mid 1 \le i \le n\}$. Then $\mathscr{G}(G, \mathscr{A})$ is a Tits geometry of rank n with diagram $C_n(2)$ on which G acts faithfully and flag-transitively so that one of the following holds:*

(i) *$\mathscr{G}(G, \mathscr{A}) \cong \mathscr{G}(Sp_{2n}(2))$ and either $G \cong Sp_{2n}(2)$, or $n = 2$ and $G \cong Alt_6$;*

(ii) *$\mathscr{G}(G, \mathscr{A}) \cong \mathscr{G}(Alt_7)$ and $G \cong Alt_7$.* □

We formulate a related result from [Tim84] which we will often use.

Theorem 1.8.2 *In the notation of* (1.8.1) *suppose that the condition* (∗) *in (WP4) is changed to*

$$P_{ij}/Q_{ij} \cong \begin{cases} Sym_3 \times Sym_3 & \text{if } j - i > 1, \\ L_3(2) & \text{otherwise.} \end{cases}$$

Then $\mathscr{G}(G, A)$ is the projective $GF(2)$-space of rank n and either $G \cong L_{n+1}(2)$ or $n = 3$ and $G \cong Alt_7$. □

1.9 From classical to sporadic geometries

Let us mimic the construction of geometries of symplectic groups for the largest sporadic simple group, the Monster.

Let $G = M$ be the Monster group. Let B be a Sylow 2-subgroup of G, whose order is 2^{46}. There are exactly five subgroups $P_1, \dots, P_5$ in M which contain B as a maximal subgroup. If we put $P_{ij} = \langle P_i, P_j \rangle$ for $1 \le i < j \le 5$, then the conditions (WP1)–(WP5) from the previous section hold with (∗) in (WP4) being changed to the following (where n is assumed to be 5).

$$(\#) \quad P_{ij}/Q_{ij} \cong \begin{cases} Sym_3 \times Sym_3 & \text{if } j - i > 1, \\ L_3(2) & \text{if } i = j - 1 \text{ and } j < n, \\ 3 \cdot Sp_4(2) \text{ or } 3 \cdot Alt_6 & \text{if } i = n - 1,\ j = n. \end{cases}$$

Here $3{\cdot}Sp_4(2)$ and $3{\cdot}Alt_6$ are non-split extensions by subgroups of order 3 of $Sp_4(2) \cong Sym_6$ and Alt_6, respectively. In fact $P_{45}/O_2(P_{45}) \cong 3 \cdot Sp_4(2)$

in the Monster group. The rank 2 geometry $\mathcal{G}(P_{45}, \{P_4, P_5\})$ is a triple cover $\mathcal{G}(3 \cdot Sp_4(2))$ of the generalized quadrangle $\mathcal{G}(Sp_4(2))$ of order $(2, 2)$. This cover is denoted by the diagram

$$T_2 : \underset{2}{\circ} \overset{\sim}{=\!=\!=} \underset{2}{\circ}$$

and called the *tilde geometry* of rank 2.

Let $G_i = \langle P_k \mid 1 \leq k \leq 5,\ k \neq i \rangle$, $\mathcal{A} = \{G_i \mid 1 \leq i \leq 5\}$ and $\mathcal{G}(M) = \mathcal{G}(G, \mathcal{A})$. It was shown in [RSt84] that $\mathcal{G}(M)$ is a geometry on which M acts flag-transitively and which has the following diagram:

$$T_5 : \underset{2}{\circ} \text{---} \underset{2}{\circ} \text{---} \underset{2}{\circ} \text{---} \underset{2}{\circ} \overset{\sim}{=\!=\!=} \underset{2}{\circ}$$

Thus the Monster group M acts flag-transitively on a geometry $\mathcal{G}(M)$ whose local properties (rank 2 residues) are similar to those of $\mathcal{G}(Sp_{10}(2))$. In view of (1.8.1) it is natural to pose the following question: "To what extent is the Monster characterized by the property that it is a flag-transitive automorphism group of a tilde geometry of rank 5?"

Notice that $\mathcal{G}(M)$ contains, as residues, tilde geometries of rank 3 and 4 associated with sporadic simple groups Mat_{24} and Co_1. It also contains a number of other nice subgeometries and one such subgeometry can be described in the following way.

Let τ be a Baby Monster involution in M, in which case $C_M(\tau) \cong 2 \cdot BM$ where BM is the Baby Monster sporadic simple group. Then a subset of the set of elements in $\mathcal{G}(M)$ fixed by τ (here we do not define this subset precisely) forms a geometry $\mathcal{G}(BM)$ on which $BM \cong C_M(\tau)/\langle \tau \rangle$ acts flag-transitively and which has the following diagram:

$$P_5 : \underset{2}{\circ} \text{---} \underset{2}{\circ} \text{---} \underset{2}{\circ} \text{---} \underset{2}{\circ} \overset{P}{\text{---}} \underset{1}{\circ}$$

Here the rightmost edge denotes the geometry $\mathcal{G}(Sym_5)$ of edges and vertices (the left and the right type on the diagram) of the Petersen graph. The vertices of the *Petersen graph* are the 2-element subsets of a set of size 5 with two vertices subsets being adjacent if they are disjoint. In other terms let $S_1 \cong D_8$ and $S_2 \cong Sym_3 \times 2$ be subgroups in $S \cong Sym_5$ such that $S_1 \cap S_2 \cong 2^2$. Then $\mathcal{G}(Sym_5) \cong \mathcal{G}(S, \{S_1, S_2\})$. By (1.6.2) there are no thick generalized 5-gons. Speaking informally the Petersen graph is as close as one can get to such a 5-gon in terms of girth (which is 10) and diameter (which is 6).

We say that $\mathcal{G}(BM)$ is a *Petersen geometry* of rank 5. As residues $\mathcal{G}(BM)$ contains Petersen geometries of rank 3 and 4 associated with sporadic simple groups Mat_{22} and Co_2.

1.10 The main results

The main aim of this monograph is to present the detailed exposition of the classification of flag-transitive Petersen and tilde geometries whose completion was announced in [ISh94b]. In this section we formulate and briefly discuss the final result of the classification.

Theorem 1.10.1 *There exist exactly eight flag-transitive Petersen geometries of rank at least* 3*, whose diagrams and full automorphism groups are the following:*

P_3 : o—2—o—2—P—o 1   Aut Mat_{22}, $3 \cdot$ Aut Mat_{22};

P_4 : o—2—o—2—o—2—P—o 1 Mat_{23}, Co_2, $3^{23} \cdot Co_2$, J_4;

P_5 : o—2—o—2—o—2—o—2—P—o 1 BM, $3^{4371} \cdot BM$. □

In what follows we will write $\mathscr{G}(G)$ for a geometry of which G is the commutator subgroup of the automorphism group.

The geometries $\mathscr{G}(Mat_{22})$, $\mathscr{G}(Co_2)$ and $\mathscr{G}(BM)$ in (1.10.1) are not 2-simply connected and their universal 2-covers are, respectively, $\mathscr{G}(3 \cdot Mat_{22})$, $\mathscr{G}(3^{23} \cdot Co_2)$ and $\mathscr{G}(3^{4371} \cdot BM)$.

Theorem 1.10.2 *There exist an infinite family of flag-transitive tilde geometries (which contains one geometry of rank n for every* $n \geq 2$*) and four exceptionals. Every flag-transitive tilde geometry is 2-simply connected. The diagrams and full automorphism groups are the following:*

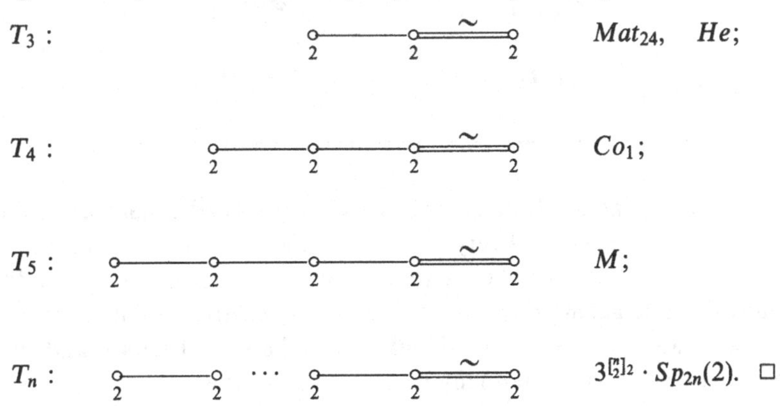

It follows from the classification that whenever H is a flag-transitive automorphism group of a Petersen or tilde geometry $\mathcal{G}$, of rank at least 3, then either $H = \mathrm{Aut}\,\mathcal{G}$, or $\mathcal{G}$ is a rank 3 Petersen geometry and H is the commutator subgroup of $\mathrm{Aut}\,\mathcal{G}$.

Geometry	Subgeometry
$\mathcal{G}(Mat_{22})$	$\mathcal{G}(Sp_4(2))$
$\mathcal{G}(3 \cdot Mat_{22})$	$\mathcal{G}(3 \cdot Sp_4(2))$
$\mathcal{G}(Mat_{23})$	$\mathcal{G}(Alt_7)$
$\mathcal{G}(Co_2)$	$\mathcal{G}(Sp_6(2))$
$\mathcal{G}(3^{23} \cdot Co_2)$	$\mathcal{G}(3^7 \cdot Sp_6(2))$
$\mathcal{G}(J_4)$	$\mathcal{G}(Mat_{24})$
$\mathcal{G}(BM)$	$\mathcal{G}(Sp_8(2))$
$\mathcal{G}(3^{4371} \cdot BM)$	$\mathcal{G}(3^{35} \cdot Sp_8(2))$
$\mathcal{G}(Mat_{24})$	$\mathcal{G}(Mat_{22})$
$\mathcal{G}(Co_1)$	$\mathcal{G}(Co_2)$
$\mathcal{G}(M)$	$\mathcal{G}(BM)$

The Petersen and tilde geometries are closely related to each other and also to $C_n(2)$-geometries. Every Petersen geometry of rank $n \geq 3$ contains either a $C_{n-1}(2)$-geometry or a tilde geometry of rank $n-1$. In addition some tilde geometries contain Petersen geometries as subgeometries (see table above). These mutual embeddings between Petersen and tilde geometries were used essentially in their classification.

By (1.10.2) the Monster group is indeed rather strongly characterized by the property that it is a flag-transitive automorphism group of a rank 5 tilde geometry, since we have the following.

Theorem 1.10.3 *Let G be a group for which conditions (WP1)–(WP5) in Section 1.8 hold for $n = 5$ with (*) in (WP4) being changed to (#) from Section 1.9. Then G is a flag-transitive automorphism group of a rank 5 tilde geometry and either $G \cong M$ or $G \cong 3^{155} \cdot Sp_{10}(2)$.* □

The next characterization involves terms more common for the Monster group.

Theorem 1.10.4 *The Monster is the only group which is generated by three subgroups C, N and L which satisfy the following:*

(i) $C \sim 2_+^{1+24}.Co_1$ *and $O_2(C)$ contains its centralizer in C;*
(ii) $N/O_2(N) \cong Sym_3 \times Mat_{24}$, $L/O_2(L) \cong L_3(2) \times 3 \cdot Sym_6$;
(iii) $[N : N \cap C] = 3$, $[L : L \cap N] = [L : L \cap C] = 7$, $[L : L \cap N \cap C] = 21$. □

In the above statement the subgroups C, N and L correspond to the stabilizers in M of pairwise incident elements in $\mathcal{G}(M)$ of type 1, 2 and 3, respectively.

In what follows P-geometry or T-geometry will mean, respectively, *flag-transitive* Petersen or tilde geometry. In case we want to specify the rank n, we talk about P_n- and T_n-geometries. Thus all our P- and T-geometries are flag-transitive unless explicitly stated otherwise.

1.11 Representations of geometries

We say that a geometry $\mathcal{G}$ of rank n belongs to a *string diagram* if all rank 2 residues of type $\{i, j\}$ for $|i - j| > 1$ are generalized digons. In this case the types on the diagram usually increase rightward from 1 to n. The elements which correspond, respectively, to the leftmost, the second left, the third left and the rightmost nodes on the diagram will be called *points, lines, planes* and *hyperplanes*:

```
      X        Y                  Z
o--------o--------o  . . .  o--------o
points   lines    planes         hyperplanes
```

A graph on the set of points of $\mathcal{G}$ in which two points are adjacent if and only if they are incident to a common line is called the *collinearity graph* of $\mathcal{G}$.

When constructing the geometry $\mathcal{G}$ associated with a classical group G we started by treating the (totally singular) subspaces in the natural module of G as abstract elements. We have seen in Section 1.6 that in many cases $\mathcal{G}$ is uniquely determined by the structure of its rank 2 residues. This means that in these cases there is a strong possibility of reconstructing from $\mathcal{G}$ the vector space V. It is natural to look for a more direct way to recover from $\mathcal{G}$ the vector space V.

In a more general setting the question can be posed in the following form. Given a geometry $\mathcal{G}$ and a vector space V, is it possible to define a mapping φ from the element set of $\mathcal{G}$ onto the set of proper subspaces of V, such that dim $\varphi(x)$ is uniquely determined by the type of x and whenever x and y are incident, either $\varphi(x) \leq \varphi(y)$ or $\varphi(y) \leq \varphi(x)$? This question leads to a very important and deep theory of presheaves on geometries which was introduced and developed in [RSm86] and [RSm89]. A special class of the presheaves, described below, has played a crucial rôle in the classification of P- and T-geometries.

Let $\mathcal{G}$ be a geometry with elements of one type called points and elements of some other type called lines. Unless stated otherwise, if $\mathcal{G}$ has a string diagram, the points and lines are as defined above. Suppose that $\mathcal{G}$ is of $GF(2)$-type which means that every line is incident to exactly three points. Let P and L denote, respectively, the point set and the line set of $\mathcal{G}$. In order to simplify the notation we will assume that every line is uniquely determined by the triple of points it is incident to. Let V be a vector space over $GF(2)$. A *natural representation* of (the point–line incidence system associated with) $\mathcal{G}$ is a mapping φ of $P \cup L$ into the set of subspaces of V such that

(NR1) V is generated by Im φ,

(NR2) $\dim p = 1$ for $p \in P$ and $\dim l = 2$ for $l \in L$,

(NR3) if $l \in L$ and $\{p, q, r\}$ is the set of points incident to l, then $\{\varphi(p), \varphi(q), \varphi(r)\}$ is the set of 1-dimensional subspaces in $\varphi(l)$.

If $\mathcal{G}$ possesses at least one natural representation then it possesses the *universal natural representation* φ_0 such that any other natural representation is a composition of φ_0 and a linear mapping. The $GF(2)$-vector–space underlying the universal natural representation (considered as an abstract group with additive notation for the group operation) has the presentation

$$V(\mathcal{G}) = \langle v_p,\ p \in P \mid 2v_p = 0;\ \ v_p + v_q = v_q + v_p \text{ for } p, q \in P;$$
$$v_p + v_q + v_r = 0, \text{ if } \{p, q, r\} = l \in L\rangle$$

and the universal representation itself is defined by

$$\varphi_0 : p \mapsto v_p \text{ for } p \in P$$

and

$$\varphi_0 : l \mapsto \langle v_p, v_q, v_r \rangle \text{ for } \{p,q,r\} = l \in L.$$

In this case $V(\mathcal{G})$ will be called *the universal representation module of* $\mathcal{G}$. The following statement is rather obvious.

Lemma 1.11.1 *Let $\mathcal{G} = \mathcal{G}(L_n(2))$ be the projective space of rank $n-1$ over $GF(2)$. Then $V(\mathcal{G})$ is the n-dimensional natural module for $L_n(2)$.* □

The next result, which is also rather standard, was originally proved in [Ti74] in a more general form.

Lemma 1.11.2 *Let $\mathcal{G} = \mathcal{G}(Sp_{2n}(2))$ be the symplectic geometry of rank n. Then $V(\mathcal{G})$ is the $(2n+1)$-dimensional orthogonal module of $Sp_{2n}(2) \cong O_{2n+1}(2)$.* □

Natural representations of geometries usually provide a nice model for geometries and "natural" modules for their automorphism groups. Besides that, in a certain sense natural representations control extensions of geometries. Below we explain this claim.

Let $\mathcal{G}$ be a geometry of rank at least 3 with a string diagram such that the residue of type $\{1,2\}$ is a projective plane of order 2, so that the diagram of $\mathcal{G}$ has the following form:

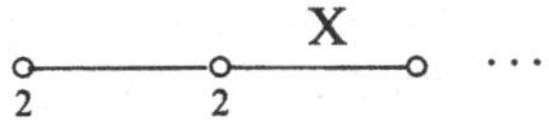

X

o———o———o ...
2 2

Let G be a flag-transitive automorphism group of $\mathcal{G}$. Let x_1 be a point of $\mathcal{G}$ (an element of type 1), G_1 be the stabilizer of x_1 in G and $\mathcal{H} = \mathrm{res}_{\mathcal{G}}(x_1)$. Then the points and lines of $\mathcal{H}$ are the lines and planes of $\mathcal{G}$ incident to x_1. Let Q_1 be the kernel of the action of G_1 on $\mathcal{H}$. Then clearly G_1/Q_1 is a flag-transitive automorphism group of $\mathcal{H}$. Let R_1 be the kernel of the action of Q_1 on the set of points collinear to x_1 (incident with x_1 to a common line) and suppose that $U = Q_1/R_1$ is non-trivial. Let x_2 be a line incident to x_1 and $\{x_1, y_1, z_1\}$ be the points incident to x_2. Since every $q \in Q_1$ stabilizes both x_1 and x_2, it either stabilizes x_2 pointwise permutes y_1 and z_1. Moreover the latter possibility must hold for some q since G is flag-transitive and $U \neq 1$. Thus U is an elementary abelian 2-group and the module U^* dual to U is generated by 1-dimensional subspaces, one for every point in $\mathcal{H}$. Now if x_3 is a plane incident to x_1, then the points and lines of $\mathcal{G}$ incident to x_3 form a projective plane π of order 2. It is easy to deduce from the flag-transitivity and the condition $U \neq 1$ that the stabilizer of x_3 in G induces the automorphism group $L_3(2)$ of π. In its turn this implies that the action induced by Q_1 on the set of points incident to x_3 is of order 4 and we have the following.

Proposition 1.11.3 *In the above notation if $U = Q_1/R_1$ is non-trivial, then it is an elementary abelian 2-group and the module U^* dual to U supports a natural representation of $\mathcal{H} = \mathrm{res}_{\mathcal{G}}(x_1)$, in particular U^* is a quotient of the universal representation module $V(\mathcal{H})$.* □

When we follow an inductive approach to classification of geometries, we can assume that $\mathcal{H}$ and its flag-transitive automorphism groups are known and we are interested in geometries $\mathcal{G}$ which are extensions of $\mathcal{H}$ by the projective plane edge in the diagram. Then the section Q_1/R_1 is either trivial or related to a natural representation of $\mathcal{H}$. In particular this section is trivial if $\mathcal{H}$ does not possess a natural representation. In practice it often happens that in this case there are no extensions of $\mathcal{H}$ at all.

For various reasons it is convenient to consider a non-abelian version of natural representations. The *universal representation group* of a geometry $\mathcal{G}$ with 3 points on every line has the following definition in terms of generators and relations:

$$R(\mathcal{G}) = \langle z_p,\ p \in P \mid z_p^2 = 1,\ z_p z_q z_r = 1 \text{ if } \{p,q,r\} = l \in L\rangle.$$

It is easy to observe that $V(\mathcal{G}) = R(\mathcal{G})/[R(\mathcal{G}), R(\mathcal{G})]$. Notice that generators z_p and z_q of $R(\mathcal{G})$ commute whenever p and q are collinear. It is straightforward from this observation that $R(\mathcal{G}(L_n(2))) = V(\mathcal{G}(L_n(2)))$. Less trivial but still not difficult to prove is the equality $R(\mathcal{G}(Sp_{2n}(2))) = V(\mathcal{G}(Sp_{2n}(2)))$. There are geometries whose universal representation groups are non-abelian. In particular the geometries $\mathcal{G}(J_4)$, $\mathcal{G}(BM)$ and $\mathcal{G}(M)$ have non-trivial representation groups while their representation modules are trivial.

1.12 The stages of classification

Our interest in P- and T-geometries originated from the classification of distance-transitive graphs of small valencies. In [FII86] within the classification of distance-transitive graphs of valency 7 we came across the intersection arrays $i(1)$ and $i(2)$:

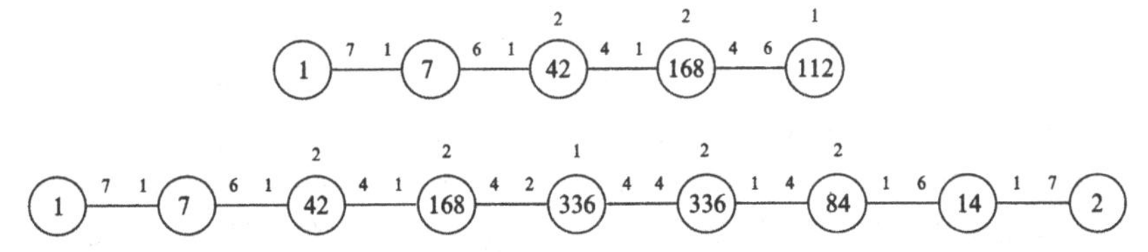

A graph $\Gamma(Mat_{22})$ with intersection array $i(1)$ was discovered in [Big75]. The vertices of this graph are the blocks of the Steiner system $S(5,8,24)$ missing a pair of points; two vertices are adjacent if the corresponding blocks are disjoint. The automorphism group of $\Gamma(Mat_{22})$ is isomorphic to $\mathrm{Aut}\, Mat_{22}$ where Mat_{22} is the sporadic Mathieu group of degree 22. Every graph with the intersection array $i(2)$ must be a 3-fold antipodal cover of a graph with intersection array $i(1)$. An example $\Gamma(3 \cdot Mat_{22})$ of such a graph was constructed in [FII86]; its automorphism group is a non-split extension of $\mathrm{Aut}\, Mat_{22}$ by a normal subgroup of order 3.

Let $H = Mat_{22}$ or $3 \cdot Mat_{22}$. It can be deduced directly from the intersection arrays $i(1)$ and $i(2)$ (Section 4 in [Iv87]) that $\Gamma(H)$ contains a Petersen subgraph. Let $\mathscr{G}(H)$ be the incidence system whose elements of type 3, 2 and 1 are respectively vertices, edges and Petersen subgraphs in $\Gamma(H)$ and where incidence relation is via inclusion. Then $\mathscr{G}(H)$ is a geometry with the diagram

P_3 : o——o——P——o (2, 2, 1)

and $\mathscr{G}(3 \cdot Mat_{22})$ is a cover of $\mathscr{G}(Mat_{22})$.

It was proved in [Sh85] that $\mathscr{G}(Mat_{22})$ and $\mathscr{G}(3 \cdot Mat_{22})$ are the only P_3-geometries (recall that in the present volume all P- and T-geometries are flag-transitive by definition). In [Sh85] for the first time ever the strategy for classification of geometries in terms of their diagrams based on consideration of amalgams of parabolic subgroups and their completions was applied. By now this is a commonly accepted strategy for studying groups and geometries.

By the result of the classification of P_3-geometries, every flag-transitive automorphism group of a P_n-geometry for $n \geq 4$ must involve Mat_{22} as a section. Using this clue rank 4 geometries $\mathscr{G}(Mat_{23})$, $\mathscr{G}(Co_2)$, $\mathscr{G}(J_4)$ and a rank 5 geometry $\mathscr{G}(BM)$ were constructed in [Iv87]. The geometries $\mathscr{G}(Mat_{23})$ and $\mathscr{G}(J_4)$ as well as $\mathscr{G}(Mat_{22})$ were mentioned in [Bue85]. The point residues in $\mathscr{G}(Mat_{23})$ and $\mathscr{G}(Co_2)$ are isomorphic to $\mathscr{G}(Mat_{22})$ while in $\mathscr{G}(J_4)$ they are isomorphic to $\mathscr{G}(3 \cdot Mat_{22})$.

Let $\mathscr{G}$ be a P_n-geometry for $n \geq 3$ and G be a flag-transitive automorphism group of $\mathscr{G}$. The *derived graph* $\Delta = \Delta(\mathscr{G})$ has the elements of type

n in $\mathscr{G}$ as vertices and two vertices are adjacent if they are incident to a common element of type $n-1$. The graph Δ is locally projective of type $(n, 2)$ (as defined below) with respect to G, the girth of Δ (the length of a shortest cycle) is 5 and unless $\mathscr{G} = \mathscr{G}(Mat_{23})$, the kernel $G_1(x)$ at every vertex x is non-trivial. A graph Γ is said to be a *locally projective graph* of type (n, q) with respect to a subgroup G in its automorphism group if G is vertex-transitive and for a vertex x the stabilizer $G(x)$ of x in G acts on the set $\Gamma(x)$ of neighbours of x in Γ as $L_n(q)$ (possibly extended by some outer automorphisms) acts on the set of 1-dimensional subspaces of an n-dimensional $GF(q)$-space. The kernel of the action of $G(x)$ on $\Gamma(x)$ is denoted by $G_1(x)$.

Locally projective graphs of girth 4 with non-trivial kernels at vertices were classified in [CPr82] and it was believed for a while that no such graphs of girth 5 exist. In [Iv88] and [Iv90] the classification problem of locally projective graphs of girth 5 with non-trivial kernels at vertices was reduced to the classification of P-geometries. This brought an additional interest in P-geometries and their derived graphs.

The local analysis needed for the classification of P_4-geometries was carried out in [Sh88]. It was shown that the amalgam of maximal parabolic subgroups associated with a flag-transitive action on a P_4-geometry is isomorphic to one of five amalgams $\mathscr{A}^j = \{G_i^j \mid 1 \leq i \leq 4\}$, $1 \leq j \leq 5$. Here $\mathscr{A}^1$, $\mathscr{A}^2$ and $\mathscr{A}^3$ are realized in the actions of Mat_{23}, Co_2 and J_4 on P-geometries associated with these groups. For $k = 1$ and 2 the amalgam $\mathscr{A}^{3+k}$ possesses a morphism onto $\mathscr{A}^k$ whose restriction to G_i^{3+k} is an isomorphism onto G_i^k for $2 \leq i \leq 4$ and whose restriction to G_1^{3+k} is a homomorphism with kernel of order 3. More precisely G_1^{3+k} is the universal completion of the amalgam $\{G_1^k \cap G_i^k \mid 2 \leq i \leq 4\}$, so that $O^2(G_1^{3+k}/O_2(G_1^{3+k})) \cong 3 \cdot Mat_{22}$ while $O^2(G_1^k/O_2(G_1^k)) \cong Mat_{22}$. This means that every geometry $\mathscr{G}$ corresponding to $\mathscr{A}^{3+k}$ (if it exists) has point residues isomorphic to $\mathscr{G}(3 \cdot Mat_{22})$ and the universal cover of $\mathscr{G}$ is the universal 2-cover of $\mathscr{G}(Mat_{23})$ or $\mathscr{G}(Co_2)$ for $k = 1$ or 2, respectively.

Thus the main result of [Sh88] reduces the classification of P_4-geometries to calculation of the universal 2-covers of $\mathscr{G}(Mat_{23})$, $\mathscr{G}(Co_2)$ and $\mathscr{G}(J_4)$. The former of the geometries was treated in [ISh90a]. This geometry contains a subgeometry $\mathscr{H} \cong \mathscr{G}(Alt_7)$. Using the simple connectedness result for the subgeometry it was shown that the geometry itself is simply connected. Furthermore, if $\mathscr{G}$ is a proper 2-cover of $\mathscr{G}(Mat_{23})$ (*i.e.* a 2-cover which is not a cover) then a connected component of the preimage of $\mathscr{H}$ in $\mathscr{G}$ is a T_3-geometry which possesses a morphism onto $\mathscr{G}(Alt_7)$. Using coset enumeration with a group given in terms of generators and relations it was shown that no tilde analogue of $\mathcal{G}(Alt_7)$ exists and hence no proper 2-cover of $\mathcal{G}(Mat_{23})$ exists either.

The universal 2-cover of $\mathcal{G}(Co_2)$ was determined in [Sh92]. First, by triangulating cycles in the collinearity graph of the geometry the latter was proved to be simply connected. Let $\varphi : \widetilde{\mathcal{G}} \to \mathcal{G}(Co_2)$ be a proper 2-covering of geometries. Then a connected component $\widetilde{\mathcal{H}}$ of the preimage in $\widetilde{\mathcal{G}}$ of a subgeometry $\mathcal{H} \cong \mathcal{G}(Sp_6(2))$ in $\mathcal{G}(Co_2)$ is a T_3-geometry and φ induces its morphism onto $\mathcal{H}$. The amalgam $\mathcal{D}$ of maximal parabolics in a flag-transitive automorphism group of $\widetilde{\mathcal{H}}$ (particularly in the action on $\widetilde{\mathcal{H}}$ of its stabilizer in a flag-transitive automorphism group of $\widetilde{\mathcal{G}}$) is specified up to isomorphism. It is possible to write down an explicit presentation for the universal completion $U(\mathcal{D})$ by modifying the Steinberg presentation for $Sp_6(2)$. By means of coset enumeration on a computer it was shown that $U(\mathcal{D}) \cong 3^7 \cdot Sp_6(2)$. Based on this result a 2-cover $\mathcal{G}(3^{23} \cdot Co_2)$ of $\mathcal{G}(Co_2)$ was constructed and its 2-simple connectedness was established.

The geometry $\mathcal{G}(J_4)$ contains the T_3-geometry $\mathcal{G}(Mat_{24})$ as a subgeometry and the simple connectedness question for $\mathcal{G}(J_4)$ heavily depends on that for $\mathcal{G}(Mat_{24})$. First the amalgams of maximal parabolics associated with flag-transitive actions on T_3-geometries have been classified and then by means of coset enumeration on a computer the universal completions of these amalgams were found. The result (presented in an unpublished preprint [ISh89b]) was the complete list of T_3-geometries: $\mathcal{G}(3^7 \cdot Sp_6(2))$ as in the above paragraph and the sporadic geometries $\mathcal{G}(Mat_{24})$ and $\mathcal{G}(He)$ constructed in [RSt84]. In the case when the stabilizer of a point induces Sym_6 on the corresponding residue an independent classification was achieved in [Hei91]. Earlier it was shown in [Row89] and independently in [Tim89] that in this case the order of the Borel subgroup is either 2^9, realized in $\mathcal{G}(3^7 \cdot Sp_6(2))$ or 2^{10}, realized in $\mathcal{G}(Mat_{24})$ and $\mathcal{G}(He)$. If the stabilizer of a point induces Alt_6 on the corresponding residue, then the geometry must have a 1-covering onto the exceptional C_3-geometry $\mathcal{G}(Alt_7)$. It was proved in [ISh90a] (see also [GM93]) that tilde analogues of $\mathcal{G}(Alt_7)$ do not exist.

After $\mathcal{G}(Mat_{24})$ was proved to be simply connected, the simple connectedness question for $\mathcal{G}(J_4)$ was attacked in [Iv92b]. In that paper instead of trying to triangulate cycles in the collinearity graph of the geometry a different graph Σ called the *intersection graph of subgeometries* was considered. The vertices of Σ are the $\mathcal{G}(Mat_{24})$-subgeometries in $\mathcal{G}(J_4)$ with two of them being adjacent if they have the maximal possible number (namely 7) of common points. Using the simple connectedness of the subgeometries it was shown that every covering φ of $\mathcal{G}(J_4)$ induces a covering φ^i of Σ with respect to which all triangles are contractible. Finally it was shown that every cycle in Σ can be triangulated, which means that φ^i and φ must be isomorphisms. About the same time the triangulability of Σ was established in [ASeg91] within the uniqueness proof for J_4.

The idea of studying triangulability of cycles in intersection graphs of various families of simply connected subgeometries turned out to be rather fruitful. In [Iv92d] the simple connectedness of the T_4-geometry $\mathcal{G}(Co_1)$ was established using its simply connected subgeometry $\mathcal{G}(Co_2)$. The geometry was constructed in [RSt84] based on the maximal parabolic geometry $\mathcal{H}(Co_1)$ of the same group constructed in [RSm80]. The simple connectedness of $\mathcal{H}(Co_1)$ was shown in [Seg88]. The Baby Monster geometry $\mathcal{G}(BM)$ contains $C_4(2)$-subgeometries $\mathcal{G}(Sp_8(2))$ and also F_4-buildings associated with the groups ${}^2E_6(2)$. Consideration of the intersection graph Σ with respect to the latter family of subgeometries led to the simple connectedness proof for $\mathcal{G}(BM)$ in [Iv92c]. Detailed information about the structure of Σ from [Seg91] has played an important rôle in the proof. Finally the simple connectedness of the T_5-geometry $\mathcal{G}(M)$ of the Monster was shown in [Iv91a] via consideration of the intersection graph with respect to the $\mathcal{G}(BM)$-subgeometries. The triangulability of that graph was established in [ASeg92].

A question which for a while looked rather intractable is the one about the universal 2-cover of the Baby Monster geometry $\mathcal{G}(BM)$. If $\widetilde{\mathcal{G}} \to \mathcal{G}(BM)$ is a proper 2-covering, then a connected component $\widetilde{\mathcal{H}}$ of the preimage in $\widetilde{\mathcal{G}}$ of a $\mathcal{G}(Sp_8(2))$-subgeometry from $\mathcal{G}(BM)$ must be a T_4-geometry possessing a morphism onto $\mathcal{G}(Sp_8(2))$. Motivated by this observation all T-geometries which possess morphisms onto $C_n(2)$-geometries were classified in [ISh93a]. It turned out that there is one family of such geometries containing one T_n-geometry for every $n \geq 2$. The full automorphism group of this T_n-geometry is a non-split extension of an elementary abelian 3-group of rank $\left[{n \atop 2}\right]_2 = (2^n-1)(2^{n-1}-1)/3$ by $Sp_{2n}(2)$. Thus the only possibility for $\widetilde{\mathcal{H}}$ is to be isomorphic to $\mathcal{G}(3^{35} \cdot Sp_8(2))$. It was decided to try to construct a 2-cover of $\mathcal{G}(BM)$ similarly to the way $\mathcal{G}(3^{23} \cdot Co_2)$ was constructed. The following question turned out to be crucial for the construction. Let $G_5 \cong 2^{5+10+10+5}.L_5(2)$ be the stabilizer in BM of an element of type 5 in $\mathcal{G}(BM)$ and $E \cong 2{\cdot}{}^2E_6(2).2$ be the stabilizer of an F_4-subgeometry. Is there always an element $q \in O_2(G_5)$ such that $q \in E \setminus E'$? The affirmative answer to this question was given in [ISh93b] and was independently checked in [Wil93] using computer calculations. This enabled to construct a proper 2-cover $\widetilde{\mathcal{G}}$ of $\mathcal{G}(BM)$. After that a very tight bound on the order of the automorphism groups of any such 2-cover was established and it became possible to deduce that $\widetilde{\mathcal{G}} \cong \mathcal{G}(3^{4371} \cdot BM)$ is in fact the universal 2-cover.

Let us turn to the local structure (the structure of amalgams of maximal parabolics) in P_n- and T_n-geometries for $n \geq 4$. In [Sh88] important information on the structure of the subgroup G_n, including a bound on its order, was deduced for the case of P-geometries. Later it was realized that these results can be extended word for word to the case of T-geometries. The structure of amalgams of maximal parabolic subgroups of flag-transitive T_n-geometries was studied in [Row91], [Row92] and [Par92]. In [Row91] it was shown that if G is a flag-transitive automorphism group of a T_4-geometry such that the Borel subgroup of the action of G_1 on the T_3-residue has order 2^{10}, then the Borel subgroup of G has order 2^{21} (which is the order of a Sylow 2-subgroup of Co_1) or 2^{25}. In [Row92] it was shown that if G is a flag-transitive automorphism group of a T_5-geometry and the Borel subgroup of the action of G_1 on the residual T_4-geometry has order 2^{21}, then the Borel subgroup of G is of order 2^{46} (which is the order of a Sylow 2-subgroup of the Monster). Proceeding by induction and assuming that all P- and T-geometries of smaller rank are known, in view of (1.11.3), certain information on the possible structure of a G_1-parabolic can be deduced from the knowledge of natural representations of residual P- and T-geometries.

The universal natural representations of P- and T-geometries were studied even before their importance for the local analysis was noticed. It was shown in [ISh89a] that $\mathcal{G}(Mat_{23})$ has no natural representation, $V(\mathcal{G}(Mat_{22}))$ is the 11-dimensional Golay code module while $V(\mathcal{G}(3 \cdot Mat_{22}))$ is the direct sum of $V(\mathcal{G}(Mat_{22}))$ and the natural 6-dimensional $GF(4)$-module for $SU_6(2)$ (the latter contains $3 \cdot Mat_{22}$). It was shown in [RSm89] that $V(\mathcal{G}(Mat_{22}))$ is also the universal representation module for $\mathcal{G}(Mat_{24})$. In [ISh94a] the universal representation module for $\mathcal{G}(Co_2)$ was identified with a 23-dimensional section of $\bar{\Lambda} = \Lambda/2\Lambda$ where Λ is the Leech lattice. Using this result it was not difficult to show in [Iv92a] that $\bar{\Lambda}$ itself is the universal representation module for $\mathcal{G}(Co_1)$. The equality $V(\mathcal{G}(Co_1)) = \bar{\Lambda}$ was independently proved in [Sm92]. After that it was shown that P- and T-geometries associated with "large" sporadic groups do not have natural representations. In [ISh90b] it was proved that there are no such representations for $\mathcal{G}(J_4)$. In [ISh94a] $\mathcal{G}(BM)$ was proved to have no natural representation. Since $\mathcal{G}(BM)$ is a subgeometry in $\mathcal{G}(M)$, the latter has no natural representations either (the question about existence of such representations was posed in [Str84]). The universal representations of the T-geometries from the symplectic series were determined in [Sh93], where the equality $V(\mathcal{G}(3^{23} \cdot Co_2)) = V(\mathcal{G}(Co_2))$ was also established. The latter equality was used in [ISh94a] to show that $\mathcal{G}(3^{4371} \cdot BM)$ does not have a natural representation. For a long time the geometry $\mathcal{G}(He)$ was known to possess a natural representation in a 51-dimensional irreducible $GF(2)$-module [MSm82]. It was established by B. McKay (private communication) using computer calculation that $V(\mathcal{G}(He))$ is in fact 52-dimensional.

Let G be one of J_4, BM and M, and $\mathcal{G} = \mathcal{G}(G)$. Then the elements of type i in $\mathcal{G}$ are certain elementary abelian 2-subgroups in G of rank i with the incidence relation defined via inclusion. From this description it is immediate that G is a quotient of $R(\mathcal{G})$. The non-triviality of the representation group of $\mathcal{G}$ explains in a sense why the original proofs for the triviality of $V(\mathcal{G})$ were rather complicated. It turned out to be easier to work with the whole representation group $R(\mathcal{G})$ and to show the equality $[R(\mathcal{G}), R(\mathcal{G})] = R(\mathcal{G})$ (which is of course equivalent to the triviality of $V(\mathcal{G})$). Recently the precise structure of $R(\mathcal{G})$ was determined in [IPS96] and [ISh97]: $R(\mathcal{G}(G))$ is J_4, $2 \cdot BM$ and M for $G \cong J_4$, BM and M, respectively.

Let $\mathcal{A} = \{G_i \mid 1 \leq i \leq n\}$ be the amalgam of maximal parabolics associated with a flag-transitive action of a group G on a P- or T-geometry $\mathcal{G}$. Since the lists in (1.10.1) and (1.10.2) are known to be closed under taking universal covers, in order to complete the classification it is sufficient to show that $\mathcal{A}$ is isomorphic to the amalgam associated with an action on a geometry from these lists. This statement was proved in [ShSt94] for the case when $\mathcal{G}$ is a T-geometry having $\mathcal{G}(3^7 \cdot Sp_6(2))$ as a residue. Using the results from [Sh88], their generalizations for T-geometries and the results from [Row91], [Row92], [Par92], one can deduce some detailed information on the structure of a G_n-parabolic, which restricts considerably the possibilities for its chief factors. Let us assume that the residue $\mathcal{H} = \mathrm{res}_{\mathcal{G}}(x_1)$ is known (as well as its natural representations) and let Q_1 and R_1 be as in (1.11.3). Since the dual of Q_1/R_1 supports a natural representation of $\mathcal{H}$ we have only a few possibilities for the structure of this factor. In particular $Q_1 = R_1$ if $\mathcal{H}$ does not have natural representations (which happens when $\mathcal{H}$ is one of $\mathcal{G}(Mat_{23})$, $\mathcal{G}(J_4)$, $\mathcal{G}(BM)$, $\mathcal{G}(3^{4371} \cdot BM)$ and $\mathcal{G}(M)$). The equality $Q_1 = R_1$ must also hold in the case $\mathcal{H} \cong \mathcal{G}(He)$ since otherwise the Borel subgroup of G would be of order at least 2^{61} which contradicts [Row91]. On the other hand the equality $Q_1 = R_1$ leads to a contradiction with the information on the chief factors of G_n. Thus Q_1/R_1 is non-trivial and we are left with the following four possibilities: (a) $\mathcal{G}$ is a T_4-geometry and $G_1 \sim 2^{11}.Mat_{24}$; (b) $\mathcal{G}$ is a P_5-geometry and $G_1 \sim 2_+^{1+22}.Co_2$; (c) $\mathcal{G}$ is a P_5-geometry and $G_1 \sim (2_+^{1+22} \times 3^{23}).Co_2$; (d) $\mathcal{G}$ is a T_5-geometry and $G_1 \sim 2_+^{1+24}.Co_1$. For the former three cases the uniqueness of the amalgam $\mathcal{A}$ was proved in an unpublished work of S.V. Shpectorov while for the latter case it was established in [Iv92a].

1.13 Consequences and development

In this section we discuss some results which were proved either using the classification of P- and T-geometries or under the inspiration of this classification.

Maximal parabolic geometries

Let G be one of the following sporadic simple groups: Mat_{24}, Co_1, M, $M(24)$, Mat_{22}, Co_2, BM and J_4. Let $\mathcal{H}(G)$ be the maximal parabolic geometry of G as introduced in [RSm80] with one of the following diagrams:

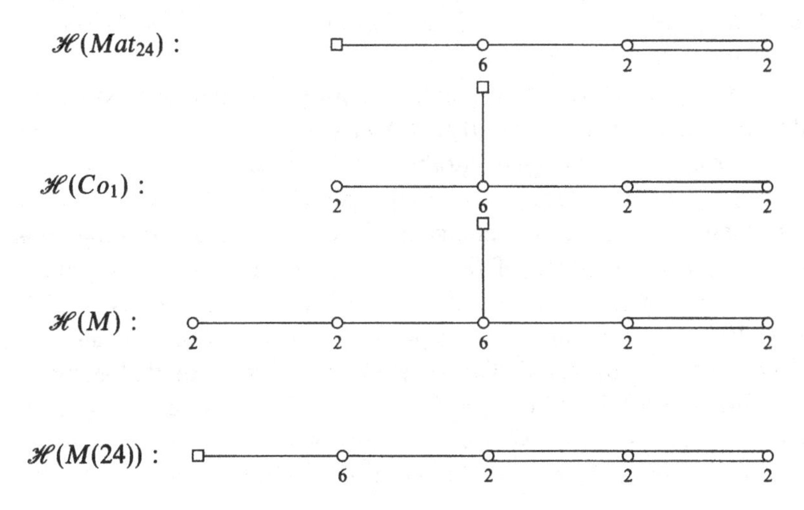

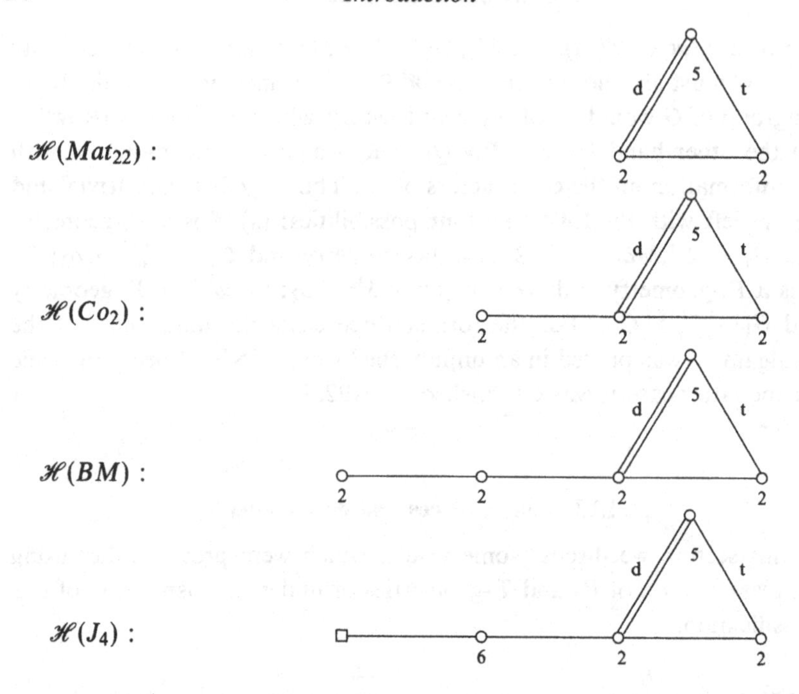

Here $\underset{5}{\circ}\!\overset{t}{\text{———}}\!\underset{2}{\circ}$ denotes the geometry of 2- and 1-element subsets in a set of size 5 in which two subsets of different size are incident if they are disjoint. The elements of type 1 in the geometry with the diagram $\underset{2}{\circ}\!\overset{d}{=\!=\!=}\!\underset{5}{\circ}$ are the maximal totally isotropic subspaces in the natural symplectic module V of $Sp_4(2)$, the elements of type 2 are the cosets of the hyperplanes in V, an element S of type 1 and an element $H+v$ of type 2 are incident if $S < H$. The semidirect product $V : Sp_4(2)$ induces on this geometry a flag-transitive action.

Let $\mathscr{A}$ be the amalgam of maximal parabolic subgroups associated with the action of G on $\mathscr{H}(G)$. If $G \cong Mat_{24}$, Co_1 or M then $\mathscr{A}$ contains the amalgam $\mathscr{B}$ of maximal parabolic subgroups associated with the action of G on the T-geometry $\mathscr{G}(G)$ and if $G \cong Mat_{22}$, Co_2, BM or J_4 then $\mathscr{A}$ contains the amalgam of maximal parabolic subgroups associated with the action of G on the P-geometry $\mathscr{G}(G)$. Furthermore in both cases $\mathscr{B}$ generates the universal completion of $\mathscr{A}$. By (1.10.1) and (1.10.2) G is the universal completion of $\mathscr{B}$ and hence it is also the universal completion of $\mathscr{A}$. The simple connectedness of the geometry $\mathscr{H}(M(24))$ was established in [Iv95]). Thus we have the following result which was proved in [Ron82] for $G \cong Mat_{24}$ and in [Seg88] for $G \cong Co_1$ and which answers the question posed at the end of [RSm80].

Theorem 1.13.1 *Let G be one of the following groups:*

$$Mat_{24},\ Co_1,\ M,\ M(24),\ Mat_{22},\ Co_2,\ BM,\ J_4.$$

Then the maximal parabolic geometry $\mathscr{H}(G)$ of G is simply connected. □

Locally projective graphs of girth 5

As we mentioned in the previous section, our interest in P-geometries originated from the classification problem of locally projective graphs of girth 5. As a direct consequence of the reduction results in [Iv88], [Iv90] together with (1.10.1) we have the following.

Theorem 1.13.2 *Let Γ be a locally projective graph of type (n,q), $n \geq 2$, with respect to a group G. Suppose that the girth of Γ is 5 and $G_1(x) \neq 1$ at every $x \in \Gamma$. Then $q = 2$ and one of the following holds:*

(i) *Γ is the derived graph $\Delta(\mathscr{G})$ of a P-geometry $\mathscr{G}$ of rank n and G is a flag-transitive automorphism group of $\mathscr{G}$, where $\mathscr{G}$ is one of the following $\mathscr{G}(Sym_5)$, $\mathscr{G}(Mat_{22})$, $\mathscr{G}(3 \cdot Mat_{22})$, $\mathscr{G}(Co_2)$, $\mathscr{G}(3^{23} \cdot Co_2)$, $\mathscr{G}(J_4)$, $\mathscr{G}(BM)$ and $\mathscr{G}(3^{4371} \cdot BM)$;*

(ii) *$n = 5$, $G \cong J_4$, the vertices of Γ are the imprimitivity blocks of size 31 of G on the vertex set of $\Delta = \Delta(\mathscr{G}(J_4))$ in which two blocks are adjacent in Γ if their union contains a pair of vertices adjacent in Δ.* □

The locally projective graphs of girth 5 with $G_1(x) = 1$ were studied in [IP98] (see (9.11.6)) and it turns out that the graph $\Delta(\mathscr{G}(Mat_{23}))$ occupies a very specific position in the class of such graphs.

Uniqueness of sporadics and their extensions

The existence of the geometry $\mathscr{G}(G)$ for $G \cong J_4$, BM or M was proved starting with very basic local properties of the group G. The information on G needed is the structure of the centralizer $C = C_G(\tau)$ of a central involution τ in G and the fact that τ is conjugate in G to involutions from $O_2(C)\setminus\{\tau\}$. The maximal parabolic geometries of J_4 and M were predicted to exist in [RSm80] even before the groups themselves were constructed. For this reason the classification of P- and T-geometries immediately provides a uniform uniqueness proof of G as a group possessing the properties needed to deduce the existence of $\mathscr{G}(G)$ [Iv91b].

Theorem 1.13.3 *Let G be a non-abelian simple group containing an involution τ such that $C := C_G(\tau)$ is of the shape*

$$2_+^{1+12}.3 \cdot \mathrm{Aut}\, Mat_{22},\ \ 2_+^{1+22}.Co_2 \ \ or \ \ 2_+^{1+24}.Co_1.$$

Suppose that $C_G(O_2(C)) \leq O_2(C)$ and that $\tau^G \cap O_2(C) \neq \{\tau\}$. Then in each of the three cases G is uniquely determined up to isomorphism and is isomorphic to J_4, BM or M, respectively. □

Other uniqueness proofs can be found in [Nor80], [ASeg91] for J_4, [LSi77], [Seg91] for BM and [Nor85], [GMS89] for M.

Within the classification of P- and T-geometries and their natural representations we have proved some interesting facts about linear representations and non-split extensions of sporadic groups. As an illustration we formulate the following result which can be deduced from [ISh93b].

Theorem 1.13.4 *Let F be a field whose characteristic is not 2. Then BM has a unique faithful representation of dimension 4371 over F which is irreducible. If the characteristic of F is not 3 then the extension of BM by the corresponding 4371-dimensional F-module always splits, and for $F = GF(3)$ there is a unique non-split extension.* □

Generators and relations

The classification of P- and T-geometries enabled us to obtain characterizations of certain sporadic simple groups, stronger than the characterization by the centralizer of an involution. The groups were proved to coincide with the universal completions of certain of their subamalgams. This provided us with presentations of the groups involved (the *geometric presentations* as they were termed in [Iv91a]). In the case of J_4 the geometric presentation was proved in [Iv92b] to be equivalent to a presentation for J_4 conjectured by G. Stroth and R. Weiss in [StW88].

In the case of BM and M the result establishes the correctness of the so-called Y-presentations for these groups. The Y-presentations ([CCNPW], [CNS88], [Nor90]) describe groups as specific factor groups of Coxeter groups with diagrams having three arms originating in a common node. The most famous is the Y_{555} diagram (below).

After the announcement of the geometric presentation of M at the Durham symposium "Groups, Combinatorics and Geometry" in July 1990 [Iv92a] S.P. Norton [Nor92] proved its equivalence to the corresponding Y-presentation. This resulted in the proof of the following theorem conjectured by J.H. Conway [Con92].

Theorem 1.13.5 *The Coxeter group corresponding to the Y_{555} diagram subject to a single additional relation*

$$(ab_1c_1ab_2c_2ab_3c_3)^{10} = 1$$

is isomorphic to the wreath product $M \wr 2$ of the Monster group and a group of order 2 (this wreath product is known as the Bimonster). □

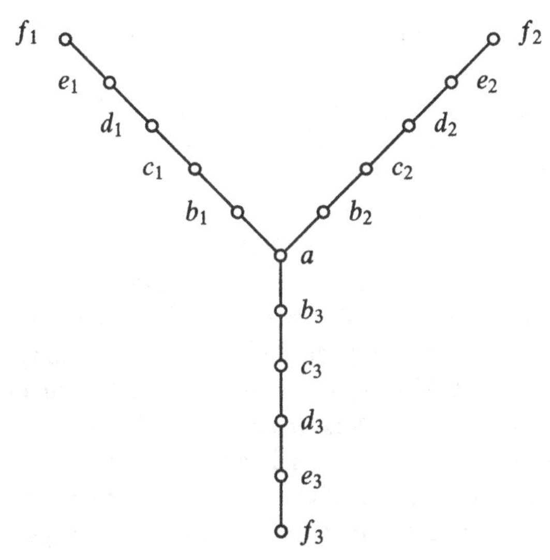

The correctness of the Y-presentation for BM is proved in [Iv94].

Construction of J_4

As a consequence of the classification of P- and T-geometries we have the following. Let G be one of the sporadic simple groups Mat_{23}, Mat_{24}, He, Co_2, Co_1, J_4, BM and M. Then G is the universal completion (and because of its simplicity it is the unique faithful completion) of the amalgam $\mathscr{A} = \{G_i \mid 1 \leq i \leq n\}$ of maximal parabolic subgroups associated with the action of G on $\mathscr{G}(G)$. In addition, unless $G = Mat_{24}$ or He, the isomorphism type of $\mathscr{A}$ is uniquely determined by the chief factors of the G_i and by the indices $k_{ij} = [G_i : G_i \cap G_j]$. This means that G can be defined as "a faithful completion of an amalgam $\mathscr{A}$ with given chief factors and indexes k_{ij}". Hence (at least in principle) one can start from this definition to establish the existence of G and to deduce all its properties including the simplicity. This would give an independent construction of G together with its uniqueness proof. In its full extent the approach was realized in [IMe93] for the fourth Janko group J_4.

In [IMe93] the amalgams $\mathcal{M} = \{M_1, M_2, M_3\}$ such that

$$M_1 \sim 2^{11}.Mat_{24}, \quad M_2 \sim 2^{10}.L_5(2), \quad M_3 \sim [2^{15}].(Sym_5 \times L_3(2))$$

and

$$[M_2 : M_{12}| = 31, \ [M_3 : M_{13}] = 5, \ [M_3 : M_{23}] = 10, \ [M_{23} : B] = 3$$

where $M_{ij} = M_i \cap M_j$ and $B = M_1 \cap M_2 \cap M_3$ have been studied. An amalgam with these properties is contained in J_4, where M_1 is the stabilizer of a $\mathcal{G}(Mat_{24})$-subgeometry in $\mathcal{G}(J_4)$, M_2 is the stabilizer of a vertex in the graph Γ as in (1.13.2 (ii)) and M_3 is the stabilizer of a Petersen subgraph in Γ. These subgroups are universal completions of their intersections with the amalgam of maximal parabolic subgroups corresponding to the action of J_4 on $\mathcal{G}(J_4)$.

It was shown that up to isomorphism there exists exactly one amalgam $\mathcal{M}$ as above. After that it was shown that $\mathcal{M}$ has a faithful completion. This was done by constructing an isomorphic embedding of $\mathcal{M}$ into $GL_{1333}(\mathbb{C})$ (notice that 1333 is the dimension of the smallest faithful complex representation of J_4). Next it was shown that if G is a completion of $\mathcal{M}$, then the subgroup M_1 when it acts naturally on the set of its right cosets in G has exactly seven orbits, whose lengths l_i, $1 \le i \le 7$, were explicitly calculated (so that these lengths are independent of the particular choice of the completion G). Thus every completion G of $\mathcal{M}$ has the same order $|M_1| \cdot \sum_{i=1}^{7} l_i$ which turns out to be

$$2^{21} \cdot 3^3 \cdot 5 \cdot 11^3 \cdot 23 \cdot 29 \cdot 31 \cdot 37 \cdot 43.$$

Moreover G is non-abelian and simple and $C_G(z) \sim 2_+^{1+12}.3 \cdot \mathrm{Aut}\, Mat_{22}$ for an involution z in G. This means that G is the fourth Janko group J_4 according to the standard definition of the latter. Originally J_4 was constructed on a computer as a subgroup of $GL_{112}(2)$ [Nor80].

Extended dual polar spaces

Another interesting class of geometries admitting flag-transitive actions of sporadic simple groups is formed by extended (classical) dual polar spaces. An *extended dual polar space* (EDPS for short) of order (s, t) has diagram

```
     c
o---------o=========o---------o ... o---------o
1         s         t         t     t         t
```

where the leftmost edge denotes the geometry of all 1- and 2-element subsets of an $(s+2)$-element set with respect to the natural incidence relation; the residue of a point is a classical thick dual polar space.

Flag-transitive EDPS's of rank 3 were classified in [BH77], [DGMP], [WY90] and [Yos91]. The representations of rank 3 flag-transitive EDPS's were described in [Yos92]. Rank 4 EDPS's with the property that the stabilizer of a point in a flag-transitive automorphism group acts faithfully on the residue of that point were classified in [Yos94]. Further progress in the classification of flag-transitive EDPS's was achieved in [Iv95], [ISt96], [IMe97], [Iv98b], [Iv98a] and [Iv97]. An EDPS $\mathscr{E}$ is called *affine* if it possesses a flag-transitive automorphism group G which contains a normal subgroup T acting regularly on the set of points of $\mathscr{E}$. In this case T is called the *translation group* of $\mathscr{E}$ with respect of G. The following characterization of affine EDPS's is proved in [Iv98a].

Theorem 1.13.6 *Let $\mathscr{E}$ be an affine EDPS of rank $n \geq 3$ and $\mathscr{D}$ be the residue in $\mathscr{E}$ of a point. Then*

- (i) *$s = 2$, so that $\mathscr{D}$ is the dual polar space associated with $Sp_{2n-2}(2)$ or $U_{2n-2}(2)$,*
- (ii) *there is a 2-covering $\widetilde{\mathscr{E}} \to \mathscr{E}$ where $\widetilde{\mathscr{E}}$ is an EDPS which is 2-simply connected and affine,*
- (iii) *the translation group of $\widetilde{\mathscr{E}}$ (as above) with respect to its full automorphism group is the universal representation group $R(\mathscr{D})$ of the residual dual polar space $\mathscr{D}$.* □

Thus the above result reduces the classification of affine EDPS's to the calculation of the universal representation groups of the dual polar spaces associated with $Sp_{2m}(2)$ and $U_{2m}(2)$. The precise structure of these representation groups is known only for $m = 2$ and 3 but some partial results are available also for larger m (see [Iv98a] for details).

In [Iv97] it is shown that there are exactly 19 flag-transitive EDPS's which are not 2-covered by affine ones.

Theorem 1.13.7 *Let $\mathscr{E}$ be an EDPS of rank at least 3 which possesses a flag-transitive automorphism group whose Borel subgroup is finite. Then one of the following holds:*

- (i) *there is a 2-covering $\widetilde{\mathscr{E}} \to \mathscr{E}$ where $\widetilde{\mathscr{E}}$ is affine;*
- (ii) *$\mathscr{E}$ is isomorphic to one of the 19 exceptional EDPS's whose diagrams and full automorphism groups are given in the table.*

o —c— o ═══ o (1, 2, 2)	Sym_8, $U_4(2).2$
o —c— o ═══ o (1, 2, 4)	$Sp_6(2) \times 2$, $Sp_6(2)$
o —c— o ═══ o (1, 4, 2)	$3 \cdot U_4(3).2^2$, $U_4(3).2^2$
o —c— o ═══ o (1, 3, 3)	$U_5(2).2$
o —c— o ═══ o (1, 3, 9)	$McL.2$
o —c— o ═══ o (1, 9, 3)	$HS.2$, $Suz.2$
o —c— o ═══ o —— o (1, 2, 2, 2)	$Sp_8(2)$, $3 \cdot M(22).2$, $M(22).2$
o —c— o ═══ o —— o (1, 2, 4, 4)	$Co_2 \times 2$, Co_2
o —c— o ═══ o —— o (1, 4, 2, 2)	$M(24)$
o —c— o ═══ o —— o (1, 3, 3, 3)	$M(24)$
o —c— o ═══ o —— o (1, 9, 3, 3)	M
o —c— o ═══ o —— o —— o (1, 2, 2, 2, 2)	$M(23)$ □

In what follows $\mathscr{E}(E)$ will denote the EDPS for which E is the commutator subgroup of the automorphism group. In case we need to specify the order (s, t) of the EDPS, we write $\mathscr{E}_{s,t}(E)$. Let $\mathscr{E}$ be an exceptional EDPS from (1.13.7), G be the automorphism group of $\mathscr{E}$ and H be a subgroup of G. Unless $\mathscr{E} = \mathscr{E}(HS)$ the action of H is flag-transitive if and only if H contains the commutator subgroup of G; Aut HS is the only flag-transitive automorphism group of $\mathscr{E}(HS)$.

The classification strategy implemented in [Iv97] is the one developed within the classification project for P- and T-geometries. This strategy is based on studying the amalgams of maximal parabolics, representations of residual geometries and the simple connectedness question. In addition the following direct connections (A), (B) and (C) between EDPS's and P- and T-geometries have played essential rôles in the classification.

(A) Let $\mathscr{E}$ be an EDPS of rank $n \geq 4$ in which the residual rank 3 EDPS's are all isomorphic to $\mathscr{E}(U_4(2))$, so that $s = t = 2$. Let Δ be the graph on the set of elements of type 2 in $\mathscr{E}$ in which two elements are adjacent if they are incident to a common element of type 3 but not to a common element of type 1. Then every connected component of Δ is the derived graph of a Petersen geometry of rank $n-1$. In this way the EDPS's $\mathscr{E}(M(22))$, $\mathscr{E}(3 \cdot M(22))$ and $\mathscr{E}(M(23))$ are related to P-geometries $\mathscr{G}(Mat_{22})$, $\mathscr{G}(3 \cdot Mat_{22})$ and $\mathscr{G}(Mat_{23})$, respectively.

(B) The EDPS $\mathscr{E}_{4,2}(M(24))$ is closely related to a geometry $\mathscr{H}(M(24))$ with the same automorphism group and diagram

2 2 2 2

The point residues in $\mathscr{H}(M(24))$ are isomorphic to the T_3-geometry $\mathscr{G}(Mat_{24})$. In [Iv95] the simple connectedness of $\mathscr{H}(M(24))$ was proved first and then use this result in the simple connectedness proof for $\mathscr{E}_{4,2}(M(24))$. It is worth mentioning that there is a geometry $\mathscr{H}(3 \cdot M(24))$ with diagram

possessing a morphism onto $\mathscr{H}(M(24))$ whose simple connectedness was also established in [Iv95].

(C) The EDPS $\mathscr{E}(M)$ of the Monster group was constructed in [BF83] and [RSt84]. The simple connectedness proof for $\mathscr{E}(M)$ in [IMe97] reduces the problem to the simple connectedness of the T_5-geometry of the Monster. That is, we show that the universal completion U of the amalgam of maximal parabolic subgroups corresponding to the action of M on $\mathscr{E}(M)$ contains subgroups C, N and L as in (1.10.4) which implies that $U = M$. It was noted in [Iv96] that 3-local and 2-local parabolics in the Monster are related via a subgeometry $\mathscr{H}(2^{24} \cdot Co_1)$ in $\mathscr{E}(M)$ with diagram

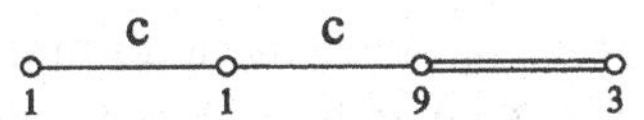

acted on flag-transitively by $C/Z(C)$. In [IMe97] it was shown that the universal cover of $\mathscr{E}(M)$ also contains $\mathscr{H}(2^{24} \cdot Co_1)$ as a subgeometry. At this stage we applied the main result of [ISt96] which proves the simple connectedness of the EDPS $\mathscr{E}_{3,3}(M(24))$ which is also a subgeometry in $\mathscr{E}(M)$. It is worth mentioning that the geometry $\mathscr{H}(2^{24} \cdot Co_1)$ is not simply connected since it possesses a double cover [ISh98] (which might or might not be universal).

1.14 Terminology and notation

In this section we fix our terminology and notation concerning groups and their actions on graphs and geometries.

Let G be a group. Then 1 is the identity element, $G^{\#} = G \setminus \{1\}$, $Z(G)$ is the centre, $G' = [G, G]$ is the commutator subgroup and $\mathrm{Aut}\, G$ is the automorphism group of G. Let g be an element in G (written $g \in G$) and H be a subgroup of H (written $H \leq G$). Then $\langle g \rangle$ is the subgroup of G generated by g, $C_G(g)$ and $C_G(H)$ are the centralizers in G of g and H, $N_G(H)$ is the normalizer of H in G. If $N_G(H)/C_G(H) \cong \mathrm{Aut}\, H$ then H is said to be *fully normalized* in G. By G/H we denote the set of right cosets of H in G. In the case where H is normal in G (written $H \trianglelefteq G$), by G/H we also denote the corresponding factor group. The core of H in G, denoted by $core_G(H)$ is the largest normal subgroup of G contained in H, so that $core_G(H)$ is the intersection of the conjugates of H in G. The smallest normal subgroup in G which contains H is the *normal closure* of H in G. Let p and q be different primes. Then $O_p(G)$ is the largest normal subgroup in G which is a p-group (*i.e.* whose order is p^k for some k), $O_{p,q}(G)$ is the preimage in G of $O_q(G/O_p(G))$, $O^p(G)$ is the smallest normal subgroup in G with the property that the corresponding factor group is a p-group. By p^n we denote the elementary abelian group of this order, so that p also denotes the cyclic group of order p. When we write $[p^n]$, we mean a group of order p^n; 2_ε^{1+2n} denotes the extraspecial group of order 2^{2n+1} of type $\varepsilon \in \{+, -\}$. Recall that a maximal abelian subgroup of 2_ε^{1+2n} has rank $n+1$ if $\varepsilon = +$ and n if $\varepsilon = -$. The symmetric and alternating groups of degree n we denote by Sym_n and Alt_n, respectively. If we want to specify the underlying set X of size n, we write $Sym(X)$ and $Alt(X)$, respectively. In the case where H is a group, naturally identified with a subgroup in Sym_n (say if $H \cong Sym_m \times Sym_{n-m}$ for some $1 \leq m \leq n$), then H^e denotes the subgroup of even permutations in H, *i.e.* $H^e = H \cap Alt_n$.

If we write $G \sim A.B$ we mean that G has a normal subgroup isomorphic to (and identified with) A such that G/A is isomorphic to B (written $G/A \cong B$). When we write $G \sim A \cdot B$ we mean that G does not split over A, *i.e.* there is no subgroup H in G such that $H \cap A = 1$ and $H \cong B$. Finally by writing $G \cong A : B$ we mean that G is a semidirect product of A and B with respect to a homomorphism of B into $\mathrm{Aut}\, A$ which is either clear from the context or irrelevant. If the homomorphism is trivial we have the direct product $A \times B$.

If V is a vector space of dimension n over the field $GF(q)$ of q elements, where $q = p^a$ for a prime p, then $\Gamma L(V)$, $GL(V)$ and $SL(V)$ denote respectively the group of semilinear, linear and linear with determinant 1 bijections of V onto itself. A group G is said to be a *linear group* of V if

$$SL(V) \trianglelefteq G \leq \Gamma L(V),$$

in which case V is said to be the *natural module of* G. When interested in groups only we write $\Gamma L_n(q)$, $GL_n(q)$ and $SL_n(q)$. By $A\Gamma L_n(q)$, $AGL_n(q)$ and $ASL_n(q)$ we denote the affine groups which are semidirect products of V (considered as an elementary abelian p-group) and the corresponding linear group. The actions of the groups $\Gamma L_n(q)$, $GL_n(q)$ and $SL_n(q)$ on the set of subspaces of the underlying vector space V are denoted by $P\Gamma L_n(q)$, $PGL_n(q)$ and $L_n(q)$, respectively. By $P\Sigma L_n(q)$ we denote the extension of $L_n(q)$ by the field automorphisms. A group H is said to be a *projective linear group* if H is the image in $P\Gamma L_n(q)$ of a linear group of V. The (doubly transitive) permutation action of a projective linear group on the set of 1-dimensional subspaces of the underlying vector space is said to be the *natural permutation representation* of the projective linear group. By $Sp_{2m}(q)$ and $U_m(q^{1/2})$ we denote the projective special symplectic and unitary groups. By $O_m^{\varepsilon}(q)$ we denote the orthogonal group of dimension m and type $\varepsilon \in \{+,-\}$; its commutator subgroup is denoted by $\Omega_m^{\varepsilon}(q)$.

Let Σ be a set on which a group G acts by permutations (*i.e.* there is a homomorphism of G into the symmetric group of Σ clear from the context). The image of $x \in \Sigma$ under $g \in G$ is denoted by x^g. If $\Xi = \{x, y, ...\}$ is a subset of Σ then $G[\Xi] = G[x, y, ...]$ and $G(\Xi) = G(x, y, ...)$ denote the setwise and the elementwise stabilizers of Ξ in G. If $g \in G(X)$ then g is said to *fix* X and if $g \in G[X]$ then g is said to *stabilize* X. Similar terminology applies to subgroups of G. If $H \leq G[\Xi]$ then H^{Ξ} denotes the permutation group induced by H in Ξ. Suppose that G acts transitively on Σ. Then an orbit of G on the set of ordered pairs of elements of Σ is called an *orbital*. The orbitals containing (x, y) and (y, x) are said to be dual to each other. An orbital which coinsides with its dual is called self-dual.

For a set Σ let 2^{Σ} be the *power set* of Σ, *i.e.* the set of all subsets of Σ. The symmetric difference operator

$$X \triangle Y = (X \cup Y) \setminus (X \cap Y)$$

provides 2^Σ with a $GF(2)$-vector-space structure. If G is a permutation group on Σ then 2^Σ is the permutational $GF(2)$-module of G. Let F be the binary function on 2^Σ taking values in $GF(2)$ and defined by

$$F(X,Y) = \begin{cases} 0 & \text{if } |X \cap Y| \text{ is even,} \\ 1 & \text{otherwise.} \end{cases}$$

Then F is non-singular and bilinear and will be called the *parity form* on 2^Σ. The parity form is invariant under every permutation group G on Σ and if $G = Sym(\Sigma)$ then F is the unique invariant form which is non-singular and bilinear. A subset of Σ will be called *even* if it contains an even number of elements.

A *partition* of Σ is a set $\mathcal{S} = \{S_1, ..., S_n\}$ of subsets of Σ such that every $\sigma \in \Sigma$ belongs to exactly one S_i. If $\mathcal{T} = \{T_1, ..., T_m\}$ is a partition of Σ such that every S_i is the union of some T_j then we say that $\mathcal{T}$ *refines* $\mathcal{S}$ and that $\mathcal{S}$ is refined by $\mathcal{T}$. Sometimes we identify a subset $S \subseteq \Sigma$ with the partition $\{S, \Sigma \setminus S\}$. We write $(n_1^{k_1} n_2^{k_2} ...)$ for a multiset of integers in which n_1 appears k_1 times, n_2 appears k_2 times *etc.*, where normally $n_1 > n_2 > ...$ and n_i is written instead of n_i^1.

We will follow the terminology and notation concerning actions of groups on geometries as introduced in Section 1.3 with the following addition. Let $\mathcal{G}$ be a geometry of rank n with a string diagram on which types increase rightward from 1 to n and let x_i be an element of type i where $1 \leq i \leq n$. Then $res^+_{\mathcal{G}}(x_i)$ and $res^-_{\mathcal{G}}(x_i)$ are the residues in $\mathcal{G}$ of a flag of type $\{1, 2, ..., i\}$ and a flag of type $\{i, i+1, ..., n\}$ containing x_i, respectively. If G is a (flag-transitive) automorphism group of $\mathcal{G}$ and $G_i = G(x_i)$ is the stabilizer of x_i in G then $\bar{G}_i$ denotes the action induced by G_i on $res_{\mathcal{G}}(x_i)$, G_i^ε denotes the kernel of the action of G_i on $res^\delta_{\mathcal{G}}(x_i)$ for $\{\varepsilon, \delta\} = \{+, -\}$.

Let Γ be a graph, which is assumed to be undirected and locally finite. The latter means that every vertex is adjacent to a finite number of other vertices. The vertex set of Γ will be denoted by the same letter Γ and we will write $E(\Gamma)$ and $\mathrm{Aut}\,\Gamma$ for the edge set of Γ and for its automorphism group, respectively. The vertices incident to an edge will be called the *ends* of the edge. For a positive integer s an s-arc (or an arc of length s) in Γ is a sequence $(x_0, x_1, ..., x_s)$ of $s+1$ vertices such that $\{x_{i-1}, x_i\} \in E(\Gamma)$ for $1 \leq i \leq s$ and $x_i \neq x_{i-2}$ for $2 \leq i \leq s$. Such an arc is said to *originate* at x_0, to *terminate* at x_s and to *join* x_0 and x_s. If $x_0 = x_s$ then the arc is called a *cycle* of length s or simply an s-cycle. The *girth* of a graph is the length of its shortest cycle. A graph whose vertices are the edges of

Γ with two of them adjacent if they are incident to a common vertex of Γ is called the *line graph* of Γ.

The graph is *connected* if any two its vertices can be joined by an arc. A graph Γ is called *n-partite* or *multipartite* if its vertex set possesses a partition $\{\Gamma^i \mid 1 \le i \le n\}$ whose members are called *parts* such that if $\{x, y\} \in E(\Gamma)$ with $x \in \Gamma^i$ and $y \in \Gamma^j$ then $i \neq j$. 2-Partite graphs are also called *bipartite*. It is a standard fact that a graph is bipartite if and only if it does not contain cycles of odd length.

For $x, y \in \Gamma$ let $d(x, y)$ denote the distance between x and y in the natural metric of Γ, that is, the number of edges in a shortest arc joining x and y. If Ξ and Λ are subsets in the vertex set of Γ then the distance $d(\Xi, \Lambda)$ between Ξ and Λ is the minimum among the distances $d(x, y)$ for $x \in \Xi$ and $y \in \Lambda$. The diameter d of Γ is the maximum of distances between its vertices. Put

$$\Gamma_i(x) = \{y \mid y \in \Gamma, d(x, y) = i\} \quad \text{for} \quad 0 \le i \le d.$$

We will usually write $\Gamma(x)$ instead of $\Gamma_1(x)$.

The number of vertices adjacent to x, that is $|\Gamma(x)|$, is called the valency of x. If $k = |\Gamma(x)|$ is independent of the choice of $x \in \Gamma$ then Γ is called *regular* of *valency* k.

Let Ξ be a subset in the vertex set of Γ. The subgraph of Γ *induced* on Ξ has Ξ as set of vertices and its edges are all the edges of Γ with both ends contained in Ξ. A subgraph in which any two vertices are adjacent is called a *clique*; a subgraph in which no two vertices are adjacent is called a *coclique*. Let Δ be a graph. Then a graph Γ is said to be *locally* Δ if for every $x \in \Gamma$ the subgraph in Γ induced by $\Gamma_1(x)$ is isomorphic to Δ.

A cycle $(x_0, x_1, ..., x_s)$ is said to be *non-degenerate* if for every i, j with $0 \le i < j \le s$, the distance $d(x_i, x_j)$ in Γ is equal to the distance between x_i and x_j in the cycle, which is

$$\min\{j - i, i + s - j\}.$$

Let Ξ be an induced subgraph in Γ. Then Ξ is said to be *geodetically closed* if whenever $x, y \in \Xi$ with $d(x, y) = i$, all arcs of length i joining x and y in Γ are contained in Ξ. If in addition Ξ contains all arcs of length $i+1$ joining x and y then Ξ is called *strongly geodetically closed*. A graph Γ can be considered as a 1-dimensional simplicial complex [Sp66]. The *fundamental group* of Γ is by definition the fundamental group of the corresponding complex.

If Γ and Γ' are graphs then a surjective mapping $\varphi : \Gamma' \to \Gamma$ is called a *covering of graphs* if for every $x' \in \Gamma'$ the restriction of φ to $\Gamma'(x')$ is a bijection onto $\Gamma(\varphi(x'))$. Let $x \in \Gamma$. Then for every $x' \in \varphi^{-1}(x)$ and an s-arc $X = (x_0 = x, x_1, ..., x_s)$ originating at x there is a unique s-arc $X^{-1}(x')$ originating at x' which maps onto X. Here $X^{-1}(x') = (x'_0 = x', x'_1, ..., x'_s)$ and x'_i is the unique vertex in $\varphi^{-1}(x_i) \cap \Gamma'(x'_{i-1})$ for $1 \leq i \leq s$. If X is a cycle then $X^{-1}(x')$ might or might not be a cycle. If $X^{-1}(x')$ is a cycle for every $x' \in \varphi^{-1}(x)$ then X is said to be *contractible* with respect to φ. A covering is characterized by the subgroup in the fundamental group of the graph generated by the contractible cycles [Sp66].

Suppose that $X = (x_0, x_1, ..., x_s = x_0)$ is an s-cycle and for $0 \leq i \leq j < s$ let $Y = (y_0 = x_i, y_1, ..., y_t = x_j)$ be an arc joining x_i and x_j. Then we say that X *splits* into $X_1 = (x_0, ...x_i, y_1, ..., y_{t-1}, x_j, x_{j+1}, ..., x_s)$ and $X_2 = (x_i, x_{i+1}, ..., x_j, y_{t-1}, ..., y_1, x_i)$ (in this case X is the sum of X_1 and X_2 modulo 2). It is easy to see that if both X_1 and X_2 are contractible then X is also contractible. If in its turn X_1 splits into cycles X_3 and X_4, then we say that X splits into the cycles X_2, X_3 and X_4. Thus inductively we can define the splittings of X into any number of cycles. If X splits into a set of triangles then X is said to be *triangulable*. A graph is triangulable if every of it cycles is triangulable. If Γ is triangulable and every triangle in Γ is contractible with respect to φ then φ must be an isomorphism and hence Γ is triangulable if and only if its fundamental group is generated by the triangles. The following sufficient condition for triangulability is a straightforward generalization of Lemma 5 in [Ron81a].

Lemma 1.14.1 *Let Γ be a graph of diameter d and suppose that for every i, $2 \leq i \leq d$, the following two conditions hold:*

(i) *if $y \in \Gamma_i(x)$ then the subgraph in Γ induced by $\Gamma(x) \cap \Gamma_{i-1}(y)$ is connected;*

(ii) *if $y, z \in \Gamma_i(x)$ and $z \in \Gamma(y)$ then $d(\Gamma(x) \cap \Gamma_{i-1}(y), \Gamma(x) \cap \Gamma_{i-1}(z)) \leq 1$.*

Then Γ is triangulable, which means that its fundamental group is generated by the triangles. The condition (ii) is implied by the following:

(iii) *if $y \in \Gamma_i(x)$ then every vertex from $\Gamma(x) \setminus \Gamma_{i-1}(y)$ is adjacent to a vertex from $\Gamma_{i-1}(y) \cap \Gamma(x)$.* □

Let $G \leq \mathrm{Aut}\,\Gamma$ be an automorphism group of Γ. Then G is said to be, respectively, vertex-transitive, edge-transitive or s-arc-transitive if it acts transitively on the vertex set, edge set or set of s-arcs. In these respective cases Γ is also called *vertex-transitive, edge-transitive* and *s-arc-transitive.*

If the action of a group is s-arc-transitive but not $(s+1)$-arc-transitive, we will say that it is *strictly* s-arc-transitive. Notice that if G acts 1-arc-transitively on Γ then Γ can be identified with a self-dual orbital of the action of G on the vertex set of the graph. For $x \in \Gamma$ let $G(x)$ be the stabilizer of x in G. The permutation group $G(x)^{\Gamma(x)}$ is known as the *subconstituent* of G on Γ. For an integer i define

$$G_i(x) = \bigcap_{d(x,y)\leq i} G(y)$$

which clearly is a normal subgroup of $G(x)$. Then $G(x)^{\Gamma(x)}$ is abstractly isomorphic to the factor group $G(x)/G_1(x)$. If $\{x,y\}$ is an edge, we put $G_i(x,y) = G_i(x) \cap G_i(y)$.

Let Γ be a graph and G be a vertex-transitive automorphism group of Γ. Suppose that G preserves on Γ an imprimitivity system $\mathscr{B}$. Define Δ to be a graph whose vertices are the imprimitivity blocks from $\mathscr{B}$ and two such blocks B_1, B_2 are adjacent if there is $\{x,y\} \in E(\Gamma)$ such that $x \in B_1$ and $y \in B_2$. Then Δ is said to be constructed from Γ by *factorizing over the imprimitivity system* $\mathscr{B}$.

The *standard double cover* $2\cdot\Gamma$ of Γ is a graph on $2\cdot\Gamma := \{(x,\alpha) \mid x \in \Gamma, \alpha \in \{0,1\}\}$ with vertices (x,α) and (y,β) being adjacent if and only if $\{x,y\} \in E(\Gamma)$ and $\alpha \neq \beta$. Then $\varphi : (x,\alpha) \mapsto x$ is a covering of $2\cdot\Gamma$ onto Γ. If C is a cycle of length m in Γ then $\varphi^{-1}(C)$ is a disjoint union of two cycles of length m when m is even and it is a cycle of length $2m$ if m is odd. This means that a cycle in Γ is contractible with respect to φ if and only if it has even length. In particular Γ and $2\cdot\Gamma$ have the same girth if the girth of Γ is even. Furthermore, if Γ is bipartite and connected then $2\cdot\Gamma$ consists of two connected components, each isomorphic to Γ. If Γ is non-bipartite and connected then $2\cdot\Gamma$ is bipartite and connected. If G is a vertex-transitive automorphism group of Γ then the group $\widetilde{G}$ generated by the automorphisms $\widetilde{g} : (x,\alpha) \mapsto (x^g,\alpha)$ for every $g \in G$ together with the automorphism $\delta : (x,\alpha) \mapsto (x, 1-\alpha)$ is isomorphic to $G \times 2$ and it acts vertex-transitively on $2\cdot\Gamma$. The pairs $\{(x,0),(x,1)\}$ form an imprimitivity system of $\widetilde{G}$ and Γ can be reconstructed from $2\cdot\Gamma$ by factorizing over this system.

The action of a group $G \leq \operatorname{Aut}\Gamma$ on Γ is *distance-transitive* if for every $0 \leq i \leq d$ the group G acts transitively on the set

$$\Gamma_i = \{(x,y) \mid x,y \in \Gamma, d(x,y) = i\}.$$

A graph which possesses a distance-transitive action is called a *distance-transitive graph.* If Γ is distance-transitive then for every i, $0 \leq i \leq d$, the parameters

$$c_i = |\Gamma_{i-1}(y) \cap \Gamma(x)|, \ a_i = |\Gamma_i(y) \cap \Gamma(x)|, \ b_i = |\Gamma_{i+1}(y) \cap \Gamma(x)|$$

are independent of the choice of the pair $x, y \in \Gamma$ satisfying $d(x, y) = i$. Clearly in this case Γ is regular and $c_i + a_i + b_i = |\Gamma(x)| = k$ is the valency of Γ. The sequence

$$i(\Gamma) = \{b_0 = k, b_1, ..., b_{d-1}; c_1 = 1, c_2, ..., c_d\}$$

is called the *intersection array* of the distance-transitive graph Γ. If we put $k_i = |\Gamma_i(x)|$ for $1 \leq i \leq d$ then

$$k_i = \frac{k \cdot b_1 \cdot ... \cdot b_{i-1}}{c_1 \cdot c_2 \cdot ... \cdot c_i}.$$

To represent the decomposition of a distance-transitive graph with respect to a vertex we draw the following distance diagram:

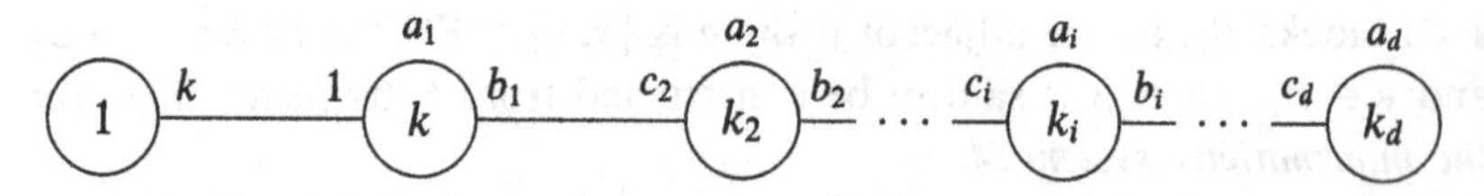

We draw similar diagrams for non-distance-transitive actions. Let G be a group acting on a graph Γ and x be a (basic) vertex. The *suborbit diagram* (with respect to x) consists of ovals (or circles) joined by curves (or lines). The ovals represent the orbits of $G(x)$ on the vertex set of Γ. Inside the oval which represents an orbit Σ_i (the Σ_i-oval) we show the size of Σ_i or place its name explained in the context. Next to the Σ_i-oval we show the number n_i (if non-zero) of vertices in Σ_i and adjacent to a given vertex $y_i \in \Sigma_i$. On the curve joining the Σ_i- and Σ_j-ovals we put the numbers n_{ij} and n_{ji} (called *valencies*.) Here n_{ij} (appearing closer to the Σ_i-oval) is the number of vertices in Σ_j adjacent to y_i. Clearly

$$|\Sigma_i| \cdot n_{ij} = |\Sigma_j| \cdot n_{ji}$$

and we draw no curve if $n_{ij} = n_{ji} = 0$. Normally we present the valencies n_i and n_{ij} as sums of lengths of orbits of $G(x, y_i)$ on the vertices in Σ_i and Σ_j adjacent to y_i. When the orbit lengths are unknown or irrelevant, we put the valencies into square brackets. Generally the suborbit diagram depends on the orbit of G on Γ from which the basic vertex x is taken. Even if a graph is not necessarily distance-transitive, we use the notation c_i, a_i, b_i if the corresponding parameters are independent of the choice of a pair of vertices at distance i.

2
Mathieu groups

In this chapter we construct the Mathieu groups and study their basic properties. We construct the largest Mathieu group Mat_{24} as the automorphism group of the (binary extended) Golay code defined in Section 2.1. In Section 2.2 we construct a Golay code as the quadratic residue code over $GF(23)$. In Section 2.3 we show that a minimal non-empty subset in a Golay code has size 8 (called an octad). Moreover the set of all octads in a Golay code forms the block set of a Steiner system of type $S(5,8,24)$. The residue of a 3-element subset of elements in a Steiner system of type $S(5,8,24)$ is a projective plane of order 4. In Section 2.4 we review some basic properties of the linear groups and in Sections 2.5 and 2 6 we define the generalized quadrangle of order $(2,2)$ and its triple cover which is the tilde geometry of rank 2. In Section 2.7 we prove uniqueness of the projective plane of order 4 and analyse some properties of the plane and its automorphism group. This analysis enables us to establish the uniqueness of the Steiner system of type $S(5,8,24)$ in Section 2.8. The Mathieu group Mat_{24} of degree 24 is defined in Section 2.9 as the automorphism group of the unique Golay code. The uniqueness proof implies rather detailed information about Mat_{24} and two other large Mathieu groups Mat_{23} and Mat_{22}. In Section 2.10 we study the stabilizers in Mat_{24} of an octad, a trio and a sextet. In Section 2.11, analysing dodecads in the Golay code and their stabilizers in Mat_{24}, we introduce the little Mathieu groups. In Sections 2.12, 2.13 and 2.14 we classify the subgroups in Mat_{24} of order 2 and 3 and determine octads, trios and sextets stabilized by such a subgroup. In Section 2.15 we study the action of the large Mathieu groups on the Golay code and on its cocode. Finally in Section 2.16 we describe the generalized quadrangle of order (3,9) in terms of the projective plane of order 4.

2.1 The Golay code

Let X be a finite set of elements. A *binary linear code* $\mathscr{C}$ based on X is a subspace of the power set 2^X (considered as a $GF(2)$-vector–space).

In general a linear code over $GF(q)$ is a triple $(V, X, \mathscr{C})$ where V is a $GF(q)$-vector–space, X is a basis of V and $\mathscr{C}$ is a subspace in V. It is obvious that in the case $q = 2$ this definition is equivalent to the above one. In what follows unless explicitly stated otherwise, when talking about codes we always mean binary linear codes.

The size of X is called the *length* of a code $\mathscr{C}$ based on X. A code is *even* (respectively *doubly even*) if the number of elements in every non-empty subset in $\mathscr{C}$ is even (respectively divisible by 4). The *minimal weight* of $\mathscr{C}$ is the number of elements in a smallest non-empty subset in $\mathscr{C}$. The dual code $\mathscr{C}^*$ of $\mathscr{C}$ is the orthogonal complement of $\mathscr{C}$ with respect to the parity form:

$$\mathscr{C}^* = \{A \mid A \in 2^X, |A \cap B| \text{ is even for all } B \in \mathscr{C}\}.$$

Since $|A \cap B| = \frac{1}{2}(|A| + |B| - |A \triangle B|)$, every doubly even code is contained in its dual.

If $\mathscr{C} = \mathscr{C}^*$ then the code is called *self-dual.* The following characterization of self-dual codes is immediate since the parity form is non-singular and bilinear.

Lemma 2.1.1 *A code $\mathscr{C}$ is self-dual if it is totally singular with respect to the parity form and* $\dim \mathscr{C} = |X|/2$. □

By the above lemma the length of a self-dual code is always even.

Definition 2.1.2 *A code $\mathscr{C}$ is called a* Golay code *(extended binary Golay code) if $\mathscr{C}$ is self-dual of length* 24 *with minimal weight greater than or equal to* 8.

By (2.1.1) a Golay code is 12-dimensional and we will denote it by $\mathscr{C}_{12}$.

We will show that up to isomorphism there exists a unique Golay code whose automorphism group is the sporadic Mathieu group Mat_{24} and that the Golay code is doubly even. We start in the next section by constructing a Golay code as the quadratic residue code over $GF(23)$.

2.2 Constructing a Golay code

In this section we give an elementary construction of a Golay code. That is, we will construct such a code as the quadratic residue code over $GF(23)$. The construction is elementary since quadratic residue codes over $GF(q)$ can be constructed for all $q = -1 \bmod 8$.

Consider the field $GF(23)$ of size 23 whose elements will be denoted by integers from 0 to 22 so that the field operations are the usual addition and multiplication modulo 23. The set $GF(23)^*$ of non-zero elements in $GF(23)$ is the union of the set Q of squares and the set N of non-squares, where

$$Q = \{1,2,3,4,6,8,9,12,13,16,18\},$$

$$N = \{5,7,10,11,14,15,17,19,20,21,22\}.$$

Let $\mathscr{P}$ be the projective line over $GF(23)$ considered as the union of $GF(23)$ and the formal symbol ∞. The group $L \cong L_2(23)$ acts on $\mathscr{P}$ by means of transformations

$$x \mapsto \frac{ax+b}{cx+d},$$

where $ad - bc \in Q$. We introduce three elements of L:

$$t : x \mapsto x+1; \quad s : x \mapsto 2x; \quad \tau : x \mapsto -1/x.$$

Let T and S be the subgroups of L generated by t and s, respectively.

The following statement is immediate.

Lemma 2.2.1

(i) *T is cyclic of order 23 acting regularly on $GF(23) = \mathscr{P} \setminus \{\infty\}$;*
(ii) *S is cyclic of order 11, it acts regularly on Q and N, normalizing T;*
(iii) *S is the elementwise stabilizer in L of the pair $\{\infty, 0\}$;*
(iv) *the stabilizer $L(\infty)$ of ∞ in L is the semidirect product of T and S and it acts transitively on the set of unordered pairs of elements of $GF(23)$;*
(v) *τ is an involution which normalizes S, maps ∞ onto 0 and Q onto N.* □

The following result is also rather standard, but we present a brief proof for the sake of completeness.

Lemma 2.2.2

(i) *L is generated by t, s and τ;*

(ii) *L acts doubly transitively on $\mathcal{P}$ and has order* $24 \cdot 23 \cdot 11$;

(iii) *L acts transitively on the set of* 3*-element subsets of $\mathcal{P}$, and the elementwise stabilizer in L of every such subset is trivial;*

(iv) *the setwise stabilizer in L of a* 3*-element subset of $\mathcal{P}$ is cyclic of order* 3 *acting fixed-point freely on $\mathcal{P}$.*

Proof. Since $L(\infty)$ acts transitively on $\mathcal{P} \setminus \{\infty\}$ and τ does not fix ∞ we have (i) and (ii). By the double transitivity every L-orbit on 3-element subsets of $\mathcal{P}$ contains a triple $\{\infty, 0, a\}$ for some $a \in GF(23)^*$. Under the action of S the triples of this shape split into two orbits depending on whether $a \in Q$ or $a \in N$. Since τ stabilizes $\{\infty, 0\}$ and permutes Q and N, these two orbits are fused and we obtain (iii). Finally (iv) follows from (iii) and the order of L. □

For $a \in GF(23)$ put $N_a = \{n + a \mid n \in N\} \cup \{a\}$ and let $\mathcal{N} = \{N_a \mid a \in GF(23)\}$.

Lemma 2.2.3

(i) *$L(\infty)$ preserves $\mathcal{N}$ as a whole and acts on $\mathcal{N}$ as it acts on $GF(23)$, in particular $L(\infty)$ acts transitively on the set of unordered pairs of subsets in $\mathcal{N}$;*

(ii) *every element $b \in GF(23)$ is contained in exactly* 12 *subsets from $\mathcal{N}$;*

(iii) *any two subsets from $\mathcal{N}$ have intersection of size* 6.

Proof. It is straightforward that $N_a^t = N_{a+1}$ and $N_a^s = N_{2a}$, which imply (i). Since $L(\infty)$ acts transitively on $GF(23)$ and preserves $\mathcal{N}$, the number of subsets in $\mathcal{N}$ containing a given element $b \in GF(23)$ is independent of the choice of b and (ii) follows. Now counting in two ways the number of configurations $(a, \{A, B\})$ where $a \in GF(23)$, $A, B \in \mathcal{N}$ and $a \in A \cap B$, we obtain (iii). □

Let $\mathcal{C}$ be the code based on $\mathcal{P}$ generated by $\mathcal{N}$ and the whole set $\mathcal{P}$.

Lemma 2.2.4 *$\mathcal{C}$ is stable under L.*

Proof. By (2.2.3 (i)) $\mathcal{C}$ is stable under $L(\infty)$. Hence by (2.2.2) it is sufficient to show that $\mathcal{C}$ is stable under τ. Clearly $\mathcal{P}^\tau = \mathcal{P}$ and $N_0^\tau = \mathcal{P} \triangle N_0$. Now one can check directly or consult [MS77], p. 492 for a general argument that for $a \neq 0$ we have

$$N_a^\tau = N_{-1/a} \triangle N_0 \triangle \mathcal{P}.$$

Lemma 2.2.5

(i) $\mathscr{C}$ *is totally singular with respect to the parity form;*
(ii) $\dim \mathscr{C} = 12$*;*
(iii) *a non-empty subset in* $\mathscr{C}$ *has size at least* 8*;*
(iv) $\mathscr{C}$ *is a Golay code.*

Proof. By (2.2.3 (ii), (iii)) the intersection of any two subsets from $\mathscr{N} \cup \{\mathscr{P}\}$ has an even number of elements and (i) follows. The code $\mathscr{C}$ is a faithful $GF(2)$-module for the cyclic group T of order 23 and $\mathscr{P}$ generates in $\mathscr{C}$ a 1-dimensional submodule. On the other hand 11 is the smallest number m such that $2^m - 1$ is divisible by 23. Hence $\mathscr{C}/\langle\mathscr{P}\rangle$ is at least 11-dimensional. By (i) $\mathscr{C}$ is at most 12-dimensional and we obtain (ii).

In view of (i) in order to prove (iii) we have to show that $\mathscr{C}$ does not contain subsets of size 2, 4 and 6. Let D be a subset in $\mathscr{C}$ of size d.

(a) Let $d = 2$. Then by (2.2.2 (ii)) $\mathscr{C}$ contains all 2-element subsets of $\mathscr{P}$ and hence all even subsets. This is impossible since the dimension of $\mathscr{C}$ is only 12.
(b) Let $d = 4$ and E be a 3-element subset of D. Let g be an element of order 3 in L which stabilizes E as a whole (compare (2.2.2 (iv))). Since g acts fixed-point freely on $\mathscr{P}$ it cannot stabilize D as a whole. Hence $|D \triangle D^g| = 2$ which is impossible by (a).
(c) Let $d = 6$. Let E and F be any two 3-element subsets in D and g be an element from L which maps E onto F (2.2.2 (iii)). Then $D \cap D^g$ contains F and must be of even size. Hence it is of size 4 or 6 and in the latter case $D^g = D$. If the intersection is of size 4 then $|D \triangle D^g| = 4$ which is impossible by (b). Thus D must be stable under all such elements g. This means that the setwise stabilizer of D in L acts transitively on the set of 3-element subsets of D. Since the number of such subsets is divisible by 5, by (2.2.2 (ii)) this contradicts the Lagrange theorem.

Now (iv) follows from (i), (ii), (iii) and (2.1.1). □

2.3 The Steiner system $S(5,8,24)$

In this section $\mathscr{P}$ is an arbitrary set of size 24 and $\mathscr{C}_{12}$ is a Golay code based on $\mathscr{P}$. Eventually we will show that $\mathscr{C}_{12}$ is unique up to isomorphism, in particular it is isomorphic to the code constructed in the previous section.

Let $\bar{\mathscr{C}}_{12}$ be the set of cosets of $\mathscr{C}_{12}$ in $2^{\mathscr{P}}$, so that $\bar{\mathscr{C}}_{12}$ is a 12-dimensional

$GF(2)$-space. For $0 \leq i \leq 24$ let $\mathcal{P}_i$ be the set of i-element subsets in $\mathcal{P}$, $|\mathcal{P}_i| = \binom{24}{i}$, and let $\bar{\mathcal{C}}_{12}(i)$ be the set of cosets in $\bar{\mathcal{C}}_{12}$ having non-empty intersection with $\mathcal{P}_i$. If D and E are distinct subsets from $2^{\mathcal{P}}$ contained in the same coset from $\bar{\mathcal{C}}_{12}$ then $D \Delta E$ is a non-empty subset from $\mathcal{C}_{12}$ and hence its size is at least 8. This immediately gives the following.

Lemma 2.3.1 *Let D and E be distinct subsets of $\mathcal{P}$ contained in the same coset from $\bar{\mathcal{C}}_{12}$, such that $D \in \mathcal{P}_i$, $E \in \mathcal{P}_j$ with $0 \leq i, j \leq 4$. Then $i = j = 4$ and $D \cap E = \emptyset$.* □

Since there can be at most six pairwise disjoint 4-element subsets of $\mathcal{P}$ we obtain the next result.

Lemma 2.3.2 *If $0 \leq i < j \leq 4$ then $\bar{\mathcal{C}}_{12}(i) \cap \bar{\mathcal{C}}_{12}(j) = \emptyset$; $|\bar{\mathcal{C}}_{12}(i)| = \binom{24}{i}$ and $|\bar{\mathcal{C}}_{12}(4)| \geq \binom{24}{4}/6$.* □

Now one can easily verify the following remarkable equality:

$$1 + \binom{24}{1} + \binom{24}{2} + \binom{24}{3} + \binom{24}{4}/6 = 2^{12}.$$

Since the right hand side in the above equality is the total number of cosets in $\bar{\mathcal{C}}$, and the summands in the left hand size are the lower bounds for $|\bar{\mathcal{C}}_{12}(i)|$, $i = 0, 1, 2, 3$ and 4, given by (2.3.2), we have the following.

Lemma 2.3.3 *$\bar{\mathcal{C}}_{12}$ is the disjoint union of the $\bar{\mathcal{C}}_{12}(i)$ for $i = 0, 1, 2, 3$ and 4; $|\bar{\mathcal{C}}_{12}(4)| = \binom{24}{4}/6$, which means that for every $S \in \mathcal{P}_4$ there is a partition of $\mathcal{P}$ into six subsets $S_1 = S, S_2, ..., S_6$ from $\mathcal{P}_4$ such that $S_i \cup S_j \in \mathcal{C}$ for $1 \leq i < j \leq 6$.* □

By the above lemma the minimal weight of a Golay code is exactly 8. A subset of size 8 in a Golay code will be called an *octad*. A partition of $\mathcal{P}$ into six 4-element subsets such that the union of any two is an octad will be called a *sextet*. The elements from $\mathcal{P}_4$ will be called *tetrads*. In these terms by (2.3.3) every tetrad is a member of a unique sextet.

Lemma 2.3.4 *Every element $F \in \mathcal{P}_5$ is contained in a unique octad.*

Proof. Let S be a tetrad contained in F and $\{S_1 = S, S_2, ..., S_6\}$ be the unique sextet containing S. Let $\{x\} = F \setminus S$ and let j, $2 \leq j \leq 6$, be such that $x \in S_j$. Then $O = S_1 \cup S_j$ is the octad containing F. If there were another octad O' containing F then $O \triangle O'$ would be a non-empty subset in $\mathscr{C}_{12}$ of size at most 6, which is impossible. □

Definition 2.3.5 *Let t, k, v be integers with $1 \leq t < k < v$. A Steiner system of type $S(t,k,v)$ is a pair $(\mathscr{X}, \mathscr{B})$ where $\mathscr{X}$ is a set of v elements and $\mathscr{B}$ is a collection of k-element subsets of $\mathscr{X}$ called* blocks *such that every t-element subset of $\mathscr{X}$ is contained in a unique block.*

Lemma 2.3.6

(i) *The minimal weight of a Golay code $\mathscr{C}_{12}$ is 8 and the subsets of size 8 (called octads) are the blocks of a Steiner system of type $S(5,8,24)$;*

(ii) *$\mathscr{C}_{12}$ is generated by its octads as a $GF(2)$-vector–space.*

Proof. Immediately from (2.3.4) and (2.3.5) we have (i). Let $\mathscr{D}$ be the subspace in $\mathscr{C}_{12}$ generated by the octads. To prove (ii) it is sufficient to show that $|2^{\mathscr{P}}/\mathscr{D}| \leq 2^{12}$. Let $E \subseteq \mathscr{P}$. We claim that the coset of $\mathscr{D}$ containing E contains a subset of size at most 4. In fact, suppose that $|E| \geq 5$ and D is a 5-element subset in E. Then the symmetric difference of E and the (unique) octad containing D is smaller than E and the claim follows by induction. Also it is clear that the tetrads from a sextet are equal modulo $\mathscr{D}$ and the result follows from the equality given after (2.3.2). □

By the above lemma, to prove the uniqueness of the Golay code it is sufficient to establish the uniqueness of the Steiner system of type $S(5,8,24)$.

It is easy to see that the number of blocks in a Steiner system of type $S(t,k,v)$ is $\binom{v}{t}/\binom{k}{t}$. The next two lemmas are standard and easy to prove.

Lemma 2.3.7 *Let $(\mathscr{X}, \mathscr{B})$ be a Steiner system of type $S(t,k,v)$ with $t \geq 2$ and Y be an m-element subset of $\mathscr{X}$ where $m < t$. Let*

$$\mathscr{B}(Y) = \{B \setminus Y \mid B \in \mathscr{B}, Y \subseteq B\}.$$

Then $(\mathscr{X} \setminus Y, \mathscr{B}(Y))$ is a Steiner system of type $S(t-m, k-m, v-m)$ called the residual system of $(\mathscr{X}, \mathscr{B})$ with respect to Y. □

Lemma 2.3.8 *Let $(\mathscr{X}, \mathscr{B})$ be a Steiner system of type $S(2, n+1, n^2+n+1)$ and let $\mathscr{G}$ be the incidence system whose points are the elements of $\mathscr{X}$, whose* *lines are the elements of $\mathcal{B}$ and where incidence relation is via inclusion. Then $\mathcal{G}$ is a projective plane of order n.* □

Thus the residue of a 3-element subset in a Steiner system of type $S(5,8,24)$ is a projective plane of order 4. In the next section we prove uniqueness of this plane and study its basic properties to be prepared for the uniqueness proof for the Steiner system of type $S(5,8,24)$.

2.4 Linear groups

In this section we summarize some standard properties of linear and projective linear groups. We refer the reader to [AB95], [Tay92] and Section 9.3 in [BCN89] for proofs and further details.

Let $V = V_n(q)$ be an n-dimensional $GF(q)$-vector–space where $n \geq 2$ and $q = p^m$ where p is a prime. Let G be a linear group on V and put $G^0 = G \cap GL(V)$. First of all we have the following.

Lemma 2.4.1 *The group $SL_n(q)$ is perfect and the group $L_n(q)$ is non-abelian and simple, unless $(n,q) = (2,2)$ or $(2,3)$; $GL_2(2) \cong Sym_3$ and $GL_2(3) \cong 2_-^{1+2} : Sym_3$.* □

Let $\mathcal{P} = \mathcal{P}(V)$ be the projective geometry of rank $n-1$ associated with V, *i.e.* the set of all proper subspaces of V with type function being the dimension and incidence relation defined via inclusion. Let $\mathcal{P}^i$ be the set of i-dimensional subspaces in $\mathcal{P}$. Then $|\mathcal{P}^i| = \left[{n \atop i}\right]_q$ where the latter is the q-ary Gaussian binomial coefficient:

$$\begin{bmatrix} n \\ i \end{bmatrix}_q = \frac{(q^n-1)\cdot(q^{n-1}-1)\cdot \ldots \cdot(q^{n-i+1}-1)}{(q^i-1)\cdot(q^{i-1}-1)\cdot \ldots \cdot(q-1)}.$$

An isomorphism between two projective geometries is also called a *collineation*; a *correlation* is a product of a collineation and a diagram automorphism. Let W be a hyperplane in V and w be a non-zero vector in W. A *transvection* $t = t(w, W)$ with *centre* w and *axis* W is a linear transformation defined as follows:

$$v^t = \begin{cases} v & \text{if } v \in W, \\ v + w & \text{otherwise.} \end{cases}$$

Lemma 2.4.2 *Let $\{V_i \mid 1 \le i \le n-1\}$ be a maximal flag in $\mathcal{P}$, where $V_i \in \mathcal{P}^i$, G_i be the stabilizer of V_i in G and $B = \bigcap_{i=1}^{n} G_i$ be the Borel subgroup. Then*

(i) *the G_i are pairwise different maximal subgroups in G,*
(ii) *V_i is the only proper subspace of V stabilized by G_i,*
(iii) *G_i induces linear groups on both V_i and V/V_i,*
(iv) *the G-orbit containing a given pair (U, W) of subspaces in V is uniquely determined by the dimensions of U, W and $U \cap W$,*
(v) *if $G_i \cap G_j < H < G$ then $H = G_k$ for $k = i$ or j,*
(vi) *$B = N_G(S)$ where S is a Sylow p-subgroup in G and the index of S in B divides $(q-1)^n \cdot m$,*
(vii) *G induces non-equivalent doubly transitive actions on $\mathcal{P}^1$ and on $\mathcal{P}^{n-1}$.* □

The next lemma contains important information on the structure of the parabolics G_i. Notice that since $\mathcal{P}$ possesses a diagram automorphism, we have $G_i \cong G_{n-i}$.

Lemma 2.4.3 *For $1 \le i \le (n-1)/2$ let U_{n-i} be an $(n-i)$-dimensional subspace of V with trivial intersection with V_i and L_i be the stabilizer of U_{n-i} in G_i. Then*

(i) *$G_i = Q_i : L_i$ where $Q_i = O_p(G_i)$,*
(ii) *if $i \ge 2$ then L_i contains a characteristic subgroup $K_i \cong SL(V_i) \times SL(U_{n-i})$ and Q_i is a $GF(q)K_i$-module isomorphic to the dual of $V_i \otimes U_{n-i}$,*
(iii) *L_1 contains a characteristic subgroup $K_1 \cong SL(U_{n-1})$ and Q_1 is a $GF(q)K_1$-module isomorphic to the dual of U_{n-1},*
(iv) *if R_i is the kernel of the action of G_i on $\mathrm{res}_{\mathcal{G}}(V_i)$ then $Q_i \le R_i$, $R_i \le G^0$ and R_i/Q_i is a cyclic group whose order divides $q-1$.* □

The next two lemmas provide further details on the structure of G_1. We follow notation introduced in (2.4.3); in addition for an element $U \in \mathcal{P}^2$ incident to V_1 let $L(U)$ denote the set of elements from $\mathcal{P}^1$ other then V_1 incident to U, so that $L(U)$ is a q-element subset of $\mathcal{P}^1$.

Lemma 2.4.4 *The subgroup $G_1 \cap G^0$ induces $PGL_{n-1}(q)$ on $\mathrm{res}_{\mathcal{G}}(V_1)$. If U_1, U_2 are different elements from $\mathcal{P}^2$ incident to V_1 then Q_1 induces a regular action of order q on both $L(U_1)$ and $L(U_2)$ and an action of order q^2 on $L(U_1) \cup L(U_2)$.* □

Lemma 2.4.5 *Let $N \neq 1$ be a normal subgroup in G_1. Then N contains Q_1. If N is not contained in R_1 then either $\bar{N} := NR_1/R_1$ contains $L_{n-1}(q)$, or $(n,q) = (3,2)$ and $\bar{N} \cong 3$, or $(n,q) = (3,3)$ and $\bar{N} \cong 2^2$.* □

In order to simplify the notation put V_0 to be the zero subspace in V and V_n to be the whole space V. Let $\{e_1, e_2, ..., e_n\}$ be a basis of V such that $V_i = \langle e_1, e_2, ..., e_i \rangle$ for $0 \leq i \leq n$. Let $\bigwedge^i V$ be the i-th exterior power of V turned into a $GF(q)G$-module, $0 \leq i \leq n$. Recall that $\bigwedge^i V$ is of dimension $\binom{n}{i}$ and has a basis

$$\{e_{k_1} \wedge e_{k_2} \wedge ... \wedge e_{k_i} \mid 1 \leq k_1 < k_2 ... < k_i \leq n\}.$$

This shows that the elements of $\mathcal{P}^i$ correspond to certain 1-dimensional subspaces of $\bigwedge^i V$ so that V_i corresponds to $\langle e_1 \wedge e_2 \wedge ... \wedge e_i \rangle$. We will use the following characterization of exterior powers which generalizes Lemma 1 in [CPr82] and Lemma 2.10 in [IMe93].

Proposition 2.4.6 *Let G be a linear group of an n-dimensional $GF(q)$-space V where $q = p^m$. Let W be a $GF(q)G$-module. Suppose that for some i, $1 \leq i \leq n-1$, there is an injective mapping φ of $\mathcal{P}^i$ into the set of 1-dimensional subspaces of W such that*

(i) *W is generated by the image of φ,*
(ii) *φ commutes with the action of G,*
(iii) *if $E_1, ..., E_{q+1}$ are the subspaces from $\mathcal{P}^i$ contained in V_{i+1} and containing V_{i-1}, then $\langle \varphi(E_i) \mid 1 \leq i \leq q+1 \rangle$ is a 2-dimensional subspace in W which is the natural module for the chief factor $SL(V_{i+1}/V_{i-1})$ of $G_{i-1} \cap G_{i+1}$.*

Then W is isomorphic to $\bigwedge^i V$.

From the above proposition it is straightforward to deduce the structure of the permutational $GF(2)$-module of $L_n(2)$ acting on the set of 1-dimensional subspaces (on the set of non-zero vectors) of the natural module.

Lemma 2.4.7 *Let V be the natural module of $G = SL_n(2) \cong L_n(2)$ and $\mathcal{P}$ be the projective geometry of V. Let P be the point set of $\mathcal{P}$ (the set of 1-dimensional subspaces) and $V_n = V > V_{n-1} > ... > V_1$ be a maximal flag in $\mathcal{P}$ where V_i is identified with the set of points it is incident to. Let W be the power set of P. Then W, as a $GF(2)G$-module, possesses the decomposition*

$$W = W^1 \oplus W^e,$$

where $W^1 = \{\emptyset, P\}$ *and* W^e *consists of the even subsets of* P. *Moreover,* W^e *is uniserial,*

$$W^e = W_1 > W_2 > ... > W_{n-1} > W_n = \{\emptyset\},$$

where W_i *is spanned by the images under* G *of* $P \setminus V_i$ *and* $W_i/W_{i+1} \cong \bigwedge^i V$, $1 \leq i \leq n-1$. □

2.5 The quad of order (2, 2)

In this section we present a description of the classical generalized quadrangle (or simply quad) $\mathcal{G}(Sp_4(2)) \cong \mathcal{B}_2(2)$ of order (2,2) and study its basic properties.

Let Ω be a set of size 6. Let $\mathcal{S} = (P, L)$ be a point–line incidence system whose point set P is the set of all 2-element subsets of Ω and three such points form a line if they are pairwise disjoint. Thus the line set L is the set of partitions of Ω into three pairs. It is an easy combinatorial exercise to check that $\mathcal{S}$ is a generalized quadrangle of order $(2, 2)$.

The symmetric group H of Ω isomorphic to Sym_6 acts naturally on $\mathcal{S}$. Moreover the points can be identified with the transpositions in H and the lines can be identified with the fixed-point free involutions (which are products of three pairwise commuting transpositions). In these terms a point and a line are incident if and only if they commute and the action of H on $\mathcal{S}$ is by conjugation. It is well known [Tay92] that $H \cong Sym_6$ possesses an outer automorphism τ which maps transpositions onto fixed-point free involutions. Since τ is an automorphism of H it maps commuting pairs of involutions onto commuting ones. This means that τ preserves the incidence in $\mathcal{S}$ and hence it performs a diagram automorphism. It is well known and straightforward to check that $\langle H, \tau \rangle$ acts distance-transitively on the incidence graph Γ of $\mathcal{S}$ and that Γ has the following distance diagram:

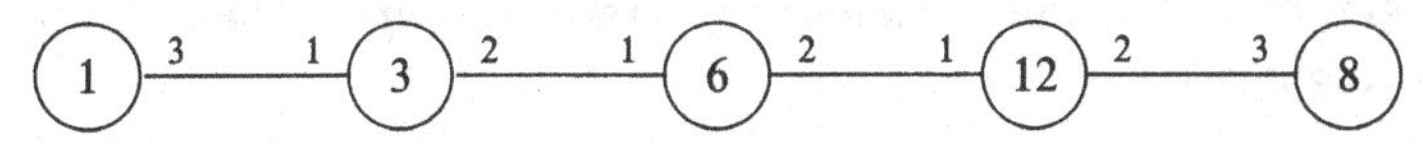

By the construction Γ is bipartite with parts P and L. Let Ξ^1 and Ξ^2 be graphs on P and L, respectively, in which two vertices are adjacent if they are at distance 2 in Γ. In other terms Ξ^1 is the collinearity graph of $\mathcal{S}$. Since $\mathcal{S}$ possesses a diagram automorphism, Ξ^1 and Ξ^2 are isomorphic. The group H acts distance-transitively on Ξ^i for $i = 1$ and 2 and the distance diagram of Ξ^i is the following:

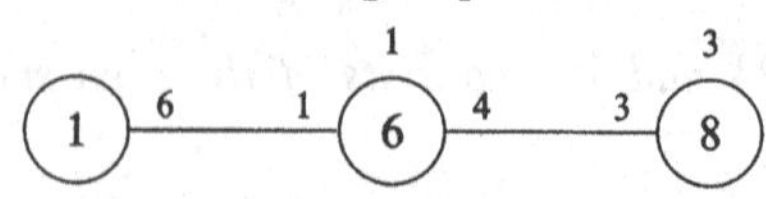

The vertices of Ξ^1 are the 2-element subsets of Ω and two vertices are adjacent if they are disjoint. For $\alpha \in \Omega$ there are five vertices in Ξ^1 which contain α and these vertices form a *coclique* (a maximal induced subgraph with no edges). Such a coclique will be called *standard*. Define a coclique in Ξ^2 to be standard if it is an image of a standard coclique in Ξ^1 under a diagram automorphism.

Lemma 2.5.1 *For $i = 1$ and 2 we have the following:*

(i) *every 5-vertex coclique in Ξ^i is standard;*
(ii) *$H \cong Sym_6$ is the full automorphism group of Ξ^i;*
(iii) *the stabilizer in H of a standard 5-vertex coclique is isomorphic to Sym_5 and acts transitively on the vertex set of Ξ^{3-i} and on the set of standard 5-vertex cocliques in Ξ^{3-1}.*

Proof. Since $\Xi^1 \cong \Xi^2$ we can assume that $i = 1$. Then (i) is an elementary combinatorial exercise. Let A be the automorphism group of Ξ^1. There are six standard cocliques and by (i) they are permuted by A. Let Σ be a standard coclique and B be the stabilizer of Σ in A. Then a vertex from $\Xi^1 \setminus \Sigma$ is adjacent to exactly three vertices in Σ and different vertices are adjacent to different triples. This shows that the action of B on Σ is faithful, hence

$$|A| \leq |B| \cdot 6 \leq |Sym_5| \cdot 6 = |Sym_6|$$

and (ii) follows. Finally it is an easy exercise to check (iii). □

Notice that if Σ is a standard 5-coclique in Ξ^i for $i = 1$ or 2 stabilized by $B \cong Sym_5$ then the subgraph in Ξ^i induced on $\Xi^i \setminus \Sigma$ is the Petersen graph and B acts on it as the full automorphism group. Such subgraphs will be called *standard Petersen subgraphs*. The following result is straightforward.

Lemma 2.5.2 *For $i = 1$ and 2 every 2-path (x, y, z) in Ξ^i with $d(x, z) = 2$ is contained in exactly two non-degenerate 5-cycles and in a unique standard Petersen subgraph.* □

Let $V_6(2) = 2^\Omega$ considered as a 6-dimensional $GF(2)$-space. Then it is easy to see that $V_6(2)$ contains exactly two proper H-submodules:

$V_1(2) = \{\emptyset, \Omega\}$ and $V_5(2) = \{X \mid X \subseteq \Omega, |X| = 0 \bmod 2\}$ (of dimension 1 and 5, respectively) with $V_1(2) \leq V_5(2)$.

Let φ be a mapping of $P \cup L$ into the set of subspaces of $V_5(2)$ defined as follows: if $p \in P$ (here p is a 2-element subset of Ω) then $\varphi(p) = \Omega \setminus p$ and if $l = \{p, q, r\} \in L$ then $\varphi(l) = \langle \varphi(p), \varphi(q), \varphi(r) \rangle$. It is easy to check that $\varphi(p) + \varphi(q) + \varphi(r) = 0$ and hence φ is a natural representation of $\mathcal{S}$.

Consider $V_4(2) = V_5(2)/V_1(2)$ and let χ be the natural homomorphism of $V_5(2)$ onto $V_4(2)$. The parity form induces on $V_4(2)$ a non-singular symplectic form Ψ. The composition of φ and χ is a natural representation of $\mathcal{S}$ in $V_4(2)$. Moreover the images of points are all 1-dimensional subspaces in $V_4(2)$ and a 2-dimensional subspace in $V_4(2)$ is an image of a line if and only if it is totally singular with respect to Ψ. Hence $\mathcal{S}$ is the classical generalized quadrangle $\mathcal{G}(Sp_4(2)) \cong \mathcal{B}_2(2)$ and H is the set of all linear transformations of $V_4(2)$ which preserve Ψ, reflecting the remarkable isomorphism $Sp_4(2) \cong Sym_6$. Notice that $V_5(2)$ is the orthogonal module for $Sp_4(2) \cong O_5(2)$ and by (1.11.2) φ is the universal natural representation of $\mathcal{G}(Sp_4(2))$. Applying (1.6.5) or analysing the maximal subgroups in Sym_6 one can see that $Sym_6 \cong Sp_4(2)$ and Alt_6 are the only flag-transitive automorphism groups of $\mathcal{G}(Sp_4(2))$. The following lemma is easy to deduce from elementary properties of symmetric groups.

Lemma 2.5.3 *Let $\Xi = \Xi^1$ be the collinearity graph of $\mathcal{G} = \mathcal{G}(Sp_4(2))$ and G be a flag-transitive automorphism group of $\mathcal{G}$. Let $x \in \Xi$ be a vertex, $G(x)$ be the stabilizer of x in G, $K(x) = O_2(G(x))$ and $Z(x) = Z(G(x))$. Then $G \cong Alt_6$ or $G \cong Sym_6 \cong Sp_4(2)$, and the following assertions hold:*

(i) *$G(x) \cong Sym_4 \times 2 \cong 2^3 : Sym_3$ if $G \cong Sym_6$ and $G(x) \cong Sym_4 \cong 2^2 : Sym_3$ if $G \cong Alt_6$;*

(ii) *$G(x)$ induces on $\mathrm{res}_{\mathcal{G}}(x)$ the natural action of Sym_3 with kernel $K(x)$;*

(iii) *$G(x)$ induces on $\Xi(x)$ the group Sym_4 in its action of degree 6 with kernel $Z(x)$;*

(iv) *the pointwise stabilizer in $G(x)$ of any two lines from $\mathrm{res}_{\mathcal{G}}(x)$ is contained in $Z(x)$;*

(v) *$K(x)$ is elementary abelian of order 2^2 or 2^3 and $K(x) \cap G(y) = 1$ for $y \in \Xi_2(x)$;*

(vi) *$Z(x) = 1$ if $G \cong Alt_6$ and $Z(x)$ is of order 2 acting fixed-point freely on $\Xi_2(x)$ if $G \cong Sym_6$;*

(vii) *every subgroup of index 15 in G is the stabilizer of either a vertex or a triangle in Ξ.* □

There are 16 quadratic forms f on $V_4(2)$ associated with Ψ in the sense that

$$\Psi(u,v) = f(v+u) + f(v) + f(u)$$

[Tay92]; 10 of these forms are of plus and 6 are of minus type. If f is of minus type then the set of 1-subspaces totally singular with respect to f is a standard coclique in Ξ^1. This reflects another important isomorphism $Sym_5 \cong O_4^-(2)$.

Let B be a Sylow 2-subgroup in $G \cong Sym_6$, so that $B \cong D_8 \times 2$. There are exactly two proper subgroups in G (say P_1 and P_2) properly containing B. Here $P_1 \cong P_2 \cong Sym_4 \times 2$ and $P_i = N_G(R_i)$ for $i = 1, 2$ where R_1 and R_2 are two elementary abelian subgroups of order 8 in B. Then in accordance with the standard principle we have the following.

Lemma 2.5.4 $\mathcal{G}(G, \{P_1, P_2\}) \cong \mathcal{G}(Sp_4(2))$. □

2.6 The rank 2 T-geometry

The incidence graph Γ of the generalized quadrangle $\mathcal{G}(Sp_4(2))$ possesses a distance-transitive antipodal triple cover $\widehat{\Gamma}$. The graph $\widehat{\Gamma}$ is known as the Foster graph ([BCN89], Theorem 13.2.1) and it has the following distance diagram:

(1) 3 1 (3) 2 1 (6) 2 1 (12) 2 1 (24) 2 2 (24) 1 2 (12) 1 2 (6) 1 3 (2)

Define $\mathcal{G}(3 \cdot Sp_4(2))$ to be a geometry whose points are the vertices in one of the parts of $\widehat{\Gamma}$, whose lines are the vertices in another part and the incidence relation corresponds to the adjacency in $\widehat{\Gamma}$. Then $\mathcal{G}(3 \cdot Sp_4(2))$ is the rank 2 tilde geometry (or simply T_2-geometry) with the diagram

o═~═o
2 2

The collinearity graph $\widehat{\Xi}$ of $\mathcal{G}(3 \cdot Sp_4(2))$ is an antipodal triple cover of the collinearity graph Ξ of $\mathcal{G}(Sp_4(2))$ and it has the following distance diagram:

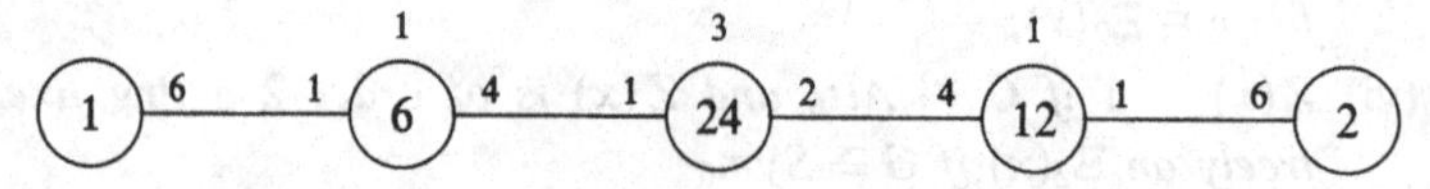

The following information comes from Theorem 13.2.1 in [BCN89].

Lemma 2.6.1 *Let $\widehat{G}$ be the full automorphism group of $\mathcal{G}(3 \cdot Sp_4(2))$. Then*

(i) *$\widehat{G}$ acts distance-transitively on $\widehat{\Xi}$,*
(ii) *$O^2(\widehat{G})$ is a perfect central extension of Alt_6 by a subgroup Y of order 3,*
(iii) *every element from $\widehat{G} \setminus O^2(\widehat{G})$ inverts Y and $\widehat{G}/Y \cong Sym_6 \cong Sp_4(2)$,*
(iv) *$\widehat{H} \leq \widehat{G}$ acts flag-transitively on $\mathcal{G}(3 \cdot Sp_4(2))$ if and only if $\widehat{H}$ contains $O^2(\widehat{G})$,*
(v) *$\mathcal{G}(3 \cdot Sp_4(2))$ possesses a diagram automorphism $\widehat{\tau}$ and $\langle \widehat{G}, \widehat{\tau} \rangle / Y \cong$ Aut Sym_6.* □

An explicit incidence matrix of $\widehat{\Gamma}$ can be found in [Ito82]. We will give constructions of $\mathcal{G}(3 \cdot Sp_4(2))$ in terms of the projective plane of order 4 (2.7.13), in terms of the Steiner system $S(5, 8, 24)$ (2.10.2 (v)) and in terms of subamalgams in a group $3^6 : Sym_6$ (6.2.2). The last construction will also provide us with a characterization of $\mathcal{G}(3 \cdot Sp_4(2))$ and of its automorphism group. The following lemma specifies $\mathcal{G}(3 \cdot Sp_4(2))$ as a coset geometry.

Lemma 2.6.2 *Let $\widehat{G}$ be as in (2.6.1) and let $\widehat{B}$ be a Sylow 2-subgroup of $\widehat{G}$. Then $\widehat{B} \cong D_8 \times 2$; there are exactly two subgroups $\widehat{P}_1$ and $\widehat{P}_2$ such that $\widehat{B} < \widehat{P}_i < \widehat{G}$ and $\widehat{P}_i \cap Y = 1$. Moreover, $\widehat{P}_1 \cong \widehat{P}_2 \cong Sym_4 \times 2$ and $\mathcal{G}(3 \cdot Sp_4(2)) \cong \mathcal{G}(\widehat{G}, \{\widehat{P}_1, \widehat{P}_2\})$.* □

It follows directly from the distance diagrams of Ξ and $\widehat{\Xi}$ that every non-degenerate 5-cycle is contractible with respect to the natural antipodal covering $\varphi : \widehat{\Xi} \to \Xi$. Since the fundamental group of the Petersen graph is clearly generated by its 5-cycles, we have the following.

Lemma 2.6.3 *If Π is a (standard) Petersen subgraph in Ξ, then $\varphi^{-1}(\Pi)$ is a disjoint union of three isomorphic copies of Π (called the standard Petersen subgraphs of $\widehat{\Xi}$).* □

The next lemma is also a direct consequence of the distance diagram of $\widehat{\Xi}$.

Lemma 2.6.4 *Every 2-path $(\widehat{x}, \widehat{y}, \widehat{z})$ in $\widehat{\Xi}$ with $d(\widehat{x}, \widehat{y}) = 2$ is contained in exactly two non-degenerate 5-cycles and in a unique standard Petersen subgraph.* □

Proposition 2.6.5 *The fundamental group of $\widehat{\Xi}$ is generated by the cycles of length 3 and by the non-degenerate cycles of length 5.*

Proof. The distance diagram of $\widehat{\Xi}$ shows that if n is the length of a non-degenerate cycle then $n \in \{3,5,6,7,8\}$. Hence, proceeding by induction, it is sufficient to show that every non-degenerate n-cycle for $6 \leq n \leq 8$ is decomposable into shorter cycles. Let $x \in \widehat{\Xi}$, $y \in \widehat{\Xi}_3(x)$, $\widehat{\Xi}(y) = \{z_i \mid 1 \leq i \leq 6\}$. We assume that the $\{z_i, z_{i+1}\}$ are edges for $i = 1,3,5$ and that $z_6 \in \widehat{\Xi}_4(x)$, which forces $z_5 \in \widehat{\Xi}_3(x)$. Put $\{u_i\} = \widehat{\Xi}(x) \cap \widehat{\Xi}(z_i)$ for $1 \leq i \leq 4$ and let C_{ij} denote the 6-cycle $(y, z_i, u_i, x, u_j, z_j, y)$ for $1 \leq i < j \leq 4$. Since the $\{z_i, z_{i+1}\}$ are edges, it is easy to see that for $i = 1$ and 3 the cycle $C_{i,i+1}$ is degenerate. Let Θ be the unique (standard) Petersen subgraph containing the 2-path (y, z_1, u_1). Since Θ has diameter 2, valency 3 and does not contain triangles, it contains z_5 as well as one of z_3 and z_4. Without loss of generality we assume that $z_3 \in \Theta$. In this case $d(u_1, z_i) = 2$ for $i = 3$ and 5, which implies that C_{13} and all 7-cycles are degenerate. The cycle C_{14} is the sum of the cycles C_{13} and C_{34}, both degenerate. In view of the obvious symmetry we conclude that every 6-cycle splits into triangles and pentagons. By similar arguments it is easy to prove decomposability of the 8-cycles and we suggest this as an exercise. □

As a direct consequence of (2.6.3), (2.6.4) and (2.6.5) we obtain the following.

Corollary 2.6.6 *The subgroup of the fundamental group of Ξ corresponding to the covering $\varphi : \widehat{\Xi} \to \Xi$ is generated by the cycles of length* 3 *and by the non-degenerate cycles of length* 5. □

2.7 The projective plane of order 4

Let $\Pi = (P, L)$ be a projective plane of order 4 where P is the set of points and L is the set of lines. We follow [Beu86] to show that Π is unique up to isomorphism. Since there exists the Desarguesian plane formed by 1- and 2-dimensional subspaces in a 3-dimensional $GF(4)$-space we only have to show that up to isomorphism there exists at most one projective plane of order 4. Recall that the automorphism group of the Desarguesian plane of order 4 is isomorphic to $P\Gamma L_3(4)$.

First of all we have that $|P| = |L| = 4^2 + 4 + 1 = 21$. As usual we identify a line of Π with the set of five points it is incident to. A subset of points is called *independent* if every line intersects it in at most two points. Dually a family of lines will be called independent if every point is on at most two lines from the family. We are going to describe the maximal independent sets of points.

An easy counting argument gives the following.

Lemma 2.7.1 *There are exactly* $(21 \cdot 20 \cdot 16 \cdot 9)/4! = 2520$ *independent 4-sets of points in* Π. □

Let $Q = \{q_1, q_2, q_3, q_4\}$ be an independent 4-set of points. Every pair of points in Q determine a unique line containing these points and since Q is independent, different pairs determine different lines. Thus the set M of lines intersecting Q in two points has size 6. Two lines from M have a common point outside Q if and only if these lines are determined by disjoint pairs of points in Q. Hence there are three points, say p_1, p_2, p_3, outside Q which are intersections of lines from M and these points correspond to partitions of Q into disjoint pairs (see the figure below).

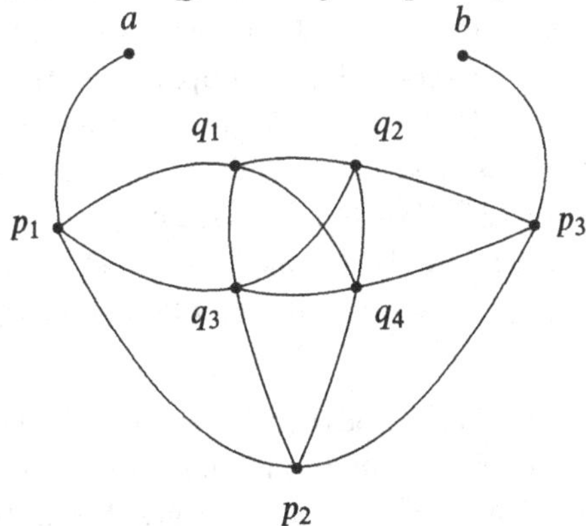

It is easy to calculate that the union of lines in M contains exactly 19 points and hence there are exactly 2 points, say a and b, which are missed by every line from M. The following lemma is an easy combinatorial exercise (see also [Beu86]).

Lemma 2.7.2 *The set* $l = \{p_1, p_2, p_3, a, b\}$ *is a line of* Π. □

Lemma 2.7.3 *The following assertions hold:*

(i) $\Omega = Q \cup \{a, b\}$ *is a maximal independent set of points, called, a* hyperoval,

(ii) *every independent 4-set of points in* Π *is in a unique hyperoval and there are exactly* 168 *hyperovals in* Π,

(iii) *if* $A = Q \cup \{p_1, p_2, p_3\}$ *and* $B = M \cup \{l\}$, *where* $l = \{p_1, p_2, p_3, a, b\}$, *then* (A, B) *is a Fano subplane in* Π *with respect to the incidence relation induced from* Π,

(iv) *every independent 4-set of points in Π is in a unique subplane and there are exactly 360 Fano subplanes in Π,*

(v) *a subset of points in Π is the point set of a Fano subplane if and only if it is the symmetric difference of a hyperoval and a line intersecting the hyperoval in 2 points.*

Proof. An independent set containing Q must not contain points outside Q on lines from M. Thus such a set is contained in $\Omega = Q \cup \{a, b\}$. By (2.7.2) the line l containing a and b misses Q and this immediately implies that Ω is independent and (i) follows. By the construction Q (which is an arbitrary 4-set) is in a unique hyperoval. By (2.7.1) there are 2520 independent 4-sets in Π and every hyperoval contains 15 of them. Hence the total number of hyperovals is $2520/15 = 168$ and we obtain (ii). It is straightforward to check the axioms of the Fano plane to see (iii). An arbitrary Fano subplane in Π must contain an independent 4-set, say Q, the lines M intersecting Q in 2 points, the intersections of the lines from M outside Q and the line through these intersection points. By (2.7.2) such a subplane can be uniquely constructed starting with an independent 4-set. Since every Fano subplane contains 7 independent 4-sets, (iv) follows. Finally (v) follows directly from the above constructions. □

We proceed with the uniqueness proof for Π. Let Ω be a hyperoval and let $\mathcal{G}$ be the generalized quadrangle constructed starting with Ω as in Section 2.5. Let Ξ^1 and Ξ^2 be the point graph and the line graph of $\mathcal{G}$, respectively, so that the vertices of Ξ^1 are the 2-element subsets of Ω and the vertices of Ξ^2 are partitions of Ω into three disjoint pairs. For $i = 1, 2$ let Θ^i be the set of (standard) 5-cocliques in Ξ^i. Then $|\Theta^i| = 6$ and the cliques in Θ^1 are identified with the elements of Ω. Notice that every vertex of Ξ^i is in exactly two cocliques from Θ^i.

Let L_i be the set of lines intersecting Ω in i points. For a point $p \in \Omega$ each of the five points in $\Omega \setminus \{p\}$ determines a line from L_2 incident to p and these lines are pairwise different since Ω is independent. Hence all five lines containing p are in L_2 and we have $L = L_0 \cup L_2$. Put $P_0 = \Omega$ and $P_2 = P \setminus \Omega$. Then

$$|P_0| = |L_0| = 6, \quad |P_2| = |L_2| = 15.$$

Lemma 2.7.4 *There exists a unique bijective mapping φ of $P \cup L$ onto $\Xi^1 \cup \Theta^1 \cup \Xi^2 \cup \Theta^2$ satisfying the following:*

(i) *φ restricted to $P_0 = \Omega$ is the identity mapping onto $\Theta^1 = \Omega$;*

(ii) $\varphi(L_2) = \Xi^1$, $\varphi(P_2) = \Xi^2$, $\varphi(L_0) = \Theta^2$;

(iii) *two distinct elements* $x, y \in P \cup L$ *are incident in* Π *if and only if either* $\varphi(x) \in \Xi^i$ *and* $\varphi(y) \in \Xi^{3-i}$ *are incident in* $\mathcal{G}$, *or* $\varphi(x) \in \Xi^i$ *is contained in* $\varphi(y) \in \Theta^i$, $i = 1, 2$.

Proof. We are going to construct φ satisfying the required conditions. By (i) φ is defined on P_0. If $l \in L_2$ we put $\varphi(l) = l \cap \Omega \in \Xi^1$. Then l is incident to $p \in \Omega$ if and only if $\varphi(l)$ is contained in the coclique corresponding to p. Let $q \in P_2$. Since L_1 is empty, every line incident to q either is disjoint from Ω or intersects Ω in exactly two points and since two lines in Π intersect in a single point the latter intersections are pairwise disjoint. Hence there are three lines, say l_1, l_2, l_3, containing q and intersecting Ω in two points. Thus $\{\Omega \cap l_i \mid 1 \leq i \leq 3\}$ is a partition of Ω and we define this partition to be the image of q under φ. In this case $l \in L_2$ and $q \in P_2$ are incident if and only if $\varphi(l)$ and $\varphi(q)$ are incident in $\mathcal{G}$. Let $m \in L_0$ and let $\{r_1, ..., r_5\}$ be the points (from P_2) incident to m. We claim that $\varphi(r_i)$ and $\varphi(r_j)$ are non-adjacent in Ξ^2 whenever $i \neq j$. In fact if they are adjacent then there is a pair $\alpha \in \Xi^1$ adjacent to them both. In this case $\varphi^{-1}(\alpha)$ is a line from L_2 incident to r_i and r_j. This is a contradiction since m is the unique line incident to r_i and r_j. Hence $\{\varphi(r_1), ..., \varphi(r_5)\}$ is a (standard) coclique in Ξ^2 which we define to be the image of m under φ. Now it is easy to see that φ is bijective, the conditions (i)–(iii) are satisfied and the uniqueness follows from the construction. □

Lemma 2.7.5 *The following assertions hold:*

(i) *let* Π *and* Π' *be projective planes of order* 4, *let* Ω *and* Ω' *be hyperovals in* Π *and* Π', *respectively, and let* ψ *be a bijection of* Ω *onto* Ω'; *then there is a unique isomorphism of* Π *onto* Π' *whose restriction to* Ω *coincides with* ψ;

(ii) *all projective planes of order* 4 *are isomorphic;*

(iii) $G = \mathrm{Aut}\,\Pi$ *acts transitively on the set of hyperovals in* Π *and the stabilizer* H *of a hyperoval is isomorphic to* Sym_6;

(iv) H *acting on* P *has two orbits with lengths* 6 *and* 15;

(v) *there is a correlation* τ *of* Π *such that* $\langle H, \tau \rangle \cong \mathrm{Aut}\,Sym_6$.

Proof. (i) follows directly from (2.7.4) while (ii) and (iii) are straightforward from (i). It is easy to see that in the terms introduced before (2.7.4) the set L_0 is a hyperoval in the dual of Π. Let τ be a correlation of Π which sends P_0 onto L_0 (such a correlation exists by (i)). Since H is also the stabilizer of L_0 in G, τ normalizes H. It is easy to observe that τ induces a diagram automorphism of the generalized quadrangle $\mathcal{G}$ of order (2,2) and (v) follows. □

Thus up to isomorphism the projective plane Π of order 4 is unique and it is Desarguesian. By (2.7.5 (iii)) and (2.7.3 (ii)) we have

$$|\mathrm{Aut}\,\Pi| = 168 \cdot 6! = 2^7 \cdot 3^3 \cdot 5 \cdot 7$$

which is of course the order of $P\Gamma L_3(4)$.

The following result is easy to deduce from the order of $\mathrm{Aut}\,\Pi$, (2.7.3 (iv)) and (2.7.5 (iii)).

Lemma 2.7.6 *The group $G = \mathrm{Aut}\,\Pi$ acts transitively on the set of Fano subplanes in Π. The stabilizer F of such a subplane is isomorphic to $L_3(2) \times 2$ and F acting on P has two orbits with lengths* 7 *and* 14. □

Let us discuss a relationship between Π and the affine plane Φ of order 3. Recall that an affine plane of order n is the rank 2 geometry of elements and blocks of a Steiner system of type $S(2, n, n^2)$. An affine plane of order q is formed by vectors and cosets of 1-dimensional subspaces in a 2-dimensional $GF(q)$-space.

Let T be a triple of independent points in Π. Let l_1, l_2 and l_3 be the lines intersecting T in two points and let $A = (l_1 \cup l_2 \cup l_3) \setminus T$. Then A is of size 9 and every line intersecting A in at least two points intersects it in three points. It is easy to check that the set B of lines intersecting A in three points has size 12 and that $\Phi = (A, B)$ is an affine plane of order 3. This construction goes back to [Edge65] and it is called the *deleting procedure.*

Lemma 2.7.7 *Every affine subplane of order* 3 *in Π can be constructed by the deleting procedure.*

Proof. Let $A \subset P$, $B \subset L$ be such that $\Phi = (A, B)$ is an affine plane of order 3 with respect to the incidence relation induced from Π. Let l_1, l_2, l_3 be parallel lines in Φ. If these lines are independent then Φ can be obtained by the deleting procedure starting with the triple of their pairwise intersection points. Suppose this is not the case and that $p \in l_1 \cap l_2 \cap l_3$. Let $\{q_i\} = l_i \setminus (A \cup \{p\})$ and $\{p_1, p_2, p_3\} = l_2 \cap A$. Then for $i = 1$, 2 and 3 the line containing q_1 and p_i must intersect l_3 in q_3, which is impossible. □

As an implication of the proof of (2.7.7) we see that every affine subplane Φ of order 3 in Π can be constructed by the deleting procedure starting with exactly four different triples $T_1, ..., T_4$ corresponding to the classes of parallel lines in Φ. It is clear that $T_i \cup T_j$ for $i \neq j$ is an independent set of size 6, that is, a hyperoval. Moreover, by (2.7.1), (2.7.3 (ii)) and (2.7.5 (iii)) $T_1 \cup T_2$, $T_1 \cup T_3$ and $T_1 \cup T_4$ are all the hyperovals containing T_1. So we have the following.

Lemma 2.7.8

(i) *There are exactly* 280 *affine subplanes of order* 3 *in* Π *and the automorphism group of* Π *permutes them transitively;*

(ii) *the symmetric difference of* 2 *hyperovals in* Π *intersecting in* 3 *points is a hyperoval.* □

Let us turn to the automorphism group $G \cong P\Gamma L_3(4)$ of Π. Let $G^1 \cong PGL_3(4)$ and $G^2 \cong L_3(4)$ be normal subgroups in G. The cosets of G^2 in G^1 are indexed by the non-zero elements of $GF(4)$ and the unique proper coset of G^1 in G contains the field automorphism. Hence $G/G^2 \cong Sym_3$. The preimage in G of a subgroup of order 2 in G/G^2 is $P\Sigma L_3(4)$.

Lemma 2.7.9 *Let* $\mathcal{H}$ *be the set of hyperovals and* $\mathcal{F}$ *be the set of Fano subplanes in* Π. *Then*

(i) G^2 *acting on* $\mathcal{H}$ *has three orbits, each of length* 56 *with stabilizers isomorphic to* Alt_6,

(ii) G^2 *acting on* $\mathcal{F}$ *has three orbits, each of length* 120 *with stabilizers isomorphic to* $L_3(2)$,

(iii) $G/G^2 \cong Sym_3$ *permutes the orbits of* G^2 *in* $\mathcal{H}$ *and the orbits of* G^2 *on* $\mathcal{F}$ *in the natural way,*

(iv) *there is a unique bijection between the orbits of* G^2 *on* $\mathcal{H}$ *and the orbits of* G^2 *on* $\mathcal{F}$ *which is stabilized by* G.

Proof. By (2.7.5 (iii)) and (2.7.6) G acts transitively on $\mathcal{H}$ and on $\mathcal{F}$ with stabilizers isomorphic to Sym_6 and $L_3(2) \times 2$, respectively. By the fundamental theorem of projective geometry G^2 is transitive on triples of independent points and we know that each such triple is in exactly three hyperovals. Hence G^2 has on $\mathcal{H}$ one or three orbits. In the former case the stabilizer in G^2 of a hyperoval is a normal subgroup of index 6 in Sym_6. Since there are no such subgroups we obtain (i). By (i) and (2.7.3 (v)) there are one or three orbits of G^2 on $\mathcal{F}$. Since there are no index 6 subgroups in $L_3(2) \times 2$, (ii) follows. Finally (iii) follows from the paragraph before the lemma and implies (iv). □

Applying (2.4.1), (2.4.2), (2.4.3) to the case $(n, q) = (3, 4)$ we obtain the following.

Lemma 2.7.10 *Let l be a line of Π identified with the set of points it is incident to and let $p \in l$. Then the following assertions hold:*

(i) *$G^2(l)$ is elementary abelian of order 2^4, it acts transitively on $P \setminus l$; $O_2(G^2(p))$ has order 2^4; two points from $P \setminus p$ are in the same orbit of $O_2(G^2(p))$ if and only if they belong to the same line passing through p; $O_2(G^2(p)) \cap G^2(l)$ is of order 2^2 and its set of orbits on $P \setminus l$ coincides with that of $O_2(G^2(p))$;*

(ii) *$G^2[l]$ induces on l the natural action of $Alt_5 \cong L_2(4)$ and it acts transitively by conjugation on the set of non-identity elements of $G^2(l)$;*

(iii) *$G^1(l)$ is an extension of $G^2(l)$ by a group of order 3 acting fixed-point freely on $G^2(l)$;*

(iv) *the elements from $G[l] \setminus G^1[l]$ induce odd permutations on l and invert $G^1(l)/G^2(l) \cong 3$, so that $G[l] \cong 2^4.(3 \times Alt_5).2$ and $G[l]$ induces on l the natural action of $Sym_5 \cong P\Gamma L_2(4)$;*

(v) *$G^2 \cong L_3(4)$ is non-abelian and simple and all involutions in G^2 are conjugate.* □

Let us discuss the structure of the stabilizer in G of an affine subplane Φ of order 3 in Π. It follows directly from the deleting procedure that the elementwise stabilizer of Φ is trivial. It is a standard fact that the automorphism group F of Φ is isomorphic to $3^2 : GL_2(3) \cong 3^2 : 2_-^{1+2} : Sym_3$. In particular F possesses a unique homomorphism onto Sym_3 with kernel $3^2 : 2_-^{1+2}$. Since G^2 is transitive on triples of independent points, by (2.7.7) it is transitive on the affine subplanes of order 3 in Π. Now comparing the order of G and the number of subplanes given by (2.7.8) we obtain the following.

Lemma 2.7.11

(i) *G^2 acts transitively on the set of affine subplanes of order 3 in Π with stabilizer isomorphic to $3^2 : 2_-^{1+2}$;*

(ii) *the stabilizer of Φ in G induces on Φ its full automorphism group isomorphic to $3^2 : GL_2(3)$.* □

The automorphism group of Φ induces Sym_4 on the set of four parallel classes of lines in Φ and hence also on the set of triples of points in Π from which Φ can be constructed by the deleting procedure.

The next lemma contains a standard result and we present a sketch of the proof for the sake of completeness.

Lemma 2.7.12 *The group $SL_3(4)$ is a non-split extension of $L_3(4)$ by a centre of order 3. Every element from $\Gamma L_3(4) \setminus GL_3(4)$ inverts the centre of $SL_3(4)$.*

Proof. Let $D \cong 3^2 : 2_{-}^{1+2}$ be the stabilizer in G^2 of an affine subplane of order 3. It is easy to check that $D/O_3(D)$ acts transitively on four elementary abelian factor groups of $O_3(D)$ having order 3. This shows that D does not have faithful $GF(2)$-representations of dimension less than 8. On the other hand the $GF(2)$-dimension of the natural module of $SL_3(4)$ is 6. Hence $SL_3(4)$ does not split over its centre (of order 3). The second statement in the lemma is obvious. □

Since the full preimage $\widetilde{H}$ in $SL_3(4)$ of the stabilizer in G^2 of a hyperoval contains a Sylow 3-subgroup of the latter, $\widetilde{H}$ does not split over the centre of $SL_3(4)$. This, (2.7.12), (2.7.9) and (6.2.2) give the following.

Lemma 2.7.13 *The full preimage $\widehat{H}$ in $\Gamma L_3(4)$ of the stabilizer in G of a hyperoval satisfies Hypothesis* 6.2.1 *so that $\widehat{H}$ is the automorphism group of the rank* 2 *T-geometry.* □

The subgroup $\widehat{H} \cong 3 \cdot Sym_6$ preserves in the natural module of $\Gamma L_3(4)$ a code known as the *hexacode.* The natural module of $\Gamma L_3(4)$ considered as a $GF(2)$-module for $\widehat{H}$ will be called the *hexacode module.* Let us discuss the orbits of $\widehat{H}$ on the vectors of the hexacode module.

Lemma 2.7.14 *The subgroup $\widehat{H}$ as in* (2.7.13) *acting on the non-zero vectors of the natural module of $\Gamma L_3(4)$ has two orbits with lengths* 18 *and* 45 *and stabilizers isomorphic to Sym_5 and $Sym_4 \times 2$, respectively.* □

2.8 Uniqueness of $S(5,8,24)$

Let $(\mathcal{P},\mathcal{B})$ be a Steiner system of type $S(5,8,24)$. In this section we follow [Lün69] to show that $(\mathcal{P},\mathcal{B})$ is unique up to isomorphism. Since we have an example formed by the octads of the Golay code constructed in Section 2.2, all we have to show is that there is at most one possibility for the isomorphism type of such a system. Although we do not assume *a priori* that $(\mathcal{P},\mathcal{B})$ comes from a Golay code, the blocks from $\mathcal{B}$ will be called *octads*.

For a 3-element subset Y in $\mathcal{P}$ put

$$\mathcal{B}_i = \mathcal{B}_i(Y) = \{B \mid B \in \mathcal{B}, |B \cap Y| = i\}.$$

Lemma 2.8.1 $|\mathcal{B}_i| = 21, 168, 360$ *and* 210 *for* $i = 3, 2, 1$ *and* 0, *respectively.*

Proof. Since $\mathcal{B}_4 = \emptyset$, proceeding by induction for a given i we can assume that $|\mathcal{B}_j|$ is known for $j > i$. Let n_i denote the number of triples (B,K,L), where $B \in \mathcal{B}$, $K \subseteq B \cap Y$, $L \subseteq B \setminus Y$, such that $|K| = i$ and $|L| = 5 - i$. Since $K \cup L$ is contained in a unique octad, we have

$$n_i = \binom{3}{i} \cdot \binom{21}{5-i}.$$

By the inductive assumption we can calculate the number of triples with $|B \cap Y| > i$. Since every octad from $\mathcal{B}_i$ corresponds to exactly $\binom{8-i}{5-i}$ triples, it is straightforward to calculate $|\mathcal{B}_i|$. □

By (2.3.7) and (2.3.8) $\Pi(Y) := (\mathcal{P} \setminus Y, \{B \setminus Y \mid B \in \mathcal{B}_3\})$ is a projective plane of order 4.

Lemma 2.8.2 *Let* $A \subseteq \mathcal{P} \setminus Y$. *Then* $A = B \setminus Y$ *for some* $B \in \mathcal{B}_2$ *if and only if* A *is a hyperoval in* $\Pi(Y)$.

Proof. Recall that two different octads have at most four elements in common. Let $B \in \mathcal{B}_2$ and $B' \in \mathcal{B}_3$. Then $|B \cap B' \cap Y| = 2$ and hence $|B \cap B' \setminus Y| \leq 2$, which shows that $B \setminus Y$ is an independent set of points in $\Pi(Y)$. Since the size of $B \setminus Y$ is 6, it is a hyperoval by (2.7.3). Clearly, different octads from $\mathcal{B}_2$ correspond to different hyperovals. On the other hand by (2.7.3 (ii)) and (2.8.1) $|\mathcal{B}_2|$ is exactly the number of hyperovals in $\Pi(Y)$ and hence the result. □

Lemma 2.8.3 *Let* $B_1, B_2 \in \mathcal{B}$ *with* $|B_1 \cap B_2| = 4$. *Then* $B_1 \Delta B_2 \in \mathcal{B}$.

Proof. Let $y_1 \in B_1 \setminus B_2$, $y_2 \in B_2 \setminus B_1$, $y_3 \in B_1 \cap B_2$ and $Y = \{y_1, y_2, y_3\}$. Then $B_1, B_2 \in \mathcal{B}_2(Y)$ and in view of (2.8.2) $B_1 \setminus Y$ and $B_2 \setminus Y$ are hyperovals in $\Pi(Y)$ intersecting in three points. By (2.7.8) $A := B_1 \Delta B_2 \setminus Y$ is a hyperoval and by (2.8.2) there is an octad $B_3 \in \mathcal{B}_2(Y)$ such that $A = B_3 \setminus Y$. For $i = 1$ and 2 we have $|B_3 \cap B_i \setminus Y| = 3$ and hence $|B_3 \cap B_i \cap Y| \leq 1$, which is possible only if $B_3 \cap Y = B_1 \Delta B_2 \cap Y$. Hence $B_3 = B_1 \Delta B_2$. □

Lemma 2.8.4 *Let $A \subseteq \mathscr{P} \setminus Y$. Then*

(i) *an octad $B \in \mathscr{B}_1$ with $A = B \setminus Y$ exists if and only if A is the point set of a Fano subplane in $\Pi(Y)$,*
(ii) *$A \in \mathscr{B}_0$ if and only if A is the symmetric difference of a pair of lines in $\Pi(Y)$.*

Proof. Choose $B_1 \in \mathscr{B}_3$ and $B_2 \in \mathscr{B}_2$ so that $B_1 \setminus Y$ is a line intersecting the hyperoval $B_2 \setminus Y$ in two points. Then $|B_1 \cap B_2| = 4$ and by (2.8.3) $B_3 := B_1 \triangle B_2$ is an octad. Clearly $B_3 \in \mathscr{B}_1$ and by (2.7.3 (v)) $B_3 \setminus Y$ is the point set of a Fano subplane in $\Pi(Y)$. Also by (2.7.3 (v)) every Fano subplane can be obtained as the symmetric difference of a hyperoval and a line intersecting the hyperoval in two points. By (2.7.3 (iv)) and (2.8.1) $|\mathscr{B}_1|$ is exactly the number of Fano subplanes in $\Pi(Y)$, hence (i) follows.

Let $B_1, B_2 \in \mathscr{B}_3$. Since any two lines in $\Pi(Y)$ intersect in a single point, we have $|B_1 \cap B_2| = 4$ and by (2.8.3) $B_3 = B_1 \triangle B_2$ is an octad. Clearly $B_3 \in \mathscr{B}_0$ and $B_3 = (B_1 \setminus Y) \triangle (B_2 \setminus Y)$. It is easy to check that different pairs of lines in $\Pi(Y)$ have different symmetric differences. By (2.8.1) $|\mathscr{B}_0|$ is exactly the number of unordered pairs of lines in $\Pi(Y)$ and (ii) follows. □

Lemma 2.8.5 *Let B_1 and B_2 be different octads. Then $|B_1 \cap B_2| \in \{0,2,4\}$.*

Proof. Without loss of generality we can assume that $0 < |B_1 \cap B_2| \leq 4$. Let $y_1 \in B_1 \cap B_2$, $y_2, y_3 \in B_1 \setminus B_2$ and $Y = \{y_1, y_2, y_3\}$. Then $B_1 \in \mathscr{B}_3(Y)$, $B_2 \in \mathscr{B}_1(Y)$. Hence $B_1 \setminus Y$ is a line, while $B_2 \setminus Y$ is a subplane in $\Pi(Y)$. Now it is an easy exercise to show that a line intersects a Fano subplane in one or three points, which implies the result. □

As a direct consequence of (2.8.5) we obtain the following.

Lemma 2.8.6 *Let $B \in \mathscr{B}_1(Y)$ and let B' be an arbitrary octad. Then $B \cap Y \subseteq B'$ if and only if $|B \cap B' \setminus Y|$ is odd.* □

Now we are ready to prove the central result of the section.

Proposition 2.8.7 *Let $(\mathscr{P},\mathscr{B})$ and $(\mathscr{P}',\mathscr{B}')$ be Steiner systems of type $S(5,8,24)$. Let $Y \subseteq \mathscr{P}$, $Y' \subseteq \mathscr{P}'$ with $|Y| = |Y'| = 3$. Let ψ be a collineation of $\Pi(Y)$ onto $\Pi(Y')$. Then there exists a unique isomorphism φ of $(\mathscr{P},\mathscr{B})$ onto $(\mathscr{P}',\mathscr{B}')$ such that the restriction of φ to $\Pi(Y)$ coincides with ψ.*

Proof. We are going to construct the isomorphism φ satisfying the required condition. The action of φ on $\mathscr{P} \setminus Y$ is determined by that of

ψ. By (2.8.2) and (2.8.3) $\psi(A) = B' \setminus Y'$ for some $B' \in \mathcal{B}'$ if and only if $A = B \setminus Y$ for some $B \in \mathcal{B}$. Let B_1, B_2, B_3 be octads from $\mathcal{B}_1(Y)$ such that $B_i \cap Y \neq B_j \cap Y$ for $i \neq j$. It is easy to see, arguing as in the proof of (2.8.1), that such a triple exists. Let B'_1, B'_2, B'_3 be octads from $\mathcal{B}'_1(Y')$ such that $B'_i \setminus Y' = \psi(B_i \setminus Y)$. If φ is an isomorphism of Steiner systems, then $\varphi(B_i \cap Y) = B'_i \cap Y'$ for $1 \leq i \leq 3$ and this condition specifies φ uniquely. We claim that, defined in this way, φ is in fact an isomorphism. Let $B \in \mathcal{B}$ and let $B' \in \mathcal{B}'$ be such that $B' \setminus Y' = \psi(B \setminus Y)$. By (2.8.6), for $1 \leq i \leq 3$ $B_i \cap Y \subseteq B$ if and only if $|B \cap B_i \setminus Y|$ is odd. Since ψ is a bijection of $\mathcal{P}$ onto $\mathcal{P}'$ the latter condition holds if and only if $|B' \cap B'_i \setminus Y'|$ is odd, in which case $B'_i \cap Y' \subseteq B'$ by (2.8.6). Hence $\varphi(B) = B'$ and the claim follows. □

In view of (2.3.6) the above proposition immediately implies the following corollary as well as substantial information on the automorphism group of the Golay code which will be discussed in the next section.

Corollary 2.8.8 *Up to isomorphism there are a unique Steiner system of type $S(5, 8, 24)$ and a unique Golay code.* □

Notice that the above uniqueness proof could be slightly simplified if we would assume that the Steiner system comes from a Golay code. In this case (2.8.3) and (2.8.5) would be immediate. Also it is worth mentioning that for $B, B' \in \mathcal{B}$ the equality $B \cap Y = B' \cap Y$ holds if and only if $B \setminus Y$ and $B' \setminus Y$ are in the same orbit on $\{B \setminus Y \mid B \in \mathcal{B}\}$ of the $L_3(4)$-subgroup in $\mathrm{Aut}\,\Pi(Y) \cong P\Gamma L_3(4)$. Using this observation one could (if one wished to) obtain an explicit model of the Steiner system of type $S(5, 8, 24)$.

2.9 Large Mathieu groups

Let $(\mathcal{P}, \mathcal{B})$ be the unique Steiner system of type $S(5, 8, 24)$, $\mathcal{C}_{12}$ be the unique Golay code generated by the octads from $\mathcal{B}$. The automorphism group of $\mathcal{C}_{12}$ (equivalently of the Steiner system) is known as the *Mathieu group* Mat_{24} of degree 24.

Lemma 2.9.1 *Let $G = Mat_{24}$ and Y be a 3-element subset of $\mathcal{P}$. Then*

(i) *G acts transitively on the set of 3-element subsets of $\mathcal{P}$ and $G[Y] \cong \mathrm{Aut}\,\Pi(Y) \cong P\Gamma L_3(4)$,*

(ii) *$G(Y) \cong L_3(4)$ and $G[Y]/G(Y) \cong Sym_3$,*

(iii) *G acts 5-fold transitively on $\mathcal{P}$,*

(iv) $|G| = 2^{10} \cdot 3^3 \cdot 5 \cdot 7 \cdot 11 \cdot 23$,
(v) *G contains a subgroup isomorphic to $L_2(23)$.*

Proof. (i) is an immediate consequence of (2.8.7). Since $G[Y] \cong \mathrm{Aut}\,\Pi(Y)$ acts transitively on the set of hyperovals in $\Pi(Y)$, it is easy to see that $G[Y]$ acts transitively on Y. On the other hand by (2.7.10) if ψ is a homomorphism of $P\Gamma L_3(4)$ onto a transitive subgroup of Sym_3, then $\mathrm{Im}\,\psi = Sym_3$ and $\ker \psi = L_3(4)$. Hence we have (ii). By (i) and (ii) the action of G on $\mathcal{P}$ is 3-fold transitive and by (2.4.2 (vii)) $G(Y)$ acts doubly transitively on $\mathcal{P} \setminus Y$ which implies (iii). By (i) we have

$$|G| = \binom{24}{3} \cdot |P\Gamma L_3(4)|$$

and (iv) follows. Finally, $\mathcal{C}_{12}$ is isomorphic to the Golay code constructed in Section 2.2 which is invariant under $L_2(23)$. □

Let $\emptyset = Y_0 \subset Y_1 \subset Y_2 \subset Y_3 = Y \subset \mathcal{P}$ where Y_i is of size i and let Mat_{24-i} denote the elementwise stabilizer of Y_i in Mat_{24}.

Lemma 2.9.2

(i) *Mat_{24-i} acts $(5-i)$-fold transitively on $\mathcal{P} \setminus Y_i$.*
(ii) *$Mat_{21} \cong L_3(4)$; $|Mat_{22}| = 2^7 \cdot 3^2 \cdot 5 \cdot 7 \cdot 11$; $|Mat_{23}| = 2^7 \cdot 3^2 \cdot 5 \cdot 7 \cdot 11 \cdot 23$.*
(iii) *Mat_{24-i} is non-abelian and simple for $0 \le i \le 3$.*

Proof. By (2.9.1 (i), (iii), (iv)) we obtain (i) and (ii). To prove (iii) we proceed by induction. For $i = 3$ the result follows from (2.7.10). Suppose that $0 \le i \le 2$, that Mat_{24-i-1} is simple and N is a proper normal subgroup in Mat_{24-i}. Since the action of Mat_{24-i} on $\mathcal{P} \setminus Y_i$ is doubly transitive and hence primitive, the action of N on this set is transitive. If $N \cap Mat_{24-i-1} \ne 1$, then by the simplicity of Mat_{24-i-1} we have $N \ge Mat_{24-i-1}$ which implies $N = Mat_{24-i}$. This shows that the action of N on $\mathcal{P} \setminus Y_i$ is regular, in particular $|N| = 24 - i$. It is well known that if L is a regular normal subgroup in a doubly transitive group H then L is elementary abelian of order p^a, say, and $C_H(L) = L$. In particular $|H| \le p^a \cdot |GL_a(p)|$. In the situation considered $24 - i$ is a prime power only if $i = 1$ but in this case we also reach a contradiction since $|Mat_{23}| > 23 \cdot 22$. □

The groups Mat_{24}, Mat_{23} and Mat_{22} are sporadic simple groups called the Mathieu groups. Two more Mathieu groups will appear later as subgroups of Mat_{24}.

Lemma 2.9.3 *The setwise stabilizer $G[Y_2]$ of Y_2 in $G = Mat_{24}$ contains Mat_{22} as an index 2 subgroup and elements from $G[Y_2] \setminus Mat_{22}$ induce outer automorphisms of Mat_{22}.*

Proof. By (2.9.1 (iii)) $G[Y_2]$ contains Mat_{22} properly. Since Mat_{22} acts transitively on $\mathcal{P} \setminus Y_2$ there exists $g \in G[Y_2] \setminus Mat_{22}$ which stabilizes Y_3 as a whole. Then g induces on $L_3(4) \cong Mat_{21} < Mat_{22}$ a conjugate of the field automorphism which is outer. □

In fact $G[Y_2]$ is the full automorphism group Aut Mat_{22} of Mat_{22} while Mat_{23} and Mat_{24} are perfect.

By (2.3.7) the residual system of $(\mathcal{P}, \mathcal{B})$ with respect to Y_i is a Steiner system of type $S(5-i, 8-i, 24-i)$. Since the residue of Y_3 is the projective plane of order 4, it is easy to see that Mat_{23} and Aut Mat_{22} are the full automorphism groups of the residual systems of Y_1 and Y_2, respectively.

By arguments similar to those in Section 2.8 one can prove the following [Lün69].

Lemma 2.9.4 *For $i = 0$, 1, 2 and 3 a Steiner system of type $S(5-i, 8-i, 24-i)$ is isomorphic to the residue of Y_i in $(\mathcal{P}, \mathcal{B})$.* □

2.10 Some further subgroups of Mat_{24}

We continue to use notation and terminology introduced in Sections 2.8 and 2.9. Till the end of this chapter and throughout the next chapter G denotes the Mathieu group Mat_{24}.

Lemma 2.10.1 *Let $B \in \mathcal{B}$, $G_b = G[B]$ and $Q_b = G(B)$. Then*

(i) *there are 759 octads and G permutes them transitively,*
(ii) *Q_b is elementary abelian of order 2^4 and it acts regularly on $\mathcal{P} \setminus B$,*
(iii) *G_b induces on B the alternating group $Alt(B) \cong Alt_8$,*
(iv) *G_b/Q_b induces the full automorphism group $L_4(2)$ of Q_b and $Alt_8 \cong L_4(2)$,*
(v) *$G_b \cong AGL_4(2)$.*

Proof. Since every 5-element subset of $\mathcal{P}$ is in a unique octad, (i) is implied by (2.9.1 (iii)). We assume that $Y \subset B$. Then $B \setminus Y$ is a line in $\Pi(Y)$ and (ii) follows from (2.7.10) and (2.9.1 (ii)). Since two octads share at most four elements, G_b induces a 5-fold transitive action on B and by (2.7.10) the elementwise stabilizer of Y in this action is isomorphic to

Alt_5. This gives (iii). G_b induces a non-trivial action on Q_b since so does its intersection with Mat_{21} by (2.7.10). Since Alt_8 is simple the action is faithful, and we observe (iv) comparing the orders of Alt_8 and $L_4(2)$. Finally, since Q_b acts regularly on $\mathcal{P} \setminus B$, for $p \in \mathcal{P} \setminus B$ the subgroup $G_b \cap G(p)$ is a complement to Q_b in G_b and (v) follows. □

By (2.3.3) and the uniqueness of the Steiner system of type $S(5,8,24)$, every 4-element subset S of $\mathcal{P}$ is contained in a unique sextet which is a partition of $\mathcal{P}$ into six 4-element subsets $S_1 = S, S_2, ..., S_6$ called tetrads, such that $S_i \cup S_j$ is an octad for $1 \leq i < j \leq 6$. Notice that $S \cup S_i$ for $2 \leq i \leq 6$ are all the octads containing S.

Lemma 2.10.2 *Let $\Sigma = \{S_1, S_2, ..., S_6\}$ be a sextet, G_s be the stabilizer of Σ in G, K_s be the kernel of the action of G_s on the set of tetrads in Σ and $Q_s = O_2(G_s)$. Then*

(i) *there are $1771 = \binom{24}{4}/6$ sextets and G permutes them transitively,*
(ii) *G_s induces the natural action of Sym_6 on the tetrads in Σ,*
(iii) *K_s induces the natural action of Alt_4 on the elements in each S_i, $1 \leq i \leq 6$,*
(iv) *Q_s is elementary abelian of order 2^6 and K_s is an extension of Q_s by a group X_s of order 3 which acts on Q_s fixed-point freely,*
(v) *G_s/Q_s is isomorphic to the automorphism group of the rank 2 tilde geometry $\mathcal{G}(3 \cdot Sym_6)$,*
(vi) *G_s is isomorphic to the full preimage in $A\Gamma L_3(4)$ of the stabilizer in $P\Gamma L_3(4)$ of a hyperoval in the corresponding projective plane of order 4, so that Q_s is the hexacode module for G_s/Q_s,*
(vii) *if $1 \leq i < j < k \leq 6$ then Q_s acts faithfully on $S_i \cup S_j \cup S_k$.*

Proof. The 5-fold transitivity of the action of G on $\mathcal{P}$ implies (i) and also the transitivity of G_s on the tetrads in Σ. We assume that $Y \subset S_1$ and let $\{p\} = S_1 \setminus Y$. Let H be the setwise stabilizer of S_1 in G and let F be the intersection of H with the setwise stabilizer of Y, isomorphic to $P\Gamma L_3(4)$. Clearly H is contained in G_s and induces on the points of S_1 the symmetric group Sym_4. On the other hand, F is the stabilizer in $P\Gamma L_3(4)$ of the point p in $\Pi(Y)$, so that $F \cong P\Sigma L_3(4)$. Since $S_i \cup \{p\}$, $2 \leq i \leq 6$, are all the lines in $\Pi(Y)$ passing through p, we see from (2.7.10 (iv)) that F induces Sym_5 on these lines and hence (ii) follows. The subgroup K_s is contained in H and its intersection with F is the elementwise stabilizer in $P\Gamma L_3(4)$ of the lines passing through p. By (2.7.10) this intersection is an extension of an elementary abelian group R of order 2^4 by an order 3 group X_s acting on R fixed-point freely. By (2.9.1 (ii)) X_s acts transitively on Y. Since p is an arbitrary element from S_1 we obtain (iii). Now Q_s induces on each S_i an elementary abelian group of order 2^2 and the kernel is of order 2^4. Hence Q_s is elementary abelian of order 2^6. The subgroup X_s acts fixed-point freely both on R and on Q_s/R, hence (iv) follows. Clearly the image of X_s in G_s/Q_s is normal and the action of X_s on Q_s induces on the latter the structure of a 3-dimensional $GF(4)$-space. Let Π_1 denote the projective plane of order 4 associated with the dual of this space. For $1 \leq i \leq 6$ let R_i be the elementwise stabilizer of S_i in Q_s. Then $\Theta = \{R_1, R_2, ..., R_6\}$ is the set of points of Π_1 and by (ii) the setwise stabilizer of Θ in the automorphism group of Π_1 induces on Θ the action of Sym_6 which shows that Θ is independent and hence it is a hyperoval in Π_1. Moreover by (2.7.5) G_s/Q_s is the full preimage in $\Gamma L_3(4)$ of the stabilizer of this hyperoval in $\mathrm{Aut}\,\Pi_1$. Now (v) follows from (2.7.13). Since X_s acts fixed-point freely on Q_s, by the Frattini argument $N_{G_s}(X_s)$ is a complement to Q_s in G_s. This observation and (v) imply (vi). If $1 \leq i < j < k \leq 5$ then $\{R_i, R_j, R_k\}$ is a $GF(4)$-basis of Π_1 and (vii) follows. □

Since $\mathcal{B}$ is the set of all 8-element subsets in the Golay code $\mathcal{C}_{12}$, whenever B_1 and B_2 are disjoint octads, the complement B_3 of their union is an octad (disjoint from B_1 and B_2). A triple of pairwise disjoint octads is called a *trio*.

Lemma 2.10.3 *Let $T = \{B_1, B_2, B_3\}$ be a trio. Let G_t be the stabilizer of T in G, K_t be the kernel of G_t acting on the set of octads in T and let $Q_t = O_2(G_t)$. We adopt for $B = B_1$ the notation introduced in* (2.10.1) *and let $\mathcal{T}(B)$ be the set of trios containing B.*

(i) *$|\mathcal{T}(B)| = 15$ and the action of G_b on $\mathcal{T}(B)$ is doubly transitive with kernel Q_b,*
(ii) *there are 3795 trios and G permutes them transitively,*
(iii) *for $i = 2$ and 3 the subgroup K_t acts on B_i as an elementary abelian group of order 2^3 extended by $L_3(2)$,*
(iv) *G_t induces Sym_3 on the octads in T,*
(v) *Q_t is elementary abelian of order 2^6 and $Q_b \cap Q_t$ is a hyperplane in Q_b,*
(vi) *there is a subgroup X_t of order 3 in G_t which permutes the octads in T transitively, such that G_t is the semidirect product of Q_t and $N_{G_t}(X_t) \cong Sym_3 \times L_3(2)$,*

(vii) *G_t is isomorphic to the full preimage in $A\Gamma L_3(4)$ of the stabilizer in $P\Gamma L_3(4)$ of a Fano subplane in the corresponding projective plane of order 4,*

(viii) *Q_t, as a module for $G_t/Q_t \cong L_3(2) \times Sym_3$, is isomorphic to the tensor product of the natural module D_1 for $L_2(2) \cong Sym_3$ and the natural module D_2 for $L_3(2)$,*

(ix) *there is a bijective mapping χ of $\mathcal{T}(B)$ onto the set of hyperplanes in Q_b such that with T as above B_2 and B_3 are the orbits of $\chi(T)$ on $\mathcal{P} \setminus B$.*

Proof. We assume that $Y \subset B = B_1$, so that $B \in \mathcal{B}_3$. Then the octads B_2 and B_3 are in $\mathcal{B}_0$. Moreover the lines in $\Pi(Y)$ of which B_i is the symmetric difference intersect in a point on the line $l := B \setminus Y$. On the other hand the symmetric difference of 2 lines intersecting in a point on l misses B. An easy calculation now shows that there are 30 octads disjoint from B and hence $|\mathcal{T}(B)| = 15$. In $\mathrm{Aut}\,\Pi(Y)$ the setwise stabilizer D of l induces Sym_5 on the points on l and for such a point p the stabilizer of p in D induces Sym_4 on the lines passing through p other than l. This shows that G_b and even its intersection with $G[Y]$ act transitively on the octads disjoint from B. Since 15 is an odd number, Q_b is in the kernel of the action of G_b on $\mathcal{T}(B)$ and $G_b/Q_b \cong L_4(2)$ acts on $\mathcal{T}(B)$ as it acts on the cosets of a parabolic subgroup. By (2.4.2) the action is doubly transitive. An element from Q_b either stabilizes each of B_2 and B_3 or switches them. So we have (i) and (ii). The subgroup $Q_b \cap K_t$ is a hyperplane in Q_b. Hence the image of K_t in $G_b/Q_b \cong L_4(2)$ is of index 15 and by (2.4.3) it is isomorphic to $2^3 : L_3(2)$. This gives (iii). Let $\{S_1, S_2, ..., S_6\}$ be a sextet. Then $\{S_1 \cup S_2, S_3 \cup S_4, S_5 \cup S_6\}$ is a trio and (iv) follows from (ii) and (2.10.2 (ii)). The subgroup K_t contains Q_t and induces on B the action of $2^3 : L_3(2)$ with kernel $Q_b \cap K_t$ of order 2^3 so we have (v). Thus $G_t \cong 2^6.L_3(2).Sym_3$. Since $L_3(2)$ does not possess an outer automorphism of order 3 there is a subgroup X_t in G_t whose image in G_t/Q_t is normal. Then X_t induces a $GF(4)$-vector–space structure on the commutator $[X_t, Q_t]$ and this structure is preserved by G_t/Q_t. The $L_3(2)$-factor of K_t acts faithfully on $Q_t \cap Q_b$ and on $Q_t/(Q_t \cap Q_b)$. Since $L_3(2)$ is not involved in $P\Gamma L_2(4)$, this implies $[X_t, Q_t] = Q_t$; so that the action of X_t on Q_t is fixed-point free. An element inverting X_t can be found inside Q_b, in particular commuting with the $L_3(2)$-factor of K_t, which implies (vi). The action of X_t on Q_t defines on the latter a $GF(4)$-structure so G_t is a subgroup in $A\Gamma L_3(4)$. Now it is easy to check that the $GF(4)$-subspaces in Q_t having non-trivial intersection with $Q_t \cap Q_b$

form a Fano subplane in the projective plane of order 4 associated with Q_t. In view of (v), (vi) and (2.7.6) this implies (vii) and (viii). Finally (ix) is immediate from (iii) and (v). □

In what follows B, $\Sigma = \{S_1, S_2, ..., S_6\}$ and $T = \{B_1, B_2, B_3\}$ are typical octad, sextet and trio and unless explicitly stated otherwise we adopt for them and their stabilizers in G notation as in (2.10.1), (2.10.2) and (2.10.3), although *a priori* we do not assume any relationship between B, Σ and T.

Lemma 2.10.4 *Let B be an octad and let $\mathcal{O}_i$ denote the set of octads intersecting B in exactly i elements. Then*

(i) $|\mathcal{O}_0| = 30$, *G_b acts transitively on $\mathcal{O}_0$ with stabilizer $2^3.2^3.L_3(2)$ which is an index 2 subgroup in the normalizer in G_b of a hyperplane in Q_b,*

(ii) $|\mathcal{O}_4| = 280$, *G_b acts transitively on $\mathcal{O}_4$ with stabilizer $2^6.3.Sym_3$, contained in a conjugate of G_t,*

(iii) $|\mathcal{O}_2| = 448$, *G_b acts transitively on $\mathcal{O}_2$ with stabilizer isomorphic to Sym_6.*

Proof. We obtain (i) directly from (2.10.3 (i), (vii)). If $B, B' \in \mathcal{B}$ and $S := B \cap B'$ is of size 4, then we can choose notation so that $B = S \cup S_2$, $B' = S \cup S_3$ where $\{S = S_1, S_2, ..., S_6\}$ is the sextet containing S. Since G acts transitively on the set of sextets and the stabilizer of a sextet induces Sym_6 on the set of the tetrads in the sextet, the transitivity assertion follows. Thus the number of pairs of octads intersecting in 4 points is the product of the number of sextets (which is 1771) and the number of pairs of 2-element subsets of a 6-element set (of tetrads in a sextet) intersecting in a single element (which is 60). Since G acts transitively on the set of octads and there are 759 of them, we can calculate $|\mathcal{O}_4|$. There are 759 octads altogether and 311 of them are in $\mathcal{O}_8 \cup \mathcal{O}_0 \cup \mathcal{O}_4$. By (2.8.5) the remaining 448 octads are in $\mathcal{O}_2$. G_b induces Alt_8 on the elements in B and hence it acts transitively on the set of 2-element subsets of B. The stabilizer in Alt_8 of such a subset is Sym_6. Hence there are $16 = 448/\binom{8}{2}$ octads intersecting B in a given 2-element subset. Let us show that Q_b acts regularly on these 16 octads. Suppose to the contrary that a non-trivial element $q \in Q_b$ stabilizes an octad B' intersecting B in 2 elements. By (2.10.1 (ii)) q fixes all elements inside B and no elements outside B. Hence q induces on B' an odd permutation, which contradicts (2.10.1 (iii)). □

Lemma 2.10.5 *Let Σ be a sextet and $B = S_1 \cup S_2$. Then there is a unique 2-dimensional subspace U in Q_b (such that $G_b \cap G_s = N_{G_b}(U)$ and S_j, $3 \leq j \leq 6$) are the orbits of U on $\mathcal{P} \setminus B$.*

Proof. The subgroup $H := G_b \cap G_s$ is the stabilizer in G_b of the partition $\{S_1, S_2\}$ of B. Since $[G_b : H] = 35$, H contains Q_b and H/Q_b is a parabolic subgroup in $L_4(2) \cong G_b/Q_b$. Comparing the orders, we conclude that $H = N_{G_b}(U)$ for a 2-dimensional subspace U in Q_b. Let Y be a 3-element subset in S_1 and $\{p\} = S_1 \setminus Y$. Then $\{p\} \cup S_i$, $2 \leq i \leq 6$, are the lines in $\Pi(Y)$ containing p. By (2.7.10 (i)) the elementwise stabilizer W of $\{p\} \cup S_2$ in $G(Y)$ is of order 2^2 and S_j, $3 \leq j \leq 6$, are the orbits of W on $\mathcal{P} \setminus B$. Comparing the orders, we obtain the equality $W = Q_b \cap Q_s$, which shows that W is normalized by H. Since U is the unique 2-dimensional subspace in Q_b normalized by H, we have $W = U$ and the result follows. □

Now by (2.10.3 (iii), (ix)) and (2.10.5) we have the following.

Lemma 2.10.6 *In terms of* (2.10.3) *let $F \cong 2^3 : L_3(2)$ be the action induced by K_t on B. Let S be a tetrad contained in B and let Σ be the sextet containing S. Then Σ refines T if and only if S is an orbit on B of a subgroup of index* 2 *in $O_2(F)$.* □

2.11 Little Mathieu groups

From the construction of the binary Golay code in Section 2.2 we know that $\mathcal{C}_{12}$ contains 12-element subsets which will be called *dodecads*. By now we know that $\mathcal{C}_{12}$ contains the empty set, the set $\mathcal{P}$, 759 octads and 759 complements of octads. Since $2^{12} - 2 \cdot 759 - 2 = 2576$ we have the following.

Lemma 2.11.1 *$\mathcal{C}_{12}$ contains exactly* 2576 *subsets of size greater than* 8 *and less than* 16. □

We are going to show that all of the remaining subsets are dodecads. Clearly the complement of a dodecad is a dodecad. If a dodecad should contain an octad then their symmetric difference would be of size 4, which is impossible in the Golay code.

Lemma 2.11.2 *A dodecad never contains an octad.* □

Clearly, the symmetric difference of two octads intersecting in two points is a dodecad.

Lemma 2.11.3 *Let D be a dodecad.*

(i) *If B_1, B_2, B_3, B_4 are pairwise distinct octads such that $B_1 \triangle B_2 = B_3 \triangle B_4 = D$ then $B_1 \cap B_2 \neq B_3 \cap B_4$,*
(ii) *there are exactly 66 ways to present D as the symmetric difference of 2 octads intersecting in 2 elements,*
(iii) *there are exactly 2576 dodecads in $\mathscr{C}_{12}$ and $G = Mat_{24}$ permutes them transitively.*

Proof. For B_i, $1 \leq i \leq 4$, as in (i) suppose that $B_1 \cap B_2 = B_3 \cap B_4 = X$. Then X is contained in $B_1 \cap B_3$. Suppose that $B_1 \cap B_3 = X$. Then $B_3 = X \cup (D \setminus B_1) = B_2$, which is a contradiction. Hence $B_1 \cap B_3$ is of size 4 and it contains X properly. Then $B_5 := B_1 \triangle B_3$ is an octad completely contained in D, which contradicts (2.11.2) and hence (i) follows. Now we see that the number of presentations of D as the symmetric difference of two octads is at most the number of 2-element subsets in $\mathscr{P} \setminus D$, which is 66. Thus we can produce at least $(759 \cdot 448)/(2 \cdot 66) = 2576$ different dodecads as symmetric differences of pairs of octads. Now (2.11.1) implies (ii). By (2.10.4 (iii)) G acts transitively on the pairs of octads intersecting in two elements and (iii) follows. □

As a direct consequence of (2.11.1) and (2.11.3 (iii)) we have the following.

Proposition 2.11.4 *The Golay code $\mathscr{C}_{12}$ is doubly even.* □

We are going to study the setwise stabilizer in $G = Mat_{24}$ of a dodecad D. First let us define a certain structure on D. Let $\mathscr{Q}$ denote the set of 6-element subsets of D (blocks) which are intersections of D with octads. If an octad B intersects D in six points then D is the symmetric difference of B and the octad $B' = D \triangle B$, also intersecting D in six points. This and (2.11.3) imply

Lemma 2.11.5 *$\mathscr{Q}$ is of size $132 = 66 \cdot 2$ and it is closed under taking complements.* □

Two octads never share a 5-element subset and the same is certainly true for the blocks from $\mathscr{Q}$. Since $132 = \binom{12}{5} / \binom{6}{5}$, every 5-element subset of D is in a unique block from $\mathscr{Q}$ and we have the following.

Lemma 2.11.6 *$\mathscr{D} = (D, \mathscr{Q})$ is a Steiner system of type $S(5,6,12)$.* □

Let Mat_{12} denote the stabilizer in Mat_{24} of a dodecad D. Let $\emptyset = Y_0 \subset Y_1 \subset Y_2 \subset Y_3 \subset D$ where Y_i is of size i. Let Mat_{12-i} denote the elementwise stabilizer of Y_i in Mat_{12}. Let C_1 and C_2 be disjoint blocks of the Steiner system $\mathscr{D}$ of type $S(5,6,12)$ defined on D.

Lemma 2.11.7

(i) *Mat_{12} permutes transitively the blocks of $\mathscr{D}$,*

(ii) *the setwise stabilizer H of C_1 in Mat_{12} is isomorphic to Sym_6 and it induces on C_1 and C_2 two faithful inequivalent 6-fold transitive actions, an element from Mat_{12} which maps C_1 onto C_2 induces an outer automorphism of H,*

(iii) *Mat_{12} acts faithfully and 5-fold transitively on D,*

(iv) *Mat_{12-i} acts $(5-i)$-fold transitively on $D \setminus Y_i$,*

(v) *$|Mat_{12}| = 2^6 \cdot 3^3 \cdot 5 \cdot 11$; $|Mat_{11}| = 2^4 \cdot 3^2 \cdot 5 \cdot 11$; $|Mat_{10}| = 2^4 \cdot 3^2 \cdot 5$; $|Mat_9| = 2^3 \cdot 3^2$.*

Proof. By (2.10.4 (iii)) we have (i). Let X be the unique 2-element subset of $\mathscr{P} \setminus D$ such that $B = C_1 \cup X$ and $B' = C_2 \cup X$ are octads. By (2.10.4 (iii)) the subgroup of G which stabilizes each of B and B' as a whole is isomorphic to Sym_6 and induces faithful 6-fold transitive actions on C_1 and C_2. Let S be a 4-element subset of C_1 and let $\{S_1 = S, S_2, ..., S_6\}$ be the sextet containing S. Since S is in four blocks of $\mathscr{D}$, it is easy to see that for $2 \le i \le 6$ the intersection $|S_i \cap D|$ is of size 2 or 0. This means that a 4-element subset of C_1 corresponds to a partition of C_2 into three 2-element subsets and we have (ii). Every 5-element subset E of D is contained in a unique block C of $\mathscr{D}$. By (i) Mat_{12} is transitive on blocks of $\mathscr{D}$ and by (ii) the stabilizer of C acts 6-fold transitively on its points. So we have (iii) and (iv). The elementwise stabilizer of E is contained in the stabilizer of C and obviously must be trivial. In view of (iv) this means that $|Mat_{12-i}| = (12-i) \cdot ... \cdot 8$ and we obtain (v) by direct calculations. □

By (2.11.3) there is a correspondence between 2-element subsets of $\mathscr{P} \setminus D$ and pairs of complementary blocks of $\mathscr{D}$. A subset X corresponds to a pair $\{C_1, C_2\}$ if and only if $C_1 \cup X$ and $C_2 \cup X$ are octads. Since two distinct octads intersect in no, two or four points, it is easy to see the following.

Lemma 2.11.8 *Suppose that X and Z are 2-element subsets of $\mathcal{P} \setminus D$ corresponding to the pairs $\{C_1, C_2\}$ and $\{C_3, C_4\}$ of complementary blocks, respectively. Then up to renaming the blocks the following hold:*

(i) *if $|X \cap Z| = 1$ then $|C_1 \cap C_3| = |C_2 \cap C_3| = 3$,*
(ii) *if $|X \cap Z| = 0$ then $|C_1 \cap C_3| = 4$ and $|C_2 \cap C_3| = 2$.* □

We are going to study in more detail the structure of the groups Mat_{12-i} for $0 \leq i \leq 3$.

Lemma 2.11.9

(i) *$Mat_9 \cong 3^2 : Q_8$ is the stabilizer in $L_3(4)$ of an affine subplane of order 3 in $\Pi(Y)$,*
(ii) *Mat_{10} is an index 2 subgroup in* Aut $Sym_6 \cong P\Gamma L_2(9)$ *in which Mat_9 is the normalizer of a Sylow 3-subgroup,*
(iii) *Mat_{11} is a non-abelian simple group, it acts 3-fold transitively on $\mathcal{P} \setminus D$ with point stabilizer isomorphic to $L_2(11)$,*
(iv) *Mat_{12} is non-abelian and simple; it contains two conjugacy classes of Mat_{11}-subgroups permuted by an outer automorphism of Mat_{12} realized by an element of Mat_{24} which maps D onto $\mathcal{P} \setminus D$.*

Proof. The residual system of $\mathcal{D}$ with respect to Y_i is a Steiner system of type $S(5-i, 6-i, 12-i)$ by (2.3.5). For $i = 3$ we obtain an affine subplane of order 3 in the residual projective plane $\Pi(Y)$ of order 4. By (2.11.7 (v)) and (2.7.11) Mat_9 is the full stabilizer of this subplane in $Mat_{21} \cong L_3(4)$ and (i) follows. Let $D' = \mathcal{P} \setminus D$ and $\mathcal{D}'$ be the Steiner system of type $S(5, 6, 12)$ defined on D'. Let $Mat_{10}.2$ denote the setwise stabilizer of Y_2 in Mat_{12} which clearly contains Mat_{10} with index 2. Since Y_2 corresponds to a partition of D' into 2 blocks of $\mathcal{D}'$ we obtain (ii) from (2.11.7 (ii)). It is easy to see that Mat_{11} acts transitively on D' which is of size 12. Y_1 is contained in 11 2-element subsets of D. These subsets determine a collection $\mathcal{E}$ of 22 blocks of $\mathcal{D}'$ forming 11 complementary pairs. By (2.11.8) any 2 blocks from different pairs have intersection of size 3. For $p \in D'$ let $\mathcal{E}(p) = \{B \setminus \{p\} \mid B \in \mathcal{E}, p \in B\}$. Then $(D' \setminus \{p\}, \mathcal{E}(p))$ is the unique 2-$(11, 5, 2)$-design whose automorphism group is isomorphic to $L_2(11)$ [BJL86]. Computing the orders we see by (2.11.7 (v)) that the latter is the stabilizer of p in Mat_{11} and (iii) follows. Since Mat_{12} acts 5-fold transitively on the set D of size 12 with point stabilizer being Mat_{11} which is simple, Mat_{12} is also simple (see the proof of (2.9.2 (iii))). The stabilizer in Mat_{12} of any point from $\mathcal{P}$ is isomorphic to Mat_{11} but the stabilizers of $p \in D$ and $p' \in D'$ are conjugate only in the extension of Mat_{12} by an element from Mat_{24} which permutes D and D'. □

The groups Mat_{11} and Mat_{12} are two further sporadic simple Mathieu groups. The group Mat_{10} is an index 2 subgroup in $\mathrm{Aut}\, Sym_6 \cong P\Gamma L_2(9)$, distinct from two other such subgroups Sym_6 and $PGL_2(9)$. Let $Mat_{12}.2$ denote the setwise stabilizer in Mat_{24} of a pair of complementary dodecads. By (2.11.9 (iv)) it contains Mat_{12} with index 2 and it induces on it an outer automorphism. In fact $Mat_{12}.2$ is the full automorphism group of Mat_{12}.

We observed that the residual system of $\mathscr{D}$ with respect to Y_i is a Steiner system of type $S(5-i, 6-i, 12-i)$. It is not difficult to show using (2.11.9) that $Mat_{10}.2$, Mat_{11} and Mat_{12} are the full automorphism groups of these systems for $i = 2$, 1 and 0, respectively. Each of these systems is unique up to isomorphism ([Wit38], [HP85], [BJL86]).

Lemma 2.11.10 *For $0 \le i \le 3$ any Steiner system of type $S(5-i, 6-i, 12-i)$ is isomorphic to the residual system of $\mathscr{D}$ with respect to Y_i.* □

2.12 Fixed points of a 3-element

In this section we calculate the normalizer in G of a subgroup X_s of order 3 and analyse the octads, sextets and trios fixed by X_s.

We start by studying the orbits of G on the set of 6-element subsets of $\mathscr{P}$. Since such a subset might or might not be contained in an octad, clearly there are at least two orbits. On the other hand G acts on $\mathscr{P}$ 5-transitively and hence all 5-element subsets form a single orbit and if D is such a subset then $G[D]$ induces on D the full symmetric group Sym_5. There is a unique octad B which contains D and in terms of (2.10.1) we have the following.

Lemma 2.12.1 *Let D be a 5-element subset of $\mathscr{P}$ and B be the unique octad which contains D. Then*

(i) *$G[D]/G(D) \cong Sym_5$,*
(ii) *$G(D)$ is the elementary abelian group Q_b of order 2^4 extended by a fixed-point free subgroup X_s of order 3,*
(iii) *$G(D)$ acts transitively on $B \setminus D$ with Q_b being the kernel and on $\mathscr{P} \setminus B$ with X_s being the stabilizer of a point.* □

By the above lemma the G-orbit which contains a given 6-element subset E depends only on whether or not E is contained in an octad and the following lemma holds.

Lemma 2.12.2 *If E is a 6-element subset of $\mathcal{P}$, then $G[E]/G(E) \cong Sym_6$. Moreover, if E is contained in an octad then $G(E)$ is elementary abelian of order 2^4, otherwise $G(E)$ is of order 3.* □

If E is contained in an octad then the exact structure of $G[E]$ follows from (2.10.1).

Lemma 2.12.3 *If E is a 6-element subset of $\mathcal{P}$ which is not contained in an octad, then there is a sextet $\Sigma = \{S_1, ..., S_6\}$ such that $G[E] = N_{G_s}(X_s)$ where G_s is the stabilizer of Σ in G and X_s is a Sylow 3-subgroup in $O_{2,3}(G_s)$. In particular $G[E] = N_G(X_s) = N_{G_s}(X_s) \cong 3 \cdot Sym_6$ is the automorphism group of the rank 2 T-geometry.*

Proof. Let Σ be a sextet and let G_s and X_s be defined as above. Then by (2.10.2) for every $1 \le i \le 6$ the subgroup X_s stabilizes S_i as a whole and it fixes exactly one element in S_i. Thus the set E of elements in $\mathcal{P}$ fixed by X_s is of size 6. Since $N_{G_s}(X_s)$ clearly stabilizes E as a whole, the result follows from (2.12.2) and (2.10.2 (iv)). □

Let E be a 6-element subset of $\mathcal{P}$ not contained in an octad and $X_s = G(E)$, so that X_s acts fixed-point freely on $\mathcal{P} \setminus E$. Let D be a 5-element subset in E. Then the unique octad B' which contains D is stabilized by X_s and hence $B' = D \cup T$ where T is an orbit of X_s on $\mathcal{P} \setminus D$ which gives the following.

Lemma 2.12.4 *There is a bijection χ between the elements in E and the orbits of X_s on $\mathcal{P} \setminus E$ such that $(E \setminus \{p\}) \cup \chi(p)$ is an octad for every $p \in E$.* □

Let $E = \{p_1, ..., p_6\}$ and let $T_1, ..., T_6$ be the orbits of X_s on $\mathcal{P} \setminus E$ indexed so that $B_i := (E \setminus \{p_i\}) \cup T_i$ is an octad. For $1 \le i < j \le 6$ the octads B_i and B_j have four elements in common and hence their symmetric difference $B_{ij} = \{p_i, p_j\} \cup T_i \cup T_j$ is also an octad.

Lemma 2.12.5 *The subgroup X_s stabilizes exactly $21 = 6 + 15$ octads, namely the octads from the sets $\{B_i \mid 1 \le i \le 6\}$ and $\{B_{ij} \mid 1 \le i < j \le 6\}$. These sets are the orbits of $N_G(X_s)$.*

Proof. Let B' be an octad stabilized by X_s. Then X_s fixes 5 or 2 elements in B' and in any case X_s stabilizes a 5-element subset in B'. But one can check that every 5-element subset of $\mathcal{P}$ stabilized by X_s is already in one of the 21 octads counted. □

Let us turn to the trios stabilized by X_s. Since X_s is not fixed-point free on $\mathcal{P}$, whenever it stabilizes a trio, it stabilizes every octad in the trio. Thus all we have to do is to decide how many trios can be formed by the octads stabilized by X_s.

Lemma 2.12.6 *There are exactly* 15 *trios stabilized by* X_s *and permuted transitively by* $N_G(X_s)$. *These trios are indexed by the partitions of E into* 3 *subsets of size* 2, *in particular* $\{B_{12}, B_{34}, B_{56}\}$ *is such a trio.* □

Again, since X_s is not fixed-point free on $\mathcal{P}$, whenever X_s stabilizes a sextet, it stabilizes at least one tetrad in the sextet. Thus in order to describe the sextets stabilized by X_s we have to decide for each 4-element subset stabilized by X_s which sextet it belongs to. The result is in the following.

Lemma 2.12.7 *There are exactly* 16 *sextets stabilized by* X_s, *namely the sextet* $\Sigma = \{\{p_i\} \cup T_i \mid 1 \leq i \leq 6\}$ *and the sextets* Σ_{ij} *containing the tetrads* $E \setminus \{p_i, p_j\}$, $T_i \cup \{p_j\}$, $T_j \cup \{p_i\}$ *for* $1 \leq i < j \leq 6$. *Moreover* $N_G(X_s)$ *stabilizes* Σ *and permutes transitively the* Σ_{ij}. □

The final result in this section can be deduced for instance by comparing the centralizers of X_s in the Golay code and Todd modules dual to each other (Section 2.15).

Lemma 2.12.8 *There are exactly* 20 *dodecads stabilized by* X_s, *namely the dodecads*

$$(E \setminus \{p_i, p_j, p_k\}) \cup T_i \cup T_j \cup T_k$$

for $1 \leq i < j < k \leq 6$. *These dodecads are transitively permuted by* $N_G(X_s)$. □

2.13 Some odd order subgroups in Mat_{24}

In this section we determine the conjugacy classes of subgroups of order 3 in G and calculate the normalizer of a subgroup of order 7. First we recall a useful general result from [Alp65].

Lemma 2.13.1 *Let H be a finite group, F, K be subgroups of H with $K \leq F$. Put*

$$\mathscr{K} = \{h^{-1}Kh \mid h \in H, h^{-1}Kh \leq F\}$$

and let $\mathscr{K}_1, \mathscr{K}_2, ..., \mathscr{K}_m$ be the orbits of F acting on $\mathscr{K}$ by conjugation. Put

$$\Delta = \{Fh \mid h \in H, Fhk = Fh \text{ for all } k \in K\}$$

and let $\Delta_1, \Delta_2, ..., \Delta_l$ be the orbits of $N_H(K)$ on Δ. Then $l = m$ and under a suitable ordering

$$|\Delta_i| = [N_H(K_i) : N_F(K_i)]$$

for $1 \leq i \leq m$, where $K_i \in \mathscr{K}_i$ (notice that $N_H(K_i) \cong N_H(K)$). □

In the previous section we discussed a subgroup X_s of order 3 in G such that $N_G(X_s) = N_{G_s}(X_s) \cong 3 \cdot Sym_6$. The subgroups in G conjugate to X_s will be called $3a$-subgroups. Let X_t be as in (2.10.3). Since $N_{G_t}(X_t) \cong Sym_3 \times L_3(2)$, X_t is not a $3a$-subgroup. The G-conjugates of X_t will be called $3b$-subgroups. We will show that every subgroup of order 3 in G is either $3a$ or $3b$ and that $N_G(X_t) = N_{G_t}(X_t)$. We start with a preliminary lemma.

Lemma 2.13.2

(i) *All subgroups of order 3 in $L_3(4)$ are conjugate,*

(ii) *if E is a set of size 6 and $D = Sym(E) \cong Sym_6$, then D has two classes of subgroups of order 3, say $3A$ and $3B$, so that $3A$-subgroups are generated by 3-cycles; these two classes are fused in* Aut D.

Proof. If D is the stabilizer in $L_3(4)$ of an affine subplane in the projective plane of order 4, then $O_3(D)$ is a Sylow 3-subgroup in $L_3(4)$ and (i) follows from (2.7.11 (i)). In (ii) everything except possibly the fusion is obvious. Since the product of two non-commuting transpositions is always a $3A$-subgroup, the classes are not stable under an outer automorphism of D. □

Lemma 2.13.3 *The following assertions hold:*

(i) *every subgroup of order 3 in G which fixes an element from $\mathscr{P}$ is a $3a$-subgroup,*

(ii) *all $3a$-subgroups in $N_{G_s}(X_s)$ other than X_s are conjugate,*

(iii) *all $3b$-subgroups in $N_{G_s}(X_s)$ are conjugate and every subgroup of order 3 in G is either $3a$ or $3b$,*

(iv) $N_G(X_t) = N_{G_t}(X_t)$.

Proof. Put $N = N_{G_s}(X_s)$. Every subgroup of order 3 in G which is not fixed-point free on $\mathcal{P}$ is contained in $G(Y) \cong L_3(4)$ for a 3-element subset Y of $\mathcal{P}$. Since G is 5-fold transitive on $\mathcal{P}$, (i) follows from (2.13.2 (i)). Since $N \cong G_s/Q_s$ any two subgroups of order 3 are conjugate in N if and only if they are conjugate in G_s. By (2.12.7) and (2.13.1) there are exactly two classes of $3a$-subgroups in G_s and we obtain (ii). For $U = 3A$ and $3B$ let $C(U)$ be the set of subgroups of order 3 in N, which maps onto the class of U-subgroups in $N/X_s \cong Sym(E)$ where E is the 6-element subset of $\mathcal{P}$ fixed by X_s. Clearly $C(3A)$ and $C(3B)$ are unions of conjugacy classes of subgroups in N. Since every subgroup from $C(3A)$ fixes an element from E, it must be a $3a$-subgroup by (i). Hence by (ii) every $3a$-subgroup in N is contained in $X_s \cup C(3A)$, all subgroups in $C(3A)$ are conjugate and $C(3B)$ consists of $3b$-subgroups. Since N is the automorphism group of the rank 2 T-geometry, we know from (2.6.1) that it possesses an outer automorphism τ which induces an outer automorphism of $N/X_s \cong Sym_6$. By (2.13.2 (ii)) τ permutes $C(3A)$ and $C(3B)$. In view of (ii) this means that all subgroups in $C(3B)$ are conjugate in N and (iii) follows. Thus G acts transitively by conjugation on the set of pairs (A, B) where A and B are respectively $3a$- and $3b$-subgroups and $[A, B] = 1$. Let us calculate the number of $3a$-subgroups commuting with a given $3b$-subgroup, say X_t. When acting on $\mathcal{P}$, X_t has eight orbits of length 3, in particular it does not stabilize an octad. Hence if E is the union of any two X_t-orbits, then E is a 6-element subset not contained in an octad. Clearly X_t commutes with the $3a$-subgroup which is the elementwise stabilizer of E. This shows that X_t commutes with $\binom{8}{2} = 28$ $3a$-subgroups. Since $|C(3A)| = |C(3B)| = 60$ we obtain

$$|N_G(X_t)| = \frac{|N_G(X_s)| \cdot 28}{60} = 2^4 \cdot 3^2 \cdot 7 = |N_{G_t}(X_t)|$$

and (iv) follows. □

Lemma 2.13.4 *Let X_t be a $3b$-subgroup in G. Then*

(i) *X_t fixes 7 sextets transitively permuted by $N_G(X_t)$,*
(ii) *there are 15 trios fixed by X_t; $N_G(X_t)$ fixes one of them and permutes transitively the remaining ones.*

Proof. It is easy to check (for instance diagonalizing the corresponding matrices) that all non-central subgroups of order 3 in $SL_3(4)$ are conjugate and the centralizer in the natural module of such a subgroup is of order 2^2. By (2.10.2 (vi)) and (2.10.3 (vii)) this means that for $x = t$ or s

if A is a subgroup of order 3 in G_x but not in $O_{2,3}(G_x)$, then $C_{Q_x}(A) \cong 2^2$. We know that $N_{G_s}(X_s)$ contains 60 $3b$-subgroups forming a conjugacy class and (i) follows from (2.13.1). It is easy to see that G_t contains 3 conjugacy classes of subgroups of order 3. By (2.12.6) one of these classes consists of $3a$-subgroups and it is easy to observe that such a subgroup can be found in the $L_3(2)$-factor of a complement to Q_t. Now (2.13.1) and straightforward calculations imply (ii). □

Lemma 2.13.5 *Let $S \leq G_t$ be a subgroup of order 7. Then $N_G(S) = N_{G_t}(S) \cong Frob_7^3 \times Sym_3$. In particular S is not fully normalized in G.*

Proof. Acting on $\mathcal{P}$ the subgroup S has three orbits of length 7 and three fixed elements. Hence S fixes at most three octads and, since $S \leq G_t$, it fixes exactly three octads forming the trio stabilized by S. Hence the result follows directly from (2.10.3). □

2.14 Involutions in Mat_{24}

In this section we study subgroups of order 2 (or rather involutions) in G. We determine the conjugacy classes of involutions, their centralizers and also octads, trios and sextets fixed by a given involution. For this purpose it is helpful to know the G-orbits on octad–sextet pairs.

Let B be an octad and Σ be a sextet. Let $\nu = \nu(B, \Sigma)$ be the multiset consisting of $|B \cap S_i|$ for $1 \leq i \leq 6$. We assume that $|B \cap S_i| \geq |B \cap S_j|$ if $i < j$.

Lemma 2.14.1 *The G-orbit containing the pair (B, Σ) is uniquely determined by the multiset $\nu = \nu(B, \Sigma)$ and one of the following holds:*

(i) $\nu = (4^2 0^4)$, *$G_b \cap G_s$ contains both Q_b and K_s and has order $2^{10} \cdot 3^2$,*
(ii) $\nu = (3\,1^5)$, *$G_b \cap G_s \cong (Sym_3 \times Sym_5)^e$,*
(iii) $\nu = (2^4 0^2)$, *$G_b \cap G_s \sim [2^6].Sym_3$ and $|Q_b \cap G_s| = 2$.*

Proof. Since the union of any two tetrads in Σ is an octad and any two octads have four, two or no elements in common, it is easy to see that the possibilities for the multiset ν are those given in the lemma. (i) follows directly from (2.10.5). In case (ii) clearly $Q_b \leq G_b \cap G[B \cap S_1]$ acts transitively on $\mathcal{P} \setminus B$ and the transitivity assertion follows. Furthermore, $G_b \cap G_s = G_b \cap G[B \cap S_1] \cap G(p)$ where $\{p\} = S_1 \setminus B$ and we obtain (ii). In case (iii) $H := G_b \cap G[B \cap S_1]$ contains Q_b and $H/Q_b \cong Sym_6$ acts transitively by conjugation on the set of non-trivial elements of Q_b. Hence H acts doubly transitively on $\mathcal{P} \setminus B$ which implies the transitivity assertion. There is exactly one involution in Q_b which stabilizes $S_1 \setminus B$ as a whole and this involution generates $Q_b \cap G_s$. Since Σ is uniquely determined by any of the tetrads S_1, S_2, S_3 and S_4 the number n of sextets Σ with $\nu(B, \Sigma) = (2^4 0^2)$ for a fixed B is given by

$$n = \binom{8}{2} \cdot \binom{16}{2} / 4$$

and (iii) follows. □

By (2.10.1) all involutions from Q_b are conjugate in G_b. The G-conjugates of the involutions from Q_b will be called $2a$-involutions.

Lemma 2.14.2 *Let q be a $2a$-involution contained in Q_b and let $C = C_G(q)$. Then*

(i) $C = C_{G_b}(q) \cong 2_+^{1+6} : L_3(2)$,
(ii) *an involution which fixes an element from $\mathcal{P}$ is a $2a$-involution,*
(iii) *q fixes 71 octads, 99 trios and 91 sextets.*

Proof. Since B is the set of elements from $\mathcal{P}$ fixed by q, $C_G(q) \leq G_b$ and (i) follows from (2.10.1). Let r be an involution which fixes an element from $\mathcal{P}$. Then there is a 5-element subset D of $\mathcal{P}$ which r stabilizes as a whole (and fixes an element in D). Without loss of generality we assume that B is the unique octad containing D. Since G_b induces Alt_8 on B, r fixes 4 or 8 elements in B and in any case r is contained in $G(Y) \cong L_3(4)$ for a 3-element subset $Y \subseteq B$ and (ii) follows from (2.7.10 (v)). By (2.10.4 (iii)) Q_b acts fixed-point freely on $\mathcal{O}_2$. By (2.10.3) an octad from $\mathcal{O}_0$ is an orbit on $\mathcal{P} \setminus B$ of a hyperplane R in Q_b. Clearly such an octad is fixed by q if and only if $q \in R$. By (2.10.5) if $B' \in \mathcal{O}_4$ then $B' \setminus B$ is an orbit of a subgroup U of order 2^2 in Q_b. Then B' is fixed by q if and only if $q \in U$. These observations and basic properties of G_b show that q fixes $1 + 2 \cdot 7 + 8 \cdot 7 = 71$ octads. If $T = \{B_1, B_2, B_3\}$ is a trio stabilized by q, then since q is an involution, q stabilizes at least one of the octads in T, say B_1. Furthermore, if $B_2^q = B_3$, then $B_2 \cap B_2^q = \emptyset$ and since q fixes B elementwise, this means that $B_2 \cap B = \emptyset$. Hence either $B_1 = B$ or every octad in T is fixed by q. Now an easy calculation shows that q fixes 99 trios. Since Q_b is the kernel of the action of G_b on the set of sextets Σ such that $\nu(B, \Sigma) = (4^2, 0)$, q fixes each of these 35 sextets. By (2.14.1) if $\nu(B, \Sigma) = (2^4 0^2)$ then Σ is fixed by exactly one involution from Q_b. Since there are 840 sextets with $\nu(B, \Sigma) = (2^4 0^2)$ (transitively permuted by G_b) and 15 involutions in Q_b, altogether we have 91 sextets fixed by q. □

By (2.7.13) and (2.10.2 (vi)) G_s/Q_s acting on the set of involutions in Q_s by conjugation has two orbits with lengths 18 and 45 and if t is contained in the former of the orbits then $C_{G_s}(t) \cong 2^6.Sym_5$. Hence t is not a $2a$-involution. The G-conjugates of t will be called $2b$-involutions.

Lemma 2.14.3 *Let t be a $2b$-involution and $C = C_G(t)$. Then*

(i) *every involution in G is either $2a$ or $2b$,*
(ii) *all $2b$-involutions in G_b are conjugate and there are 630 of them,*
(iii) *there is a unique sextet $\Sigma(t)$ such that $t \in Q_s$ where $Q_s = O_2(G_s)$ and G_s is the stabilizer of $\Sigma(t)$ in G,*
(iv) *$C = C_{G_s}(t) \cong 2^6 : Sym_5$ and C acts transitively on $\mathcal{P}$,*
(v) *t fixes 15 octads, 75 trios and 51 sextets.*

Proof. Since G_b contains a Sylow 2-subgroup of G, every involution in G is a conjugate of an involution from G_b. If $s \in G_b$ is an involution which is not a $2a$-involution, then by (2.14.2 (ii)) s acts fixed-point freely on $\mathcal{P}$ and particularly on B. This means that the image of s in $G_b/Q_b \cong L_4(2)$ is one of the 105 transvections. Let $p \in \mathcal{P} \setminus B$ and let $H = G_b \cap G(p)$ be the corresponding complement to Q_b in G_b. Then every element $r \in \mathcal{P} \setminus B$ is identified with the unique element $q \in Q_b$ such that $p^q = r$. In this way p is identified with the identity element, Q_b acts on $\mathcal{P} \setminus B$ by translation and H acts by conjugation. Let s_0 be a transvection in H with centre r and axis R. Then $C_H(s_0)$ coincides with the stabilizer in H of the pair (r, R) and it permutes transitively the involutions in $R \setminus \{r\}$. If $q \in Q_b$, then $s_0 q$ is an involution if and only if s_0 centralizes q, *i.e.* if $q \in R$. The elements s_0 and r induce the same action on $\mathcal{P} \setminus (B \cup R)$ which means that sr is a $2a$-involution. On the other hand if u is an involution from $R \setminus \{r\}$, then $s_0 u$ acts fixed-point freely on $\mathcal{P}$. By the transitivity of $C_H(s_0)$ on the involutions in $R \setminus \{r\}$ all these involutions are conjugate in G_b, hence they are $2b$-involutions and we obtain (i) and (ii).

Let σ be the partition of B into the orbits of s_0. There is a unique orbit of length 3 of $C_H(s_0)$ on the set of 2-dimensional subspaces in Q_b. This orbit consists of the subspaces containing r and contained in R. These 3 subspaces correspond to the partitions of B into 2 4-sets refined by σ. Let u be an involution from $R \setminus r$, so that $t = s_0 u$ is a $2b$-involution. Then t acts on R as u and on $Q_b \setminus R$ as ur. Hence $U = \langle r, u \rangle$ is the unique 2-dimensional subspace in Q_b whose orbits on $\mathcal{P} \setminus B$ are unions of the orbits of t on this set. Let $\Sigma(t) = \{S_1, S_2, ..., S_6\}$ be the partition of $\mathcal{P}$ such that $\{S_1, S_2\}$ is the partition of B which corresponds to U and S_i, $3 \le i \le 6$, are the orbits of U on $\mathcal{P} \setminus B$. Then $\Sigma(t)$ is a sextet by (2.10.5) and by the construction t stabilizes every tetrad in $\Sigma(t)$. We claim that $\Sigma(t)$ is the unique sextet with this property. Let $\Sigma' = \{S'_1, S'_2, ..., S'_6\}$ be a sextet, let $v = v(B, \Sigma')$ be the multiset as in (2.14.1) and suppose that every S'_i is stabilized by t. If $v = (4^2 0^4)$ then $\Sigma = \Sigma(t)$ by the above construction of $\Sigma(t)$. Notice that the set $S'_i \cap B$, if non-empty, must be a union of subsets from σ. By this observation $v \neq (3\,1^5)$ and if $v = (2^4 0^2)$ then we can assume that $\{S'_i \cap B \mid 1 \leq i \leq 4\} = \sigma$. Let Λ be the set of all sextets Σ' such that $\{S'_i \mid 1 \leq i \leq 4\} = \sigma$ and let M be the set of all $2b$-involutions in $s_0 Q_b$. Then $|\Lambda| = 8$, $|M| = 6$ and $\langle Q_b, H \rangle$ acts transitively on both Λ and M. Let Θ be a graph on $\Lambda \cup M$ in which $\Sigma' \in \Lambda$ is adjacent to $t \in M$ whenever t stabilizes every tetrad in Σ'. By the above mentioned transitivity, if Θ contains at least one edge then every $\Sigma' \in \Lambda$ is adjacent to at least 3 involutions in M, which is impossible since $|Q_b \cap G'_s| = 2$ by (2.14.1 (iii)) where G'_s is the stabilizer of Σ' in G. Hence $\Sigma' = \Sigma(t)$ and (iii) follows. By (2.7.14) we obtain $C \cong 2^6 : Sym_5$. Let Π be the projective plane of order 4 formed by the subgroups in Q_s normal in K_s. Then by (2.10.2) and the proof of (2.7.5) G_s/K_s stabilizes a hyperoval Ω in Π and a hyperoval Ω^* in the dual of Π. The points on Ω^* are the kernels of Q_s on the S_i for $1 \leq i \leq 6$. The subgroup CK_s/K_s is the stabilizer in G_s/K_s of a point on Ω which is the normal closure of t in K_s. By (2.5.1 (iii)) C permutes transitively the tetrads in $\Sigma(t)$. In addition C contains Q_s which acts transitively on every tetrad in $\Sigma(t)$, so (iv) follows. By (2.13.1), (ii) and (iv) there are 15 octads stabilized by t and transitively permuted by $C_G(t)$. Clearly they are exactly the octads refined by $\Sigma(t)$. Since t is an involution, whenever it stabilizes a trio T, it stabilizes at least one octad in T. If t stabilizes every octad in T, then by the above T is one of the 15 trios refined by $\Sigma(t)$. Consider the trios containing B and stabilized by t. By (2.10.3 (ix)) the trios containing B are in a bijection with the hyperplanes in Q_b. Since tQ_b/Q_b is a transvection, t fixes 7 hyperplanes and 3 of them correspond to trios refined by $\Sigma(t)$. Hence altogether t stabilizes $15 + 4 \cdot 15 = 75$ trios. Since there are 15 octads refined by a given sextet and t is an involution, whenever t stabilizes a sextet Σ', it stabilizes at least one octad, refined by Σ'. By (2.10.5) the sextets refining B are in a bijection with the 2-dimensional subspaces in Q_b. Since tQ_b/Q_b is a transvection, it stabilizes 11 2-dimensional subspaces. One of these subspaces, say U, corresponds to $\Sigma(t)$. Let Σ' corresponds to one of the remaining 10 subspaces, say to W. Since U and W are stabilized by a transvection, $\langle U, W \rangle$ is a hyperplane which determines a trio containing B and refined by $\Sigma(t)$ and Σ'. Hence there are $1 + (10 \cdot 15)/3 = 51$ sextets stabilized by t. $\square$

In the table below we summarize the normalizers, the numbers of elements in $\mathcal{P}$, octads, trios and sextets stabilized for various subgroups of order 2 and 3 in Mat_{24}.

Class	Normalizer	$\mathcal{P}$	Octads	Trios	Sextets
$2a$	$2_+^{1+6} : L_3(2)$	8	71	99	91
$2b$	$2^6 : Sym_5$	0	15	75	51
$3a$	$3 \cdot Sym_6$	6	21	15	16
$3b$	$Sym_3 \times L_3(2)$	0	0	15	7

Lemma 2.14.4 *Let X be a subgroup of order* 3 *and s be an involution in G such that $[Y, s] = 1$. Then one of the following holds:*

(i) *X is a $3a$-subgroup and s is a $2a$-involution,*
(ii) *X is a $3b$-subgroup and s is a $2b$-subgroup.*

Proof. One can observe from the above table that whenever s is an involution, a Sylow 3-subgroup of $C_G(s)$ is of order 3 and hence all these subgroups are conjugate. With X and s as above if s is a $2a$-involution then X stabilizes the octad formed by the elements of $\mathcal{P}$ fixed by s. Hence X is a $3a$-subgroup. On the other hand if X is a $3a$-subgroup, then all involutions in $C_G(X) \cong 3 \cdot Alt_6$ are conjugate. □

The following lemma describes the distribution of involutions inside Q_b, Q_t and Q_s.

Lemma 2.14.5

(i) *Q_b is $2a$-pure,*
(ii) *Q_t contains* 21 *$2a$-involutions and* 42 *$2b$-involutions,*
(iii) *Q_s contains* 45 *$2a$-involutions and* 18 *$2b$-involutions.*

Proof. The $2a$-involutions are central while $2b$-involutions are not. For $x = b$, t and s the subgroup G_x contains a Sylow 2-subgroup of G

and hence every involution from an odd-length orbit of G_x on $Q_x^\#$ must be a 2*a*-involution. In addition Q_s contains a 2*b*-involution by (2.14.3 (iii)) and we obtain (i) and (iii). In view of the above to prove (ii) it is sufficient to show that Q_t contains a 2*b*-involution. It follows from the proof of (2.13.4 (ii)) that there is a 3*b*-subgroup X in G_t not contained in $O_{2,3}(G_t)$. Then $C_{Q_t}(X)$ is of order 2^2 and it is 2*b*-pure by (2.14.4). □

2.15 Golay code and Todd modules

As above let $\mathcal{C}_{12}$ be the unique Golay code based on a set $\mathcal{P}$ of size 24 and $\bar{\mathcal{C}}_{12} = 2^{\mathcal{P}}/\mathcal{C}_{12}$ be the cocode. When considering $\mathcal{C}_{12}$ and $\bar{\mathcal{C}}_{12}$ as $GF(2)$-modules for Mat_{24} and its subgroups we will call them 12-dimensional *Golay code* and *Todd modules.* The modules $\mathcal{C}_{12}$ and $\bar{\mathcal{C}}_{12}$ are dual to each other. In fact, if V is a 1-dimensional subspace in $\mathcal{C}_{12}$ then the orthogonal complement $V^\perp$ of V in $2^{\mathcal{P}}$ with respect to the parity form has dimension 23 and $\mathcal{C}_{12} \leq V^\perp$ since $\mathcal{C}_{12}$ is totally singular. This means that the image of $V^\perp$ in $\bar{\mathcal{C}}_{12}$ is a hyperplane.

There is a 1-dimensional submodule $V_1 = \{\emptyset, \mathcal{P}\}$ in $\mathcal{C}_{12}$; the quotient $\mathcal{C}_{12}/V_1$ is called the 11-dimensional Golay code module and is denoted by $\mathcal{C}_{11}$. Dually $\bar{\mathcal{C}}_{12}$ contains a submodule $\bar{\mathcal{C}}_{11}$ of codimension 1 known as the 11-dimensional Todd module. Since G does not stabilize non-zero vectors in $\bar{\mathcal{C}}_{12}$, both $\mathcal{C}_{12}$ and $\bar{\mathcal{C}}_{12}$ are indecomposable.

Arguing as in (2.2.5 (ii)) one can see that if H is a subgroup of Mat_{24} whose order is divisible by 23, then $\mathcal{C}_{11}$ and $\bar{\mathcal{C}}_{11}$ are irreducible $GF(2)H$-modules.

We can now describe the orbits of $G = Mat_{24}$ on the non-zero vectors in $\mathcal{C}_{12}$, $\mathcal{C}_{11}$, $\bar{\mathcal{C}}_{12}$ and $\bar{\mathcal{C}}_{11}$. By (2.3.3), (2.11.3) and (2.11.9) we have the following.

Lemma 2.15.1 *The actions of G on $\mathcal{C}_{12}$ and $\bar{\mathcal{C}}_{12}$ are indecomposable and*

(i) *G acting on $\mathcal{C}_{12}^\#$ has four orbits: the octads, the complement of octads, the dodecads and the set $\mathcal{P}$ itself,*

(ii) *G acting on $\mathcal{C}_{11}^\#$ has two orbits indexed by the partitions of $\mathcal{P}$ into an octad and its complement and by the partitions of $\mathcal{P}$ into two dodecads; the stabilizers are $G_b \cong 2^4 : L_4(2)$ and $Mat_{12}.2$, respectively,*

(iii) *G acting on $\bar{\mathcal{C}}_{12}^\#$ has four orbits: $\bar{\mathcal{C}}_{12}(i)$ for $i = 1, 2, 3$ and 4 with stabilizers Mat_{23}, $\mathrm{Aut}\, Mat_{22}$, $P\Gamma L_3(4) \cong Mat_{21}.Sym_3$ and $G_s \cong 2^6 : 3 \cdot Sym_6$ (the vectors in $\bar{\mathcal{C}}_{12}(4)$ are naturally indexed by the sextets),*

(iv) *the submodule* $\bar{\mathcal{C}}_{11}$ *of* $\bar{\mathcal{C}}_{12}$ *contains the zero vector, the orbit* $\bar{\mathcal{C}}_{12}(2)$ *(the pairs from* $\mathcal{P}$*) and the orbit* $\bar{\mathcal{C}}_{12}(4)$ *(the sextets).* □

Let Γ_t be a graph on the set of vectors in $\bar{\mathcal{C}}_{12}$ in which two vectors are adjacent if their sum is contained in $\bar{\mathcal{C}}_{12}(1)$. Then Γ_t is a quotient of the 24-dimensional cube, $\bar{\mathcal{C}}_{12} : Mat_{24}$ acts distance-transitively on Γ_t and the distance diagram of the graph is the following:

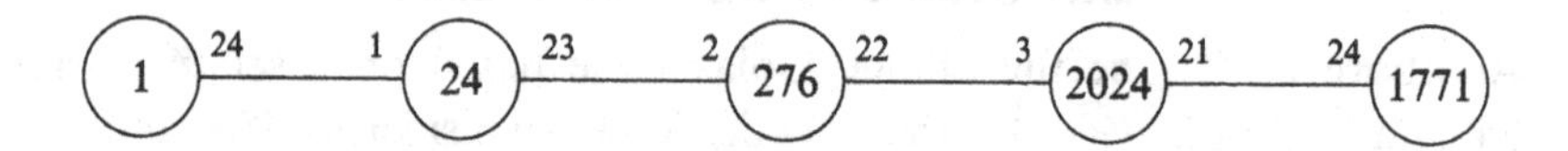

Let Γ_g be the graph on the set of vectors of $\mathcal{C}_{12}$ in which two vectors are adjacent if their sum is an octad. The suborbit diagram corresponding to the action of $\mathcal{C}_{12} : Mat_{24}$ on Γ_g is the following:

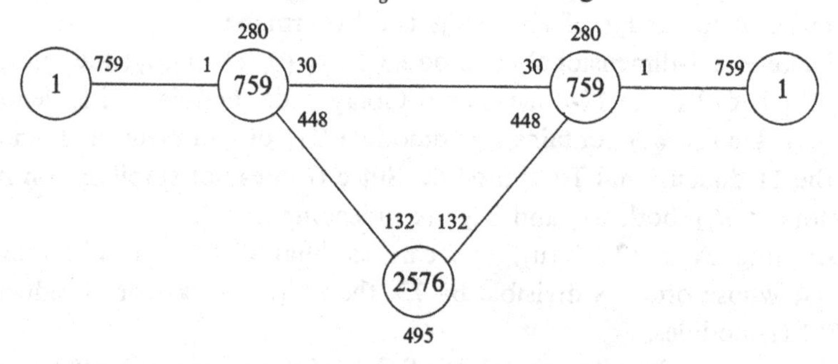

Let $Y_1 \subset Y_2 \subset Y_3 \subset \mathcal{P}$ where Y_i is of size i and let Mat_{23}, Aut Mat_{22}, $P\Gamma L_3(4)$ be the setwise stabilizers in Mat_{24} of Y_1, Y_2 and Y_3, respectively. We are going to specify the structure of $\mathcal{C}_{11}$ and $\bar{\mathcal{C}}_{11}$ considered as $GF(2)$-modules for these subgroups of Mat_{24}.

By the above remark $\mathcal{C}_{11}$ and $\bar{\mathcal{C}}_{11}$ are irreducible under Mat_{23}.

Lemma 2.15.2

(i) Mat_{23} *acting on* $\mathcal{C}_{11}^{\#}$ *has three orbits with lengths* 253 *(the octads containing* Y_1*),* 506 *(the octads not containing* Y_1*) and* 1288 *(the dodecads containing* Y_1*) with stabilizers* $2^4 : Alt_7$, Alt_8 *and* Mat_{11},

(ii) Mat_{23} *acting on* $\bar{\mathcal{C}}_{11}^{\#}$ *has three orbits with lengths* 23 *(pairs containing* Y_1*),* 253 *(pairs not containing* Y_1*) and* 1771 *(the sextets) and with stabilizers* Mat_{22}, $P\Sigma L_3(4)$ *and* $2^4 : (3 \times Alt_5).2$. □

As a $GF(2)$-module for $\mathrm{Aut}\,Mat_{22}$, $\bar{\mathscr{C}}_{11}$ is indecomposable; it contains a 1-dimensional submodule W_1 generated by the image of Y_2. The quotient $\bar{\mathscr{C}}_{11}/W_1$ is irreducible, called the 10-dimensional Todd module and denoted by $\bar{\mathscr{C}}_{10}$. Dually $\mathscr{C}_{11}$ has a $\mathrm{Aut}\,Mat_{22}$-submodule $\mathscr{C}_{10}$ of codimension 1 known as the 10-dimensional Golay code module.

Lemma 2.15.3

(i) $\mathrm{Aut}\,Mat_{22}$ *acting on* $\mathscr{C}_{10}^{\#}$ *has three orbits with lengths* 77 *(octads containing* Y_2*),* 330 *(octads disjoint from* Y_2*) and* 616 *(dodecads containing* Y_2*) and with stabilizers* $2^4 : Sym_6$, $2^4 : L_3(2)$ *and* $\mathrm{Aut}\,Sym_6$*;*

(ii) $\mathrm{Aut}\,Mat_{22}$ *acting on* $\bar{\mathscr{C}}_{10}^{\#}$ *has three orbits with lengths* 22, 231 *and* 770 *with stabilizers* $P\Sigma L_3(4)$, $2^5 : Sym_5$ *and* $2^4.(Sym_3 \wr Sym_2)$. □

As a $GF(2)$-module for $P\Gamma L_3(4)$ $\bar{\mathscr{C}}_{11}$ is indecomposable; it contains a 2-dimensional submodule W_2 generated by the pairs contained in Y_3. The quotient $\bar{\mathscr{C}}_{11}/W_2$ is irreducible, called the 9-dimensional Todd module and denoted by $\bar{\mathscr{C}}_9$. Dually $\mathscr{C}_{11}$ has a $P\Gamma L_3(4)$-submodule $\mathscr{C}_9$ of codimension 2 known as the 9-dimensional Golay code module. $\mathscr{C}_9$ is isomorphic to the module of Hermitian forms in a 3-dimensional $GF(4)$-space.

Lemma 2.15.4 *Let* $\Pi = \Pi(Y_3)$*. Then*

(i) $P\Gamma L_3(4)$ *acting on* $\mathscr{C}_9^{\#}$ *has three orbits with lengths* 21 *(the lines of* Π*),* 210 *(the pairs of lines) and* 280 *(the affine subplanes of order* 3*),*

(ii) $P\Gamma L_3(4)$ *acting on* $\bar{\mathscr{C}}_9^{\#}$ *has three orbits with lengths* 21 *(the points of* Π*),* 210 *(the pairs of points) and* 280 *(the affine subplanes of order* 3*).* □

2.16 The quad of order (3,9)

In this section we establish a relationship between the projective plane Π of order 4 and the generalized quadrangle of order (3,9) associated with the group $U_4(3) \cong \Omega_6^-(3)$. This relationship is reflected in (1.6.5 (vi)).

We consider Π as the residue $\Pi(Y)$ of the Steiner system $(\mathscr{P}, \mathscr{B})$, where $Y = \{a, b, c\}$ is a 3-element subset of $\mathscr{P}$. For $x = a, b$ and c put

$$\mathscr{H}^x = \{H \subseteq \mathscr{P} \mid H \cup Y \setminus \{x\} \in \mathscr{B}\}.$$

Then $\mathscr{H}^a$, $\mathscr{H}^b$, $\mathscr{H}^c$ are the orbits of $G(Y) \cong L_3(4)$ on the set of hyperovals in Π of length 56 each (2.7.9). If H and H' are two hyperovals then by (2.8.5) $|H \cap H'|$ is 0 or 2 if H and H' are from the same $G(Y)$-orbit and it is 1 or 3 otherwise.

Let $H \in \mathscr{H}^x$ for $x = a, b$ or c. Then by counting arguments as in the proof of (2.8.1) one can see that $\mathscr{H}^x$ contains 45 hyperovals intersecting H in 2 points and 10 hyperovals disjoint from H. Let Γ^x be the graph on $\mathscr{H}^x$ in which 2 hyperovals are adjacent if they are disjoint. This graph is known as the *Gewirtz graph*.

Let Φ be an affine subplane of order 3 in Π and $T_1, ..., T_4$ be the triples of points in Π from which Φ can be constructed by the deleting procedure (the paragraph before (2.7.8)). Then the set

$$\mathscr{H}(\Phi) = \{D_{ij} := T_i \cup T_j \mid 1 \leq i < j \leq 4\}$$

consists of hyperovals such that D_{ij} and D_{kl} are in the same $G(Y)$-orbit if and only if $\{i, j\} \cap \{k, l\} = \emptyset$. Hence $\mathscr{H}(\Phi)$ contains a single edge of Γ^x for $x = a, b$ and c. By (2.7.11) this means that the action of $G(Y)$ on the Gewirtz graph is edge-transitive. In fact it is well known and easy to check that the action is distance-transitive and the distance diagram is the following:

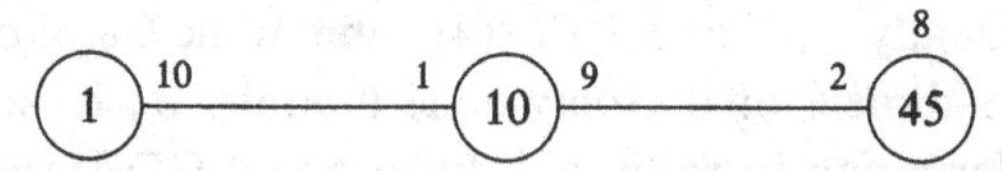

By the basic axiom of $(\mathscr{P}, \mathscr{B})$ a 3-element subset from $\mathscr{P} \setminus Y$ is contained in a unique hyperoval from $\mathscr{H}^x$ for $x = a, b$ and c. This means that for $H \in \mathscr{H}^a$ the set $\Gamma^b(H)$ of hyperovals in Γ^b intersecting H in three elements has size 20 and it is in the natural bijection with the set of 3-element subsets in H. Hence two vertices in $\Gamma^b(H)$ are adjacent in Γ^b if and only if they correspond to disjoint subsets in H. Let Δ be the graph on $\mathscr{H}^a \cup \mathscr{H}^b$ in which two hyperovals are adjacent if they either are disjoint or intersect in 3-elements and let F be the setwise stabilizer of Δ in $G[Y] \cong \mathrm{Aut}\,\Pi$, so that $F \cong P\Sigma L_3(4)$. Then the action of F on Δ is vertex-transitive and it is easy to deduce from the above that the suborbit diagram is the following:

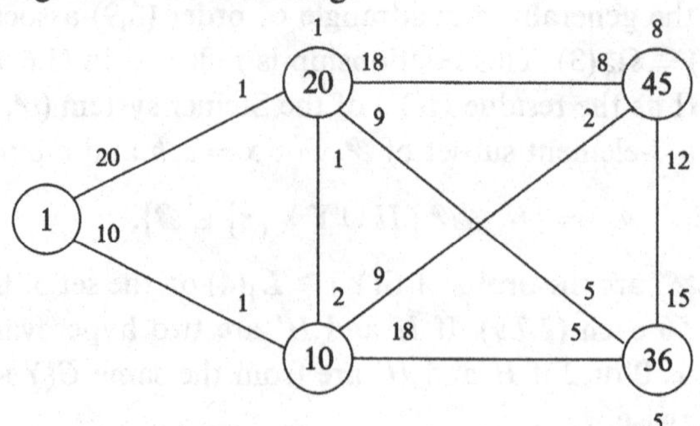

One can see from the diagram that if K is a maximal clique in Δ then $|K| = 4$ and there is an affine subplane Φ such that

$$K = \mathcal{H}(\Phi) \cap (\mathcal{H}^a \cup \mathcal{H}^b).$$

By the remark after (2.7.11) one can see that the stabilizer of Φ in F induces on K a transitive action of D_8. This and the suborbit diagram of Δ show that Δ is the point graph of a generalized quadrangle of order (3,9) on which F induces a flag-transitive action. It is known (5.3.2 (iii) in [PT84]) that every such quadrangle is isomorphic to the classical one associated with the group $U_4(3) \cong \Omega_6^-(3)$ and we have the following.

Proposition 2.16.1 *Let Π be a projective plane of order 4 and $F_0 = \mathrm{Aut}\,\Pi$. Let $\mathcal{H}^a$, $\mathcal{H}^b$, $\mathcal{H}^c$ be the orbits of the $L_3(4)$-subgroup in F_0 on the set of hyperovals in Π, let $\Delta = \mathcal{H}^a \cup \mathcal{H}^b$ and let $F \cong P\Sigma L_3(4)$ be the setwise stabilizer of Δ in F_0. Let $\mathcal{U}$ be a rank 2 incidence system whose points are the hyperovals in Δ, whose lines are the affine subplanes of order 3 in Π, a hyperoval H and a subplane Φ being incident if $H \in \mathcal{H}(\Phi)$. Then $\mathcal{U}$ is a generalized quadrangle of order (3,9) on which F induces a flag-transitive action. Furthermore, $\mathcal{U}$ is isomorphic to the classical generalized quadrangle associated with the group $U_4(3) \cong \Omega_6^-(3)$.* □

3
Geometry of Mathieu groups

In this chapter we study geometries of the Mathieu groups. Construction of the Mathieu groups as automorphism groups of Steiner systems which are extensions of the projective plane of order 4 (for large Mathieu groups) and of the affine plane of order 3 (for little Mathieu groups) leads to geometries discussed in Section 3.1. In Section 3.2 we construct and study the geometry $\mathcal{H}(Mat_{24})$ whose elements are octads, trios and sextets with the incidence relation defined via refinement. We investigate this geometry via the octad graph Γ whose vertices are the octads and two of them are adjacent if they are disjoint. The geometry $\mathcal{H}(Mat_{24})$ belongs to a locally truncated diagram and this reflects the fact that Γ does not contain a complete family of geometrical subgraphs. In Section 3.3 we follow a strategy developed in Chapter 2 to construct a graph of smaller valency with a complete family of geometrical subgraphs and the same abstract automorphism group Mat_{24}. In this way we obtain the rank 3 tilde geometry $\mathcal{G}(Mat_{24})$. In Section 3.4 considering the actions on the octad graph of Mat_{23} and $\mathrm{Aut}\, Mat_{22}$ we construct the P-geometries $\mathcal{G}(Mat_{23})$ and $\mathcal{G}(Mat_{22})$ of rank 4 and 3, respectively. In Section 3.5 we show that $\mathcal{G}(Mat_{22})$ possesses a triple cover $\mathcal{G}(3 \cdot Mat_{22})$ which is simply connected. In Section 3.6 we establish the 2-simple connectedness of $\mathcal{G}(Mat_{23})$. In Section 3.7 we calculate the suborbit diagrams corresponding to the action of Mat_{24} on its maximal parabolic geometry. In Section 3.8 we analyse the structure of the Golay code as $GF(2)$-module for the maximal parabolics associated with the action of Mat_{24} on $\mathcal{H}(Mat_{24})$. In Section 3.9 we calculate the suborbit diagrams of the maximal parabolic geometry of $\mathrm{Aut}\, Mat_{22}$. In Section 3.10 we calculate the lengths of orbits of the parabolic subgroups of Mat_{24} on the set of sextets.

3.1 Extensions of planes

In the previous chapter the large Mathieu groups have been constructed as automorphism groups of extensions of the projective plane of order 4 and the little Mathieu groups as automorphism groups of extensions of the affine plane of order 3. These constructions can be interpreted in terms of diagram geometries as follows.

Let $\mathcal{S} = (\mathcal{P}, \mathcal{B})$ be the Steiner system of type $S(5,8,24)$. Define $\mathcal{E}(Mat_{24})$ to be an incidence system of rank 5 whose elements of type i are the i-element subsets of $\mathcal{P}$ for $1 \leq i \leq 4$, the elements of type 5 are the octads from $\mathcal{B}$ and the incidence is by inclusion.

Lemma 3.1.1 *The incidence system $\mathcal{E} = \mathcal{E}(Mat_{24})$ is a geometry with the diagram*

$$\mathcal{E}(Mat_{24}): \quad \underset{1}{\circ}\!\!-\!\!-\!\!-\!\!\underset{1}{\circ}\!\!-\!\!-\!\!-\!\!\underset{1}{\circ}\!\!\overset{c^*}{-\!\!-\!\!-}\!\!\underset{4}{\circ}\!\!-\!\!-\!\!-\!\!\underset{4}{\circ}$$

The group Mat_{24} acts on $\mathcal{E}(Mat_{24})$ flag-transitively with the stabilizer of an element of type i being Mat_{23}, $\mathrm{Aut}\, Mat_{22}$, $P\Gamma L_3(4)$, $2^6 : 3 : Sym_5$ and $2^4 : L_4(2)$ for $i = 1,2,3,4$ and 5, respectively.

Proof. Since every 4-element subset of $\mathcal{P}$ is contained in an octad, it is easy to see that every flag is contained in a maximal one. Since the incidence relation is via inclusion, $\mathcal{E}$ belongs to a string diagram. Let $Y_1 \subset Y_2 \subset Y_3 \subset Y_4 \subset B$ be a maximal flag in $\mathcal{E}$ where $|Y_i| = i$ and $B \in \mathcal{B}$. Then the elements of type 4 incident to Y_3 are in the natural correspondence with the elements in $\mathcal{P} \setminus Y_3$ and the elements of type 5 incident to Y_3 are the blocks from $\mathcal{B}_3(Y_3)$. Hence $\mathrm{res}^+_{\mathcal{E}}(Y_3) = (\mathcal{P} \setminus Y_3, \mathcal{B}(Y_3))$ which is the projective plane $\Pi(Y_3)$ of order 4. The remaining rank 2 residues in $\mathcal{E}$ are even more obvious. Since Mat_{24} acts transitively on $\mathcal{B}$ and the stabilizer of an octad induces Alt_8 on the elements in the octad, the flag-transitivity of Mat_{24} on $\mathcal{E}$ follows. The structure of stabilizers follows from the definition of Mat_{23} and $\mathrm{Aut}\, Mat_{22}$, together with (2.9.1) and (2.10.2) (notice that the stabilizer of Y_4 is of index 6 in the stabilizer of the sextet determined by Y_4). □

Let $\mathcal{D} = (D, \mathcal{Q})$ be the unique Steiner system of type $S(5,6,12)$. Define $\mathcal{F}(Mat_{12})$ to be an incidence system of rank 5 whose elements of type i are the i-element subsets of D for $1 \leq i \leq 4$, the elements of type 5 are the blocks from $\mathcal{Q}$ and the incidence relation is by inclusion. The proof of the following statement is similar to that of (3.1.1).

Lemma 3.1.2 *The incidence system $\mathscr{F}(Mat_{12})$ is a geometry with the diagram*

$$\mathscr{F}(Mat_{12}): \quad \overset{}{\underset{1}{\circ}} \text{———} \underset{1}{\circ} \text{———} \underset{1}{\circ} \overset{c^*}{\text{———}} \underset{2}{\circ} \overset{\mathrm{Aff}^*}{\text{———}} \underset{3}{\circ}$$

The group Mat_{12} acts on $\mathscr{F}$ flag-transitivity with the stabilizer of an element of type i being Mat_{11}, $Mat_{10}.2$, $Mat_9.Sym_3$, $Mat_8.Sym_4$ and Sym_6 for $i = 1, 2, 3, 4$ and 5, respectively. □

Let $Y_1 \subset Y_2 \subset Y_3 \subset Y_4 \subset Q$ be a maximal flag in $\mathscr{F}$. The uniqueness results for $\mathscr{S}$, $\mathscr{D}$ and their residues can be reformulated in the following way.

Lemma 3.1.3 *Each of the following geometries is characterized by its diagram including indices: $\mathscr{E} = \mathscr{E}(Mat_{24})$, $\mathrm{res}_{\mathscr{E}}(Y_1)$, $\mathrm{res}^+_{\mathscr{E}}(Y_2)$, $\mathscr{F}(Mat_{12})$, $\mathrm{res}_{\mathscr{F}}(Y_1)$, $\mathrm{res}^+_{\mathscr{F}}(Y_2)$.* □

3.2 Maximal parabolic geometry of Mat_{24}

In this section we study the geometry $\mathscr{H}(Mat_{24})$ whose elements are sextets, trios and octads of the Steiner system $\mathscr{S}$ in which two elements are incident if one refines the other one as partitions. It is convenient to define first a graph associated with the geometry.

Let $\Gamma = \Gamma(Mat_{24})$ be the *octad graph* which is a graph on the set of octads in which two octads are adjacent if they are disjoint. Then the octads contained in a trio form a triangle while the octads refined by a sextet induce a 15-vertex subgraph (called a *quad*) isomorphic to the collinearity graph of the generalized quadrangle $\mathscr{G}(Sp_4(2))$.

Lemma 3.2.1

(i) *Γ has diameter 3 and two of its vertices are at distance 1, 2 and 3 if and only if as octads they have intersection of size 0, 4 and 2, respectively,*

(ii) *Mat_{24} acts distance-transitively on Γ and the distance diagram of Γ is*

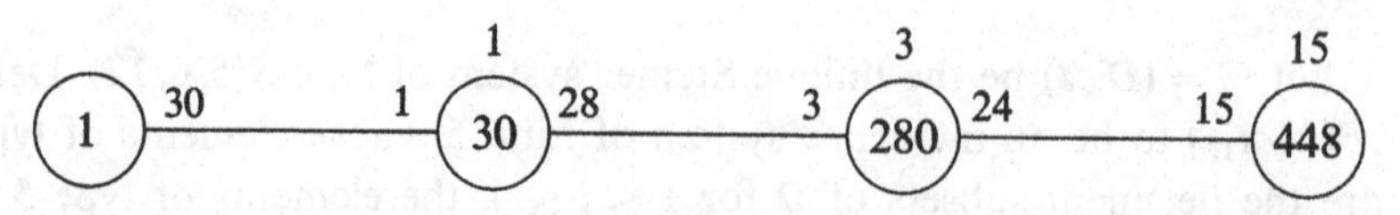

(iii) *every pair of vertices at distance 2 is contained in a unique quad and every quad is strongly geodetically closed,*

(iv) *whenever x is a vertex and T is a triangle, there is a unique vertex in T nearest to x.*

Proof. First of all the action of G on Γ is vertex-transitive. Let B be an octad. By (2.10.4) $\mathcal{O}_8 = \{B\}$, $\mathcal{O}_0$, $\mathcal{O}_4$ and $\mathcal{O}_2$ are the orbits of G_b on the vertex set of Γ. By definition $\mathcal{O}_0$ consists of the octads adjacent to B. Let B_1, B_2 be a pair of octads from $\mathcal{O}_0$. For $i = 1$ and 2 let H_i be the hyperplane in Q_b such that B_i is an orbit of H_i (compare (2.10.3 (ix)). Then $B_1 \cap B_2 = \emptyset$ if $H_1 = H_2$ and $|B_1 \cap B_2| = 4$ otherwise. Hence $a_1 = 1$, $b_1 = 28$ and $\mathcal{O}_4$ is the set of vertices at distance 2 from B. Since G_b acts transitively on $\mathcal{O}_4$ this implies $c_2 = 3$. Let $B_3 \in \mathcal{O}_4$ and U be the 2-dimensional subspace in Q_b such that $B_3 \setminus B$ is an orbit of U and let B_1, H_1 be as above. Then either $U < H_1$, in which case B_1 is refined by the sextet determined by $B \cap B_3$, *i.e.* B_1 is contained in the quad containing B and B_3 or $U \cap H_1$ is of order 2 and $B_1 \cap B_3$ is an orbit of this intersection. This implies (i) and (ii). Now (iii) and (iv) follow directly from the distance diagram of Γ. □

Because of the properties of Γ in (3.2.1 (iii), (iv)) it is a so-called near hexagon [ShY80].

For a vertex $x \in \Gamma$ define π_x to be the geometry whose points are the triangles containing x, whose lines and planes are 3- and 7-element subsets of points whose setwise stabilizers in $G(x)$ contain Sylow 2-subgroups of the latter. Since there are 35 quads containing x and transitively permuted by $G(x)$, it is easy to see that 3 triangles containing x form a line in π_x if and only if they are contained in a common quad. By (2.10.3 (ix)) and (2.10.5) the trios containing B correspond to the hyperplanes and the sextets refining B correspond to the 2-dimensional subspaces in Q_b. This shows that π_x is isomorphic to the rank 3 projective geometry associated with the dual of $O_2(G(x))$.

Let $T = \{x, y, z\}$ be a triangle in Γ. A quad Σ containing T determines a line $l_x(\Sigma)$ in π_x and a line $l_y(\Sigma)$ in π_y. The mapping ψ_{xy} defined by

$$\psi_{xy} : l_x(\Sigma) \mapsto l_y(\Sigma)$$

for every quad Σ containing T is the unique mapping of the residue of T in π_x onto the residue of T in π_y which commutes with the action of $G(x, y)$.

The local projective space structures π_x enable us to introduce the notion of *geometrical subgraphs.* A subgraph Ξ in Γ is called geometrical if

(a) whenever Ξ contains a pair of adjacent vertices it contains the unique triangle containing this pair,
(b) if $x \in \Xi$ then the set Ψ of triangles containing x and contained in Ξ is a subspace in π_x,
(c) $G[\Xi] \cap G(x) = G[\Psi] \cap G(x)$.

The geometrical subgraph as above will be denoted by $\Xi(x, \Psi)$. Since $G(x)$ induces the automorphism group of π_x and in view of the mapping ψ_{xy}, arguing as in Section 9.5, one can easily show that if $\Xi(x, \Psi)$ exists then it is unique. Moreover, a quad is a geometrical subgraph $\Xi(x, \Psi)$ where Ψ is a line in π_x.

Lemma 3.2.2 *Let Ψ be a plane in π_x. Then the geometrical subgraph $\Xi(x, \Psi)$ does not exist.*

Proof. Suppose to the contrary that $\Xi(x, \Psi)$ exists. Since G acts on Γ vertex-transitively and $G(x)$ induces the automorphism group of π_x, for every $x' \in \Gamma$ and every plane Ψ' in $\pi_{x'}$ the geometrical subgraph $\Xi(x', \Psi')$ exists. Let $\mathcal{M}$ be the set of all such geometrical subgraphs. Let Σ be a quad containing x, so that $\Sigma = \Xi(x, \Phi)$ for a line Φ of π_x. Let Ψ_i, $1 \leq i \leq 3$, be the planes in π_x containing Φ. Then $\Xi(x, \Psi_i)$, $1 \leq i \leq 3$, are the geometrical subgraphs from $\mathcal{M}$ containing Σ. Since $G(x) \cap G[\Phi]$ induces Sym_3 on $\{\Psi_i \mid 1 \leq i \leq 3\}$ we conclude that $G[\Sigma]$ induces Sym_3 on $\{\Xi(x, \Psi_i) \mid 1 \leq i \leq 3\}$, which is impossible, since $G[\Sigma] \cong 2^6 : 3 \cdot Sym_6$ does not possess a homomorphism onto Sym_3. □

One may notice a similarity between the local structure of $\Gamma(Mat_{24})$ and that of the dual polar graph of $\mathcal{G}(Sp_8(2))$; the difference is that the latter graph contains a complete family of geometrical subgraphs.

Define $\mathcal{H} = \mathcal{H}(Mat_{24})$ to be the rank 3 geometry whose elements of type 1, 2 and 3 are the sextets, trios and octads, where two elements are incident if one refines another one as partitions of $\mathcal{P}$. Equivalently the elements of $\mathcal{H}$ are the quads, triangles and vertices of Γ and the incidence relation is via inclusion. It is easy to deduce from the above that $\mathcal{H}$ is in fact a geometry and to prove the following.

Lemma 3.2.3 *Let $\{x_1, x_2, x_3\}$ be a maximal flag in $\mathcal{H}$ where x_i is of type i. Then $\text{res}_{\mathcal{H}}(x_1) \cong \mathcal{G}(Sp_4(2))$, $\text{res}_{\mathcal{H}}(x_2) \cong K_{3,7}$ and $\text{res}_{\mathcal{H}}(x_3)$ is the geometry of 1- and 2-dimensional subspaces in a 4-dimensional $GF(2)$-space.* □

By the above lemma $\mathcal{H}(Mat_{24})$ belongs to the following diagram:

The leftmost node is a convention to indicate that the residue of an element of type 3 (that is of an octad) is a truncation of the rank 3 projective geometry over $GF(2)$. The elements corresponding to this node can be defined locally (in the residue) and by (3.2.2) it is not possible to define them globally.

The following result was established in [Ron81a].

Proposition 3.2.4 *The geometry $\mathcal{H}(Mat_{24})$ is simply connected.*

Proof. Let $\varphi : \widetilde{\mathcal{H}} \to \mathcal{H}$ be the universal covering of $\mathcal{H} = \mathcal{H}(Mat_{24})$ and let $\widetilde{\Gamma}$ be a graph on the set of elements of type 3 in $\widetilde{\mathcal{H}}$, in which two elements are adjacent if they are incident to a common element of type 2. Since φ is a covering, the vertices in $\widetilde{\Gamma}$ incident to a given element of type 2 form a triangle and the vertices incident to a given element of type 1 induce a quad (the collinearity graph of $\mathcal{G}(Sp_4(2))$). Since a pair of elements of type 3 in $\mathcal{H}$ are incident to at most one element of type 2, arguing as in (6.3.3), we conclude that φ induces a covering of $\widetilde{\Gamma}$ onto $\Gamma = \Gamma(Mat_{24})$ (denoted by the same letter φ) and in view of (3.2.1 (iii)) all cycles of length 3, 4 and 5 in Γ are contractible with respect to φ. Since the diameter of Γ is 3, to prove the proposition it is sufficient to show that every non-degenerate cycle in Γ of length 6 or 7 is decomposable into shorter cycles. Let $x \in \Gamma$, $y \in \Gamma_3(x)$ and $\{z_i \mid 1 \leq i \leq 15\} = \Gamma(y) \cap \Gamma_2(x)$. For $1 \leq i, j \leq 15$, $i \neq j$, let $C_{ij} = (x, u_i, z_i, y, z_j, u_j, x)$ be a 6-cycle. It is easy to see that the decomposability of C_{ij} is independent of the particular choice of u_i and u_j. Moreover, whenever $C(z_i, z_j)$ and $C(z_j, z_k)$ are decomposable, so is C_{ik}. By (2.10.4 (iii)) $G(x, y) \cong Sym_6 \cong Sp_4(2)$. This means that $G(x, y)$ acts primitively on the point set of π_y and in view of (3.2.1 (iv)) also on $\Gamma(y) \cap \Gamma_2(x)$. Thus to show that all 6-cycles are decomposable it is sufficient to show that at least one such cycle is decomposable. Let (y, z_1, u_1) be a 2-path, where $u_1 \in \Gamma(x)$, and Σ be the unique quad containing this path. Then there is another path, say (y, z_2, u_1), joining y and u_1 in Σ. This shows that C_{12} is decomposable into shorter cycles and hence so is any other 6-cycle. Finally, by (3.2.1 (iv)), every cycle of length 7 splits into a triangle and two (possibly degenerate) 6-cycles, so the result follows. □

3.3 Minimal parabolic geometry of Mat_{24}

By (3.2.2) the octad graph $\Gamma = \Gamma(Mat_{24})$ does not contain a complete family of geometrical subgraphs. In Section 9.5 we will learn in general how to construct in such circumstances a graph of smaller valency with a complete family of geometrical subgraphs with the same abstract group of automorphisms. We start with a helpful lemma.

Lemma 3.3.1 *Let Σ be a sextet identified with the quad in Γ induced by the octads refined by Σ, $x \in \Sigma$, Φ_x be the line in π_x such that $\Sigma = \Xi(x, \Phi_x)$ and let Ψ_x be a plane containing Φ_x. Then $G[\Sigma] = G_s$, $G(\Sigma) = K_s$ and $G(\Sigma) \cap G[\Psi_x] = Q_s$.*

Proof. Everything except the last equality follow from (2.10.2) and the definition of the octad graph. The group $G(x)$ induces $L_4(2)$ on π_x and the stabilizer in $L_4(2)$ of the line Φ_x induces $Sym_3 \times Sym_3$ on the set of points and planes incident to Φ_x. On the other hand $G(x) \cap G[\Phi_x]$ induces $Sym_4 \times 2$ on Σ. Hence $G(\Sigma)$ permutes transitively the three planes in π_x containing Φ_x and the result follows. □

Let Ω be a graph whose vertices are the pairs (x, Ψ) where $x \in \Gamma$ and Ψ is a plane in π_x, two such vertices (x, Ψ) and (x', Ψ') are adjacent if x and x' are adjacent in Γ, the triangle determined by the edge $\{x, x'\}$ is contained in both Ψ and Ψ' and $\psi_{xx'}(\Psi) = \Psi'$. By (3.2.2) and a statement analogous to (9.6.4) Ω is connected of valency 14 and $G \cong Mat_{24}$ acts naturally on Ω.

We use notation as in (3.3.1); in addition let $T = \{x, y, z\}$ be a triangle contained in Φ_x. Let Φ_y and Ψ_y be the images under ψ_{xy} of Φ_x and Ψ_x and let Φ_z and Ψ_z be defined similarly.

Then $R = \{(x, \Psi_x), (y, \Psi_y), (z, \Psi_z)\}$ is a triangle in Ω and every edge of Ω is in a unique triangle. Put

$$H_1 = G(x) \cap G[\Phi_x] \cap G[\Psi_x],$$

$$H_2 = G[T] \cap G[\Phi_x \cup \Phi_y \cup \Phi_z] \cap G[\Psi_x \cup \Psi_y \cup \Psi_z].$$

Lemma 3.3.2

(i) *$H_1 = G_s \cap G((x, \Psi_x))$; $H_2 = G_s \cap G[R]$,*

(ii) *$H_1 \cap H_2$ contains a Sylow 2-subgroup of G_s, in particular it contains Q_s,*

(iii) *H_1/Q_s is a complement to K_s/Q_s in $(G_s \cap G(x))/Q_s$ and H_2/Q_s is a complement to K_s/Q_s in $(G_s \cap G[T])/Q_s$,*

(iv) $\langle H_1, H_2 \rangle = G_s$,

(v) *the subgraph* Θ *in* Ω *induced by the images of* (x, Ψ_x) *under* G_s *is a geometrical subgraph isomorphic to the collinearity graph of* $\mathcal{G}(3 \cdot Sp_4(2))$.

Proof. Since Σ is a geometrical subgraph, we have $G(x) \cap G[\Sigma] = G(x) \cap G[\Phi_x]$ and (i) follows. Now (ii) and (iii) follow directly from (3.3.1). Since G_s/Q_s does not split over K_s/Q_s, we obtain (iv). By (2.10.2 (v)) G_s/Q_s is the automorphism group of $\mathcal{G}(3 \cdot Sp_4(2))$. By (i), (ii) and (iii) $\{H_1K_s/K_s, H_2K_s/K_s\}$ is the amalgam of maximal parabolics of the action of $G_s/K_s \cong Sp_4(2)$ on $\mathcal{G}(Sp_4(2))$. By (2.6.2) this implies that $\{G_s/Q_s, \{H_1/Q_s, H_2/Q_s\}\}$ is isomorphic to $\mathcal{G}(3 \cdot Sp_4(2))$ and (v) follows. □

Below we present the suborbit diagram of the action of $G \cong Mat_{24}$ on Ω computed by D.V. Pasechnik.

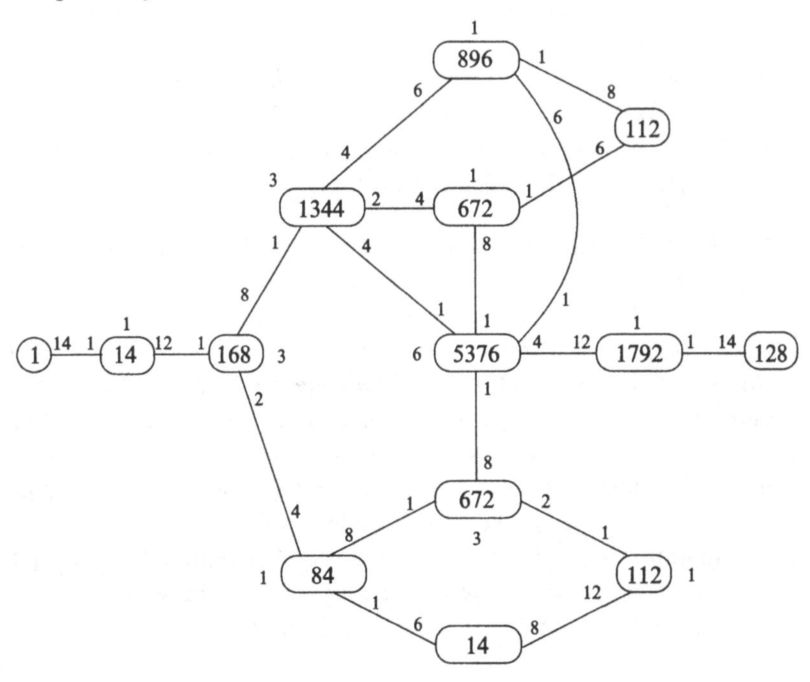

Let $\mathcal{G} = \mathcal{G}(Mat_{24})$ be the rank 3 incidence system whose elements of type 1 are the geometrical subgraphs in Ω as in (3.3.2 (v)), the elements of type 2 and 3 are the triangles and vertices of Ω; the incidence relation is via inclusion. We obtain the following result directly from (3.3.2).

Lemma 3.3.3 *$\mathcal{G}(Mat_{24})$ is a geometry with the diagram*

$$\overset{}{\underset{2}{\circ}}\!\!-\!\!\!-\!\!\!-\!\!\!-\!\!\underset{2}{\circ}\overset{\sim}{=\!=\!=\!=}\underset{2}{\circ}$$

and $G \cong Mat_{24}$ is a flag-transitive automorphism group of $\mathcal{G}(Mat_{24})$. □

Let us analyse the amalgam of maximal parabolics corresponding to the action of G on $\mathcal{G}$. We continue to use notation as in (3.3.1) and (3.3.2); in addition assume that $x = B$, $T = \{B_1 = B, B_2, B_3\}$, $B = S_1 \cup S_2$. Then $\mathcal{B} = \{G_s, G_t, G_b\}$ is the amalgam of maximal parabolics corresponding to the action of G on $\mathcal{H}(Mat_{24})$ and associated with the flag $\{\Sigma, T, x\}$. Put $G_1 = G[\Theta]$, $G_2 = G[R]$, $G_3 = G((x, \Psi_x))$. Then $\mathcal{A} = \{G_1, G_2, G_3\}$ is the amalgam of maximal parabolics corresponding to the action of G on $\mathcal{G}$ and associated with the flag $\{\Theta, R, (x, \Psi_x)\}$.

Lemma 3.3.4 *The following assertions hold:*

(i) $G_1 = G_s$,
(ii) $G_3 = C_{G_b}(\tau) = C_G(\tau)$ *where τ is an involution from $Q_b \cap Q_t \cap Q_s$, so that $G_3 \cong 2_+^{1+6} : L_3(2)$ and $[G_b : G_3] = 15$,*
(iii) $G_2 = N_{G_t}(E)$ *where $E \cong 2^2$ is the normal closure of τ in $O_{2,3}(G_t)$, so that $G_2 \cong 2^6 : (Sym_3 \times Sym_4)$ and $[G_t : G_2] = 7$.*

Proof. (i) follows directly from (3.3.2 (iv)). There is a bijection φ of the set of planes in π_x onto $Q_b^{\#}$ such that $\varphi(\Psi)$ is the unique involution in Q_b which fixes elementwise every triangle in Ψ. By (2.14.2 (i)) this implies that there is a mapping χ from the vertex set of Ω onto the set of $2a$-involutions in G such that $G(v) = C_G(\chi(v))$ for $v \in \Omega$. Let U be the subgroup of order 2^2 in Q_b which corresponds to the partition $\{S_1, S_2\}$ of B (2.10.5). It is easy to see that $U = Q_b \cap Q_s = Q_b \cap Q_t \cap Q_s$ and $\varphi(\Psi_x) \in U$ which implies (ii). Since $G_2 = G[R]$, we have $G_2 = N_G(E)$ where E is generated by the images under χ of the vertices in R. Let X be a subgroup of order 3 in $O_{2,3}(G_t)$. Since X permutes transitively the vertices in R, it acts transitively by conjugation on the set of generators of E. Since $\tau \in Q_t$ and X acts on Q_t fixed-point freely, $E \cong 2^2$ and (iii) follows. □

By the above lemma we could first define $\mathcal{A}$ as an abstract subamalgam in $\mathcal{B}$ and then define $\mathcal{G}$ as the coset geometry $\mathcal{G}(G, \mathcal{A})$.

There is yet another way to construct the minimal parabolic geometry $\mathcal{G}$ of Mat_{24} from its maximal parabolic geometry $\mathcal{H}$. The point is that the elements of type 1 in both $\mathscr{H}$ and $\mathscr{G}$ are the sextets. Moreover, an element of either of these two geometries can be identified with the set of sextets it is incident to, so that the incidence relation is via inclusion. In their turn the sextets are identified with the quads in the octad graph, so an element of type 2 or 3 in $\mathscr{H}$ is the set of quads containing a triangle or a vertex of Γ, respectively. In order to define the collections of sextets corresponding to elements of type 2 and 3 in $\mathscr{G}$ it is convenient first to describe the orbits of G on the pairs of sextets.

Let $\Sigma = \{S_1, S_2, ..., S_6\}$ and $\Sigma' = \{S_1', S_2', ..., S_6'\}$ be distinct sextets. For a tetrad S let $\mu(S)$ be a non-increasing sequence consisting of the non-zero values from the set $\{|S \cap S_j| \mid 1 \le j \le 6\}$ and let $\mu = \mu(\Sigma, \Sigma')$ be the lexicographically largest among the sequences $\mu(S_i')$ for $1 \le i \le 6$. For a sequence μ_0 let $n(\mu_0)$ be the number of sextets Σ' with $\mu(\Sigma, \Sigma') = \mu_0$ for a given sextet Σ.

Lemma 3.3.5 *The G-orbit containing (Σ, Σ') is uniquely determined by $\mu = \mu(\Sigma, \Sigma')$ and one of the following holds:*

(i) $\mu = (2, 2)$, $n(\mu) = 90$, *there is a unique trio refined by Σ and Σ',*

(ii) $\mu = (3, 1)$, $n(\mu) = 240$, *there is a unique octad refined by Σ and Σ',*

(iii) $\mu = (2, 1, 1)$, $n(\mu) = 1440$, *there are no octads refined by Σ and Σ'.*

Proof. We claim that $\mu = (1, 1, 1, 1)$ is not possible. Suppose the contrary. Then $\mu(S_i') = (1, 1, 1, 1)$ for every $1 \le i \le 6$. We can assume without loss of generality, that for $1 \le j \le 4$ we have $|S_1' \cap S_j| = 1$ and that $|S_2' \cap S_1| = |S_2' \cap S_5| = 1$, in which case $|(S_1' \cup S_2') \cap (S_1 \cup S_5)| = 3$, a contradiction with (2.8.5). Hence μ is one of the sequences in (i), (ii) and (iii). We assume that $\mu = \mu(S_1') = (|S_1' \cap S_1|, ..., |S_1' \cap S_r|)$ where $r = 2$ or 3. Let $B = S_1 \cup S_2$. Suppose first that $r = 2$, so we are in case (i) or (ii). Since B is an octad containing S_1', $B \setminus S_1'$ is another tetrad from Σ' which we assume to be S_2'. It is easy to see that in this case $\mu(S_2') = \mu$. Since $G_b \cap G_s$ induces $(Sym_4 \times Sym_4)^e$ on the set of elements in B, it is easy to see that it acts transitively on the set of tetrads S with $|S \cap S_1| = |S \cap S_2| = 2$ and on the set of such tetrads with $|S \cap S_1| = 3$, $|S \cap S_2| = 1$. Hence the transitivity assertions for (i) and (ii) follow. Let U and U' be the subgroups of order 2^2 in Q_b corresponding to the partitions $\{S_1, S_2\}$ and $\{S_1', S_2'\}$, respectively. It follows in particular that $U \cap U'$ is of order 2 in case (i) and trivial in case (ii). Thus in case (i) if an orbit of U and an orbit of U' have a non-empty intersection then the intersection is of size 2; Σ and Σ' refine the trio containing B and corresponding to the hyperplane $\langle U, U' \rangle$ in Q_b; $\mu(S_i') = \mu$ for $1 \leq i \leq 6$ which gives $n((2,2)) = 90$. In case (ii) if an orbit of U and an orbit of U' have a non-empty intersection, the intersection is of size 1. Hence $\mu(S_i') = (1,1,1,1)$ for $3 \leq i \leq 6$ and $n((3,1)) = 240$. Let us turn to the case (iii). Clearly $G_b \cap G_s$ acts transitively on the set of triples Y with $|Y \cap S_1| = 2$ and $|Y \cap S_2| = 1$. In addition Q_b stabilizes B elementwise and permutes transitively the elements in $\mathscr{P} \setminus B$. This implies the transitivity assertion in (iii). Since the total number of sextets is 1771, by (i) and (ii) we obtain $n((2,1,1)) = 1440$. □

It follows from the proof of the above lemma that G_s acts transitively on the set of tetrads S with $\mu(S) = (2,2)$, $(3,1)$ and $(2,1,1)$.

Definition 3.3.6 *The* sextet graph *is a graph on the set of sextets with two sextets Σ and Σ' being adjacent if $\mu(\Sigma, \Sigma') = (2,2)$.*

Let $\mathscr{X}$ be the sextet graph. Then the vertices of $\mathscr{X}$ can be considered as quads in the octad graph and in these terms two quads are adjacent if they intersect in a triangle. Also, in view of (2.15.1 (iii), (iv)) the vertex set of $\mathscr{X}$ can be identified with a G-orbit on the set of non-zero vectors in $\bar{\mathscr{C}}_{11}$. Let Σ, Σ' be adjacent sextets as in the proof of (3.3.5). Then $S_1'' = S_1 \Delta S_1'$ is of size 4 and the sextet Σ'' determined by S_1'' is adjacent to both Σ and Σ'. Thus there is a binary operation (denoted by $\star$) defined on the pairs of adjacent vertices in $\mathscr{X}$ such that $x \star y$ is a vertex adjacent to x and y. Here $x \star y$ is the sum of x and y, considered as vectors in $\bar{\mathscr{C}}_{11}$. A clique (a complete subgraph) K in $\mathscr{X}$ will be called $\star$-closed if whenever K contains x and y, it contains $x \star y$. Since a $\star$-closed clique is the set of non-zero vectors in a subspace in $\bar{\mathscr{C}}_{11}$, it contains $2^n - 1$ vertices for an integer n. Clearly a $\star$-closed clique of size 3 is of the form $\{x, y, x \star y\}$, where x and y are adjacent vertices in $\mathscr{X}$. We are going to classify all $\star$-closed cliques in $\mathscr{X}$ and start with the following.

Lemma 3.3.7 *For every clique in the sextet graph $\mathscr{X}$ the corresponding quads in the octad graph contain a common vertex.*

Proof. It is sufficient to show that whenever three quads pairwise have common triangles then all three of them have a common vertex. Let Σ_1, Σ_2 be quads such that $\Sigma_1 \cap \Sigma_2 = T$ where T is a triangle and let $x \in T$. Let Σ_3 be a quad such that $T_i = \Sigma_i \cap \Sigma_3$ is a triangle for $i = 1$ and 2. Then by (3.2.1) T_i contains a vertex, say y_i, adjacent to x. Since Σ_3

contains y_1 and y_2, it contains the path (y_1, x, y_2). Hence $x \in \Sigma_3$ and the result follows. □

By the above lemma in order to classify the cliques in $\mathscr{X}$ it is sufficient to consider the quads containing a given vertex in the octad graph. The 35 quads containing $x = B$ can be identified with the lines in π_x or, equivalently, with the 2-dimensional subspaces in Q_b. Two such subspaces U_1 and U_2 are adjacent (as vertices of $\mathscr{X}$) if $U_1 \cap U_2$ is 1-dimensional, equivalently if $\langle U_1, U_2 \rangle$ is a hyperplane. In this case $U_1 \star U_2$ is the unique 2-subspace other than U_1 and U_2 which contains $U_1 \cap U_2$ and is contained in $\langle U_1, U_2 \rangle$.

Lemma 3.3.8 *A maximal clique K in $\mathscr{X}$ has size 7, it is $\star$-closed and $G[K]$ induces on K the natural action of $L_3(2)$. The group G acting on the set of maximal cliques in $\mathscr{X}$ has two orbits $\mathscr{K}_v$ and $\mathscr{K}_t$ such that $|\mathscr{K}_v| = 3 \cdot |\mathscr{K}_t|$. Moreover,*

(i) *if $K \in \mathscr{K}_v$ then there are a unique vertex x in the octad graph and a plane Ψ_x in π_x such that K corresponds to the lines of π_x contained in Ψ_x,*

(ii) *if $K \in \mathscr{K}_t$ then there is a unique triangle T in the octad graph (a trio) such that K corresponds to the quads containing T.*

Proof. It follows from the general description of the cliques in the Grassmann graphs [BCN89] or can be checked directly, that a maximal set of pairwise intersecting lines in π_x is of size 7 and consists either of the lines containing a given point or of the lines contained in a given plane. In view of (3.3.7) this implies the result. □

Now directly from (3.3.2), (3.3.3) and (3.3.8) we obtain the following combinatorial characterization of $\mathscr{G}(Mat_{24})$.

Lemma 3.3.9 *Let $\mathscr{D}$ be a rank 3 incidence system whose elements of type 1 are the vertices of $\mathscr{X}$ (the sextets), the elements of type 2 are the $\star$-closed cliques of size 3 and the elements of type 3 are the cliques from the G-orbit $\mathscr{K}_v$, where the incidence relation is via inclusion. Then $\mathscr{D}$ is isomorphic to $\mathscr{G}(Mat_{24})$.* □

On the following page we present the suborbit diagram of the sextet graph with respect to the action of Mat_{24}. It is straightforward to deduce this diagram from the diagram $D_s(Mat_{24})$ proved in (3.7.3).

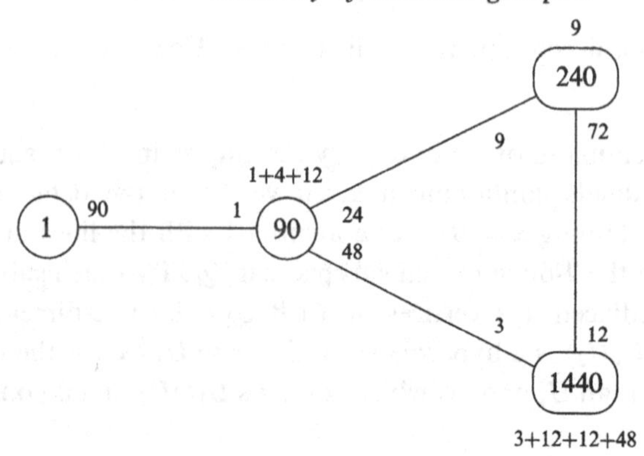

By (3.3.9) the elements of type i in $\mathcal{G}(Mat_{24})$ can be identified with certain i-dimensional subspaces in $\bar{\mathcal{C}}_{11}$ so that the incidence relation is via inclusion. This defines a natural representation of $\mathcal{G}(Mat_{24})$ in $\bar{\mathcal{C}}_{11}$. The following result was established in [RSm89].

Proposition 3.3.10 *The respesentation of $\mathcal{G}(Mat_{24})$ in $\bar{\mathcal{C}}_{11}$ is universal and it is the only representation which is invariant under the action of Mat_{24}.* □

In fact $\bar{\mathcal{C}}_{11}$ is the universal representation group of $\mathcal{G}(Mat_{24})$ [IPS96].

In was shown in [Hei91] and independently in [ISh89a] using coset enumeration on a computer that Mat_{24} is the universal completion of the amalgam of maximal parabolics corresponding to its action on $\mathcal{G}(Mat_{24})$. By the standard principle (1.5.2) this is equivalent to the following.

Proposition 3.3.11 *The geometry $\mathcal{G}(Mat_{24})$ is simply connected.* □

3.4 Petersen geometries of the Mathieu groups

In this section we construct a rank 4 P-geometry $\mathcal{G}(Mat_{23})$ possessing the Mathieu group Mat_{23} as a flag-transitive automorphism group and a rank 3 P-geometry $\mathcal{G}(Mat_{22})$ as a residue. The latter residue possesses $\mathrm{Aut}\, Mat_{22}$ as a flag-transitive automorphism group.

Let Y_1 be a 1-element subset of $\mathcal{P}$ and let $H = Mat_{23}$ be the stabilizer of Y_1 in $G = Mat_{24}$. Since the stabilizer in G of an octad B acts transitively both on B and on $\mathcal{P} \setminus B$, we conclude that H, acting on the vertex set of the octad graph Γ, has two orbits Γ^0 and Γ^1 with lengths 253 and 506, consisting of the octads containing and not containing Y_1, respectively. It is easy to deduce from (2.10.1) that the stabilizer in H of an octad from Γ^0 or Γ^1 is isomorphic to $2^3 : L_3(2)$ or $Alt_8 \cong L_4(2)$, respectively.

Let Γ^1 denote also the subgraph in Γ induced by that orbit. We assume that $x = B$ is a vertex-octad contained in Γ^1. Then for a trio $T = \{B_1 = B, B_2, B_3\}$ containing B exactly one of B_2 and B_3 is contained in Γ^1. Hence the valency of Γ^1 is 15 and $H(x) \cong L_4(2)$ acts doubly transitively on $\Gamma^1(x)$ as on the point set of π_x. Let π_x denote also the projective space having $\Gamma^1(x)$ as its point set and preserved by $H(x)$. Let Σ be a quad containing x. Then by the remark before (2.5.2) $\Sigma \cap \Gamma^0$ is a standard 5-coclique and $\Sigma \cap \Gamma^1$ is a Petersen subgraph. It is easy to deduce from the structure of the octad graph that $\Sigma \cap \Gamma^1$ is a geometrical subgraph, corresponding to a line in π_x. We are going to show that Γ^1 contains a complete family of geometrical subgraphs.

Let Y_2 be a 2-element subset of $\mathcal{P}$ containing Y_1 and disjoint from B. Let $F = \mathrm{Aut}\, Mat_{22}$ and $F^\flat = Mat_{22}$ be the setwise and elementwise stabilizers of Y_2 in G, respectively. Let Γ^2 be the subgraph in Γ induced by the octads disjoint from Y_2. One can easily deduce from (2.8.1) that $|\Gamma^2| = 330$. Since G_b acts doubly transitively on $\mathcal{P} \setminus B$, both F and $F^\flat$ act transitively on Γ^2 with stabilizers isomorphic to $2^3 : L_3(2) \times 2$ and $2^3 : L_3(2)$, respectively. The direct factor of order 2 in $F[B]$ is the unique subgroup W of order 2 in Q_b such that Y_2 is a W-orbit. Let $T = \{B_1 = B, B_2, B_3\}$ be a trio containing B and corresponding to a hyperplane D in Q_b. Then $Y_2 \subset B_i$ for $i = 2$ or 3 if and only if $W \leq D$. This shows that $\Gamma^2 \cap \Gamma^1(x)$ is a plane in π_x corresponding to W, so that

$$H(x) \cap H[\Gamma^2 \cap \Gamma^1(x)] = F^\flat(x)$$

and hence Γ^2 induces in Γ^1 a geometrical subgraph corresponding to a plane in π_x. Thus Γ^1 contains a complete family of geometrical subgraphs and by (9.8.1) we obtain the following.

Lemma 3.4.1 *Let $\mathcal{G}(Mat_{23})$ be a rank 4 incidence system whose elements of type 1 and 2 are the geometrical subgraphs in Γ^1 of valency 7 and 3, respectively, whose elements of type 3 and 4 are the edges and vertices of Γ^1 with the incidence relation via inclusion. Then $\mathcal{G}(Mat_{23})$ is a geometry with the diagram*

```
                          P
o-----------o-----------o-----------o
2           2           2           1
```

and Mat_{23} is a flag-transitive automorphism group of this geometry. □

The geometry $\mathcal{G}(Mat_{23})$ can be defined directly in terms of the Steiner system $S(5, 8, 24)$: the elements of type 1 are the 1-element subsets X of $\mathcal{P} \setminus Y_1$, the elements of type 2 are the 3-element subsets Z of $\mathcal{P} \setminus Y_1$, the elements of type 3 are the pairs $\{B_1, B_2\}$ of disjoint octads both missing Y_1 and the elements of type 4 are the octads B_3 disjoint from Y_1. The incidence relation between these elements is given by the following conditions where Σ is the sextet containing $Y_1 \cup Z$:

$$X \subset Z; \quad X \cap (B_1 \cup B_2) = \emptyset; X \cap B_3 = \emptyset;$$

$$\Sigma \text{ refines } B_1 \text{ and } B_2; \quad \Sigma \text{ refines } B_3; \quad B_3 \in \{B_1, B_2\}.$$

It is an easy combinatorial exercise to check that this definition of $\mathcal{G}(Mat_{23})$ is equivalent to the one given in (3.4.1). Notice that $G[B_1] \cap G[B_2] \cap G[Y_1]$ induces on $\mathcal{P} \setminus (B_1 \cup B_2 \cup Y_1)$ the natural action of $L_3(2)$ on the point set of a projective plane π of order 2. By the above and in view of (2.10.6) an element Z of type 2 is incident to $\{B_1, B_2\}$ if and only if Z is a line of π.

The description of $\mathcal{G}(Mat_{23})$ in terms of the Steiner system $S(5, 8, 24)$ enables us to define an important subgeometry. Let B be an octad *containing* Y_1. Define a rank 3 subgeometry $\mathcal{S}$ in $\mathcal{G}(Mat_{23})$ containing the elements of type 1 and 2 which are subsets of B and the elements $\{B_1, B_2\}$ of type 3 such that $\{B, B_1, B_2\}$ is a trio. The incidence relation is induced by that in $\mathcal{G}(Mat_{23})$.

Lemma 3.4.2 *The subgeometry $\mathcal{S}$ is isomorphic to the C_3-geometry $\mathcal{G}(Alt_7)$; the group $G[B] \cap G[Y_1] \cong 2^4 : Alt_7$ acts flag-transitively on $\mathcal{S}$ with $G(B) \cong 2^4$ being the kernel.*

Proof. A trio T containing B corresponds by (2.10.3) to a hyperplane in Q_b and T induces on $B \setminus Y_1$ a structure $\pi(T)$ of a projective plane of order 2. The Alt_7-subgroup in $G_b/Q_b \cong L_4(2)$ acts flag-transitively on the projective geometry associated with Q_b (compare (1.6.5)). In particular it permutes transitively the 15 trios containing B. Thus the corresponding 15 projective plane structures $\pi(T)$ form an Alt_7-orbit. Now the result follows directly from the definition of the Alt_7-geometry given in Section 1.7. □

For $i = 1, 2, 3$ and 4 let H_i be the stabilizer in $H \cong Mat_{23}$ of an element of type i in $\mathcal{G}(Mat_{23})$. In view of the description of $\mathcal{G}(Mat_{23})$ in terms of the Steiner system we have the following:

$$H_1 \cong Mat_{22}; \; H_2 \cong 2^4 : (Alt_5 \times 3).2; \; H_3 \cong 2^4 : L_3(2); \; H_4 \cong Alt_8.$$

Here H_2 is the stabilizer in G_s of an element from $\mathcal{P}$. The structure of H_3 deserves a further comment.

Lemma 3.4.3 *$H_3 \cong \mathrm{Aut}\, AGL_3(2)$, in particular $O_2(H_3)$ is an indecomposable module for $H_3/O_2(H_3) \cong L_3(2)$.*

Proof. Let $\{B_2, B_3\}$ be the element of type 3 stabilized by H_3. Put $B = B_1 = \mathcal{P} \setminus (B_2 \cup B_3)$. For $i = 2$ and 3 K_t induces the group $AGL_3(2) \cong 2^3 : L_3(2)$ on the elements in B_i and the kernel, which is elementary abelian of order 2^3, acts regularly on B. Hence $K_t \cap H \cong AGL_3(2)$ acts faithfully on B_i. Since B_2 and B_3 are orbits of a hyperplane D in Q_b, an element from D stabilizes both B_2 and B_3 while an element from $Q_b \setminus D$ switches them. This shows that $H_3 = \langle K_t \cap H, Q_b \rangle$ and since H_3 stabilizes B, $Q_b \trianglelefteq H_3$. On the other hand $H[B] \cong 2^4 : Alt_7$ and $H[B]/Q_b \cong Alt_7$ acts flag-transitively on the projective space associated with Q_b. It follows from elementary properties of Alt_7 that the stabilizer of a point and the stabilizer of a plane are non-conjugate subgroups isomorphic to $L_3(2)$ (and conjugated in Sym_7). Since $L_3(2) \cong (K_t \cap H)Q_b/Q_b \leq H[B]/Q_b \cong Alt_7$ and $K_t \cap H$ normalizes the hyperplane D, it does not normalize subgroups of order 2 and hence Q_b is indecomposable under H_3/Q_b. Thus an element from $Q_b \setminus D$ permutes the classes of complements to D in $K_t \cap H$. Since there are two classes of such complements [JP76], H_3 is the automorphism group of $AGL_3(2)$. □

The residue in $\mathcal{G}(Mat_{23})$ of an element of type 1 is a rank 3 P-geometry with the diagram

$$\overset{}{\underset{2}{\circ}}\!\!-\!\!-\!\!-\!\!\underset{2}{\circ}\!\!\overset{P}{-\!\!-\!\!-}\!\!\underset{1}{\circ}$$

denoted by $\mathcal{G}(Mat_{22})$. This geometry can also be described directly in terms of the Steiner system $S(5, 8, 24)$. Specifically, the elements of type 1 are sextets $\Sigma = \{S_1, S_2, ..., S_6\}$ such that $Y_2 \subseteq S_i$ for some $1 \leq i \leq 6$, the elements of type 2 are the pairs $\{B_1, B_2\}$ of disjoint octads, both disjoint from Y_2 and the elements of type 3 are the octads B_3 disjoint from Y_2. The incidence between these elements is given by the following conditions:

$$\Sigma \text{ refines } B_1 \text{ and } B_2; \quad \Sigma \text{ refines } B_3; \quad B_3 \in \{B_1, B_2\}.$$

There is a natural bijection φ between the set of 2-element subsets Z of $\mathcal{P} \setminus Y_2$ and the set of elements of type 1 in $\mathcal{G}(Mat_{22})$ where $\varphi(Z)$ is the unique sextet containing $Y_2 \cup Z$. An element B_3 of type 3 in $\mathcal{G}(Mat_{22})$ is incident to $\varphi(Z)$ if and only if Y_2 and Z are different orbits of a subgroup of order 2 in $G(B_3) \cong 2^4$.

The stabilizer in Mat_{23} of an element of type 1 in $\mathcal{G}(Mat_{23})$ induces on $\mathcal{G}(Mat_{22})$ the group $F^b \cong Mat_{22}$ and it is clear from the above description that $\mathcal{G}(Mat_{22})$ also admits Aut Mat_{22} as a flag-transitive automorphism group.

For $i = 1, 2$ and 3 let $F_i^\flat$ and F_i be the stabilizers in $F^\flat$ and F of an element of type i in $\mathcal{G}(Mat_{22})$. Then from the above we can easily deduce the following:

$$F_1^\flat \cong 2^4 : Sym_5; \quad F_2^\flat \cong 2^4 : Sym_4; \quad F_3^\flat \cong 2^3 : L_3(2);$$

$$F_1 \cong 2^5 : Sym_5; \quad F_2 \cong 2^4 : (Sym_4 \times 2); \quad F_3 \cong 2^3 : L_3(2) \times 2.$$

The residue of an element of type 1 in the Alt_7-subgeometry $\mathcal{S}$ in $\mathcal{G}(Mat_{23})$ is a subgeometry $\mathcal{Q}$ in $\mathcal{G}(Mat_{22})$ isomorphic to the generalized quadrangle of order (2, 2). This subgeometry can be described in the following way. Let B be an octad containing Y_2. Then $\mathcal{Q}$ consist of the elements of type 1 contained in B and of the elements $\{B_1, B_2\}$ of type 2 such that $\{B, B_1, B_2\}$ is a trio. The stabilizers $Q^\flat$ and Q of $\mathcal{Q}$ in $F^\flat$ and F are the elementwise and the setwise stabilizers of Y_2 in G_b isomorphic to $2^4 : Alt_6$ and $2^4 : Sym_6$, respectively. Clearly $O_2(Q) = Q_b$ is the kernel of the action of Q on $\mathcal{Q}$. We formulate this as follows.

Lemma 3.4.4 *Every octad B containing Y_2 corresponds to a rank 2 subgeometry $\mathcal{Q}$ in $\mathcal{G}(Mat_{22})$ isomorphic to the generalized quadrangle $\mathcal{G}(Sp_4(2))$.* □

Recall that if $\mathcal{G}$ is a P-geometry of rank n then the *derived graph* $\Delta(\mathcal{G})$ of $\mathcal{G}$ is a graph on the set of elements of type n in $\mathcal{G}$ in which two elements are adjacent if they are incident to a common element of type $n-1$. In these terms the subgraphs in the octad graph Γ induced by the orbits Γ^1 and Γ^2 are the derived graphs $\Delta(\mathcal{G}(Mat_{23}))$ and $\Delta(\mathcal{G}(Mat_{22}))$, respectively. It is well known [BCN89] and can be easily deduced from (3.2.1) that these two graphs are distance-transitive with the following respective distance diagrams.

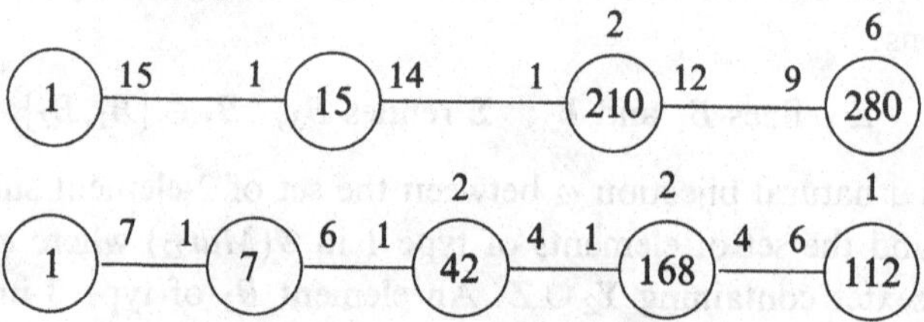

We conclude this section with the following.

Lemma 3.4.5 *$\mathcal{G}(Mat_{22})$ is a subgeometry of $\mathcal{G}(Mat_{24})$.*

Proof. We consider $\mathcal{G}(Mat_{24})$ as described in (3.3.9). Then the set of elements of type 1 in $\mathcal{G}(Mat_{22})$ is a subset of the set of elements of type 1 in $\mathcal{G}(Mat_{24})$. Let B be an element of type 3 in $\mathcal{G}(Mat_{22})$, *i.e.* an octad disjoint from Y_2. Let W be the setwise stabilizer of Y_2 in $Q_b = G(B)$, so that W is the unique subgroup of order 2 in Q_b, such that Y_2 is a W-orbit. If Σ is a sextet which refines B then Σ is determined by a tetrad which is an orbit on $\mathcal{P} \setminus B$ of a subgroup U of order 2^2 in Q_b. Furthermore, Y_2 is contained in a tetrad of Σ if and only if $W \leq U$. In terms of (3.3.8) this means that the set L of elements of type 1 incident to B is a maximal clique in the sextet graph from the orbit $\mathcal{K}_v$. Hence L is an element of type 3 in $\mathcal{G}(Mat_{24})$. Clearly B is uniquely determined by L. In a very similar way one can show that an element of type 2 in $\mathcal{G}(Mat_{22})$ is uniquely determined by the set M of sextets incident to it and that M is an element in $\mathcal{G}(Mat_{24})$. □

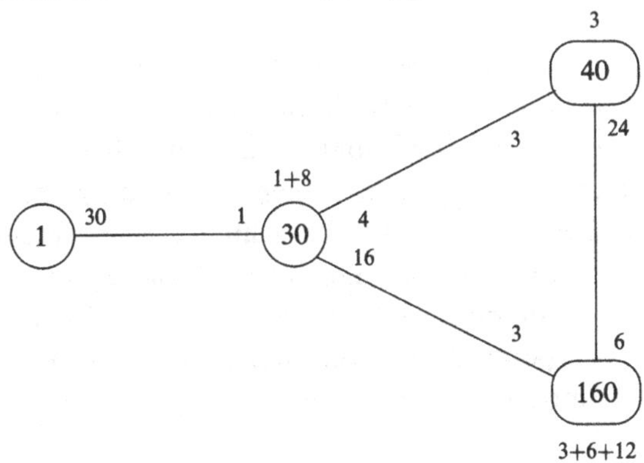

By (3.4.5) the sextet graph contains the collinearity graph of $\mathcal{G}(Mat_{22})$ as a subgraph. The suborbit diagram of the latter graph as given above is easy to deduce from the diagram $D_p(Mat_{22})$ proved in (3.9.6).

3.5 The universal cover of $\mathcal{G}(Mat_{22})$

In this section we show that $\mathcal{G}(Mat_{22})$ possesses a triple cover $\mathcal{G}(3 \cdot Mat_{22})$ which is simply connected.

We start by proving a result established by S.V. Shpectorov in [Sh85] which gives an upper bound 1898 for the number of elements of type 3 in a P-geometry of rank 3.

Let $\mathcal{G}$ be a P-geometry of rank 3 and $\Delta = \Delta(\mathcal{G})$ be the derived graph of $\mathcal{G}$. An element y of type 2 in $\mathcal{G}$ corresponds to an edge in Δ whose ends are the elements of type 3 incident to y. We claim that different elements of type 2 correspond to different edges. In fact, suppose that y_1 and y_2 are different elements of type 2 incident to the same pair $\{x_1, x_2\}$ of elements of type 3. Since y_1 and y_2 are lines in the projective plane $\text{res}_{\mathcal{G}}(x_1)$, there is a point z in this plane (an element of type 1 in $\mathcal{G}$) incident to both y_1 and y_2. Since $\text{res}_{\mathcal{G}}(z)$ is the geometry of edges and vertices of the Petersen graph, which does not contain multiple edges, we reach a contradiction. Thus the elements of type 2 in $\mathcal{G}$ are identified with the edges of Δ. Similarly an element z of type 1 corresponds to a subgraph Ξ (which is not necessarily an induced subgraph) formed by the vertices and edges incident to z. Clearly Ξ is isomorphic to the Petersen graph. Since a point in a projective plane is uniquely determined by the set of lines it is incident to, different elements of type 1 correspond to different subgraphs. Throughout this section when talking about a Petersen subgraph in Δ we always mean a subgraph formed by vertices and edges incident to an element of type 1.

For $x \in \Delta$ let π_x denote the projective plane structure having $\Delta(x)$ as the point set and dual to $\text{res}_{\mathcal{G}}(x)$. A triple $\{u, v, w\}$ of vertices from $\Delta(x)$ is a line of π_x if there is a Petersen subgraph Ξ containing x, such that $\{u, v, w\} = \Xi(x)$. If (y, x, z) is a 2-arc in Δ then the points y and z in π_x determine a unique line and hence there is a unique Petersen subgraph $\Xi(y, x, z)$ in which (y, x, z) is a 2-arc. Dually, if Ξ and Ξ' are Petersen subgraphs containing a common vertex x, then $\Xi(x)$ and $\Xi'(x)$ are lines in π_x which must have a point in common. Hence whenever two Petersen subgraphs have a common vertex, they have a common edge incident to this vertex.

As usual, for a vertex x of Δ and an integer i by $\Delta_i(x)$ we denote the set of vertices at distance i from x in Δ and $\Delta(x) = \Delta_1(x)$. In addition put

$$\Delta^i(x) = \{x\} \cup \Delta_1(x) \cup ... \cup \Delta_i(x)$$

In this section the meaning of the parameters b_i and c_i slightly differs from that in the rest of the book. To wit, we put

$$b_i = \max \{|\Delta_{i+1}(x) \cap \Delta(y)| \quad | \; x \in \Delta, \; y \in \Delta_i(x)\},$$

$$c_i = \min \{|\Delta_{i-1}(x) \cap \Delta(y)| \quad | \; x \in \Delta, \; y \in \Delta_i(x)\}.$$

The following lemma generalizes some well-known properties of distance-transitive graphs.

Lemma 3.5.1 *Let $1 \leq i \leq d$ where d is the diameter of Δ. Then*

(i) $b_i + c_i \le 7$ *and* $c_i \ge 1$,
(ii) *if* $1 \le j < i$ *then* $b_j \ge b_i$ *and* $c_i \ge c_j$,
(iii) $|\Delta_{i+1}(x)| \le (b_i/c_{i+1}) \cdot |\Delta_i(x)|$.

Proof. Since $|\Delta(y)|$ is the number of points in the projective plane π_y of order 2, which is 7, (i) follows. Let $x \in \Delta$, $y \in \Delta_i(x)$ and $z \in \Delta_{i-j}(x) \cap \Delta_j(y)$. Then

$$\Delta_{i+1}(x) \cap \Delta(y) \subseteq \Delta_{j+1}(z) \cap \Delta(y),$$

$$\Delta_{j-1}(z) \cap \Delta(y) \subseteq \Delta_{i-1}(x) \cap \Delta(y),$$

and we obtain (ii). Let D be the set of edges $\{u, v\}$ such that $u \in \Delta_i(x)$, $v \in \Delta_{i+1}(x)$. Then every vertex from $\Delta_i(x)$ is incident to at most b_i edges from D while every vertex from $\Delta_{i+1}(x)$ is incident to at least c_{i+1} edges from D. Hence

$$|\Delta_i(x)| \cdot b_i \ge |D| \ge |\Delta_{i+1}(x)| \cdot c_{i+1}$$

and (iii) follows. □

Lemma 3.5.2 $b_2 \le 4$.

Proof. Let (x, y, z) be a 2-arc such that $z \in \Delta_2(x)$ and $\Xi = \Xi(x, y, z)$. Since Ξ is of diameter 2, $\Xi(z) \subseteq \Delta^2(x)$ and $|\Delta_3(x) \cap \Delta(z)| \le 7 - |\Xi(z)| = 4$. □

Lemma 3.5.3 *If* $y \in \Delta_4(x)$ *then there is a Petersen subgraph* Ξ *containing* y *such that* $\Xi(y) \subseteq \Delta^4(x)$. *Furthermore* $|\Xi(y) \cap \Delta_3(x)| \ge 2$, *in particular* $c_4 \ge 2$.

Proof. Let (x, u, z, v, y) be a 4-arc in Δ joining x and y. Put $\Xi = \Xi(z, v, y)$ and $\Phi = \Xi(x, u, z)$. Since Ξ and Φ have a common vertex z, they have a common edge, say $\{z, w\}$. Since both Ξ and Φ have diameter 2, $w \in \Delta_2(x) \cap \Delta_2(y)$ and $\Xi \subseteq \Delta^4(x)$. Both $\Xi(y) \cap \Xi(z)$ and $\Xi(y) \cap \Xi(w)$ are non-empty while $\Xi(z) \cap \Xi(w)$ is empty since there are no triangles in Ξ. Hence $|\Xi(y) \cap \Delta_3(x)| \ge 2$. □

Lemma 3.5.4 $b_5 \le 1$.

Proof. Let $y \in \Delta_5(x)$ and suppose to the contrary that z_1 and z_2 are different vertices from $\Delta(y) \cap \Delta_6(x)$. Let $a \in \Delta(y) \cap \Delta_4(x)$. Let $\Lambda = \Xi(z_1, y, z_2)$, $\Phi_i = \Xi(a, y, z_i)$ for $i = 1$ and 2 and let Ξ be the Petersen subgraph containing a and contained in $\Delta^4(x)$ whose existence is guaranteed by (3.5.3). Since a Petersen subgraph has diameter 2, for $i = 1$ and 2 the subgraph

Φ_i intersects Ξ in an edge which joins a to a vertex x_i in $\Xi(a) \cap \Delta_4(x)$. By (3.5.3) $|\Xi(a) \cap \Delta_3(x)| \geq 2$ and hence $x_1 = x_2$. This means that the 2-arc (x_1, a, y) is contained in Φ_1 and in Φ_2, so $\Phi_1 = \Phi_2$. Since $\{z_1, y\}$ is an edge of Φ_1 and $\{y, z_2\}$ is an edge of Φ_2, we obtain $\Phi_1 = \Phi_2 = \Lambda$ and hence $\{y, a\}$ is an edge in Λ. Since a was taken to be an arbitrary vertex from $\Delta(y) \cap \Delta_4(x)$, we have $\Delta(y) \cap \Delta_4(x) \subseteq \Lambda(y)$. Since the valency of Λ is 3 and $z_1, z_2 \in \Lambda(y) \cap \Delta_6(x)$ we must have $|\Delta(y) \cap \Delta_4(x)| = 1$, which is impossible, since $|\Delta(y) \cap \Delta_4(x)| \geq c_5 \geq c_4$ and $c_4 \geq 2$ by (3.5.3). □

Now we are ready to establish the upper bound.

Proposition 3.5.5 *A P-geometry of rank* 3 *(possibly not flag-transitive) contains at most* 1898 *elements of type* 3.

Proof. Using (3.5.1), (3.5.2) and (3.5.3) we obtain $|\Delta_0(x)| = 1$, $|\Delta(x)| = 7$, $|\Delta_2(x)| \leq 42$, $|\Delta_3(x)| \leq 168$, $|\Delta_4(x)| \leq 336$ and $|\Delta_5(x)| \leq 672$. By (3.5.1 (iii)), (3.5.3) and (3.5.4) $|\Delta_{i+1}(x)| \leq \frac{1}{2} \cdot |\Delta_i(x)|$ for $i \geq 5$. Hence $\Sigma_{i=6}^{d} |\Delta_i(x)| \leq |\Delta_5(x)| \leq 672$ and the result follows. □

The existence of a triple cover $\mathcal{G}(3 \cdot Mat_{22})$ of $\mathcal{G}(Mat_{22})$ can be established using the following result (see [Maz79] and references therein).

Lemma 3.5.6 *There exists a group* $\widetilde{F} \sim 3 \cdot \mathrm{Aut}\, Mat_{22}$ *having a normal subgroup* Y *of order* 3, *such that* $\widetilde{F}/Y \cong \mathrm{Aut}\, Mat_{22}$. *The commutator subgroup* $\widetilde{F}^\flat$ *of* $\widetilde{F}$ *is a perfect central extension of* $F^\flat \cong Mat_{22}$ *by* Y *and* $\widetilde{F}^\flat = C_{\widetilde{F}}(Y)$. *If* $\widetilde{D}^\flat$ *is the preimage in* $\widetilde{F}$ *of a subgroup* $Mat_{21} \cong L_3(4)$ *in* $F^\flat$, *then* $\widetilde{D}^\flat \cong SL_3(4)$ *and an element from* $N_{\widetilde{F}}(\widetilde{D}^\flat) \setminus \widetilde{D}^\flat$ *induces on* $\widetilde{D}^\flat$ *a field automorphism.* □

Let $\widetilde{F} \sim 3 \cdot \mathrm{Aut}\, Mat_{22}$ and $\widetilde{F}^\flat \sim 3 \cdot Mat_{22}$ be as in (3.5.6) and let $\varphi : \widetilde{F} \to F$ be the canonical homomorphism. Let $\mathscr{A} = \{F_i^\flat \mid 1 \leq i \leq 3\}$ be the amalgam of maximal parabolics corresponding to the action of $F^\flat$ on $\mathcal{G}(Mat_{22})$. For $i = 1, 2$ and 3 let $\widetilde{F}_i^\flat = \varphi^{-1}(F_i^\flat)$ be the preimage of $F_i^\flat$ in $\widetilde{F}^\flat$. Since a Sylow 3-subgroup of $F_i^\flat$ is of order 3 and every such subgroup is inverted in its normalizer, we conclude that $\widetilde{F}_i^\flat$ splits over Y and since $O^3(F_i^\flat) = F_i^\flat$ we obtain the following direct product decomposition:

$$\widetilde{F}_i^\flat = O^3(\widetilde{F}_i^\flat) \times Y,$$

where $O^3(\widetilde{F}_i^\flat) \cong F_i^\flat$. Let $\widetilde{\mathscr{A}} = \{O^3(\widetilde{F}_i^\flat) \mid 1 \leq i \leq 3\}$. It is easy to check that for $1 \leq i < j \leq 3$

$$O^3(\widetilde{F}_i^\flat) \cap O^3(\widetilde{F}_j^\flat) = O^3(\widetilde{F}_i^\flat \cap \widetilde{F}_j^\flat) \cong F_i^\flat \cap F_j^\flat$$

and hence φ induces an isomorphism of $\widetilde{\mathscr{A}}$ onto $\mathscr{A}$.

Lemma 3.5.7 *Let $\mathscr{G}(3 \cdot Mat_{22})$ be the coset geometry $\mathscr{G}(\widetilde{F}^{\flat}, \widetilde{\mathscr{A}})$. Then φ induces a (3-fold) covering of $\mathscr{G}(3 \cdot Mat_{22})$ onto $\mathscr{G}(Mat_{22})$ which is universal.*

Proof. Since φ induces an isomorphism of $\widetilde{\mathscr{A}}$ onto $\mathscr{A}$, it induces a covering

$$\varphi : \mathscr{G}(\widetilde{F}^{\flat}, \widetilde{\mathscr{A}}) \to \mathscr{G}(F^{\flat}, \mathscr{A}) \cong \mathscr{G}(Mat_{22})$$

(denoted by the same letter φ). Notice that $\mathscr{G}(3 \cdot Mat_{22})$ is connected since $\widetilde{F}^{\flat}$ does not split over Y. Since $\mathscr{G}(3 \cdot Mat_{22})$ is a triple cover of $\mathscr{G}(Mat_{22})$, it contains $990 = 330 \cdot 3$ elements of type 3. A flag-transitive (in particular the universal) cover of $\mathscr{G}(3 \cdot Mat_{22})$ has $990 \cdot n$ elements of type 3, where n is either an integer or infinity. Since $990 \cdot 2 > 1898$, (3.5.5) implies the simple connectedness of $\mathscr{G}(3 \cdot Mat_{22})$. □

We could define $\mathscr{G}(3 \cdot Mat_{22})$ to be the coset geometry $\mathscr{G}(\widetilde{F}, \{N_{\widetilde{F}}(O^3(\widetilde{F}_i^{\flat})) \mid 1 \leq i \leq 3\})$ and this shows that it admits $\widetilde{F}$ as a flag-transitive automorphism group.

The derived graph of $\mathscr{G}(3 \cdot Mat_{22})$ is a triple antipodal cover of the derived graph of $\mathscr{G}(Mat_{22})$ and it has the following distance diagram:

2 2 1 2 2
(1) 7 1 (7) 6 1 (42) 4 1 (168) 4 2 (336) 4 4 (336) 1 4 (84) 1 6 (14) 1 7 (2)

Let $\mathscr{Q} \cong \mathscr{G}(Sp_4(2))$ be the subgeometry in $\mathscr{G}(Mat_{22})$ and $Q \cong 2^4 : Sym_6$ be the stabilizer of $\mathscr{Q}$ in F. Recall that Q is the stabilizer in $G \cong Mat_{24}$ of an octad B and a 2-element subset Y_2 in B. In particular $O_2(Q) = Q_b$ acts regularly on $\mathscr{P} \setminus B$ and hence for $p \in \mathscr{P} \setminus B$ the subgroup $Q(p) \cong Sym_6$ is a complement to Q_b in Q. Since $Q(p)$ acts naturally on the 6-element set $B \setminus Y_2$, it is easy to see that $B \setminus Y_2$ is a hyperoval in $\Pi(Y_2 \cup \{p\})$ and $Q(p)$ is its full stabilizer in $G[Y_2 \cup \{p\}] \cong P\Gamma L_3(4)$.

Lemma 3.5.8 *Let $\widetilde{\mathscr{Q}}$ be the preimage of $\mathscr{Q}$ in $\mathscr{G}(3 \cdot Mat_{22})$ and $\widetilde{Q}$ be the preimage of Q in $\widetilde{F}$. Then $\widetilde{\mathscr{Q}}$ is the rank 2 tilde geometry and $\widetilde{Q}$ induces the automorphism group of $\widetilde{\mathscr{Q}}$.*

Proof. It is clear that $O_2(\widetilde{Q}) \cong O_2(Q) \cong 2^4$. Let $\widetilde{Q}(p)$ be the preimage of $Q(p)$ in $\widetilde{F}$, which is clearly a complement to $O_2(\widetilde{Q})$ in $\widetilde{Q}$. By (3.5.6), (2.7.13) and in view of the discussion before the lemma, $\widetilde{Q}(p) \cong 3 \cdot Sym_6$ is the automorphism group of the rank 2 T-geometry. Let $\{\widetilde{x}_1, \widetilde{x}_2\}$ be a flag in $\widetilde{\mathscr{Q}}$ and $\widetilde{Q}_i$ be the stabilizer of $\widetilde{x}_i$ in $\widetilde{F}$, $i = 1, 2$. Then $O_2(\widetilde{Q}) \trianglelefteq \widetilde{Q}_i$ and $\widetilde{Q}_i / O_2(\widetilde{Q}) \cong Sym_4 \times 2$ is a complement to Y in the stabilizer in $\widetilde{F}$ of $\widetilde{x}_i$. Now the result follows directly from (2.6.2). □

It is clear that the stabilizer of $\widetilde{\mathcal{Q}}$ in $\widetilde{F}^{\flat} \cong 3 \cdot Mat_{22}$ is isomorphic to $2^4 : 3 \cdot Alt_6$ and it induces on $\widetilde{\mathcal{Q}}$ a flag-transitive action.

3.6 $\mathcal{G}(Mat_{23})$ is 2-simply connected

Let $\mathcal{H} = \mathcal{G}(Mat_{23})$ be the rank 4 P-geometry of the Mathieu group $H \cong Mat_{23}$ with the diagram

$$\underset{2}{\circ}\!\!-\!\!-\!\!-\!\!-\underset{2}{\circ}\!\!-\!\!-\!\!-\!\!-\underset{2}{\circ}\overset{P}{-\!\!-\!\!-\!\!-}\underset{1}{\circ}$$

Let $\mathcal{A} = \{H_i \mid 1 \leq i \leq 4\}$ and $\mathcal{B} = \{P_{ij} \mid 1 \leq i < j \leq 4\}$ be the amalgams of maximal and rank 2 parabolics corresponding to the action of H on $\mathcal{H}$ and associated with a maximal flag $\{x_1, x_2, x_3, x_4\}$ where x_i is of type i. In this section we follow [ISh90a] to show that $\mathcal{H}$ is 2-simply connected by proving that H coincides with the universal completion U of the amalgam $\mathcal{B}$. First we show that U is a completion and hence the universal completion of $\mathcal{A}$. After that we prove the simple connectedness of $\mathcal{H}$.

For $k = 2, 3$ and 4 the residue $\mathrm{res}_{\mathcal{H}}(x_k)$ is 2-simply connected which means that H_k is the universal completion of the amalgam $\mathcal{B}^{(k)} = \{P_{ij} \mid 1 \leq i < j \leq 4, i \neq k, j \neq k\}$ of rank 2 parabolics corresponding to the action of H_k on $\mathrm{res}_{\mathcal{H}}(x_k)$. Since the amalgam $\mathcal{A}^1 = \{H_k \mid 2 \leq k \leq 4\}$ contains $\mathcal{B}$ and is generated by the elements in $\mathcal{B}$, we have the following.

Lemma 3.6.1 *U is the universal completion of the amalgam $\mathcal{A}^1$.* □

The amalgam $\mathcal{D} = \{P_{34}, P_{24}, P_{23}\}$ is the amalgam of maximal parabolics corresponding to the action of $H_1 \cong Mat_{22}$ on $\mathrm{res}_{\mathcal{H}}(x_1) \cong \mathcal{G}(Mat_{22})$ and by (3.5.6) the universal completion of $\mathcal{D}$ is isomorphic to $3 \cdot Mat_{22}$.

Lemma 3.6.2 *Let D be the subgroup in U generated by $\mathcal{D}$. Then $D \cong H_1 \cong Mat_{22}$.*

Proof. We know that D is isomorphic either to H_1 or to the universal completion of $\mathcal{D}$ which is $3 \cdot Mat_{22}$. Let $\mathcal{S} \cong \mathcal{G}(Alt_7)$ be the subgeometry in $\mathcal{H}$ as in (3.4.2) and $S \cong 2^4 : Alt_7$ be the stabilizer of $\mathcal{S}$ in H. Assuming without loss of generality that $\{x_1, x_2, x_3\}$ is a maximal flag in $\mathcal{S}$, let $S_i = S \cap H_i$ denote the stabilizer of x_i in S. Let $\mathcal{E} = \{S_i \cap S_j \mid 1 \leq i < j \leq 3\}$ be the amalgam of minimal parabolics corresponding to the action of S on $\mathcal{S}$. Since $S_i \cap S_j \leq P_{kl}$, whenever $\{i, j, k, l\} = \{1, 2, 3, 4\}$, $\mathcal{E}$ is contained in $\mathcal{B}$ and hence also in U. Let E be the subgroup in U generated by

$\mathcal{E}$ and for $i = 1, 2, 3$ let T_i be the subgroup in E generated by $S_i \cap S_j$ and $S_i \cap S_k$ where $\{i, j, k\} = \{1, 2, 3\}$. Since H_k is the universal completion of $\mathcal{B}^{(k)}$ for $k = 2$ and 3 we have $T_2 \cong S_2$ and $T_3 \cong S_3$. If $D \cong H_1$ then $T_1 \cong S_1$ and if $D \cong 3 \cdot Mat_{22}$, then by (3.5.8) $T_1 \cong 2^4 : 3 \cdot Alt_7$. In the latter case $\mathcal{T} = \mathcal{G}(E, \{T_1, T_2, T_3\})$ is a flag-transitive rank 3 tilde geometry possessing a morphism onto $\mathcal{S} \cong \mathcal{G}(Alt_7)$ which commutes with the flag-transitive action. By (6.11.4) there are no such geometries and the result follows. □

Since $\mathcal{G}(Alt_7)$ is simply connected by (1.7.1), the proof of (3.6.2) has the following implication.

Lemma 3.6.3 *In the above terms $E \cong S \cong 2^4 : Alt_7$ and $\mathcal{T} \cong \mathcal{S} \cong \mathcal{G}(Alt_7)$.* □

By (3.6.2) U contains $\mathcal{A}$ and hence it is the universal completion of $\mathcal{A}$. This means that the universal 2-cover of $\mathcal{H}$ coincides with its universal cover.

Lemma 3.6.4 *The geometry $\mathcal{H} = \mathcal{G}(Mat_{23})$ is simply connected.*

Proof. Let $\psi : \widetilde{\mathcal{H}} \to \mathcal{H}$ be the universal covering of $\mathcal{H}$. We are going to show that $\widetilde{\mathcal{H}}$ and $\mathcal{H}$ have the same number of elements of type 1, which is 23. By (3.6.3) a connected component of the preimage in $\widetilde{\mathcal{H}}$ of the subgeometry $\mathcal{S} \cong \mathcal{G}(Alt_7)$ is isomorphic to $\mathcal{S}$. Let Θ be a graph on the elements of type 1 and 2 in $\mathcal{H}$ in which 2 distinct elements are adjacent if they are incident in $\mathcal{H}$ and let $\widetilde{\Theta}$ be the analogous graph associated with $\widetilde{\mathcal{H}}$. Since ψ is a covering of geometries, it induces a covering of $\widetilde{\Theta}$ onto Θ (denoted by the same letter ψ). Recall that the elements of type 1 and 2 in $\mathcal{H}$ are the elements and the 3-element subsets of $\mathcal{P} \setminus Y_1$, respectively, with the incidence relation via inclusion. Let $\tilde{x}$ be an element of type 1 in $\widetilde{\mathcal{H}}$ and $x = \psi(\tilde{x})$ (where x is also considered as an element from $\mathcal{P} \setminus Y_1$). Since $\text{res}_{\mathcal{H}}(x) \cong \mathcal{G}(Mat_{22})$, there are $231 \cdot (3 - 1) = 462$ 2-arcs in Θ originating in x and, since ψ is a covering, the same number of 2-arcs in $\widetilde{\Theta}$ originate in $\tilde{x}$. Let

$$C = (x, \{x, y, a\}, y, \{x, y, b\}, x)$$

be a 4-cycle in Θ, where y, a, b are distinct elements from $\mathcal{P} \setminus Y_1$. There is an octad B which contains $Y_1 \cup \{x, y, a, b\}$. Hence C is contained in the $\mathcal{G}(Alt_7)$-subgeometry associated with B and C is contractible with respect to ψ. Since C was taken to be an arbitrary 4-cycle containing

x, we have $|\widetilde{\Theta}_2(\widetilde{x})| = |\Theta_2(x)| = 462/21 = 22$. This implies in particular that for any $\widetilde{y}, \widetilde{z} \in \widetilde{\Theta}_2(\widetilde{x})$ there is an element $\{\widetilde{x}, \widetilde{y}, \widetilde{z}\}$ of type 2 in $\widetilde{\mathscr{H}}$. Let $y, z \in \Theta_2(x)$ and let

$$C' = (x, \{x, y, a\}, y, \{y, z, a\}, z, \{x, z, a\}, x)$$

be a 6-cycle where $a \in \mathscr{P} \setminus (Y_1 \cup \{x, y, z\})$. Then C' is contained in the $\mathscr{G}(Alt_7)$-subgeometry determined by the octad containing $Y_1 \cup \{x, y, z, a\}$. Hence C' is contractible, which means the following. Whenever 2 elements of type 1 in $\widetilde{\mathscr{H}}$ are joined by a 4-arc in $\widetilde{\Theta}$, they are joined by a 2-arc. Since $\widetilde{\Theta}$ is connected, in view of the above this means that altogether there are 23 elements of type 1 in $\widetilde{\mathscr{H}}$ and the result follows. □

Now combining (3.6.2) and (3.6.4) we obtain the final result of the section.

Proposition 3.6.5 *The geometry $\mathscr{G}(Mat_{23})$ is 2-simply connected.* □

3.7 Diagrams for $\mathscr{H}(Mat_{24})$

Consider the maximal parabolic geometry $\mathscr{H} = \mathscr{H}(Mat_{24})$ as a 3-partite graph with the partition $\{\mathscr{H}_b, \mathscr{H}_t, \mathscr{H}_s\}$ where $\mathscr{H}_x$ is the set of octads, trios and sextets for $x = b, t$ and s, respectively. In this section we calculate the suborbit diagrams of $\mathscr{H}$ with respect to the action of $G \cong Mat_{24}$. For $x, y \in \{b, t, s\}$, by N_y we denote the orbit of G_x on $\mathscr{H}_y$ of length N. It turns out that in all cases x is uniquely determined by the pair (N, y), so that there is no need to mention x explicitly. The suborbit diagram with the base vertex taken from $\mathscr{H}_x$ will be denoted by $D_x(Mat_{24})$. In $D_x(Mat_{24})$ the valencies of N_y will always be given as sums of lengths of orbits of $G_x \cap G(Y)$ for $Y \in N_y$. The diagrams $D_x(Mat_{24})$ together with similar diagrams for the maximal parabolic geometry of Aut Mat_{22} (to be calculated in Section 3.9) are of crucial importance for studying geometries of larger sporadic groups, especially of J_4. We identify the elements of $\mathscr{H}$ with vertices, triangles and quads in the octad graph Γ and when talking about distances we mean the distances in Γ. As usual B, $T = \{B_1, B_2, B_3\}$ and $\Sigma = \{S_1, S_2, ..., S_6\}$ are typical octad, trio and sextet, respectively.

To calculate the diagrams $D_x(Mat_{24})$ it will be helpful to analyse the action of Q_x on the $\mathscr{H}_y$. Since the distribution of classes of involutions in Q_x is given in (2.14.5) and for an involution its permutation character on

$\mathcal{H}_y$ can be taken from (2.14.2 (iii)) and (2.14.3 (v)), it is straightforward to calculate the number of orbits of Q_x on $\mathcal{H}_y$. In order to find out how many Q_x-orbits are contained in a given G_x-orbit, we will use the following rather obvious lemma.

Lemma 3.7.1 *Let F be a group acting transitively on a set Φ of size $n = 2^r \cdot m$ where m is odd and suppose that $|O_2(F)| = 2^s$. Then there is $t \le \min\{r, s\}$ such that every $O_2(F)$-orbit on Φ has length 2^t and $F/O_2(F)$ permutes these orbits transitively. In particular the maximal number of $O_2(F)$-orbits is m and if $F(x)$ is the stabilizer in F of $x \in \Phi$ then $F(x)O_2(F)/O_2(F)$ is the stabilizer in $F/O_2(F)$ of the $O_2(F)$-orbit containing x.* □

We start with the easiest diagram $D_b(Mat_{24})$.

Lemma 3.7.2 *The following assertions hold:*

(i) *G_b has four orbits on $\mathcal{H}_b$ with lengths* 1, 30, 280 *and* 448,
(ii) *G_b has three orbits on $\mathcal{H}_t$ with lengths* 15, 420 *and* 3360,
(iii) *G_b has three orbits on $\mathcal{H}_s$ with lengths* 35, 840 *and* 896,
(iv) *if $\Sigma \in 840_s$ then there is a unique octad B' adjacent to Σ such that $(G_b \cap G_s)K_s = G_s \cap G[B']$,*
(v) *the diagram $D_b(Mat_{24})$ is as given below.*

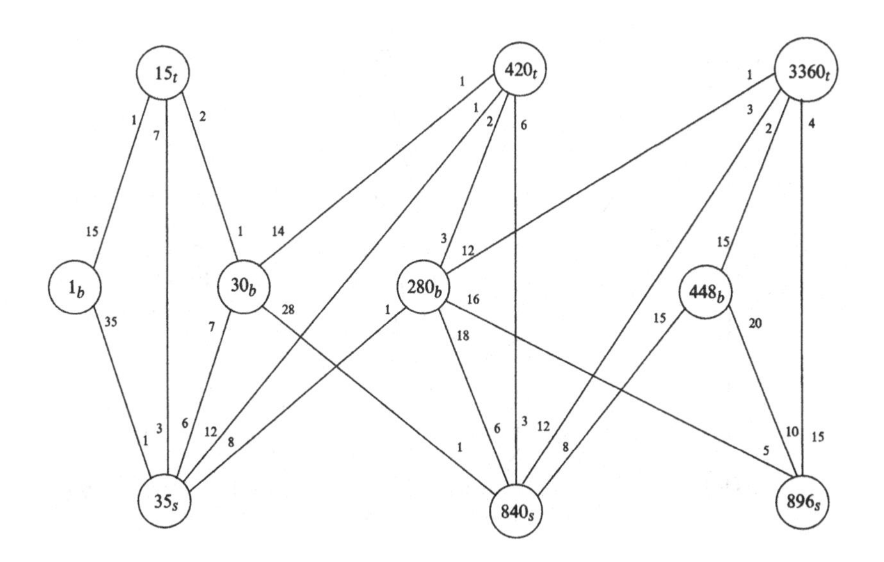

$D_b(Mat_{24})$

Proof. Part (i) follows from (3.2.1). It is easy to deduce from the proofs of (3.2.1) and (3.2.4) that if $B' \in \Gamma_i(B)$ for $i = 1, 2$ and 3 then $G_b \cap G[B']$ acts transitively on $\Gamma(B') \cap \Gamma_j(B)$ for $j = i-1$, i and $i+1$. This immediately implies (ii) together with the valencies between the N_t and N_b. By (2.14.1) and straightforward calculations we obtain (iii). To prove (iv) notice that by (2.14.1 (iii)) there is a unique octad B' adjacent to Σ and disjoint from B. Specifically, if $|B \cap S_i| = 2$ for $1 \le i \le 4$, then $B' = S_5 \cup S_6$. Hence $G_b \cap G_s \le G[B']$ and since $G_b \cap G_s \sim [2^6].Sym_3$, it is sufficient to show that $|G_b \cap K_s| \le 2^3$. Since $B \cap S_1$ is not stabilized by an element of order 3 from K_s, $G_b \cap K_s \le Q_s$. Let R_i be the kernel of the action of Q_s on S_i. Since Q_s/R_i are the points on a hyperoval on the $GF(4)$-space dual to Q_s (2.10.2), $R_1 \cap R_2 \cap R_3 = 1$. On the other hand for $i = 1$, 2 and 3 the stabilizer of $S_i \cap B$ in the action of order 2^2 induced by Q_s on S_i is of order 2, hence $|G_b \cap K_s| \le 2^3$ which implies that $(G_b \cap G_s)/(G_b \cap K_s) \cong Sym_4 \times 2$ and (iv) follows. Now the valencies of N_s are straightforward from the possible shapes of the multiset v in (2.14.1). The information on the stabilizers $G_b \cap G_s$ contained in (2.14.1) and (iv) shows that the orbit under $G_b \cap G[\Sigma]$ of an element adjacent to Σ is uniquely determined by the orbit of G_b containing this element. This gives (v). □

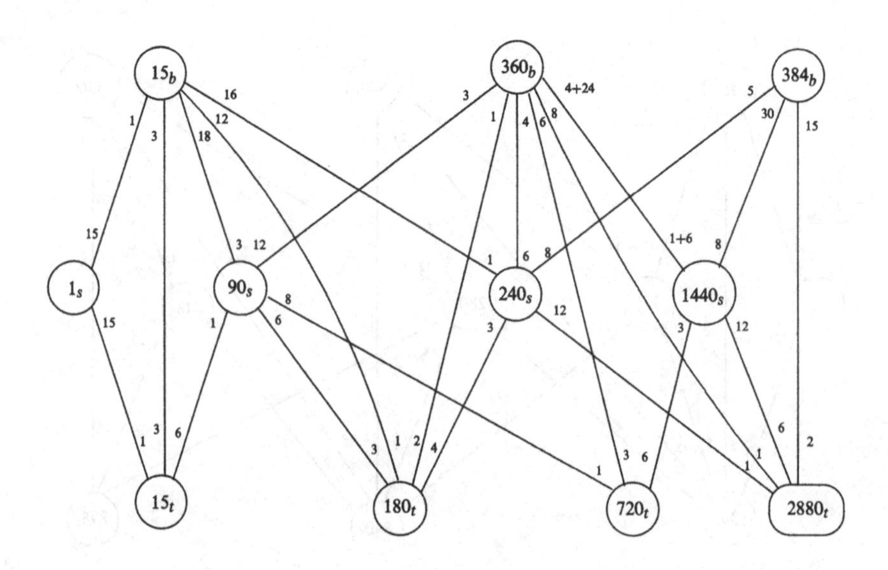

$D_s(Mat_{24})$

Lemma 3.7.3 *For $\Sigma' = \{S'_1, S'_2, ..., S'_6\} \in \mathcal{H}_s \setminus \Sigma$ let G'_s be the stabilizer of Σ' in G, $K'_s = O_{2,3}(G'_s)$ and $Q'_s = O_2(G'_s)$. Then*

(i) *G_s has four orbits on $\mathcal{H}_s$ with lengths* 1, 90, 240 *and* 1440,

(ii) *G_s has three orbits on $\mathcal{H}_b$ with lengths* 15, 360 *and* 384,

(iii) *$L(\Sigma') := (G_s \cap G'_s)K'_s/K'_s$ is the stabilizer in G'_s/K'_s of a point from $\mathrm{res}_{\mathcal{H}}(\Sigma')$ if $\Sigma' \in 90_s$ and $L(\Sigma')$ is the stabilizer of a line in this residue if $\Sigma' \in 240_s \cup 1440_s$,*

(iv) *G_s has four orbits on $\mathcal{H}_t$ with lengths* 15, 180, 720 *and* 2880,

(v) *the diagram $D_s(Mat_{24})$ is as given above.*

Proof. Part (i) follows from (3.3.5) while (ii) follows from (2.14.1). Let A be the 6×6-matrix whose (i,j)-entry is $|S_i \cap S'_j|$. It is easy to deduce from the proof of (3.3.5) or otherwise that A is of the form

$$\begin{pmatrix} 220000 \\ 220000 \\ 002200 \\ 002200 \\ 000022 \\ 000022 \end{pmatrix}, \quad \begin{pmatrix} 310000 \\ 130000 \\ 001111 \\ 001111 \\ 001111 \\ 001111 \end{pmatrix} \quad \text{or} \quad \begin{pmatrix} 200011 \\ 020011 \\ 002011 \\ 000211 \\ 111100 \\ 111100 \end{pmatrix}$$

if $\mu = (2,2)$, $(3,1)$ or $(2,1,1)$, respectively. From this it is straightforward to calculate the valencies between the N_s and N_b (but not necessarily the decompositions of the valencies into sums of orbit lengths). Direct calculation with the data in (2.14.5), (2.14.2 (iii)) and (2.14.3 (v)) shows that Q_s has 105 orbits on $\mathcal{H}_s \setminus \Sigma$. Since $90 = 45 \cdot 2$, $240 = 15 \cdot 2^4$ and $1440 = 45 \cdot 2^5$, by (3.7.1) we conclude that Q_s has 45, 15 and 45 orbits on N_s for $N = 90$, 240 and 1440, respectively. Since $K_s/Q_s \trianglelefteq G_s/Q_s$, the action of K_s/Q_s on the set of Q_s-orbits on N_s is either trivial or fixed-point free. In view of (2.12.7) we conclude that the action is trivial if and only if $N = 240$. Hence K_s has exactly 15 orbits on N_s for $N = 90$, 240 and 1440. By (3.7.1) and (2.5.3 (vii)) $L(\Sigma')$ is the stabilizer in G'_s/K'_s of an element a from $\mathrm{res}_{\mathcal{H}}(\Sigma')$. The matrices A given above now show that a is a point if $\Sigma' \in 90_s$ and a line otherwise, so (iii) follows. By (iii) G'_s acts transitively on the set of octads adjacent to Σ' and contained in N_b unless $N = 360$ in which case there are 2 orbits with lengths 1 and 6.

Recall that $B \in N_b$ for $N = 15$, 360 and 384 if $\nu(B,\Sigma) = (4^2 0^4)$; $(2^4 0^2)$ and $(3\,1^5)$, respectively, where $\nu(B,\Sigma)$ is the multiset as in (2.14.1). For a trio $T = \{B_1, B_2, B_3\}$ clearly the set $\lambda(T) = \{\nu(B_i,\Sigma) \mid 1 \le i \le 3\}$ is an invariant of the G_s-orbit containing T. Certainly every trio is refined by some sextet. Thus, considering for each of the 3 matrices A given above all possible partitions of the set of columns into 3 pairs and summing up the pairs of columns, we obtain the following exhaustive list of possibilities for $\lambda(T)$:

$$\lambda_1 = \{(4^2 0^4), (4^2 0^4), (4^2 0^4)\}, \quad \lambda_2 = \{(4^2 0^4), (2^4 0^2), (2^4 0^2)\},$$

$$\lambda_3 = \{(2^4 0^2), (2^4 0^2), (2^4 0^2)\}, \quad \lambda_4 = \{(2^4 0^2), (3\,1^5), (3\,1^5)\}.$$

For $1 \le i \le 4$ let Θ_i denote the set of trios T such that $\lambda(T) = \lambda_i$. Within the analysis of the partitions of columns of the matrices A we also obtain the numbers of trios in Θ_i adjacent to a sextet Σ' depending on i and on the G_s-orbit containing Σ'. In view of (iii) these numbers (as on $D_s(Mat_{24})$ below) show that for every i the subgroup G'_s acts transitively on the set of trios in Θ_i adjacent to Σ'. This implies that for every $1 \le i \le 4$ the set Θ_i is a G_s-orbit. In addition the number of octads in N_b adjacent to a given $T \in \Theta_i$ is readily seen from the shape of λ_i. Let us determine the sizes of the Θ_i. Since $\nu(B,\Sigma) = (4^2 0^4)$ if and only if B (as a vertex of Γ) is contained in the quad Σ, we conclude that Θ_1 consists of the trios (the triangles in Γ) contained in Σ while Θ_2 consists of the trios intersecting Σ in a single vertex. This gives $|\Theta_1| = 15$, $|\Theta_2| = 180$. All the 15 trios adjacent to an octad from 384_b are contained in Θ_4 while every trio from Θ_4 is adjacent to 2 octads from 384_b. Hence $|\Theta_4| = 2880$ and since the total number of trios is 3795, we have $|\Theta_3| = 720$ and (iv) follows. To complete the proof of (v) it remains to show that if $T \in \Theta_3 = 720_t$ then the stabilizer of T in G_s permutes transitively the octads in 360_b adjacent to T and we suggest this as an exercise. □

Lemma 3.7.4 *The following assertions hold:*

(i) *G_t has three orbits on $\mathcal{H}_b$ with lengths* 3, 84 *and* 672,

(ii) *G_t has four orbits on $\mathcal{H}_s$ with lengths* 7, 84, 336 *and* 1344,

(iii) *$L(\Sigma) := (G_t \cap G_s)K_s/K_s$ is the stabilizer in G_s/K_s of a point from $\mathrm{res}_{\mathcal{H}}(\Sigma)$ if $\Sigma \in 7_t \cup 336_t$ and $L(\Sigma)$ is the stabilizer of a line from this residue if $\Sigma \in 84_t \cup 1344_t$,*

(iv) *G_t has four orbits on $\mathcal{H}_t \setminus T$ with lengths* 42, 56, 1008 *and* 2688,

(v) *the diagram $D_t(Mat_{24})$ is as given below.*

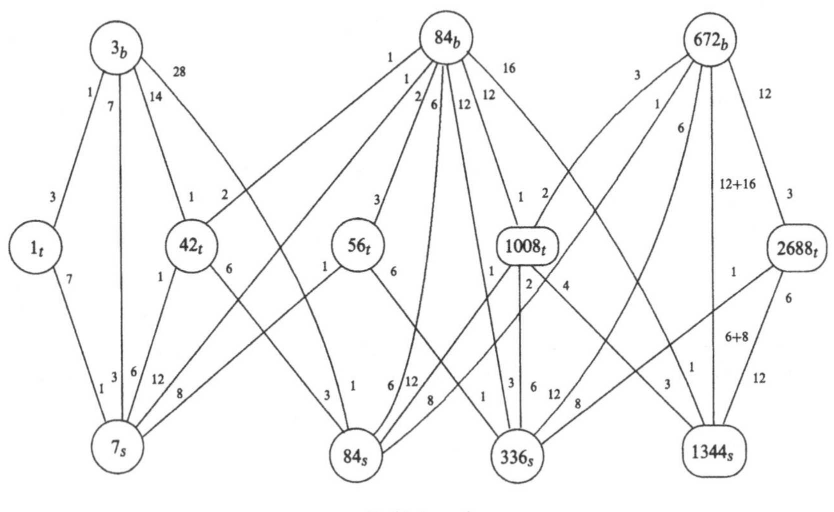

$D_t(Mat_{24})$

Proof. (i) is dual to (3.7.2 (ii)) while (ii) is dual to (3.7.3 (iv)). Notice that $B \in N_b$ for $N = 3$, 84 and 672 if and only if the distance between T and B is 0, 1 and 2, respectively. Similarly, $\Sigma \in 7_s$ if Σ contains T, $\Sigma \in 84_s$ if Σ intersects T in a vertex, $\Sigma \in 336_s$ if every octad in T is at distance 1 from Σ and $\Sigma \in 1344_s$ if exactly 1 octad in T is at distance 1 from Σ. This shows that a sextet from 7_s or 84_s is adjacent, respectively, to 3 or 1 octads in 3_b while a sextet from 336_s or 1344_s is adjacent, respectively, to 3 or 1 octads from 84_s. To prove (iii) consider the action of Q_s on $\mathscr{H}_t$ (compare $D_s(Mat_{24})$). By (2.14.5), (2.14.2 (iii)) and (2.14.3 (v)) Q_s has 150 orbits on $\mathscr{H}_t$. Since $15 = 15 \cdot 2^0$, $180 = 45 \cdot 2^2$, $720 = 45 \cdot 2^4$ and $2880 = 45 \cdot 2^6$, by (3.7.1) Q_s fixes 15_t elementwise and has 45 orbits on N_t for $N = 180$, 720 and 2880. In view of (2.12.6) we conclude that K_s fixes 15_t elementwise and acts fixed-point freely on the set of orbits of Q_s on N_t for $N = 180$, 720 and 2880. Hence K_s has 15 orbits on each of the N_t. By (2.5.3 (vii)) and the above established partial information on the valencies between the N_s and N_b on $D_t(Mat_{24})$ we obtain (iii) and easily calculate the remaining valencies between the N_s and N_b.

Consider the action of G_t on $\mathscr{H}_t \setminus T$. For $T' = \{B'_1, B'_2, B'_3\} \neq T$ put $\omega(T, T') = e^m$ where $e = d(T, T')$ (the distance in the octad graph) and m is the number of pairs (i, j) such that $d(B_i, B'_j) = e$. By (3.2.1 (iii), (iv)), for every $1 \leq i \leq 3$ there is a unique octad in T' nearest to B_i and every 4-cycle in Γ is contained in a quad. In view of these observations it is easy to see that $\omega(T, T')$ is one of the following:

$$\omega_1 = 0^1; \quad \omega_2 = 1^3; \quad \omega_3 = 1^1; \quad \omega_4 = 2^3.$$

Let Ω_i be the set of $T' \in \mathcal{H}_t \setminus T$ such that $\omega(T, T') = \omega_i$ for $1 \le i \le 4$. Since G_b acts doubly transitively on the set of trios containing B, Ω_1 is a G_t-orbit of length 42. Since every 4-cycle in Γ is in a unique quad, $\omega(T, T') = \omega_2$ if and only if T and T' are disjoint but contained in a common quad. Since G_s permutes transitively the pairs of disjoint trios in Σ, Ω_2 is a G_t-orbit of length 56. Let $\Sigma \in 1344_s$, so that exactly 1 octad from T is at distance 1 from Σ. Then 3 trios from Σ are in Ω_3 and 12 are in Ω_4. On the other hand by (iii) $G_s \cap G_t$ acting on the set of trios contained in Σ has 2 orbits with lengths 3 and 12. Hence Ω_3 and Ω_4 are G_t-orbits. It is easy to calculate that $|\Omega_3| = 1008$, so that $|\Omega_4| = 2688$ and (iv) follows. Now the rest of $D_t(Mat_{24})$ is straightforward except possibly for the transitivity of the stabilizer in G_t of $T' \in 2688_t$ on the octads contained in T' and we suggest this as an exercise. □

3.8 More on Golay code and Todd modules

In this section we analyse the structure of 11-dimensional Golay code $\mathcal{C}_{11}$ and Todd $\bar{\mathcal{C}}_{11}$ modules as $GF(2)$-modules for G_b, G_t and G_s. In order to simplify the notation, we put $X = \mathcal{C}_{11}$, $Y = \bar{\mathcal{C}}_{11}$.

Let $x = b$, t or s. Since X and Y are $GF(2)$-modules, Q_x acts trivially on each irreducible composition factor of G_x in X or Y. In particular every minimal G_x-submodule in X or Y is contained in $C_X(Q_x)$ or $C_Y(Q_x)$, respectively. In addition, since X and Y are dual to each other, $C_X(Q_x)$ is dual to $Y/[Y, Q_x]$ and $C_Y(Q_x)$ is dual to $X/[X, Q_x]$. So it is natural to calculate first the centralizers of the Q_x in X and Y. We start with the following.

Lemma 3.8.1 *If $x = b$, t or s then Q_x does not stabilize pairs of complementary dodecads.*

Proof. Let $\{D, D'\}$ be a pair of complementary dodecads stabilized by Q_x. Then a subgroup of index at most 2 in Q_x stabilizes D. It is easy to deduce from (2.10.1), (2.10.2) and (2.10.3) that for every subgroup of index 2 in Q_x a union of size 12 of its orbits always contains an octad, a contradiction with (2.11.2). □

Lemma 3.8.2 *For $x = b$, t and s put $A_x = C_X(Q_x)$ and $B_x = C_Y(Q_x)$. Then G_x acts irreducibly on A_x and B_x and the following hold:*

(i) $|A_b| = 2$, $B_b \cong \bigwedge^2 Q_b$,

(ii) *in terms of* (2.10.3 (viii)) $A_t \cong D_1$ *and* B_t *is the dual of* D_2,

(iii) A_s *is the natural symplectic module of* $G_s/K_s \cong Sp_4(2)$ *and* $|B_s| = 2$.

Proof. The dimensions of the A_x and B_x are straightforward from (2.15.1), (3.8.1) and the diagrams $D_x(Mat_{24})$. The sextets in 35_s are indexed by the 2-dimensional subspaces in Q_b. Applying (3.3.8) and (2.4.6) we obtain (i). The non-zero vectors in A_t are indexed by the octads in T and their sum is zero, hence $A_t \cong D_1$. If $B = B_1$ then $D_2 \cong Q_b \cap Q_t$. The non-zero vectors from B_t are indexed by 2-dimensional subspaces in D_2 and (ii) follows. The non-zero vectors of A_s are indexed by the octads in the quad Σ. The sum of the vectors corresponding to octads in a triangle is zero. Hence A_s supports a natural representation of $\mathrm{res}_{\mathscr{H}}(\Sigma) \cong \mathscr{G}(Sp_4(2))$ and (iii) follows by (1.11.2). □

By (3.8.2), if $Z = X$ or Y then $C_Z(Q_x)$ and $[Z, Q_x]$ are, respectively, the only minimal and the only maximal proper G_x-submodules, in particular $C_Z(Q_x) \leq [Z, Q_x]$. In addition $[X, Q_x]/C_X(Q_x)$ and $[Y, Q_x]/C_Y(Q_x)$ are dual to each other.

Lemma 3.8.3 *For $x = b$, t and s put $C_x = [X, Q_x]/C_X(Q_x)$. Then G_x acts irreducibly on C_x and the following hold:*

(i) *C_b is dual to Q_b,*
(ii) *C_t is dual to Q_t,*
(iii) *C_s is isomorphic to Q_s.*

Proof. By (3.8.2) the dimension of C_x is 4, 6 and 6 for $x = b$, t and s, respectively. It is easy to see that C_b is generated by the images of octads from 30_b and these images are indexed by the trios containing B. Since the setwise stabilizer in Q_b of such a trio is a hyperplane in Q_b, (i) follows. For $x = t$ and s let U_x be a Sylow 3-subgroup in $O_{2,3}(G_x)$. Comparing the dimensions of the centralizers of U_x in $2^{\mathscr{P}}$, $C_X(Q_x)$ and $C_Y(Q_x)$, we conclude that C_x, as a module for $N_x := N_{G_x}(U_x)$, is isomorphic to $[U_x, X]$. We claim that the action of N on C_x is faithful. In fact for $x = t$ it is immediate from the dimension of the centralizer in X of an element of order 7 and for $x = s$ it follows from the fact that N_x does not split over U_x. Hence C_x is an irreducible 3-dimensional $GF(4)$-module for N_x. It is well known and easy to check that every subgroup in $P\Gamma L_3(4)$ isomorphic to Sym_6 or $L_3(2) \times 2$ stabilizes a hyperoval or a Fano subplane in the corresponding projective plane of order 4. This shows that C_x is isomorphic to Q_x or to its dual. Since a hyperplane in C_b is contained in C_t, (i) implies (ii). From the diagram $D_s(Mat_{24})$ we observe that $[Y, Q_s]/C_Y(Q_s)$ (which is dual to C_x) is generated by the images of sextets from 90_s and these images are indexed by the $\star$-closed triangles in the sextet graph. Since the elementwise stabilizer in Q_s of such a triangle is a hyperplane in Q_s, (iii) follows. □

Thus we have the following main result of the section.

Lemma 3.8.4 *For $Z = \mathscr{C}_{11}$ or $\bar{\mathscr{C}}_{11}$ and $x = b$, t or s*

$$1 < C_Z(Q_x) < [Z, Q_x] < Z$$

is the only composition series of Z as a module for G_x; the composition factors are as in (3.8.2) *and* (3.8.3). □

In terms of (3.8.4) let X_s be a Sylow 3-subgroup in G_s. Then clearly

$$Z = C_Z(X_s) \oplus [Z, X_s]$$

and using (3.8.2) and (3.8.3) one can easily show the following.

Lemma 3.8.5 *$C_Z(X_s)$ is an reducible indecomposable 5-dimensional module for $G_s/O_{2,3}(G_s) \cong Sp_4(2)$. Moreover $C_{\mathscr{C}_{11}}(X_s)$ contains the natural 4-dimensional symplectic submodule while $C_{\bar{\mathscr{C}}_{11}}(X_s)$ contains a 1-dimensional submodule.* □

3.9 Diagrams for $\mathscr{H}(Mat_{22})$

An element of type 2 in $\mathscr{G}(Mat_{22})$ is a pair $\{B_1, B_2\}$ of disjoint octads, both disjoint from Y_2. Such an element determines a unique octad

$$B = \mathscr{P} \setminus (B_1 \cup B_2)$$

containing Y_2. In its turn B determines a $\mathscr{G}(Sp_4(2))$-subgeometry containing $\{B_1, B_2\}$. Thus an element of type 2 is contained in a unique such subgeometry. The maximal parabolic geometry $\mathscr{H}(Mat_{22})$ introduced in [RSm80] can be defined as follows. The elements of type 1 and 3 together with the incidence between them are as in $\mathscr{G}(Mat_{22})$; the elements of type 2 are the $\mathscr{G}(Sp_4(2))$-subgeometries with an element of type 1 or 3 being incident to an element of type 2 if in $\mathscr{G}(Mat_{22})$ it is incident to an element of type 2 in the subgeometry. Notice that an element of type 1 is incident to an element of type 2 in a subgeometry if and only if it is contained in the subgeometry. In order to distinguish between $\mathscr{G}(Mat_{22})$ and $\mathscr{H}(Mat_{22})$, the elements of type 1, 2 and 3 in the latter geometry will be called *pairs*, *hexads* and *octets*, respectively. The diagram of $\mathscr{H}(Mat_{22})$ is the following:

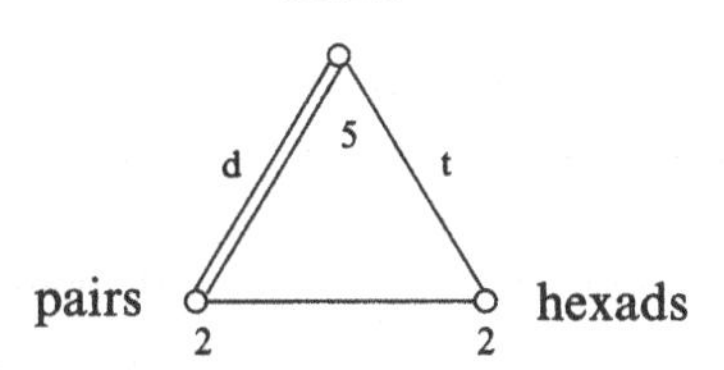

In this section we calculate the suborbit diagrams of $\mathcal{H}$ as a 3-partite graph with the partition

$$\mathcal{H} = \mathcal{H}_p \cup \mathcal{H}_h \cup \mathcal{H}_o$$

where $\mathcal{H}_p$, $\mathcal{H}_h$ and $\mathcal{H}_o$ are the pairs, hexads and octets, respectively. The diagrams are with respect to the group $F \cong \mathrm{Aut}\, Mat_{22}$ acting naturally on $\mathcal{H}$. If $\{F_p, F_h, F_o\}$ is the amalgam of maximal parabolics associated with the action, then

$$F_p \cong 2^5 : Sym_5, \quad F_h \cong 2^4 : Sym_6, \quad F_o \cong 2 \times 2^3 : L_3(2).$$

A typical pair, hexad or octet will be denoted by P, H or O, respectively. For $x = p$, h or o put $Q_x = O_2(F_x)$.

To calculate the diagrams it is helpful to know the conjugacy classes of involutions in F and their distributions inside the Q_x.

Lemma 3.9.1 *The group $F \cong \mathrm{Aut}\, Mat_{22}$ has 3 classes of involutions: 2a, 2b and 2c; an involution from F is contained in $F^\flat$ if and only if it is a 2a-involution. Furthermore,*

(i) *a 2a-involution s can be chosen so that $s \in Q_h$ and $C_F(s) = C_{F_h}(s) \cong 2^4 : (Sym_4 \times 2)$, s fixes 23 pairs, 13 hexads and 26 octets,*
(ii) *a 2b-involution t can be chosen so that $t \in Q_o$ and $C_F(t) = F_o$, t fixes 35 pairs, 21 hexads and 50 octets,*
(iii) *a 2c-involution u can be chosen so that $u \in Q_p$ and $C_F(u) = C_{F_p}(u) \cong 2^5 : Frob_5^4$, u fixes 11 pairs, 5 hexads and 10 octets.*

Proof. It is clear that the classification of involutions in F (resp. in $F^\flat$) is equivalent to classification of the orbits of $C_G(\tau)$ on the set of 2-element subsets of $\mathcal{P}$ stabilized (resp. fixed) by τ for various involutions τ in $G \cong Mat_{24}$. In view of this observation the classes of involutions in F and the corresponding centralizers are immediate from (2.14.2) and (2.14.3). The number of pairs stabilized by an involution $v \in F$ follows directly from the cyclic shape of v on $\mathcal{P} \setminus Y_2$. Let τ be a 2a-involution in G and let B be the octad formed by the elements of $\mathcal{P}$ fixed by τ. Then the set of octads stabilized by τ consists of B, 14 octads disjoint from B which are orbits of the hyperplanes in Q_b containing τ and 56 octads B' such that $|B \cap B'| = 4$ and B' is refined by a sextet containing an orbit on $\mathcal{P} \setminus B$ of a subgroup of order 4 on Q_b containing τ (compare the proof of (2.14.2)). Using this description it is straightforward to calculate the numbers of hexads and octets stabilized by the involutions s and t. The octads stabilized by u (which is a $2b$-involution in G) are exactly those refined by the sextet which contains the tetrad $Y_2 \cup P$ where P is the pair stabilized by F_p. Since 5 of these octads contain Y_2 and 10 do not, (iii) follows. □

Lemma 3.9.2

(i) *Q_h is 2a-pure,*
(ii) *Q_o contains 7 and 8 2a- and 2b-involutions, respectively,*
(iii) *Q_p contains* 15, 10 *and* 6 *2a-, 2b- and 2c-involutions, respectively.*

Proof. Since Q_h is the natural symplectic module for $F_h/Q_h \cong Sp_4(2)$, (i) follows. The structure of F_o implies that F_o/Q_o acting on the set of involutions in Q_o by conjugation has three orbits with lengths 1, 7 and 7. One of the orbits of length 7 is formed by the involutions contained in $Q_o \cap F^\flat$. By (3.9.1) a $2c$-involution never commutes with an element of order 3 and hence (ii) follows. One can deduce from (2.7.14) or otherwise that the orbits of F_p/Q_p on the set of involutions in Q_p are of lengths 15, 10 and 6 (in particular Q_p is indecomposable). The former of the orbits consists of the involutions contained in $Q_p \cap F^\flat$. Since an involution from the orbit of length 10 is centralized by a 3-element, (iii) follows. □

In Chapter 7 we will make use of the following result.

Lemma 3.9.3 *For $x = p$, h and o put $F_x^\flat = F^\flat \cap F_x$. Then the following assertions hold:*

(i) *$O_2(F_p)$ acting on $\mathcal{P} \setminus Y_2$ has one orbit of length 2 (the pair) and five orbits of length 4, $O_2(F_p^\flat)$ fixes every element in the pair and acts transitively on every $O_2(F_p)$-orbit of length 4,*
(ii) *$O_2(F_h)$ and $O_2(F_h^\flat)$ has the same orbits on $\mathcal{P} \setminus Y_2$, namely, one orbit of length 16 (the complement of the hexad) and three orbits of length 2,*
(iii) *$O_2(F_o)$ and $O_2(F_o^\flat)$ have the same orbits on $\mathcal{P} \setminus Y_2$, namely, one orbit of length 8 (the octet) and seven orbits of length 2,*
(iv) *$F_x = N_F(O_2(F_x))$.*

Proof. In view of the definition of $\mathscr{G}(Mat_{22})$ in terms of subsets of $\mathscr{P} \setminus Y_2$ the assertions (i)–(iii) follow directly from (3.9.1) and (3.9.2). These assertions immediately imply (iv). □

For $x = p$, h and o by $D_x(Mat_{22})$ we denote the suborbit diagram corresponding to the action of F on $\mathscr{H}(Mat_{22})$ with the base point taken from $\mathscr{H}_x$.

Lemma 3.9.4 *The following assertions hold:*

(i) *F_h has two orbits on $\mathscr{H}_h \setminus H$ with lengths* 16 *and* 60*; if $H' \in N_h$ then $|(H \cap H') \setminus Y_2| = 0$ and* 2 *for $N = 16$ and* 60*, respectively,*
(ii) *F_h has three orbits on $\mathscr{H}_o$ with lengths* 30, 60 *and* 240*,*
(iii) *F_h has three orbits on $\mathscr{H}_p$ with lengths* 15, 96 *and* 120*,*
(iv) *$(F_h \cap F_p)Q_p/Q_p$ is isomorphic to Sym_4, Alt_5 and $Sym_3 \times Sym_2$ for P taken from 15_p, 96_p and 120_p, respectively,*
(v) *the diagram $D_h(Mat_{22})$ is as given below.*

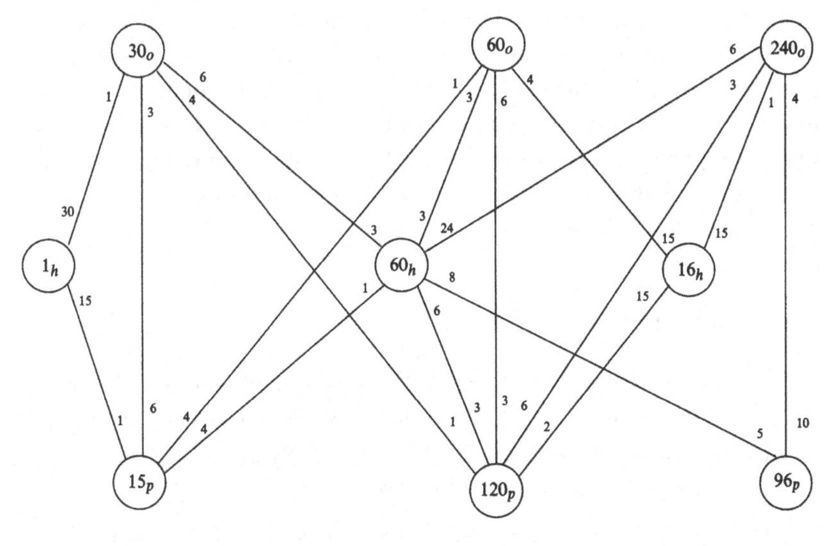

$D_h(Mat_{22})$

Proof. Suppose that $Y_2 \in B$ so that $H = B$ is a hexad and consider B as a vertex of the octad graph Γ. Then F_h is the stabilizer of Y_2 in G_b, in particular it contains Q_b. By (2.10.4) and its proof Q_b has 15 orbits of length 2 on $\Gamma(B)$ and these orbits are indexed by the hyperplanes in Q_b. If $B' \in \Gamma_i(B)$ for $i = 2$ or 3, then the Q_b-orbit of B' (of length 4 or 16) is uniquely determined by $B' \cap B$ (of size 4 or 2). Hence F_h acts transitively on the set of 30 octets in $\Gamma(B)$ and the orbit under F_h of $B' \in \Gamma_i(B)$ for $i = 2$ or 3 is uniquely determined by i and $|B' \cap Y_2|$. This means that there are 60 hexads and 60 octets in $\Gamma_2(B)$, 16 hexads and 240 octets in $\Gamma_3(B)$ and F_h acts transitively on each of these 4 sets. From this information we easily deduce (i), (ii) and the valencies between the N_h and M_o on $D_h(Mat_{22})$.

Since the actions of F_h on $B \setminus Y_2$ and $\mathscr{P} \setminus B$ are doubly transitive while the action of Q_b on $\mathscr{P} \setminus B$ is transitive, we conclude that the orbit of a pair under F_h is uniquely determined by the size of its intersection with H which implies (iii). Dualizing (iii) we obtain that F_p has 3 orbits on $\mathscr{H}_h$ with lengths 5, 32 and 40. By (3.9.1) and (3.9.2) Q_p has 17 orbits on $\mathscr{H}_h$ and it is easy to deduce from (3.7.1) that Q_p fixes 5_h elementwise, has 2 orbits on 32_h and 10 orbits on 40_h. Since $Sym_5 = F_p/Q_p$ has a single class of subgroups of index 5 (isomorphic to Sym_4) and a unique subgroup of index 2 (isomorphic to Alt_5), to complete (iv) we have to show that for $P \in 120_p$ $(F_h \cap F_p)Q_p/Q_p$ is isomorphic to $Sym_3 \times Sym_2$ rather than to Alt_4. If H' is a hexad from 16_h and D is the stabilizer of H' in F_h, then D is a complement to Q_b in F_h and hence it permutes transitively the 15 pairs in $H' \setminus Y_2$. This means that a pair from 120_p is incident in $\mathscr{H}(Mat_{22})$ to $2 = 16 \cdot 15/120$ hexads from 16_h and (iv) follows from the obvious fact that Alt_4 does not have orbits of length 2 in the natural action of Sym_5 of degree 5. Now using the divisibility it is straightforward to reconstruct the remainder of $D_h(Mat_{22})$. □

Lemma 3.9.5 *The following assertions hold:*

(i) *F_o has three orbits on $\mathscr{H}_h$ with lengths* 7, 14 *and* 56 *consisting of the hexads intersecting O in no, four and two elements, respectively,*

(ii) *for a hexad H we have $(F_o \cap F_h)Q_h/Q_h \cong Sym_4$ and $F_o \cap F_h$ stabilizes a pair in H if and only if $H \in 14_h \cup 56_h$,*

(iii) *F_o has four orbits on $\mathscr{H}_o \setminus O$ with lengths* 7, 42, 112 *and* 168,

(iv) *if $O' \in N_o$ then $F_o[O']Q_o/Q_o$ is isomorphic to Sym_4, D_8, Alt_4 and D_8 for $N = 7, 42, 112$ and 168, respectively,*

(v) *if $O' \in 7_o \cup 112_o$ then $F_o[O']$ stabilizes a hexad incident to O',*

(vi) *F_o has four orbits on $\mathscr{H}_p$ with lengths* 7, 28, 84 *and* 112,

(vii) *if $P \in N_p$ then $(F_o \cap F_p)Q_p/Q_p$ is isomorphic to $Sym_3 \times Sym_2$, $Sym_3 \times Sym_2$, D_8 and Sym_4 for $N = 7, 28, 84$ and 112, respectively,*

(viii) *the diagram $D_o(Mat_{22})$ is as given below.*

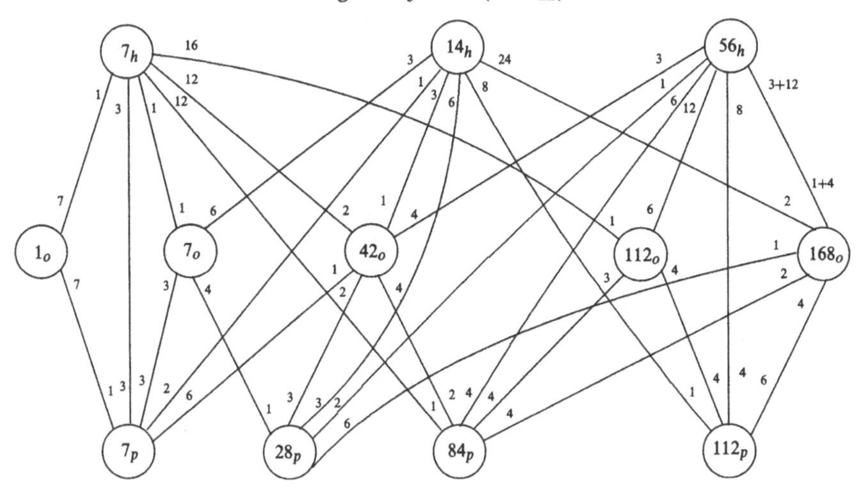

$D_o(Mat_{22})$

Proof. (i) is dual to (3.9.4 (ii)). By (3.9.1), (3.9.2) and (3.7.1) Q_h has 15 orbits on N_o for $N = 30$, 60 and 240 (the diagram $D_h(Mat_{22})$) hence $(F_o \cap F_h)Q_h/Q_h \cong Sym_4$ for every hexad H. If $H \in 56_h$ then $F_o \cap F_h$ stabilizes the pair $H \cap O$ and if $H \in 14_h$ then $F_o \cap F_h$ stabilizes the pair $H \setminus ((H \cap O) \cup Y_2)$. On the other hand if $H \in 7_h$ then $F_o \cap F_h$ permutes transitively the 3 pairs incident to both O and H, so we have (ii). Part (iii) follows from the distance diagram of $\Delta = \Delta(\mathscr{G}(Mat_{22}))$ given before (3.4.5). The proof of (vi) is similar to that of (ii). Since on the distance diagram of Δ we have $c_1 = a_4 = 1$, if $O' \in 7_o \cup 112_o$ then $F_o[O']$ stabilizes an octet O'' adjacent to O' in the derived graph. Hence it also stabilizes the hexad $\mathscr{P} \setminus (O' \cup O'')$ and (v) follows. The group F_o acts triply transitively on the elements in O; $F_o/Q_o \cong L_3(2)$ permutes doubly transitively the 7 orbits of $Z(F_o)$ on $\mathscr{P} \setminus (O \cup Y_2)$ and hence Q_o stabilizes each of these 7 orbits as a whole. This implies that F_o acts transitively on the pairs in O; on the pairs intersecting O in one element; on the orbits of $Z(F_o)$ on $\mathscr{P} \setminus (O \cup Y_2)$ and on the remaining pairs in the latter set, so that (vi) follows. The proof of (vii) is similar to that of (ii). Now with this information in hand it is straightforward to reconstruct $D_o(Mat_{22})$. □

Every statement in the next lemma either is dual to a statement in (3.9.4) and (3.9.5) or can be deduced by similar methods.

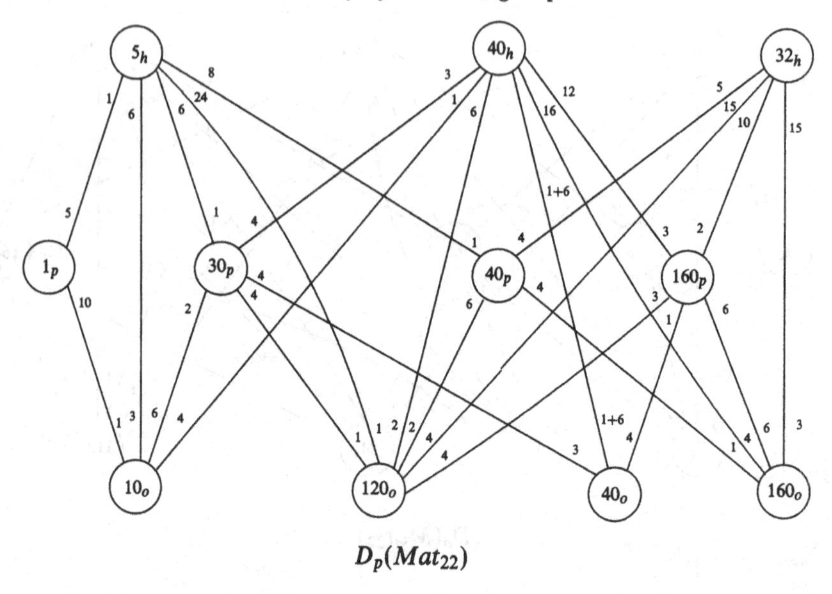

$D_p(Mat_{22})$

Lemma 3.9.6 *The following assertions hold:*

(i) F_p *has three orbits on* $\mathscr{H}_h$ *with lengths* 5, 32 *and* 40,

(ii) *if* $H \in N_p$ *then* $(F_p \cap F_h)Q_h/Q_h$ *is isomorphic to* Sym_4, Alt_5 *and* Sym_4 *for* $N = 5$, 32 *and* 40, *respectively,*

(iii) F_p *has three orbits on* $\mathscr{H}_p$ *with lengths* 30, 40 *and* 160,

(iv) F_p *has four orbits on* $\mathscr{H}_o$ *with lengths* 10, 40, 120 *and* 160,

(v) *if* $O \in N_o$ *then* $(F_p \cap F_o)Q_o/Q_o$ *is isomorphic to* Sym_4, Sym_4, D_8 *and* Alt_4 *for* $N = 10$, 40, 120 *and* 160, *respectively,*

(vi) *the diagram* $D_p(Mat_{22})$ *is as given above.* □

3.10 Actions on the sextets

In this section we describe the orbits of a few subgroups of G on $\mathscr{H}_s$, that is on the set of sextets.

Lemma 3.10.1 *Let* Y *be a 3-element subset of* $\mathscr{P}$. *Then* $G[Y] \cong P\Gamma L_3(4)$ *acting on* $\mathscr{H}_s$ *has three orbits with lengths* 21, 630 *and* 1120 *consisting of the sextets* Σ *such that* Y *intersects* i *tetrads in* Σ *for* $i = 1$, 2 *and* 3, *respectively.*

Proof. It follows from (2.10.2) that the orbit under G_s of a 3-element subset X is uniquely determined by the multiset $\{|X \cap S_i| \mid 1 \leq i \leq 6\}$.

Now the result follows from the obvious duality and straightforward calculations. □

Lemma 3.10.2 *Let* $\{D_1, D_2\}$ *be a complementary pair of dodecads and* $R \cong Mat_{12}.2$ *be the setwise stabilizer in* G *of this pair. Then* R *acting on* $\mathscr{H}_s$ *has three orbits with lengths* 396, 495 *and* 880.

Proof. For a sextet $\Sigma = \{S_1, S_2, ..., S_6\}$ put $m_{ji} = |D_j \cap S_i|$ for $j = 1, 2$, $1 \leq i \leq 6$, and suppose that m is the maximum of the m_{ji}. Then $m \in \{4, 3, 2\}$ and we are going to show that each value of m corresponds to a single orbit Ω_m of R on $\mathscr{H}_s$. Notice that because Golay subsets always have even intersection, all the values m_{ji} have the same parity. Suppose first that $m = 4$. By (2.11.2) D_j does not contain octads and hence for $j = 1$ and 2 at most one of the values m_{ji} can be equal to 4 and in order to meet the total balance exactly one of the m_{ji} must be 4. Since the action on D_1 of its stabilizer in R is 5-fold transitive, Ω_4 is an R-orbit and $|\Omega_4| = \binom{12}{4} = 495$.

If $m = 3$ then for $j = 1$ and 2 exactly three of the m_{ji} are equal to 3 and three to 1. By (2.11.9 (iii)) the stabilizer in R of an element from D_2 induces on D_1 a 3-fold transitive action of Mat_{11} and hence Ω_3 is an R-orbit and $|\Omega_3| = \binom{12}{3} \cdot 12 \cdot \frac{1}{3} = 880$.

Finally, if $m = 2$ then all the m_{ji} are equal to 2. For $1 \leq i \leq 6$ put $A = S_i \cap D_1$ and $B = S_i \cap D_2$. By (2.11.3) A determines a partition $\{T_1, T_2\}$ of D such that both $A \cup T_1$ and $A \cup T_2$ are octads. If $B \subset T_k$ then $T_k \setminus B$ is a tetrad in Σ, which is impossible since $m = 2$. Hence B intersects both T_1 and T_2. Since the elementwise stabilizer of A in R induces two inequivalent actions of Sym_6 on T_1 and T_2, Ω_2 is an R-orbit and $|\Omega_2| = \binom{12}{2} \cdot 6 \cdot 6 \cdot \frac{1}{6} = 396$. □

Lemma 3.10.3 *The subgroup* $G_3 \cong 2^{1+6}_+ : L_3(2)$ *acting on* $\mathscr{H}_s$ *has six orbits with lengths* 7, 28, 56, 336, 448 *and* 896.

Proof. By (3.3.4) G_3 has index 15 in G_b, it contains Q_b and G_3/Q_b is the stabilizer of a point in the natural action of $G_b/Q_b \cong L_4(2)$ on the rank 3 projective geometry π of the proper subgroups in Q_b. Thus a G_3-orbit on $\mathscr{H}_s$ is a union of Q_b-orbits and is contained in a G_b-orbit. By (2.14.1) and (3.7.2) G_b acting on $\mathscr{H}_s$ has three orbits Θ_1, Θ_2 and Θ_3 with lengths 35, 810 and 896; Q_b-orbits on Θ_i have lengths 1, 8 and 16, respectively. Furthermore, if $\bar{\Theta}_i$ is the set of Q_b-orbits on Θ_i, then $G_b/Q_b \cong L_4(2)$ acts on $\bar{\Theta}_i$ as it acts on the set of lines in π, on the incident point–hyperplane pairs in π and on 3-element subsets of B for $i = 1, 2$ and 3, respectively. This and elementary geometric arguments show that G_3/Q_b has two orbits on $\bar{\Theta}_1$ with lengths 7 and 28, three orbits on $\bar{\Theta}_2$ with lengths 7, 42 and 56, and acts transitively on $\bar{\Theta}_3$. The latter follows from the 3-fold transitivity of $G_3/Q_b \cong 2^3 : L_3(2)$ on the octad B stabilized by G_b. □

4
Conway groups

The largest Conway sporadic simple group is the quotient over the centre of order 2 of the automorphism group of the Leech lattice (the unique even unimodular lattice of dimension 24 without roots). In Section 4.1 we recall some standard results concerning construction of lattices from binary codes. In Section 4.2 we discuss some symmetries of the lattices coming from the binary code construction. In Section 4.3 we follow [Con69] to prove the uniqueness of the Leech lattice Λ; the proof immediately gives the order of the automorphism group Co_0 of Λ. In Section 4.4 we introduce the standard coordinate system for the Leech lattice and describe explicitly the Leech vectors of length up to 8. In Section 4.5 we discuss the action of Co_0 on the Leech lattice modulo 2 (denoted by $\bar{\Lambda}$) as well as on the Leech vectors of small length. In this way we introduce the sporadic Conway groups Co_1, Co_2 and Co_3. In Sections 4.6 and 4.7 we study the action of Co_1 on the images in $\bar{\Lambda}$ of the Leech vectors of length 8 and calculate the suborbit diagram of the Leech graph which is the smallest orbital graph of this action. In Section 4.8 we study the structure of the centralizer in Co_1 of a central involution which we use in Section 4.9 to construct the tilde geometry $\mathcal{G}(Co_1)$ and the Petersen geometry $\mathcal{G}(Co_2)$. In order to establish the simple connectedness of these geometries in Section 4.12, in Sections 4.10 and 4.11 we study the affine Leech and the shortest vector graphs in terms of their suborbit diagrams. In Sections 4.13 and 4.14 we discuss some further geometries possessing natural descriptions in terms of the Leech lattice.

4.1 Lattices and codes

Let n be a positive integer, let $\mathbb{R}^n$ be an n-dimensional Euclidean vector space and for $x, y \in \mathbb{R}^n$ let (x, y) denote the inner product of x and

y. A *lattice* of dimension n is a subset $L \subset \mathbb{R}^n$ with the property that there exists a basis $\mathscr{B}$ of $\mathbb{R}^n$ such that L consists of all integral linear combinations of vectors from $\mathscr{B}$. In this case $\mathscr{B}$ is said to be a *basis* of L. A lattice $L \subset V \cong \mathbb{R}^n$ and a lattice $L' \subset V' \cong \mathbb{R}^n$ are isomorphic if there is an isomorphism $\varphi : V \to V'$ of Euclidean spaces which sends L onto L'.

Let L be a lattice of *dimension n*. The *dual lattice* L^* of L is defined as follows:

$$L^* = \{y \mid y \in \mathbb{R}^n, (y, x) \in \mathbb{Z} \text{ for every } x \in L\}.$$

A lattice L is *integral* if $L \subseteq L^*$, which means that the inner product of any two vectors from L is an integer; L is *even* if $(x, x) \in 2\mathbb{Z}$ for every $x \in L$. Since

$$(x, y) = \frac{1}{2}(x + y, x + y) - (x, x) - (y, y),$$

every even lattice is integral. Let $\mathscr{B}$ be a basis of an integral lattice L and let A denote the Gram matrix of $\mathscr{B}$. The absolute value of the determinant of A is independent of the choice of $\mathscr{B}$ and it is called the *discriminant* of L, written as disc L. It is known [Ebe94] that disc $L = |L^*/L|$. A lattice L is *unimodular* if $L^* = L$, in which case disc $L = 1$. We will write x^2 for the (squared) length of a vector x, that is for (x, x). If $x^2 = 2$ then x is called a *root*. We are mainly interested in even unimodular lattices. It is known [Ebe94] that the dimension of such a lattice is divisible by 8. A *Leech lattice* is an even unimodular lattice of dimension 24 which does not contain roots. We will see in due course that there exists a unique Leech lattice which has a certain remarkable group of automorphisms.

Let L and M be integral lattices of dimension n and suppose that $M \subseteq L$. Then clearly $L \subseteq M^*$ so that L corresponds to a subgroup in the finite abelian group M^*/M. We will discuss a family of lattices which contain specific sublattices and show that this family possesses a natural description in terms of binary codes.

Let $\mathscr{R}_n$ be a basis of $\mathbb{R}^n$ consisting of pairwise orthogonal roots. This means that $\mathscr{R}_n$ is of size n, $a^2 = 2$ for every $a \in \mathscr{R}_n$ and $(a, b) = 0$ for all $a, b \in \mathscr{R}_n$ with $a \neq b$. Then for $x \in \mathbb{R}^n$ we have

$$x = \sum_{a \in \mathscr{R}_n} \frac{1}{2}(x, a)a.$$

Let $\mathscr{L} = \mathscr{L}(\mathscr{R}_n)$ be the lattice having $\mathscr{R}_n$ as a basis. It is easy to see the following.

Lemma 4.1.1 *The lattice $\mathscr{L} = \mathscr{L}(\mathscr{R}_n)$ is n-dimensional and even with discriminant 2^n.* □

For $\varepsilon = 0$ or 1 put

$$\mathscr{L}_\varepsilon = \mathscr{L}_\varepsilon(\mathscr{R}_n) = \left\{ \sum_{a \in \mathscr{R}_n} m_a a \;\middle|\; m_a \in \mathbb{Z}, \sum_{a \in \mathscr{R}_n} m_a = \varepsilon \bmod 2 \right\}.$$

Then $\mathscr{L}_0$ is a sublattice of index 2 in $\mathscr{L}$ and $\mathscr{L}_1$ is the only proper coset of $\mathscr{L}_0$ in $\mathscr{L}$. We will study the lattices which contain $\mathscr{L}_0$. In order to understand these lattices it is helpful to have a description of the duals $\mathscr{L}^*$ and $\mathscr{L}_0^*$, which is rather straightforward and given in the lemma below.

Lemma 4.1.2 *The following assertions hold:*

(i) *$\mathscr{L}^*$ is a lattice with the basis $\{\frac{1}{2}a \mid a \in \mathscr{R}_n\}$;*
(ii) *$x, y \in \mathscr{L}^*$ are in the same coset of $\mathscr{L}$ if and only if for every $a \in \mathscr{R}_n$ we have $(x, a) = (y, a) \bmod 2$;*
(iii) *$x \in \mathscr{L}_0^* \setminus \mathscr{L}^*$ if and only if $(x, a) \in \mathbb{Z} + \frac{1}{2}$ for every $a \in \mathscr{R}_n$.* □

The above lemma possesses an easy reformulation in terms of coordinates of vectors in the basis $\mathscr{R}_n$. To wit, $x \in \mathscr{L}^*$ if and only if every coordinate of x is either an integer, or a half integer, $x, y \in \mathscr{L}^*$ are in the same coset of $\mathscr{L}$ if and only if x and y have the same set of non-integer coordinates (which are half integers), $x \in \mathscr{L}_0^* \setminus \mathscr{L}^*$ if and only if every coordinate of x is an odd integer divided by 4.

For a vector $x \in \mathscr{L}^*$ put

$$X(x) = \{a \mid a \in \mathscr{R}_n,\ (x, a) \in 2\mathbb{Z} + 1\},$$

so that $a \in X(x)$ if and only if the corresponding coordinate of x in the basis $\mathscr{R}_n$ is not integral. By (4.1.2 (ii)) we have $X(x) = X(y)$ if and only if x and y are in the same coset of $\mathscr{L}$ in $\mathscr{L}^*$. Since $|\mathscr{L}^*/\mathscr{L}| = \operatorname{disc} \mathscr{L} = 2^n = |2^{\mathscr{R}_n}|$, the mapping

$$\varphi : x \mapsto X(x)$$

induces a bijection of $\mathscr{L}^*/\mathscr{L}$ onto $2^{\mathscr{R}_n}$ and for $X \subseteq \mathscr{R}_n$ we have

$$\varphi^{-1}(X) = \mathscr{L} + e_X, \quad \text{where } e_X = \frac{1}{2}\sum_{a \in X} a.$$

Lemma 4.1.3 *Let $x, y \in \mathscr{L}^*$, then*

(i) $\varphi(x+y) = \varphi(x)\Delta\varphi(y) = X(x)\Delta X(y)$,
(ii) $(x, y) \in \mathbb{Z}$ *if and only if* $|X(x) \cap X(y)| \in 2\mathbb{Z}$, *i.e. if and only if* $X(x)$ *and* $X(y)$ *are orthogonal with respect to the parity form,*
(iii) *if* $X \subseteq \mathcal{R}_n$ *then* $e_X^2 = \frac{1}{2}|X|$, *in particular* e_X *is a root if and only if* $|X| = 4$,
(iv) *if* $x \in \mathcal{L}^*$ *then* $x^2 = e_{X(x)}^2 + 2l$ *for some* $l \geq 0$.

Proof. Let $z = x + y$ and $a \in \mathcal{R}_n$. Then $(z, a) = (x, a) + (y, a)$ and hence $a \in X(z)$ if and only if a is contained in exactly one of the sets $X(x)$ and $X(y)$. So $X(z) = X(x)\Delta X(y)$ and (i) follows. To prove (ii) put $x = e_{X(x)} + a$ and $y = e_{X(y)} + b$ for $a, b \in \mathcal{L}$. Then

$$(x, y) = (e_{X(x)}, e_{X(y)}) + (e_{X(x)}, b) + (e_{X(y)}, a) + (a, b).$$

Since $e_{X(x)}, e_{X(y)} \in \mathcal{L}^*$, the last three terms on the right hand side of the above equality are integers and, since $(e_{X(x)}, e_{X(y)}) = \frac{1}{2}|X(x) \cap X(y)|$, (ii) follows. The assertion (iii) comes by direct calculations. In (iv) if we put $x = e_{X(x)} + a$ for $a \in \mathcal{L}$, then

$$x^2 = e_{X(x)}^2 + 2(e_{X(x)}, a) + a^2,$$

where a^2 is even since $\mathcal{L}$ is even. In addition it is easy to see that for $X \subseteq \mathcal{R}_n$ and $a \in \mathcal{R}_n$

$$(e_X, a) = \min_{x \in \mathcal{L} + e_X} |(x, a)|$$

and hence (iv) follows. □

For $\mathcal{C} \subseteq 2^{\mathcal{R}_n}$ put

$$\mathcal{L}^A(\mathcal{C}) = \varphi^{-1}(\mathcal{C}) = \bigcup_{X \in \mathcal{C}} (\mathcal{L} + e_X)$$

(the A-construction in [CS88]).

Lemma 4.1.4 *The set* $\mathcal{L}^A(\mathcal{C})$ *is a lattice if and only if* $\mathcal{C}$ *is a (binary linear) code. If* $\mathcal{C}$ *is a code then*

(i) $\mathcal{L}^A(\mathcal{C})$ *is integral if and only if* $\mathcal{C}$ *is contained in its dual,*
(ii) $\mathcal{L}^A(\mathcal{C})$ *is even if and only if* $\mathcal{C}$ *is doubly even,*
(iii) $\mathcal{L}^A(\mathcal{C})$ *is unimodular if and only if* $\mathcal{C}$ *is self-dual.*

Proof. If $\mathcal{L}^A(\mathcal{C})$ is a lattice then it is closed under addition and by (4.1.3 (i)) this happens exactly when $\mathcal{C}$ is closed under taking symmetric differences, *i.e.* when $\mathcal{C}$ is a code. On the other hand if $\mathcal{C}$ is a code then $\mathcal{L}^A(\mathcal{C})$ is closed under addition, and it is always closed under negation.

Hence $\mathcal{L}^A(\mathcal{C})$ is a subgroup of finite index in $\mathcal{L}^*$ which means that it is a free abelian group of rank n. This implies that $\mathcal{L}^A(\mathcal{C})$ is a lattice. Now (i) holds by (4.1.3 (ii)) while (ii) holds by (4.1.3 (iii), (iv)). If $\mathcal{C}$ is a code then the index of $\mathcal{L}^A(\mathcal{C})$ in $\mathcal{L}^*$ equals the index of $\mathcal{C}$ in $2^{\mathcal{R}_n}$. In view of (4.1.1) $\mathcal{L}^A(\mathcal{C})$ is unimodular if and only if it is of index $2^{n/2}$ in $\mathcal{L}^*$ and by (2.1.1) the index of $\mathcal{C}$ in $2^{\mathcal{R}_n}$ is $2^{n/2}$ if and only if $\mathcal{C}$ is self-dual. □

The above lemma reduces the classification of even unimodular lattices containing $\mathcal{L}(\mathcal{R}_n)$ to that of doubly even self-dual codes based on $\mathcal{R}_n$. Notice that all these lattices contain roots, since so does $\mathcal{L}(\mathcal{R}_n)$. Let us turn to the lattices which contain $\mathcal{L}_0$ but do not contain $\mathcal{L}$ starting with those contained in $\mathcal{L}^*$. For $X \subseteq \mathcal{R}_n$ the coset $\mathcal{L} + e_X$ splits into two $\mathcal{L}_0$-cosets $\mathcal{L}_0 + e_X$ and $\mathcal{L}_1 + e_X$. A lattice which contains $\mathcal{L}_0$ but does not contain $\mathcal{L}$ may contain at most one of these two $\mathcal{L}_0$-cosets. Let $\mathcal{C} \subseteq 2^{\mathcal{R}_n}$ and let β be a $\{0,1\}$-valued function on $\mathcal{C}$. Put

$$\mathcal{L}^B(\mathcal{C},\beta) = \bigcup_{X\in\mathcal{C}} (\mathcal{L}_{\beta(X)} + e_X)$$

(B-construction in [CS88]).

Lemma 4.1.5 *The set $\mathcal{L}^B(\mathcal{C},\beta)$ is a lattice if and only if $\mathcal{C}$ is a code and for all $X, Y \in \mathcal{C}$ we have*

$$\beta(X\triangle Y) = (\beta(X) + \beta(Y) + |X \cap Y|) \bmod 2.$$

If $\mathcal{L}^B(\mathcal{C},\beta)$ is a lattice then

(i) *$\mathcal{L}^B(\mathcal{C},\beta)$ is integral if and only if the code $\mathcal{C}$ is contained in its dual, in which case β is a linear function on $\mathcal{C}$,*
(ii) *$\mathcal{L}^B(\mathcal{C},\beta)$ is even if and only if $\mathcal{C}$ is doubly even,*
(iii) *if $\mathcal{L}^B(\mathcal{C},\beta)$ is integral then* disc *$\mathcal{L}^B(\mathcal{C},\beta) \geq 4$ with the equality holding if and only if $\mathcal{C}$ is self-dual,*
(iv) *if $\mathcal{L}^B(\mathcal{C},\beta)$ is integral then we can change the signs of some of the vectors in $\mathcal{R}_n$ so that β becomes the zero function.*

Proof. Since

$$e_X + e_Y = e_{X\triangle Y} + \sum_{a\in X\cap Y} a$$

and the latter sum is contained in $\mathcal{L}_0$ if $|X \cap Y|$ is even and it is in $\mathcal{L}_1$ otherwise, we obtain the condition for $\mathcal{L}^B(\mathcal{C},\beta)$ to be a lattice. The proofs of (i) and (ii) are analogous to proofs of (i) and (ii) in (4.1.4). Since $\mathcal{L}_0$ is of index 2 in $\mathcal{L}$, its discriminant is 2^{n+2} and if $\mathcal{C} \subseteq \mathcal{C}^*$, then the order of $\mathscr{C}$ is at most $2^{n/2}$ and (iii) follows. Let us turn to (iv) and suppose that $\mathscr{L}^B(\mathscr{C},\beta)$ is integral which means that $\mathscr{C}$ is totally singular with respect to the parity form and β is linear. Put $\mathscr{C}_0 = \{X \mid X \in \mathscr{C}, \beta(X) = 0\}$. If $\mathscr{C}_0 = \mathscr{C}$ then we are done, otherwise there is $Y \subseteq \mathscr{R}_n$ such that $Y^\perp \cap \mathscr{C} = \mathscr{C}_0$. Now it is easy to see that the change of signs of vectors in Y transfers β to the zero function. □

By (4.1.5 (iv)), if $\mathscr{C} \subseteq \mathscr{C}^*$, then we can (and will) assume that the signs of vectors in $\mathscr{R}_n$ are chosen in such a way that β is the zero function and write $\mathscr{L}^B(\mathscr{C})$ instead of $\mathscr{L}^B(\mathscr{C},\beta)$, so that

$$\mathscr{L}^B(\mathscr{C}) = \bigcup_{X \in \mathscr{C}} (\mathscr{L}_0 + e_X).$$

Let M be an even unimodular lattice which contains $\mathscr{L}_0$ and does not contain $\mathscr{L}$. Since $|\mathscr{L}/\mathscr{L}_0| = |\mathscr{L}_0^*/\mathscr{L}^*| = 2$, we have $|M/(M \cap \mathscr{L}^*)| \leq 2$. On the other hand disc$(M \cap \mathscr{L}^*) \geq 4$ by (4.1.5 (iii)). Hence we must have $|M/(M \cap \mathscr{L}^*)| = 2$ and disc$(M \cap \mathscr{L}^*) = 4$. By (4.1.5 (ii), (iii)) and in view of the above notational convention the latter equality and the fact that M is even imply that $M \cap \mathscr{L}^* = \mathscr{L}^B(\mathscr{C})$, where $\mathscr{C}$ is a doubly even code (of length n). Since disc $\mathscr{L}^B(\mathscr{C}) = 4$, there are three proper cosets of $\mathscr{L}^B(\mathscr{C})$ in $\mathscr{L}^B(\mathscr{C})^*$. Since $\mathscr{L}^B(\mathscr{C}) \subseteq \mathscr{L}^*$, clearly $\mathscr{L} \subseteq \mathscr{L}^B(\mathscr{C})^*$. The vector $\frac{1}{2}e_{\mathscr{R}_n}$ (whose coordinates in the basis $\mathscr{R}_n$ are all equal to $\frac{1}{4}$) is contained in $\mathscr{L}_0^* \setminus \mathscr{L}^*$ and since $\mathscr{C}$ is doubly even, $(\frac{1}{2}e_{\mathscr{R}_n}, e_X) \in \mathbb{Z}$ for all $X \in \mathscr{C}$. This shows that the proper cosets of $\mathscr{L}^B(\mathscr{C})$ in $\mathscr{L}^B(\mathscr{C})^*$ are the ones containing

$$\mathscr{L}_1, \ \ \mathscr{L}_0 + \frac{1}{2}e_{\mathscr{R}_n}, \text{ and } \mathscr{L}_1 + \frac{1}{2}e_{\mathscr{R}_n}.$$

Since M is unimodular, it must contain one of these cosets and not $\mathscr{L}_1$, since in that case it would contain the whole of $\mathscr{L}$.

Lemma 4.1.6 *If* $x \in \mathscr{L}_\varepsilon + \frac{1}{2}e_{\mathscr{R}_n}$ *for* $\varepsilon = 0$ *or* 1, *then*

$$x^2 = \left(\frac{n}{8} + \varepsilon\right) \bmod 2.$$

Proof. It is straightforward to check that $(\frac{1}{2}e_{\mathscr{R}_n})^2 = \frac{n}{8}$, $(\frac{1}{2}e_{\mathscr{R}_n} - a)^2 = \frac{n}{8} + 1$ for $a \in \mathscr{R}_n$ and $(\frac{1}{2}e_{\mathscr{R}_n}, x) \in 2\mathbb{Z}$ for $x \in \mathscr{L}_0$. Since $\mathscr{L}_1 + \frac{1}{2}e_{\mathscr{R}_n} = \mathscr{L}_0 - a + \frac{1}{2}e_{\mathscr{R}_n}$, the result follows. □

For $\varepsilon = \frac{n}{8} \bmod 2$ put

$$\mathscr{L}^C(\mathscr{C}) = \mathscr{L}^B(\mathscr{C}) \cup \left(\mathscr{L}^B(\mathscr{C}) + \mathscr{L}_\varepsilon + \frac{1}{2}e_{\mathscr{R}_n}\right),$$

(C-construction in [KKM91]). By (4.1.6) and the arguments before that lemma we have the following.

Lemma 4.1.7 *If $\mathscr{C}$ is a self-dual doubly even code of length n then $\mathscr{L}^C(\mathscr{C})$ is an n-dimensional even unimodular lattice and up to isomorphism every even unimodular lattice which contains $\mathscr{L}_0$ and does not contain $\mathscr{L}$ can be obtained in this way.* □

Using (4.1.3 (iii)) it is not difficult to check that for $n \geq 24$ $\mathscr{L}^C(\mathscr{C})$ does not contain roots if and only if the minimal weight of $\mathscr{C}$ is greater than 4. In view of (2.8.8) and (2.11.4) this gives the following.

Lemma 4.1.8 *Let $\mathscr{C}_{12}$ be the unique Golay code. Then $\mathscr{L}^C(\mathscr{C}_{12})$ is a Leech lattice and up to isomorphism it is the unique Leech lattice which contains $\mathscr{L}_0(\mathscr{R}_{24})$.* □

4.2 Some automorphisms of lattices

Recall that an automorphism of a lattice $L \subset \mathbb{R}^n$ is a linear transformation of $\mathbb{R}^n$ (*i.e.* an element of $GL_n(\mathbb{R})$) which stabilizes L as a whole. In this section we show that $\mathscr{L}$ and $\mathscr{L}_0$ have the same automorphism group isomorphic to $2^n : Sym_n$ and also calculate the stabilizers of $\mathscr{L}$ in the automorphism groups of the lattices $\mathscr{L}^A(\mathscr{C})$, $\mathscr{L}^B(\mathscr{C})$ and $\mathscr{L}^C(\mathscr{C})$ (depending on the code $\mathscr{C}$).

Let $\mathscr{R}_n$, $\mathscr{L}$ and $\mathscr{L}_0$ be as in the previous section. It is easy to see that

$$\mathscr{F} = \{\pm a \mid a \in \mathscr{R}_n\}$$

is the set of roots in $\mathscr{L}$ and such a subset in $\mathbb{R}^n$ will be called a *frame*.

Let M be an integral lattice of dimension n and $\mathscr{F}$ be a frame. Then M is said to be of type A, B or C with respect to $\mathscr{F}$ if the following conditions hold:

type A: $\mathscr{L} \subseteq M$,

type B: $\mathscr{L} \cap M = \mathscr{L}_0$ and $M \subseteq \mathscr{L}^*$,

type C: $\mathscr{L} \cap M = \mathscr{L}_0$ and $M \cap \mathscr{L}^* \neq M$.

The following proposition is a reformulation of (4.1.4), (4.1.5) and (4.1.7).

Proposition 4.2.1 *Let M be an integral lattice in $\mathbb{R}^n$ which is of type A, B or C with respect to a frame $\mathscr{F}$. Suppose also that in the type C case M*

is even and unimodular. Let

$$\mathscr{C} = \{X \subseteq \mathscr{R}_n \mid (\mathscr{L} + e_X) \cap M \neq \emptyset\}.$$

Then $\mathscr{C}$ *is a code and possibly after changing the signs of some vectors in* $\mathscr{R}_n$, *we have* $M = \mathscr{L}^A(\mathscr{C})$, $\mathscr{L}^B(\mathscr{C})$ *or* $\mathscr{L}^C(\mathscr{C})$, *respectively.* □

Lemma 4.2.2 *Let* $D = \mathrm{Aut}\,\mathscr{L}$. *Then* $D \cong 2^{\mathscr{R}_n} : Sym(\mathscr{R}_n) \cong 2^n : Sym_n$.

Proof. Let $\mathscr{F}$ be the frame formed by the roots in $\mathscr{L}$. Since $\mathscr{F}$ contains the basis $\mathscr{R}_n$ of $\mathbb{R}^n$, the action of D on $\mathscr{F}$ is faithful. Let us say that two roots in $\mathscr{F}$ are equivalent if they are scalar multiples of each other. The group D preserves this equivalence relation (with n classes of size 2 each) and induces on the set $\bar{\mathscr{F}}$ of equivalence classes a subgroup of $Sym(\bar{\mathscr{F}})$. Furthermore, the kernel of the action of D on $\bar{\mathscr{F}}$ is an elementary abelian 2-group of rank at most n. On the other hand each permutation of $\mathscr{R}_n$ can be extended to a linear transformation of $\mathbb{R}^n$ (which stabilizes $\mathscr{L}$) and for a subset $Y \subseteq \mathscr{R}_n$ the mapping which sends $a \in \mathscr{R}_n$ to $\gamma(Y,a) \cdot a$, where $\gamma(Y,a) = -1$ if $a \in Y$ and $\gamma(Y,a) = 1$ otherwise, defines a linear transformation of $\mathbb{R}^n$ which stabilizes every equivalence class in $\mathscr{F}$ as a whole. □

Lemma 4.2.3 *The lattices* $\mathscr{L}$ *and* $\mathscr{L}_0$ *have the same group of automorphisms.*

Proof. It follows from the proof of (4.2.2) that every automorphism from $D = \mathrm{Aut}\,\mathscr{L}$ stabilizes $\mathscr{L}_0$. So in order to prove the equality it is sufficient to show that there is a canonical (*i.e.* basis independent) way to reconstruct $\mathscr{L}$ from $\mathscr{L}_0$. Let Δ be the set of vectors which are sums of pairs of non-collinear vectors from $\mathscr{F}$. It is easy to check that Δ is in fact the set of all vectors of length 4 in $\mathscr{L}_0$. Define on Δ a graph (denoted by the same letter Δ) in which two vectors are adjacent if they are equal, collinear or orthogonal. Let us say that two vectors from Δ are equivalent if they have the same support in the basis $\mathscr{R}_n$. This equivalence relation can be described in the internal terms of $\mathscr{L}_0$ as follows: two vectors are equivalent if and only if in the graph Δ they are adjacent to the same set of vectors. Now it is sufficient to observe that the roots from $\mathscr{F}$ are halves of sums of pairs of equivalent but not collinear vectors from Δ.□

We would like to describe the stabilizers of $\mathscr{L}$ in the automorphism groups of the lattices $\mathscr{L}^A(\mathscr{C})$, $\mathscr{L}^B(\mathscr{C})$ and $\mathscr{L}^C(\mathscr{C})$ in terms of the code $\mathscr{C}$ and its automorphism group.

Let $V \cong V' \cong \mathbb{R}^n$ be two n-dimensional Euclidean vector spaces, $\mathscr{R}_n$ and $\mathscr{R}'_n$ be bases in V and V' consisting of pairwise orthogonal roots and let $\mathscr{L} = \mathscr{L}(\mathscr{R}_n)$ and $\mathscr{L}' = \mathscr{L}(\mathscr{R}'_n)$ be the lattices formed by integral linear combinations of vectors from $\mathscr{R}_n$ and $\mathscr{R}'_n$, respectively. By (4.2.2) the linear transformations of V onto V' which map $\mathscr{L}$ onto $\mathscr{L}'$ are parametrized by the pairs (Y, σ) where $Y \subseteq \mathscr{R}_n$ and σ is a bijection of $\mathscr{R}_n$ onto $\mathscr{R}'_n$. If $\tau(Y, \sigma)$ is the transformation corresponding to such a pair and $a \in \mathscr{R}_n$, then

$$\tau(Y, \sigma) : a \mapsto \gamma(Y, a) \cdot \sigma(a),$$

where $\gamma(Y, a) = -1$ if $a \in Y$ and $\gamma(Y, a) = 1$ if $a \notin Y$.

Lemma 4.2.4 *Let $\mathscr{C}$ be a code based on $\mathscr{R}_n$ and $\mathscr{C}'$ be a code based on $\mathscr{R}'_n$. Let $M = \mathscr{L}^A(\mathscr{C})$ and $M' = \mathscr{L}^A(\mathscr{C}')$ be the lattices obtained by A-construction using $\mathscr{C}$ and $\mathscr{C}'$, respectively. Then*

(i) *$\tau(Y, \sigma)$ maps M onto M' if and only if σ maps $\mathscr{C}$ onto $\mathscr{C}'$,*
(ii) $\mathrm{Aut}\, \mathscr{L}^A(\mathscr{C}) \cap \mathrm{Aut}\, \mathscr{L} \cong 2^{\mathscr{R}_n} : \mathrm{Aut}\, \mathscr{C}$.

Proof. Let $X \in \mathscr{C}$ so that $\mathscr{L} + e_X \subseteq M$. Then

$$(\mathscr{L} + e_X)^{\tau(Y,\sigma)} = \mathscr{L}' + e_{\sigma(X)} - \sum_{a' \in \sigma(Y) \cap \sigma(X)} a' = \mathscr{L}' + e_{\sigma(X)}$$

and this coset belongs to M' if and only if $\sigma(X) \in \mathscr{C}'$. So (i) follows and immediately implies (ii). □

Lemma 4.2.5 *Let $\mathscr{C}$ be a code based on $\mathscr{R}_n$ and $\mathscr{C}'$ be a code based on $\mathscr{R}'_n$. Let $M = \mathscr{L}^B(\mathscr{C})$ and $M' = \mathscr{L}^B(\mathscr{C}')$ be the lattices obtained by B-construction using $\mathscr{C}$ and $\mathscr{C}'$, respectively. Then*

(i) *$\tau(Y, \sigma)$ maps M onto M' if and only if σ maps $\mathscr{C}$ onto $\mathscr{C}'$ and $Y \in \mathscr{C}^*$,*
(ii) $\mathrm{Aut}\, \mathscr{L}^B(\mathscr{C}) \cap \mathrm{Aut}\, \mathscr{L} \cong \mathscr{C}^* : \mathrm{Aut}\, \mathscr{C}$.

Proof. Let $X \in \mathscr{C}$ so that $\mathscr{L}_0 + e_X \subseteq M$. Then

$$(\mathscr{L}_0 + e_X)^{\tau(Y,\sigma)} = \mathscr{L}'_0 + e_{\sigma(X)} - \sum_{a' \in \sigma(Y) \cap \sigma(X)} a'$$

and this coset belongs to M' if and only if $\sigma(X) \in \mathscr{C}'$ and $|\sigma(Y) \cap \sigma(X)| = |Y \cap X|$ is even. The latter condition holds for all $X \in \mathscr{C}$ if and only if $Y \in \mathscr{C}^*$. So (i) follows and implies (ii). □

Lemma 4.2.6 *Let $\mathscr{C}$ be a doubly even self-dual code based on $\mathscr{R}_n$ and $\mathscr{C}'$ be a code based on $\mathscr{R}'_n$. Let $M = \mathscr{L}^C(\mathscr{C})$ and $M' = \mathscr{L}^C(\mathscr{C}')$ be the lattices obtained by C-construction using $\mathscr{C}$ and $\mathscr{C}'$, respectively. Then*

(i) *$\tau(Y, \sigma)$ maps M onto M' if and only if σ maps $\mathscr{C}$ onto $\mathscr{C}'$ and $Y \in \mathscr{C}^* = \mathscr{C}$,*

(ii) $\mathrm{Aut}\,\mathscr{L}^C(\mathscr{C}) \cap \mathrm{Aut}\,\mathscr{L} \cong \mathscr{C} : \mathrm{Aut}\,\mathscr{C}$.

Proof. Let $N = M \cap \mathscr{L}^* = \mathscr{L}^B(\mathscr{C})$ and $N' = M' \cap \mathscr{L}'^* = \mathscr{L}^B(\mathscr{C}')$. Since $\tau(Y, \sigma)$ maps $\mathscr{L}$ onto $\mathscr{L}'$, it maps $\mathscr{L}^*$ onto $\mathscr{L}'^*$. Hence if $\tau(Y, \sigma)$ maps M onto M', it maps N onto N'. By (4.2.5) the latter happens exactly when σ maps $\mathscr{C}$ onto $\mathscr{C}'$ and $Y \in \mathscr{C}^*$ (in our case $\mathscr{C}^* = \mathscr{C}$). Assuming that these conditions hold let us show that $\tau(Y, \sigma)$ maps M onto M'. Since $\mathscr{C}$ is doubly even and self-dual, by the argument before (4.1.6) we conclude that $M \setminus N$ is the only coset of N in N^* which contains even vectors only and does not contain $\mathscr{L}_1$. Similarly $M' \setminus N'$ is the only coset of N' in N'^* which contains even vectors only and does not contain $\mathscr{L}'_1$. Since $\tau(Y, \sigma)$ maps $\mathscr{L}_1$ onto $\mathscr{L}'_1$ and preserves the lengths of vectors, it indeed sends M onto M' and (i) follows. Now (ii) is immediate with the remark that $\mathscr{C}^* = \mathscr{C}$ because of the self-duality assumption. □

Since the Golay code $\mathscr{C}_{12}$ has no subsets of size 4, all roots in $\mathscr{L}^A(\mathscr{C}_{12})$ are contained in the frame $\mathscr{F}$. Hence $\mathscr{F}$ is the only frame for which $\mathscr{L}^A(C_{12})$ is of type A and by (4.2.4) $\mathrm{Aut}\,\mathscr{L}^A(\mathscr{C}_{12}) \cong 2^{24} : Mat_{24}$.

The stabilizer in $\mathrm{Aut}\,\mathscr{L}^C(\mathscr{C}_{12})$ of the frame $\mathscr{F}$ is isomorphic to $2^{12} : Mat_{24}$ by (4.2.6) and we will see in the next section that it is a proper subgroup in $\mathrm{Aut}\,\mathscr{L}^C(\mathscr{C}_{12})$.

In view of the above discussion and by (4.2.6) we have the following.

Proposition 4.2.7 *Let Λ be a Leech lattice and $\mathscr{F}$ be a frame. Then Λ cannot be of type A or B with respect to $\mathscr{F}$ and if Λ is of type C then $\Lambda = \mathscr{L}^C(\mathscr{C}_{12},)$ for a basis $\mathscr{R}_{24}$ consisting of roots from $\mathscr{F}$. Furthermore, if $\mathscr{F}'$ is another frame for which Λ is of type C then the automorphism group of Λ contains an element which maps $\mathscr{F}$ onto $\mathscr{F}'$.* □

4.3 The uniqueness of the Leech lattice

In this section we follow the brilliant article [Con69] by J.H. Conway to show that the Leech lattice is unique up to isomorphism.

Let Λ be a Leech lattice. It can be deduced from the general theory of integral lattices that for every r the number N_r of vectors of length r in

Λ is the same for all Leech lattices and can be computed explicitly. The situation is the following.

Let L be an integral lattice. The theta function Θ_L of L is a power series in a formal variable q defined as follows:

$$\Theta_L(q) = \sum_{x \in L} q^{x^2/2} = \sum_{r=0}^{\infty} N_r q^{r/2}$$

where N_r is the number of vectors of length r in L.

The following fundamental result is known as Hecke's theorem ([Ser73], [Ebe94]).

Theorem 4.3.1 *Let L be an even unimodular lattice of dimension n. If q in the above expression for $\Theta_L(q)$ is replaved by $e^{2\pi i z}$ where z is a variable taking values in the complex upper half plane then $\Theta_L(z)$ is a modular form of weight $n/2$, which means that*

$$\Theta_L(z) = (cz+d)^{-n}\Theta_L\left(\frac{az+b}{cz+d}\right) \quad \textit{for} \quad \begin{pmatrix} a & b \\ c & d \end{pmatrix} \in SL_2(\mathbb{Z}).$$

Thus in the case of a Leech lattice Λ the theta function $\Theta_\Lambda(z)$ is a modular form of weight 12. It is known [Ser73] that the space of modular forms of weight 12 is 2-dimensional. On the other hand $\Theta_\Lambda(z)$ satisfies two additional conditions: $N_0 = 1$ (true for all lattices) and $N_2 = 0$ (since there are no roots in Λ). These two conditions turn out to be independent and they are satisfied by a unique modular form of weight 12. Hence this unique form is the theta function of a Leech lattice. The coefficient of q^{2m} in the power series expansion of this form is

$$N_{2m} = \frac{65\,520}{691}(\sigma_{11}(m) - \tau(m)),$$

where $\sigma_{11}(m)$ is the sum of 11th powers of the divisors of m and $\tau(m)$ is the Ramanujan function defined by

$$q\prod_{m=1}^{\infty}(1-q^m)^{24} = \sum_{m=1}^{\infty}\tau(m)q^m.$$

Let Λ_m denote the set of vectors of length $2m$ in Λ. Then $|\Lambda_m| = N_{2m}$ and by the above for a given m we can calculate the size of Λ_m explicitly. In particular we have the following.

Proposition 4.3.2 *Let Λ be a Leech lattice and let Λ_m be the set of vectors* *of length 2m in* Λ. *Then* $|\Lambda_0| = 1$, $|\Lambda_1| = 0$,

$$|\Lambda_2| = 196\,560,\ |\Lambda_3| = 16\,773\,120 \text{ and } |\Lambda_4| = 398\,034\,000.$$

Let $\bar{\Lambda} = \Lambda/2\Lambda$ be the Leech lattice Λ taken modulo 2, so that $\bar{\Lambda}$ is an elementary abelian 2-group of rank 24. If $\lambda \in \Lambda$ and $M \subseteq \Lambda$ then $\bar{\lambda}$ and $\bar{M}$ denote the images in $\bar{\Lambda}$ of λ and M, respectively.

One may notice a similarity between (2.3.1) and the following lemma.

Lemma 4.3.3 *Let* λ *and* ν *be distinct vectors in* Λ *with* $\bar{\lambda} = \bar{\nu}$ *such that* $\lambda \in \Lambda_i$, $\nu \in \Lambda_j$ *for* $0 \le i, j \le 4$. *Then either* $\lambda = -\nu$, *or* $i = j = 4$ *and* $(\lambda, \nu) = 0$.

Proof. Clearly λ and $-\lambda$ have the same image in $\bar{\Lambda}$, so we assume that $\lambda \neq -\nu$. Replacing λ by $-\lambda$, if necessary we can assume that (λ, ν) is non-negative. Since both λ and ν have length at most 8 we have

$$(\lambda - \nu)^2 = \lambda^2 - 2(\lambda, \nu) + \nu^2 \le 16$$

with the equality holding if and only if λ and ν are orthogonal vectors from Λ_4. Since $\bar{\lambda} = \bar{\nu}$ and $\lambda \neq -\nu$, there is a non-zero vector μ in Λ such that $\lambda - \nu = 2\mu$. Since Λ is a Leech lattice the length of μ is at least 4 and hence

$$(\lambda - \nu)^2 = 4\mu^2 \ge 16.$$

So $(\lambda - \nu)^2 = 16$ and the result follows. □

A maximal set of pairwise orthogonal 1-dimensional subspaces in a 24-dimensional Euclidean space is obviously of size 24 and each 1-subspace contains exactly two vectors of any given positive length (in particular of length 8). In view of this observation (4.3.3) implies the following.

Lemma 4.3.4 *If* $0 \le i < j \le 4$ *then* $\bar{\Lambda}_i \cap \bar{\Lambda}_j = \emptyset$, $|\bar{\Lambda}_i| = \frac{1}{2}|\Lambda_i|$ *and* $|\bar{\Lambda}_4| \ge \frac{1}{48}|\Lambda_4|$. □

By direct calculation with numbers in (4.3.2) one can easily check the following equality:

$$1 + \frac{|\Lambda_2|}{2} + \frac{|\Lambda_3|}{2} + \frac{|\Lambda_4|}{48} = 2^{24}.$$

Since the right hand side is exactly the order of $\bar{\Lambda}$ we have the following.

Lemma 4.3.5 *$\bar{\Lambda}$ is the disjoint union of the $\bar{\Lambda}_i$ for $i = 0, 2, 3$ and 4; $|\bar{\Lambda}_4| = \frac{1}{48}|\Lambda_4|$, which means that for every $\lambda \in \Lambda_4$ there is a unique $\mathbb{R}^{24}$-basis $\mathscr{D}$ consisting of pairwise orthogonal vectors from Λ_4 such that $\lambda \in \{\pm\alpha \mid \alpha \in \mathscr{D}\}$ and whenever α and β are distinct vectors from $\mathscr{D}$ then $\alpha + \beta = 2\mu$ for some $\mu \in \Lambda_2$, in particular $|\bar{\mathscr{D}}| = 1$.* □

Let $\lambda \in \Lambda_4$ and let $\mathscr{D}$ be as in the above lemma. Put $\mathscr{R}_{24} = \{a \mid 2a \in \mathscr{D}\}$ and $\mathscr{F} = \{\pm a \mid a \in \mathscr{R}_{24}\}$. Then $\mathscr{R}_{24}$ is a basis of $\mathbb{R}^{24}$ consisting of pairwise orthogonal roots and by (4.3.5) the sum of any two (possibly equal) vectors from $\mathscr{F}$ is contained in Λ. This immediately implies that $\mathscr{L}_0(\mathscr{R}_{24}) \subseteq \Lambda$ and by (4.2.7) we obtain the following.

Proposition 4.3.6 *Let Λ be a Leech lattice. Then*

(i) $\Lambda \cong \mathscr{L}^C(\mathscr{C}_{12})$,

(ii) *there is a one-to-one correspondence between the set $\bar{\Lambda}_4$ and the set of frames for which Λ is of type C,*

(iii) $\mathrm{Aut}\,\Lambda$ *acts transitively on the set $\bar{\Lambda}_4$ of size $|\Lambda_4|/48 = 8\,292\,375 = 3^6 \cdot 5^3 \cdot 7 \cdot 13$ with stabilizer isomorphic to $2^{12} : Mat_{24}$,*

(iv) $|\mathrm{Aut}\,\Lambda| = 2^{22} \cdot 3^9 \cdot 5^4 \cdot 7^2 \cdot 11 \cdot 13 \cdot 23$. □

4.4 Coordinates for Leech vectors

In the remainder of the volume Λ is the unique Leech lattice in $\mathbb{R}^{24}$, the vectors in Λ will be called *Leech vectors.* In order to carry out more or less explicit calculations in Λ it is convenient to choose a basis $\mathscr{P}$ in $\mathbb{R}^{24}$ such that $\Lambda = \mathscr{L}^C(\mathscr{C}_{12})$ with respect to $\mathscr{R}_{24} = \{4a \mid a \in \mathscr{P}\}$. In this case the coordinates of a Leech vector are integral. Moreover, a vector $\lambda \in \mathbb{R}^{24}$ whose coordinates $\{\lambda(a) \mid a \in \mathscr{P}\}$ in the basis $\mathscr{P}$ are integral is a Leech vector if and only if for $m = 0$ or 1 the following three conditions hold (we assume that $\mathscr{C}_{12}$ is based on $\mathscr{P}$):

(Λ1) $\lambda(a) = m \bmod 2$ for every $a \in \mathscr{P}$,

(Λ2) $\{a \mid \lambda(a) = m \bmod 4\} \in \mathscr{C}_{12}$,

(Λ3) $\sum_{a\in\mathscr{P}} \lambda(a) = 4m \bmod 8$.

Here $m = 0$ if $\lambda \in \mathscr{L}(\mathscr{R}_{24})^*$ and $m = 1$ otherwise.

Notice that if $\lambda, \nu \in \Lambda$ and $\lambda \in \Lambda_i$ then

$$(\lambda, \nu) = \frac{1}{8}\sum_{a\in\mathscr{P}} \lambda(a)\nu(a) \quad \text{and} \quad i = \frac{1}{16}\sum_{a\in\mathscr{P}} \lambda(a)^2.$$

In this chapter when talking about a Golay code and a Steiner system we always mean the code $\mathscr{C}_{12}$ based on $\mathscr{P}$ and the system formed by the octads in $\mathscr{C}_{12}$, respectively.

We write $\widehat{G}$ for the automorphism group of Λ (also denoted by .0 and Co_0). Let $\widehat{G}_1 = \widehat{G} \cap \operatorname{Aut} \mathscr{L}(\mathscr{R}_{24})$. By (4.2.6) $\widehat{G}_1$ consists of the transformations induced by the mappings

$$\widehat{\tau}(Y,\sigma) : a \mapsto \gamma(Y,a) \cdot \sigma(a) \;\text{ for }\; a \in \mathscr{P},$$

where $\sigma \in \operatorname{Aut} \mathscr{C}_{12}$, $Y \in \mathscr{C}_{12} = \mathscr{C}_{12}^*$, $\gamma(Y,a) = -1$ if $a \in Y$ and $\gamma(Y,a) = 1$ otherwise. Thus $\widehat{G}_1$ is the semidirect product of

$$\widehat{Q}_1 = \{\widehat{\tau}(Y,1) \mid Y \in \mathscr{C}_{12}\} \cong 2^{12}$$

and

$$\widehat{L}_1 = \{\widehat{\tau}(\emptyset,\sigma) \mid \sigma \in \operatorname{Aut} \mathscr{C}_{12}\} \cong Mat_{24}.$$

with respect to the natural action.

We are going to describe the orbits of $\widehat{G}_1$ on Λ_2, Λ_3 and Λ_4. For this purpose we represent the coordinates of a Leech vector λ in the basis $\mathscr{P}$ by a triple $(N(\lambda), P(\lambda), X(\lambda))$ where $N(\lambda)$ is the multiset of absolute values of coordinates of λ, $P(\lambda)$ is an ordered partition $(P_{n_1}, P_{n_2}, ..., P_{n_l})$ of $\mathscr{P}$ such that $a \in P_{n_j}$ if and only if $|\lambda(a)| = n_j$ for $1 \le i < j \le l$, and finally $X(\lambda)$ is a subset of $\mathscr{P}$ such that $\lambda(a)$ is negative if and only if $a \in X(\lambda)$. It is clear that the coordinates of λ (and hence λ itself) are uniquely determined by the triple $(N(\lambda), P(\lambda), X(\lambda))$. The multiset $N(\lambda)$ is called the *shape* of λ. Notice that if λ and v are in the same $\widehat{G}_1$-orbit, then they have the same shape; if they are in the same $\widehat{Q}_1$-orbit, then $P(\lambda) = P(v)$ as well. In the case of short vectors it often happens that the shape of λ uniquely determines the $\widehat{G}_1$-orbit containing λ. It is common to denote by Λ_i^n the vectors in Λ_i for which n is the maximum of the absolute values of coordinates. If $\lambda \in \Lambda_i^n$ then i and n are determined by the shape of λ, which means that Λ_i^n is a union of $\widehat{G}_1$-orbits. If there more than one $\widehat{G}_1$-orbit in Λ_i^n, we denote these orbits by $\Lambda_i^{na}, \Lambda_i^{nb}, \dots$.

Lemma 4.4.1 *The orbits of $\widehat{G}_1 \cong 2^{12} : Mat_{24}$ on Λ_2, Λ_3 and Λ_4, the shapes of vectors they consist of, their lengths and the corresponding stabilizers are as given in the table below.*

Λ_i	Orbits	Shapes	Lengths	Stabilizers
Λ_2	Λ_2^4	$(4^2 0^{22})$	$\binom{24}{2}\cdot 2^2$	$[2^{10}]$: Aut Mat_{22}
	Λ_2^3	$(3\,1^{23})$	$24\cdot 2^{12}$	Mat_{23}
	Λ_2^2	$(2^8 0^{16})$	$759\cdot 2^7$	$[2^5] : (2^4 : L_4(2))$
Λ_3	Λ_3^5	$(5\,1^{23})$	$24\cdot 2^{12}$	Mat_{23}
	Λ_3^4	$(4\,2^8 0^{15})$	$759\cdot 16\cdot 2^8$	$[2^4]\cdot L_4(2)$
	Λ_3^3	$(3^3 1^{21})$	$\binom{24}{3}\cdot 2^{12}$	$P\Gamma L_3(4)$
	Λ_3^2	$(2^{12} 0^{12})$	$2576\cdot 2^{11}$	$[2]\times Mat_{12}$
Λ_4	Λ_4^8	$(8\,0^{23})$	$24\cdot 2$	$[2^{11}] : Mat_{23}$
	Λ_4^6	$(6\,2^7 0^{16})$	$759\cdot 8\cdot 2^7$	$[2^5] : (2^4 : Alt_7)$
	Λ_4^5	$(5\,3^2 1^{21})$	$\binom{24}{3}\cdot 3\cdot 2^{12}$	$P\Sigma L_3(4)$
	Λ_4^{4a}	$(4^4 0^{20})$	$\binom{24}{4}\cdot 2^4$	$[2^8] : (2^6 : 3 : Sym_5)$
	Λ_4^{4b}	$(4^2 2^8 0^{14})$	$759\cdot\binom{16}{2}\cdot 2^9$	$[2^3] : (2\times 2^3 : L_3(2))$
	Λ_4^{4c}	$(4\,2^{12} 0^{11})$	$2576\cdot 12\cdot 2^{12}$	Mat_{11}
	Λ_4^3	$(3^5 1^{19})$	$\binom{24}{5}\cdot 2^{12}$	$2^4 : (Sym_3\times Sym_5)^e$
	Λ_4^{2a}	$(2^{16} 0^8)$	$759\cdot 2^{11}$	$[2] : 2^4 : L_4(2)$
	Λ_4^{2b}	$(2^{16} 0^8)$	$759\cdot 15\cdot 2^{11}$	$[2].2_+^{1+6} : L_3(2)$

Proof. Given $i\in\{2,3,4\}$ we first determine the possible shapes of vectors in Λ_i. If $\lambda\in\Lambda_i$ and $N(\lambda)=(n_1^{k_1}n_2^{k_2}...n_l^{k_l})$ is the shape of λ, then

$$k_1+k_2+...+k_l=24 \text{ and } k_1n_1^2+k_2n_2^2+...+k_ln_l^2=16\cdot i.$$

By (Λ1) the numbers n_j have the same parity and in the even case by (Λ2) the sum s of the k_j with $n_j=2$ mod 4 is the size of a Golay subset (*i.e.* a subset from $\mathscr{C}_{12}$), that is $s\in\{0,8,12,16,24\}$. Having these conditions it is not difficult to list the possible shapes as in the third column of the table. We will see below that for every shape in the table the signs can be chosen so that (Λ3) is satisfied.

If $N(\lambda)$ is as above then $P(\lambda) = (P_{n_1}, P_{n_2}, ..., P_{n_l})$ and $|P_{n_i}| = k_i$ for $1 \le i \le l$. We claim that for every shape in the table $\widehat{G}_1/\widehat{Q}_1 \cong \mathrm{Aut}\,\mathscr{C}_{12} \cong Mat_{24}$ acts transitively on the set of corresponding ordered partitions. In fact for shapes $(4^2 0^{22})$, $(3\,1^{23})$, $(5\,1^{23})$, $(3^3 1^{21})$, $(8\,0^{23})$, $(5\,3^2 1^{21})$, $(4^4 0^{20})$ and $(3^5 1^{19})$ this follows from the 5-fold transitivity of Mat_{24} on $\mathscr{P}$ (2.9.1 (iii)), for shapes $(2^8 0^{16})$, $(4\,2^8 0^{15})$, $(6\,2^7 0^{16})$, $(4^2 2^8 0^{14})$ and $(2^{16} 0^8)$ from the transitivity of Mat_{24} in the set of octads and the double transitivity of the stabilizer of an octad B on B and on $\mathscr{P} \setminus B$ (2.10.1) and finally for shapes $(2^{12} 0^{12})$ and $(4\,2^{12} 0^{11})$ from the transitivity of Mat_{24} on the set of dodecads and the (5-fold) transitivity of the stabilizer of a dodecad D on D and on $\mathscr{P} \setminus D$ (2.11.7).

Now it remains to analyse the possibilities for the signs of coordinates. For a multiset N from the table and an ordered partition P corresponding to N put

$$\Phi = \Phi(N, P) = \{\lambda \mid \lambda \in \Lambda, N(\lambda) = N, P(\lambda) = P\}.$$

We consider even and odd cases separately, starting with the latter one. Thus assume first that all the integers in N are odd (so that there are five possibilities for N from the table). Let μ_0 be a vector such that $N(\mu_0) = N$, $P(\mu_0) = P$ and $\mu_0(a) = 3 \bmod 4$ for all $a \in \mathscr{P}$. Then one easily checks that in each of the five cases we have $\sum_{a \in \mathscr{P}} \mu_0(a) = 4 \bmod 8$, which means that μ_0 is a Leech vector and hence $\mu_0 \in \Phi$. Since μ_0 has been chosen so that $|\mu_0(a)| = |\mu_0(b)|$ implies $\mu_0(a) = \mu_0(b)$, the stabilizer of μ_0 in $\widehat{G}_1$ is contained in the complement $\widehat{L}_1$. The isomorphism type of this stabilizer as in the last column of the table follows directly from (2.9.1), the definition of Mat_{23} and (2.10.1). For an arbitrary vector $\lambda \in \Phi$ put $Y = X(\mu_0) \triangle X(\lambda)$. Then λ can be obtained from μ_0 by changing signs in the coordinates in Y, *i.e.* $\lambda = \mu_0^{\widehat{\tau}(Y,1)}$. On the other hand Y is the set of coordinates of λ equal to 1 modulo 4 and by (Λ2) we have $Y \in \mathscr{C}_{12}$. Hence $\widehat{\tau}(Y, 1) \in \widehat{Q}_1$, which shows that $\widehat{Q}_1$ acts regularly on Φ and the analysis of the odd case is completed.

Let us turn to the even case which is slightly more delicate. Let N be an even multiset from the table, $P = (P_8, P_6, P_4, P_2, P_0)$ be the corresponding partition (where some of the P_j can be empty) and let $\Phi = \Phi(N, P)$ be as above. Notice that by (Λ2) $P_6 \cup P_2$ is a Golay set. Let μ_0 be a vector such that $N(\mu_0) = N$, $P(\mu_0) = P$ and $\mu_0(a) < 0$ if and only if

$a \in P_6$. Then μ_0 is a Leech vector unless $N = (4\,2^8 0^{15})$ or $(4\,2^{12} 0^{11})$. We postpone the analysis of these two cases and assume that μ_0 is a Leech vector. Then μ_0 is stabilized by $\widehat{\tau}(Y, \sigma) \in \widehat{G}_1$ if and only if $Y \subseteq P_0$ and σ stabilizes the partition P. Thus if μ_0 is stabilized by such a $\widehat{\tau}(Y, \sigma)$ then it is stabilized by both $\widehat{\tau}(Y, 1) \in \widehat{Q}_1$ and $\widehat{\tau}(\emptyset, \sigma) \in \widehat{L}_1$. Hence $\widehat{G}_1(\mu_0)$ is the semidirect product of $\widehat{Q}_1(\mu_0)$ (consisting of the Golay subsets contained in P_0) and $\widehat{L}_1(\mu_0)$ which is the stabilizer of P in $\widehat{L}_1$. The structure of $\widehat{G}_1(\mu_0)$ is given in the last column of the corresponding row in the table in the form $[\widehat{Q}_1(\mu_0)] : \widehat{L}_1(\mu_0)$. It is not difficult to deduce these structures from the properties of Mat_{24} and $\mathscr{C}_{12}$ contained in Chapter 2. A vector $\lambda = \mu_0^{\widehat{\tau}(Y,1)}$ belongs to Φ if and only if $|Y \cap (P_6 \cup P_2)|$ is even. This enables us to calculate the size of Φ. Comparing this size with $[\widehat{G}_1 : \widehat{G}_1(\mu_0)]$ we conclude that $\widehat{Q}_1 : \widehat{L}_1(\mu_0)$ is transitive on Φ in all cases under consideration except for the case $N = (2^{16}0^8)$. In the latter case the orbit Λ_4^{2a} containing μ_0 does not contain all the Leech vectors of this shape. This is seen by the following argument. Let $\lambda \in \Phi((2^{16}0^8), P)$ and $|X(\lambda)| = 2$. Then μ_0 cannot be mapped onto λ by an element from $\widehat{G}_1$. In fact, P_2 is the complement of an octad B and we know ((2.8.5) and (2.11.4)) that a Golay set cannot intersect it in two elements. Let us calculate the stabilizer in $\widehat{G}_1$ of such a vector λ. Clearly $\widehat{Q}_1(\lambda)$ is of order 2 and $\widehat{\tau}(B, 1)$ is its only non-identity element. Let c be the unique involution in $\widehat{L}_1$ which stabilizes B elementwise and $X(\lambda)$ setwise (compare (2.10.1 (ii))). We claim that $\widehat{G}_1(\lambda)\widehat{Q}_1 = C_{\widehat{L}_1}(c)\widehat{Q}_1$. In fact, if $\widehat{\tau}(Y, \sigma)$ stabilizes λ then either $\sigma(X(\lambda)) = X(\lambda)$ and $Y \subseteq P_0$ or $\sigma(X(\lambda)) \cap X(\lambda) = \emptyset$ and Y is an octad such that $Y \cap P_2 = X(\lambda) \cup \sigma(X(\lambda))$. Notice that by (2.10.5) for a 2-element subset $Z \in P_2 \setminus X(\lambda)$ an octad which intersects P_2 in $X(\lambda) \cup Z$ exists if and only if Z is an orbit of c. Since c is uniquely determined by any of its orbits on P_2, the claim follows. Now direct calculations show that $\Lambda_4^{2a} \cup \Lambda_4^{2b}$ contains all Leech vectors of the shape $(2^{16}0^8)$.

Let us turn to the pair of multisets left out before. If $N(\lambda) = (4\,2^{12}0^{11})$ and $X(\lambda)$ consists of a single element from P_2, then λ is a Leech vector. We calculate the stabilizer $\widehat{G}_1(\lambda)$. By (2.11.2) $\widehat{Q}_1(\lambda) = 1$. We claim that $\widehat{G}_1(\lambda)\widehat{Q}_1 = \widehat{L}_1(P)\widehat{Q}_1$. In fact $\widehat{L}_1(\lambda)$ is the stabilizer in $\widehat{L}_1$ of the partition $(P_4, X(\lambda), P_2 \setminus X(\lambda), P_0)$ (this stabilizer is isomorphic to $L_2(11)$ by (2.11.9 (iii))). Let B be an octad such that $|B \cap P_2| = 2$, $X(\lambda) \subseteq B$, $P_4 \cap B = \emptyset$ (such an octad exists by (2.15.1)) and let σ be an element which stabilizes (P_4, P_2, P_0) and maps $X(\lambda)$ onto $B \setminus (B \cap (P_0 \cup X(\lambda))$. Then $\widehat{\tau}(B, \sigma)$ stabilizes λ, which means that $\widehat{G}_1(\lambda)\widehat{Q}_1$ acts transitively on $\{\pm a \mid a \in P_2\}$ and the claim follows. By (2.11.7) $\widehat{G}_1(\lambda) \cong Mat_{11}$.

Finally let $N(\lambda) = (4\,2^8 0^{15})$ and suppose that $X(\lambda)$ consists of a single element from P_2. Then λ is a Leech vector. Let us calculate its stabilizer in $\widehat{G}_1$. There are exactly 15 octads disjoint from $P_4 \cup P_2$ and hence $\widehat{Q}_1(\lambda) \cong 2^4$. Clearly $\widehat{L}_1(\lambda)$ is the stabilizer in $\widehat{L}_1$ of the partition $(P_4, X(\lambda), P_2 \setminus X(\lambda), P_0)$ which is isomorphic to Alt_7 by (2.10.1). Arguing as in the previous paragraph we show that $\widehat{G}_1(\lambda)$ acts transitively on $\{\pm a \mid a \in P_2\}$, which implies that $\widehat{G}_1(\lambda) \sim 2^4.L_4(2)$. A more detailed analysis shows that the extension does not split. □

4.5 Co_1, Co_2 and Co_3

Our nearest goal is to show that $\widehat{G}$ acts transitively on Λ_2 and Λ_3. Notice that Leech vectors with all their coordinates in the basis $\mathscr{P}$ divisible by 4 form the sublattice $\mathscr{L}_0(\mathscr{R}_{24})$ and those with all their coordinates divisible by 2 form the sublattice $\Lambda \cap \mathscr{L}(\mathscr{R}_{24})^*$. We should emphasize that $\mathscr{P}$ is a basis of $\mathbb{R}^{24}$ and not a basis of the lattice Λ. By (4.2.6) and (4.3.6) $\widehat{G}_1$ is the intersection of $\widehat{G}$ and Aut $\mathscr{L}(\mathscr{R}_{24})$. In view of (4.2.3) this means that $\widehat{G}_1$ is the stabilizer in $\widehat{G}$ of the sublattice $\mathscr{L}_0(\mathscr{R}_{24})$ and also of the sublattice $\Lambda \cap \mathscr{L}(\mathscr{R}_{24})^*$. Since $\widehat{G}_1$ is a proper subgroup in $\widehat{G}$, by (4.3.6 (iii)) these two sublattices are not stable under $\widehat{G}$ and we have the following.

Lemma 4.5.1 *Let $\Omega = \mathscr{L}_0(\mathscr{R}_{24})$ or $\Omega = \Lambda \cap \mathscr{L}(\mathscr{R}_{24})^*$. Suppose that M is an orbit of $\widehat{G}_1$ on the set of Leech vectors such that $M \subseteq \Omega$ and $\Omega = \{mv \mid m \in \mathbb{Z}, v \in M\}$. Then the orbit of M under $\widehat{G}$ contains a vector outside Ω.* □

Lemma 4.5.2 *$\widehat{G}$ acts transitively on Λ_2.*

Proof. By (4.4.1) Λ_2^4, Λ_2^3 and Λ_2^2 are the orbits of $\widehat{G}_1$ on Λ_2. The vectors in Λ_2^4 are the shortest vectors in $\mathscr{L}_0(\mathscr{R}_{24})$ and hence Λ_2^4 generates $\mathscr{L}_0(\mathscr{R}_{24})$ over the integers. The orbit Λ_2^2 is contained in $(\Lambda \cap \mathscr{L}(\mathscr{R}_{24})^*) \setminus \mathscr{L}_0(\mathscr{R}_{24})$. Since the Golay code is generated by its octads as a $GF(2)$-space, Λ_2^2 generates $\Lambda \cap \mathscr{L}(\mathscr{R}_{24})^*$ over the integers. Now the result is immediate from (4.5.1). □

Lemma 4.5.3 *$\widehat{G}$ acts transitively on Λ_3.*

Proof. By (4.4.1) the orbits of $\widehat{G}_1$ on Λ_3 are Λ_3^5, Λ_3^4, Λ_3^3 and Λ_3^2. The second and the last of the orbits are contained in $\Lambda \cap \mathscr{L}(\mathscr{R}_{24})^*$. Since the Golay code is generated by its octads as well as by its dodecads as a $GF(2)$-space, each of Λ_3^4 and Λ_3^2 generates $\Lambda \cap \mathscr{L}(\mathscr{R}_{24})^*$ over the integers.

Suppose that the action of $\widehat{G}$ on Λ_3 is not transitive. Then in view of the above and by (4.5.1) the orbit of Λ_3^4 under $\widehat{G}$ is one of the following four sets: $\Lambda_3^4 \cup \Lambda_3^5$; $\Lambda_3^4 \cup \Lambda_3^3$; $\Lambda_3^4 \cup \Lambda_3^5 \cup \Lambda_3^2$; and $\Lambda_3^4 \cup \Lambda_3^3 \cup \Lambda_3^2$. But the size of each of the four sets is divisible by a prime number which does not divide the order of $\widehat{G}$ given in (4.3.6 (iv)). Hence we get the result. □

The set $\bar{\Lambda}_4^8$ consists of a single element which we will denote by $\bar{\lambda}_0$. By (4.3.5) $\widehat{G}_1$ is the stabilizer of $\bar{\lambda}_0$ in $\widehat{G}$. Let K be the kernel of the action of $\widehat{G}$ on $\bar{\Lambda} = \Lambda/2\Lambda$.

Lemma 4.5.4 $K = \langle \widehat{\tau}(\mathscr{P}, 1) \rangle$ *is of order* 2 *and* $\widehat{G}$ *does not split over* K.

Proof. Since $\widehat{G}_1 = \widehat{G}(\bar{\lambda}_0)$, clearly $K \leq \widehat{G}_1$. Since $\widehat{\tau}(\mathscr{P}, 1)$ multiplies every (Leech) vector by minus 1, it is contained in K. Every normal subgroup in $\widehat{G}_1$ which properly contains $\langle \widehat{\tau}(\mathscr{P}, 1) \rangle$ contains the whole of $\widehat{Q}_1$. We claim that $\widehat{Q}_1$ is not contained in K. Let $a, b \in \mathscr{P}$, $a \neq b$, and v be such that $v(a) = v(b) = 4$ and $v(c) = 0$ for $c \in \mathscr{P} \setminus \{a, b\}$. Then $(\Lambda 1)-(\Lambda 3)$ are satisfied and hence v is a Leech vector. It is clear that there is a subset $X \in \mathscr{C}_{12}$ which contains a and does not contain b. Since

$$\frac{1}{2}(v^{\widehat{\tau}(X,1)} - v) = -4a$$

is not a Leech vector, the claim follows. By (2.15.1) $\mathscr{C}_{12}$ is indecomposable under Mat_{24} and hence $\widehat{G}$ does not split over K. □

The action induced by $\widehat{G}$ on $\bar{\Lambda}$ is the first Conway sporadic simple group denoted by Co_1. By (4.3.6 (iv)) and (4.5.4) we have

$$|Co_1| = 2^{21} \cdot 3^9 \cdot 5^4 \cdot 7^2 \cdot 11 \cdot 13 \cdot 23.$$

The stabilizers in $\widehat{G}$ of vectors from Λ_2 and Λ_3 are the second and the third Conway sporadic simple groups denoted by Co_2 and Co_3, respectively. By (4.5.2) and (4.5.3) we have $Co_i = |\widehat{G}|/|\Lambda_i|$. Hence (4.3.2) and (4.3.6 (iv)) give

$$|Co_2| = 2^{18} \cdot 3^6 \cdot 5^3 \cdot 7 \cdot 11 \cdot 23,$$

$$|Co_3| = 2^{10} \cdot 3^7 \cdot 5^3 \cdot 7 \cdot 11 \cdot 23.$$

For the remainder of the chapter G will denote the first Conway group Co_1 isomorphic to the action induced by $\widehat{G}$ on $\bar{\Lambda}$. The image G_1 of $\widehat{G}_1$ in G is the semidirect product of $Q_1 = \widehat{Q}_1/\langle \widehat{\tau}(\mathscr{P}, 1) \rangle$ which is the irreducible 11-dimensional Golay code module and the (bijective) image L_1 of $\widehat{L}_1$ in G. The elements $\widehat{\tau}(Y, \sigma)$ and $\widehat{\tau}(\mathscr{P} \setminus Y, \sigma)$ have the same image in Q_1

and we denote this image by $\tau(Z,\sigma)$ where Z is either Y or $\mathcal{P}\setminus Y$. Since neither Co_2 nor Co_3 contains $\widehat{\tau}(\mathcal{P},1)$, they map isomorphically onto their images in Co_1 and will be identified with these images.

Directly from (4.3.5), (4.3.6 (iii)), (4.5.2) and (4.5.3) we obtain the following.

Lemma 4.5.5 *The group $G \cong Co_1$ acting on $\bar{\Lambda}^{\#}$ has three orbits $\bar{\Lambda}_2$, $\bar{\Lambda}_3$ and $\bar{\Lambda}_4$ with lengths* 98 280, 8 386 560 *and* 8 292 375 *and stabilizers Co_2, Co_3 and $G_1 \cong 2^{11}.Mat_{24}$, respectively.* □

Let θ be a mapping of Λ onto $GF(2)$ such that

$$\lambda^2 = 2 \cdot \theta(\lambda) \bmod 4$$

for $\lambda \in \Lambda$. Then θ induces on $\bar{\Lambda}$ a quadratic form (denoted by the same letter θ), which is clearly preserved by G, here

$$\theta(\bar{\lambda}) = 1 \text{ if } \bar{\lambda} \in \bar{\Lambda}_3 \text{ and } \theta(\bar{\lambda}) = 0 \text{ otherwise.}$$

Let β denote the bilinear form on $\bar{\Lambda}$ associated with θ:

$$\beta(\bar{\lambda}, \bar{v}) = \theta(\bar{\lambda}) + \theta(\bar{v}) + \theta(\bar{\lambda} + \bar{v}).$$

For a quadratic form on a 24-dimensional $GF(2)$-space the numbers of isotropic and non-isotropic vectors are known [Tay92]; comparing these numbers with the numbers in (4.5.5) we have

Lemma 4.5.6 *The form θ is the only non-trivial quadratic form on $\bar{\Lambda}$ preserved by G.* □

4.6 The action of Co_1 on $\bar{\Lambda}_4$

In this section we study the action of $G \cong Co_1$ on the set $\bar{\Lambda}_4$. By (4.3.6 (iii)) this action is transitive and G_1 is the stabilizer of the element $\bar{\lambda}_0 \in \bar{\Lambda}_4$ such that $\bar{\Lambda}_4^8 = \{\bar{\lambda}_0\}$. Thus the action under consideration is of G on the cosets of G_1 or equivalently of $\widehat{G}$ on the cosets of $\widehat{G}_1$. For every element $\bar{v} \in \bar{\Lambda}_4$ the 48 vectors in Λ_4 which map onto $\bar{v}$ under the natural homomorphism $\psi : \Lambda \to \bar{\Lambda}$ belong to 24 pairwise orthogonal lines (1-dimensional subspaces) in $\mathbb{R}^{24}$. Furthermore, $Q(\bar{v}) := O_2(G(\bar{v}))$ is the image in G of the kernel of the action of $\widehat{G}(\bar{v})$ on these 24 lines. Notice that $Q_1 = Q(\bar{\lambda}_0)$.

Lemma 4.6.1 *The group G_1 acting on $\bar{\Lambda}_4$ has six orbits $\bar{\Lambda}_4^\alpha$ for $\alpha = 8, 6, 5, 4a, 4b$ and $4c$. The preimages of these orbits in Λ_4, their lengths and element stabilizers are as given in the table below.*

Orbits	Preimages	Lengths	Stabilizers
$\bar{\Lambda}_4^8$	$\frac{1}{48}\Lambda_4^8$	1	$[2^{11}] : Mat_{24}$
$\bar{\Lambda}_4^6$	$\frac{1}{16}\Lambda_4^6 + \frac{1}{32}\Lambda_4^{2a}$	$2^6 \cdot 759$	$[2^5] : (2^4 : L_4(2))$
$\bar{\Lambda}_4^5$	$\frac{1}{6}\Lambda_4^5 + \frac{1}{42}\Lambda_4^3$	$2^{11} \cdot \binom{24}{3}$	$P\Gamma L_3(4)$
$\bar{\Lambda}_4^{4a}$	$\frac{1}{48}\Lambda_4^{4a}$	$2 \cdot 1771$	$[2^{10}] : (2^6 : 3 \cdot Sym_6)$
$\bar{\Lambda}_4^{4b}$	$\frac{1}{32}\Lambda_4^{4b} + \frac{1}{16}\Lambda_4^{2b}$	$2^7 \cdot 15 \cdot 759$	$[2^4].2_+^{1+6} : L_3(2)$
$\bar{\Lambda}_4^{4c}$	$\frac{1}{48}\Lambda_4^{4c}$	$2^{11} \cdot 1288$	$Mat_{12}.2$

In the second column corresponding to an orbit $\bar{N}$ by writing $\frac{1}{m}M + \frac{1}{l}L$ we mean that $\bar{v} \in \bar{N}$ is the image of m vectors from the orbit M of $\widehat{G}_1$ on Λ_4 and of l vectors from the orbit L.

Proof. If Λ_4^γ is an orbit of $\widehat{G}_1$ on Λ_4 then $\bar{\Lambda}_4^\gamma$ (which is the image of Λ_4^γ in $\bar{\Lambda}$) is an orbit of G_1 on $\bar{\Lambda}_4$. Furthermore, for $\bar{v} \in \bar{\Lambda}_4^\gamma$ the set

$$\psi^{-1}(\bar{v}) \cap \Lambda_4^\gamma = \{\mu \mid \mu \in \Lambda_4^\gamma, \bar{\mu} = \bar{v}\}$$

is an imprimitivity block of $\widehat{G}_1$ on Λ_4^γ and $G_1(\bar{v})$ is the image in G of the setwise stabilizer of $\psi^{-1}(\bar{v}) \cap \Lambda_4^\gamma$ in $\widehat{G}_1$. We say that two vectors in Λ_4 are *equivalent* if they have the same image in $\bar{\Lambda}$. We know that each equivalence class consists of 48 vectors and that every vector is equivalent to its negative.

We adopt the following strategy of the proof. For every $\alpha \in \{8, 6, 5, 4a, 4b, 4c\}$ we choose a representative $v \in \Lambda_4^\alpha$ and find 48 vectors in Λ_4 equivalent to v. This will show in particular that every vector $\mu \in \Lambda_4^\beta$ for $\beta \in \{3, 2a, 2b\}$ is equivalent to a vector from Λ_4^α with α as above. As in the proof of (4.4.1) we represent the coordinates of a Leech vector λ by the triple $(N(\lambda), P(\lambda), X(\lambda))$.

For $\alpha = 8$ everything is clear. Let $v \in \Lambda_4^6$ so that $P_6(v) \cup P_2(v)$ is an octad B of $\mathscr{C}_{12}$ and let $X(v) = P_6(v)$. If $\mu \in \Lambda_4^6$, $\mu \neq v$, $X(\mu) = P_6(\mu)$

and $P_6(\mu) \cup P_2(\mu) = B$ then $\lambda := \frac{1}{2}(\nu + \mu) \in \Lambda_2^2$ with $P_2(\lambda) = B$ and $X(\lambda) = P_6(\nu) \cup P_6(\mu)$. Since every vector is equivalent to its negative this gives 16 vectors in Λ_4^6 equivalent to ν. Let ω be a vector from Λ_4^{2a} such that $P_2(\omega) = \mathcal{P} \setminus B$ and $X(\omega)$ is a Golay set disjoint from B. Then $\lambda := \frac{1}{2}(\nu + \omega) \in \Lambda_2^3$ with $P_3(\lambda) = P_6(\nu)$ and $X(\lambda) = X(\omega)$. Since there are 32 possibilities for $X(\omega)$ (including the empty set), we obtain 32 vectors in Λ_4^{2a} equivalent to ν. It is now easy to see that the stabilizer of $\psi^{-1}(\bar{\nu}) \cap \Lambda_4^6$ in $\widehat{G}_1$ is the semidirect product of the stabilizer in $\widehat{Q}_1$ of the pair $\{\nu, -\nu\}$ (of order 2^6) and $\widehat{L}_1[B] \cong 2^4 : L_4(2)$ which gives the structure of $G_1(\bar{\nu})$ as in the table. Notice that $Q_1(\bar{\nu}) = [Q_1, O_2(L_1[B])]$ (3.8.4) and that $\tau(B, 1) = \tau(\mathcal{P} \setminus B, 1)$ is the only non-trivial element in $Q_1 \cap Q(\bar{\nu})$.

Let $\nu \in \Lambda_4^5$ and $\nu(a) = 3 \bmod 4$ for every $a \in \mathcal{P}$. If $\mu \in \Lambda_4^5$, $\mu \neq \nu$, $\mu(a) = 3 \bmod 4$ for all $a \in \mathcal{P}$ and $P_5(\mu) \cup P_3(\mu) = P_5(\nu) \cup P_3(\nu)$ then $\frac{1}{2}(\nu - \mu) \in \Lambda_2^4$ and in this way we obtain 6 vectors in Λ_4^5 equivalent to ν. If $\mu \in \Lambda_4^3$ with $\mu(a) = 3 \bmod 4$ for all $a \in \mathcal{P}$ and $P_3(\nu) \cup P_5(\nu) \cup P_3(\mu)$ is an octad then $\frac{1}{2}(\nu - \mu) \in \Lambda_2^2$. Since there are 21 octads containing a given 3-element subset of $\mathcal{P}$, we obtain 42 vectors in Λ_4^3 equivalent to ν. The stabilizer of $\psi^{-1}(\bar{\nu}) \cap \Lambda_4^5$ in $\widehat{G}_1$ is contained in $\widehat{L}_1$ and coincides with the stabilizer of $P_5(\nu) \cup P_3(\nu)$, isomorphic to $P\Gamma L_3(4)$.

Let $\nu \in \Lambda_4^{4a}$, $X(\nu) = \emptyset$ and $\Sigma = \{S_1 = P_4(\nu), S_2, ..., S_6\}$ be the sextet containing $P_4(\nu)$. If $\mu \in \Lambda_4^{4a}$ with $P_4(\mu) = S_i$ for some i, $1 \leq i \leq 6$, and $|X(\mu)|$ even, then $\frac{1}{2}(\nu + \mu)$ is contained in Λ_2^4 if $i = 1$ and in Λ_2^4 if $i \neq 1$. Thus all the 48 vectors equivalent to ν are in Λ_4^{4a}. The stabilizer of $\psi^{-1}(\bar{\nu}) \cap \Lambda_4^{4a}$ in $\widehat{G}_1$ is the semidirect product of the subgroup $\{\widehat{\tau}(Y, 1) \mid |Y \cap P_4(\nu)| \text{ is even}\}$ of index 2 in $\widehat{Q}_1$ and the stabilizer in $\widehat{L}_1$ of the sextet Σ. This implies the structure of $G_1(\bar{\nu})$ as given in the table. Notice that $Q_1(\bar{\nu}) = [Q_1, O_2(L_1[\Sigma])]$ and that

$$(Q_1 \cap Q(\bar{\nu}))^{\#} = \{\tau(Y, 1) \mid Y = S_i \cup S_j, 1 \leq i < j \leq 6\}.$$

Let $\nu \in \Lambda_4^{4b}$ so that $B = P_2(\nu)$ is an octad and let $X(\nu)$ be of size 1 and contained in $P_4(\nu)$. Let c be the unique involution in $L_1(B) \cong 2^4$ which stabilizes $P_4(\nu)$ as a whole and let $R_1 = P_4(\nu), R_2, ..., R_8$ be the orbits of c on $\mathcal{P} \setminus B$. If μ is a vector from Λ_4^{4b} such that $P_2(\mu) = B$, $P_4(\mu) = P_4(\nu)$ and $X(\mu) = P_4(\nu) \setminus X(\nu)$ then $\frac{1}{2}(\nu - \mu) \in \Lambda_2^4$. In addition if $P_2(\mu) = B$, for $2 \leq i \leq 8$ we have $P_4(\mu) = R_i$, $|X(\mu) \cap P_4(\mu)| = 1$ and $R_1 \cup R_i \cup (X(\mu) \cap P_2(\mu))$ is an octad, then $\frac{1}{2}(\nu + \mu) \in \Lambda_2^2$ and altogether we obtain 32 vectors in Λ_4^{4b} equivalent to ν. Let ω be the vector from Λ_4^{2b} such that $P_2(\omega) = \mathcal{P} \setminus B$ and $X(\omega) = P_4(\nu)$. Then $\lambda := \frac{1}{2}(\nu + \omega) \in \Lambda_2^3$ with $P_3(\lambda) = X(\nu)$. Similar results will be achieved if instead of ω we consider its image under $\widehat{\tau}(Y, 1)$

where Y is one of the 7 octads disjoint from $P_2(v) \cup P_4(v)$. Including the negatives this gives 16 vectors in Λ_4^{2b} equivalent to v. Comparing this with the information in the table from (4.4.1) we conclude that $Q_1(\bar{v})$ is of order 2^4, $Q_1 \cap Q(\bar{v}) = 1$ and $G_1(\bar{v})Q_1/Q_1 = C_{L_1}(c)Q_1/Q_1$.

Finally let $v \in \Lambda_4^{4c}$ so that $D = P_2(v)$ is a dodecad and suppose that $|X(v)| = 2$ with $P_4(v) \subset X(v)$. Recall that by (2.11.3) for every 2-element subset T in $\mathscr{P} \setminus D$ there is a unique partition $\{E_1(T), E_2(T)\}$ of D such that $E_1(T) \cup T$ and $E_2(T) \cup T$ are both octads. Let μ be a vector from Λ_4^{4c} such that $P_2(\mu) = P_2(v)$, $P_4(\mu) \neq P_4(v)$, $P_4(\mu) \subseteq X(\mu)$ and $X(\mu) \cap P_2(\mu) = X(v) \triangle E_i(P_4(v) \cup P_2(\mu))$ for $i = 1$ or 2. Then $\lambda := \frac{1}{2}(v - \mu) \in \Lambda_2^2$ with $P_2(\lambda) = E_i(P_4(v) \cup P_4(\mu)) \cup P_4(v) \cup P_4(\mu)$. Thus (including v and its negative) we obtain 24 vectors μ in Λ_4^{4c} with $P_2(\mu) = P_2(v)$ equivalent to v. Now let ω be a vector from Λ_4^{4c} such that $P_2(\omega) = \mathscr{P} \setminus P_2(v)$, $X(\omega) = P_4(v)$ and $P_4(\omega) = X(v) \cap P_2(v)$. Then $\lambda := \frac{1}{2}(v + \omega) \in \Lambda_2^3$ with $P_3(\lambda) = P_4(v)$ and $\lambda(a) = 1 \bmod 4$ for all $a \in \mathscr{P}$. In this way we obtain the remaining 24 vectors from Λ_4^{4c} equivalent to v. It is clear that $Q_1(\bar{v}) = 1$ and that $G_1(\bar{v})Q_1/Q_1 \leq SQ_1/Q_1$ where $S \cong Mat_{12}.2$ is the stabilizer in L_1 of the partition $\{D, \mathscr{P} \setminus D\}$. On the other hand $\widehat{G}_1(\bar{v})$ contains $\widehat{G}_1(v) \cong Mat_{11}$ with index 48 which implies that $G_1(\bar{v}) \cong S$. □

We summarize (4.6.1) and its proof in the following.

Lemma 4.6.2 *For $\alpha = 4a$, 6, $4b$, $4c$ and 5, respectively, the following two assertions hold:*

- (i) *$Q(\bar{\lambda}_0) = Q_1$ acts on $\bar{\Lambda}_4^{\alpha}$ with orbits of length 2, 2^6, 2^7, 2^{11} and 2^{11},*
- (ii) *$G_1/Q_1 \cong Mat_{24}$ acts on the set of Q_1-orbits in $\bar{\Lambda}_4^{\alpha}$ as it acts on the set of sextets, octads, elements of type 3 in $\mathscr{G}(Mat_{24})$, 3-element subsets of $\mathscr{P}$ and complementary pairs of dodecads.*

If $\bar{v} \in \bar{\Lambda}_4^{4a}$ then $Q(\bar{\lambda}_0) \cap Q(\bar{v})$ is of order 2^4 with non-identity elements being $\tau(B, 1)$ for the octads B refined by the sextet corresponding to the Q_1-orbit of $\bar{v}$; if $\bar{v} \in \bar{\Lambda}_4^6$ then $Q(\bar{\lambda}_0) \cap Q(\bar{v})$ is of order 2 containing $\tau(B, 1)$ where B is the octad corresponding to the Q_1-orbit of $\bar{v}$; $Q(\bar{\lambda}_0) \cap Q(\bar{v})$ is trivial in the remaining cases. □

4.7 The Leech graph

For $\alpha \in \{8, 6, 5, 4a, 4b, 4c\}$ and $\bar{v} \in \bar{\Lambda}_4$ let $\bar{\Lambda}_4^{\alpha}(\bar{v})$ denote the image of $\bar{\Lambda}_4^{\alpha}$ under an element $g \in G$ such that $\bar{\lambda}_0^g = \bar{v}$. Since $\bar{\Lambda}_4^{\alpha}$ is an orbit of $G_1 = G(\bar{\lambda}_0)$, this definition is independent of the particular choice of g with the above property. In this section we study a graph Γ on $\bar{\Lambda}_4$

(called *the Leech graph*) such that $\Gamma(\bar{\nu}) = \bar{\Lambda}_4^{4a}(\bar{\nu})$. We will use this graph to construct a rank 4 tilde geometry $\mathcal{G}(Co_1)$ associated with the Conway group Co_1. In fact Γ is the collinearity graph of $\mathcal{G}(Co_1)$. We are going to sketch the calculation of the suborbit diagram of the Leech graph Γ based at $\bar{\lambda}_0$ given below.

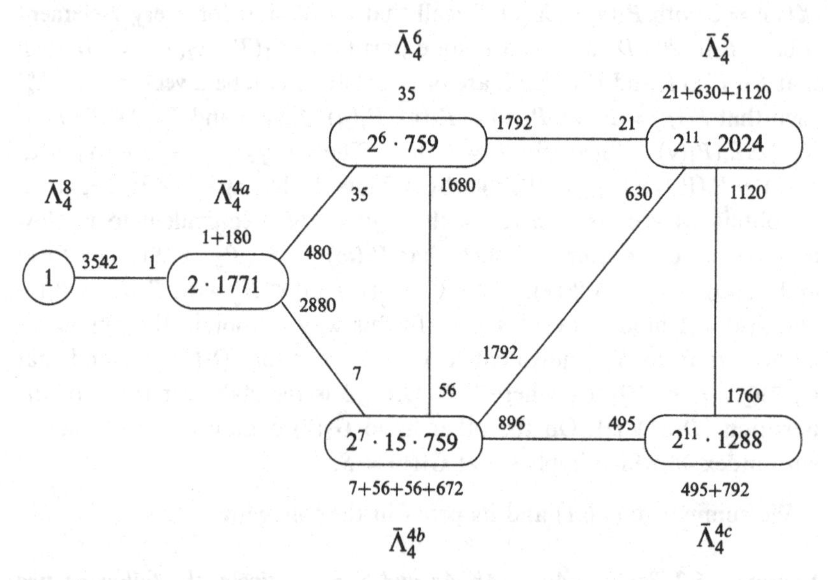

Since a Sylow 2-subgroup in G_1 has order 2^{21}, directly from (4.6.1) we obtain the following.

Lemma 4.7.1 *If* $\bar{\nu}, \bar{\mu} \in \bar{\Lambda}_4$ *with* $\bar{\nu} \neq \bar{\mu}$ *and* $|G(\bar{\nu}) \cap G(\bar{\mu})|$ *is divisible by* 2^{16} *then* $\bar{\nu}$ *and* $\bar{\mu}$ *are adjacent in* Γ. □

In view of (4.6.2) there is a mapping ϱ of $\bar{\Lambda}_4^{4a}$ onto the set of sextets which commutes with the action of G_1. Recall that the sextet graph defined in (3.3.6) is the collinearity graph of $\mathcal{G}(Mat_{24})$.

Lemma 4.7.2 *Let* $\bar{\nu}, \bar{\mu} \in \bar{\Lambda}_4^{4a}$ *with* $\bar{\nu} \neq \bar{\mu}$. *Then*

(i) $G_1(\bar{\nu})$ *acting on* $\bar{\Lambda}_4^{4a} \setminus \{\bar{\nu}\}$ *has* 4 *orbits* Φ_1, Φ_2, Φ_3 *and* Φ_4 *with lengths* 1, 180, 480 *and* 2880, *respectively,*

(ii) $\bar{\mu}$ *is adjacent to* $\bar{\nu}$ *in* Γ *if and only if* $\bar{\mu} \in \Phi_1 \cup \Phi_2$,

(iii) *the orbit of* $\bar{\mu}$ *under* $Q(\bar{\nu}) = O_2(G(\bar{\nu}))$ *has length* 2^6 *if* $\bar{\mu} \in \Phi_3$ *and* 2^7 *if* $\bar{\mu} \in \Phi_4$.

Proof. Consider the action of Q_1 on $\Gamma(\bar{\lambda}_0) = \bar{\Lambda}_4^{4a}$. Each orbit has length 2, corresponds to a sextet Σ and the kernel is $[Q_1, O_2(L_1[\Sigma])]$

which is dual to the subgroup of order 2 in the irreducible Todd module $\bar{\mathscr{C}}_{11}$. Since different sextets correspond to different subgroups in the Todd module, the kernels at different orbits are different. This implies that the $G_1(\bar{v})$-orbit of $\bar{\mu}$ is uniquely determined by the $(G_1(\bar{v})Q_1/Q_1)$-orbit of $\varrho(\bar{\mu})$ and by (3.7.3) we get (i). Furthermore, $\Phi_1 \cup \{\bar{v}\}$ is a Q_1-orbit and $\varrho(\Phi_2)$ consists of the sextets adjacent to $\varrho(\bar{v})$ in the sextet graph. If $\bar{\mu} \in \Phi_1 \cup \Phi_2$, then $|G_1(\bar{v}) \cap G_1(\bar{\mu})|$ is divisible by 2^{18}, hence $\bar{v}$ and $\bar{\mu}$ are adjacent by (4.7.1) and we have established the "if" part of (ii). By (4.6.1) and (4.6.2) we have $G_1(\bar{v}) \sim 2^{4+6}.2^6.3 \cdot Sym_6$, $|Q_1 \cap G_1(\bar{v})| = 2^{10}$ and $|Q_1 \cap Q(\bar{v})| = 2^4$. By the obvious symmetry $|Q(\bar{v}) \cap G_1(\bar{v})| = 2^{10}$ and hence $(Q(\bar{v}) \cap G_1)Q_1/Q_1 = O_2(L_1(\bar{v}))Q_1/Q_1$. By the proof of (3.7.3) the orbits of $O_2(L_1(\bar{v}))$ on $\varrho(\Phi_3)$ and $\varrho(\Phi_4)$ are of length 2^4 and 2^5, respectively. Since every orbit of $Q(\bar{v})$ on $\Gamma(\bar{v})$ must be of length 2, $\bar{v}$ is not adjacent to vertices from $\Phi_3 \cup \Phi_4$ and we obtain the "only if" part of (ii). In view of (3.8.2) it is easy to see that $Q_1 \cap Q(\bar{v})$ equals to $C_{Q_1}(O_2(L_1[\varrho(\bar{v})]))$ and it is generated by the elements $\tau(B,1)$ for the octads B refined by $\varrho(\bar{v})$. The matrices given in the proof of (3.7.3) show that for $\bar{\mu} \in \Phi_3 \cup \Phi_4$ there is an octad refined by $\varrho(\bar{v})$ which has odd intersection with a tetrad from $\varrho(\bar{\mu})$. Hence the orbits of $Q(\bar{v}) \cap G_1$ on Φ_3 and Φ_4 are of length 2^5 and 2^6, respectively. If $\Phi_1 = \{\bar{\lambda}\}$ then $\bar{\lambda}$ is fixed by $G_1 \cap G(\bar{v})$ and by the proof of (i) $\{\bar{\lambda}_0, \bar{\lambda}\}$ is an orbit of $Q(\bar{v})$. By (i) $\bar{\lambda}$ is not adjacent to vertices from $\Phi_3 \cup \Phi_4$ and hence the latter set is disjoint from its image under an element from $Q(\bar{v}) \setminus G_1$. In view of the above this gives (iii). □

By the above lemma, for every edge $\{\bar{v}, \bar{\mu}\}$ of Γ there is a unique vertex $\bar{\lambda}$ such that $T = \{\bar{v}, \bar{\mu}, \bar{\lambda}\}$ is a triangle and $T \setminus \{\bar{\alpha}\}$ is an orbit of $Q(\bar{\alpha})$ for every $\bar{\alpha} \in T$. Such triangles will be called *lines*. It follows from the tables in (4.4.1) and (4.6.1) that $\bar{\lambda}$ is the unique element in $\bar{\Lambda}^{\#}$ stabilized by $G(\bar{v}, \bar{\mu})$ which implies the following.

Lemma 4.7.3 *If $T = \{\bar{v}, \bar{\mu}, \bar{\lambda}\}$ is a line then $\bar{v} + \bar{\mu} + \bar{\lambda} = 0$.* □

In the notation introduced in the proof of (4.7.2) let $\bar{\mu} \in \Phi_i$ for $i = 3$ or 4. Then by (4.6.1) and (4.7.2 (iii)) $\bar{\mu} \in \bar{\Lambda}_4^6(\bar{v})$ if $i = 3$ and $\bar{\mu} \in \bar{\Lambda}_4^{4b}(\bar{v})$ if $i = 4$. This gives the valencies between $\bar{\Lambda}_4^{4a}$ and $\bar{\Lambda}_4^{\alpha}$ for $\alpha = 6$ and $4b$ as on the diagram.

In order to calculate the remaining data on the suborbit diagram we adopt the following strategy. Notice that the subdegrees of G on $\bar{\Lambda}_4$ are pairwise different, hence $\bar{\mu} \in \bar{\Lambda}_4^{\alpha}(\bar{v})$ if and only if $\bar{v} \in \bar{\Lambda}_4^{\alpha}(\bar{\mu})$. In order to calculate the valencies on the diagram, for all $\alpha, \beta \in \{8, 6, 5, 4a, 4b, 4c\}$ we have to determine the number $n_{\alpha\beta}$ of vertices in $\bar{\Lambda}_4^{4a}(\bar{v})$ contained in $\bar{\Lambda}_4^{\beta}$ for a given vertex $\bar{v} \in \bar{\Lambda}_4^{\alpha}$. Since the orbitals are self-paired, $n_{\alpha\beta}$ is equal to the number of vertices in $\bar{\Lambda}_4^{4a} \cap \bar{\Lambda}_4^{\beta}(\bar{v})$. Hence the numbers n_{α} can be calculated using the following lemma.

Lemma 4.7.4 *Let v be a preimage of $\bar{v}$ in Λ_4^{α} and $\mu_1, ..., \mu_{48}$ be the preimages of $\bar{\mu} \in \bar{\Lambda}_4^{4a}$ in Λ_4^{4a} such that $\mu_{24+i} = -\mu_i$ for $1 \leq i \leq 24$. Then the multiset consisting of the values $\frac{1}{8}(\mu_i, v)$ for $1 \leq i \leq 24$ is the shape of v in the basis $\{\frac{1}{8}\mu_i \mid 1 \leq i \leq 24\}$. By* (4.6.1) *this multiset determines the orbital containing $\{\bar{v}, \bar{\mu}\}$.* □

Below we apply this lemma for the remaining cases. We start with two remarks concerning the action of G_1 on $\bar{\Lambda}_4^{4a}$.

Lemma 4.7.5 *The group L_1 acting on $\bar{\Lambda}_4^{4a}$ has two orbits O_1 and O_2 such that $\varrho(O_i) = \varrho(\bar{\Lambda}_4^{4a})$ for $i = 1$ and 2.*

Proof. Clearly L_1 acts transitively on $\varrho(\bar{\Lambda}_4^{4a})$. Let v_1 and v_2 be vectors from Λ_4^{4a} such that $|X(v_1)|$ and $|X(v_2)|$ have different parities. Then $\bar{v}_1 \neq \bar{v}_2$ and since $\widehat{L}_1$ preserves multisets of coordinates, $\bar{v}_1$ and $\bar{v}_2$ are in different L_1-orbits. □

Lemma 4.7.6 *Let B be an octad, R be an orbit of Q_1 on $\bar{\Lambda}_4^{4a}$ and $\Sigma = \varrho(R)$, where $\Sigma = \{S_1, S_2, ..., S_6\}$ is a sextet. Then $\tau(B, 1)$ is not contained in $Q_1(R)$ if and only if $\mu(B, \Sigma) = (3\,1^5)$ in the notation of* (2.14.1).

Proof. Let v_1 and v_2 be vectors from Λ_4^{4a} such that $R = \{\bar{v}_1, \bar{v}_2\}$ and suppose that $v_1^{\widehat{\tau(B,1)}} = v_2$. Then $P_4(v_1) = P_4(v_2) = S_i$ for some $1 \leq i \leq 6$, $|X(v_1)|$ and $|X(v_2)|$ have different parities and hence $|B \cap S_i|$ must be odd. Now the result is immediate from (2.14.1). □

Lemma 4.7.7 *Let $\bar{v} \in \bar{\Lambda}_4^6$. Then $G_1(\bar{v})$ has four orbits on $\bar{\Lambda}_4^{4a}$ with lengths* 35, 35, 1680 *and* 1792 *which are contained in $\bar{\Lambda}_4^{\alpha}(\bar{v})$ for $\alpha = 4a$, 6, $4b$ and 5, respectively.*

Proof. Let B be the octad such that $L_1(\bar{v}) = L_1[B]$. Then by (2.14.1) and (3.7.2) $G_1(\bar{v})$ acting on $\varrho(\bar{\Lambda}_4^{4a})$ has 3 orbits $\widehat{\Psi}_1$, $\widehat{\Psi}_2$ and $\widehat{\Psi}_3$ with lengths 35, 840 and 896 containing the sextets such that $\mu(B, \Sigma) = (4^2 0^4)$, $(2^4 0^2)$ and $(3\,1^5)$, respectively. Let Ψ_i be the preimage of $\widehat{\Psi}_i$ in $\bar{\Lambda}_4^{4a}$ for $1 \leq i \leq 3$. We claim that the kernel of Q_1 on its orbit R contains $Q_1(\bar{v})$ if and only if $R \subseteq \Psi_1$. By the proof of (4.6.1) $Q_1(\bar{v})^{\#} = \{\tau(B_1, 1) \mid |B_1 \cap B| \in \{0, 8\}\}$. A sextet from $\widehat{\Psi}_1$ refines B and hence $Q_1(\bar{v})$ is in the kernel of R when

$R \subseteq \Psi_1$. If $R \subseteq \Psi_3$ then $\tau(B,1)$ is not in the kernel by (4.7.6). Suppose that $Q_1(\bar{v})$ is in the kernel of R if $R \subseteq \Psi_2$. Then by (4.7.6) the action of $\tau(B,1)$ on $\bar{\Lambda}_4^{4a}$ coincides with that of $\tau(B_1,1)$ where $|B_1 \cap B| = 0$ which is impossible and the claim follows. Thus the actions of $G_1(\bar{v})$ on Ψ_2 and Ψ_3 are transitive. On the other hand we know that $\bar{v}$ is adjacent in Γ to exactly 35 vertices in $\bar{\Lambda}_4^{4a}$. Hence there are 2 G_1-orbits in Ψ_1 of length 35 each.

Let v be a preimage of $\bar{v}$ in $\bar{\Lambda}_4^6$ so that $P_2(v) \cup P_6(v) = B$ and suppose that $X(v) = P_6(v)$. Let $\lambda \in \Lambda_4^{4a}$ with $P_6(v) \subset P_4(\lambda) \subset P_2(v) \cup P_6(v)$ and $X(\lambda) = P_6(v)$. Then $\bar{\lambda} \in \Psi_1$ and $\frac{1}{8}(v,\lambda) = 6$, which shows that one of the $G_1(\bar{v})$-orbits in Ψ_1 is contained in $\bar{\Lambda}_4^6(\bar{v})$ (the other one is known to be contained in $\bar{\Lambda}_4^{4a}$). Now let $\lambda \in \Lambda_4^{4a}$ with $P_6(v) \subset P_4(\lambda)$, $|P_4(\lambda) \cap P_2(v)| = 2$ and $X(\lambda) \cap (P_2(v) \cup P_6(v)) = P_6(v)$. Then $\bar{\lambda} \in \Psi_3$, $\frac{1}{8}(v,\lambda) = 5$ and we have $\Psi_3 \subset \bar{\Lambda}_4^5(\bar{v})$. By similar calculations we can deduce that $\Psi_2 \subset \bar{\Lambda}_4^{4b}$ but we can also apply a different argument. In the notation of (4.7.2) a sextet from $\varrho(\Phi_2)$ is adjacent in the sextet graph to a sextet from $\varrho(\Phi_3)$ (the diagram $D_s(Mat_{24})$) and hence $\bar{v}$ must be adjacent to some vertices in $\bar{\Lambda}_4^{4b}$. Since we are left with only one $G_1(\bar{v})$-orbit to locate, the result follows. □

Lemma 4.7.8 *Let v be a vector from Λ_4^5 such that $v(a) = 3 \bmod 4$ for all $a \in \mathscr{P}$ and put $Y = P_5(v) \cup P_3(v)$. Then*

(i) *$G_1(\bar{v})$ acting on $\varrho(\bar{\Lambda}_4^{4a})$ has three orbits $\widehat{\Delta}_1$, $\widehat{\Delta}_2$ and $\widehat{\Delta}_3$ with lengths* 21, 630 *and* 1120,

(ii) *$G_1(\bar{v})$ acting on $\bar{\Lambda}_4^{4a}$ has six orbits with lengths* 21, 21, 630, 630, 1120 *and* 1120 *contained in $\bar{\Lambda}_4^{\alpha}(\bar{v})$ for $\alpha = 6, 5, 4b, 5, 4c$ and 5, respectively.*

Proof. By (3.10.1) we obtain (i) and also that $\widehat{\Delta}_i$ consists of the sextets Σ such that Y intersects i tetrads in Σ. Let Δ_i be the preimage of $\widehat{\Delta}_i$ in $\bar{\Lambda}_4^{4a}$ for $1 \leq i \leq 3$. It was established in the proof of (4.6.1) that $G_1(\bar{v}) = L_1(\bar{v})$ and by (4.7.5) $\Delta_i \cap O_1$ and $\Delta_i \cap O_2$ are the orbits of $G_1(\bar{v})$ on Δ_i. Now considering representatives $\bar{\mu}$ from the orbits $\Delta_i \cap O_j$, $1 \leq i \leq 3$, $j = 1,2$, and calculating the inner products of v with the vectors μ_k, $1 \leq k \leq 48$, from the preimage of $\bar{\mu}$ in Λ_4^{4a} and using (4.7.4) we obtain (ii). □

Lemma 4.7.9 *If $v \in \Lambda_4^{4c}$, then*

(i) *$G_1(\bar{v}) \cong Mat_{12}.2$ acting on $\varrho(\bar{\Lambda}_4^{4a})$ has three orbits $\widehat{\Theta}_4$, $\widehat{\Theta}_3$ and $\widehat{\Theta}_2$ with lengths* 495, 880 *and* 396, *respectively,*

(ii) *$G_1(\bar{v})$ acting on $\bar{\Lambda}_4^{4a}$ has four orbits with lengths* 495, 495, 792 *and* 1760, *contained in $\bar{\Lambda}_4^{\alpha}(\bar{v})$ for $\alpha = 4b, 4c, 4c$ and 5, respectively.*

Proof. Let $D_1 = P_2(\nu)$, which is a dodecad, and let $D_2 = \mathcal{P} \setminus D_1$ be the complementary dodecad. By the proof of (4.6.1) the action of $G_1(\bar{\nu})$ on $\varrho(\bar{\Lambda}_4^{4a})$ is similar to that of the stabilizer F in L_1 of the partition $\{D_1, D_2\}$. Hence by (3.10.2) we obtain (i) and also that $\widehat{\Theta}_m$ consists of the sextets $\Sigma = \{S_1, S_2, ..., S_6\}$ such that

$$m = \max_{\substack{j=1,2 \\ 1 \le i \le 6}} |D_j \cap S_i|.$$

Let Θ_m be the preimage of $\widehat{\Theta}_m$ in $\bar{\Lambda}_4^{4a}$. Then by (4.7.7) Θ_3 must be a single $G_1(\bar{\nu})$-orbit contained in $\bar{\Lambda}_4^5(\bar{\nu})$. By (4.7.4) and straightforward calculations we see that Θ_4 consists of two $G_1(\bar{\nu})$-orbits contained in $\bar{\Lambda}_4^{4b}(\bar{\nu})$ and $\bar{\Lambda}_4^{4c}(\bar{\nu})$, respectively. Similarly (4.7.4) shows that Θ_2 is contained in $\bar{\Lambda}_4^{4c}(\bar{\nu})$ and it only remains to show that $G_1(\bar{\nu})$ acts on Θ_2 transitively. Let λ be a vector from Λ_4^{4c} such that $P_2(\lambda) = P_2(\nu)$, $P_4(\lambda) \neq P_4(\nu)$, $P_4(\lambda) \subseteq X(\lambda)$ and

$$B := (X(\lambda) \cap P_2(\lambda)) \cup P_4(\nu) \cup P_4(\lambda)$$

is an octad. Then by the proof of (4.6.1), or otherwise, one can see that $\bar{\lambda} = \bar{\nu}$. Let $\mu \in \Lambda_4^{4a}$ with $P_4(\nu) \cup P_4(\lambda) \subseteq P_4(\mu) \subseteq P_4(\nu) \cup P_4(\lambda) \cup P_2(\nu)$, $|P_4(\mu) \cap P_2(\lambda) \cap X(\lambda)| = 1$ and $X(\mu) = \emptyset$. Then $\bar{\mu} \in \Theta_2$. It can be shown (the proof of (3.10.2)) that there is $\widehat{\delta} = \widehat{\tau}(B, \sigma) \in \widehat{G}_1$ such that $P_4(\mu)^\sigma = P_4(\mu)$ and $\nu^{\widehat{\delta}} = \lambda$. Then $|B \cap P_4(\mu)| = 3$ and hence for $\mu_1 = \mu^{\widehat{\delta}}$ we have $\bar{\mu}_1 \neq \bar{\mu}$ and the result follows. □

Lemma 4.7.10 *If $\nu \in \Lambda_4^{4b}$, then*

(i) *$G_1(\bar{\nu})$ has six orbits on $\varrho(\bar{\Lambda}_4^{4a})$ with lengths* 7, 28, 56, 336, 448 *and* 896,

(ii) *$G_1(\bar{\nu})$ has eight orbits on $\bar{\Lambda}_4^{4a}$ with lengths* 7, 7, 56, 56, 56, 672, 896 *and* 1792 *contained in $\bar{\Lambda}_4^{\alpha}(\bar{\nu})$ for $\alpha = 4a, 4b, 4b, 6, 4b, 4b, 4c$ and* 5, *respectively.*

Proof. (i) is immediate from the proofs of (4.6.1) and (3.10.3). Let $\{\widehat{\Omega}_i \mid 1 \le i \le 6\}$ be the set of $G_1(\bar{\nu})$-orbits on $\varrho(\bar{\Lambda}_4^{4a})$ assuming that $|\widehat{\Omega}_i| < |\widehat{\Omega}_j|$ for $i < j$ and let Ω_i be the preimage of $\widehat{\Omega}_i$ in $\bar{\Lambda}_4^{4a}$. By (4.7.8) Ω_6 is a $G_1(\bar{\nu})$-orbit contained in $\bar{\Lambda}_4^5(\bar{\nu})$; by (4.7.9) Ω_5 is a $G_1(\bar{\nu})$-orbit contained in $\bar{\Lambda}_4^{4c}(\bar{\nu})$. By (4.7.7) and its proof $\bar{\nu}$ is adjacent in Γ to 56 vertices from $\bar{\Lambda}_4^6$. Furthermore, if $\bar{\mu}$ is one of them and $\{\bar{\nu}, \bar{\mu}, \bar{\lambda}\}$ is a line then $\bar{\lambda} \in \bar{\Lambda}_4^{4b}$. This shows that Ω_3 consists of 2 $G_1(\bar{\nu})$-orbits contained in $\bar{\Lambda}_4^6(\bar{\nu})$ and $\bar{\Lambda}_4^{4b}(\bar{\nu})$, respectively. By the paragraph after (4.7.3) Ω_1 consists of 2 $G_1(\bar{\nu})$-orbits contained in $\bar{\Lambda}_4^{4a}(\bar{\nu})$ and $\bar{\Lambda}_4^{4b}(\bar{\nu})$, respectively. By the above Ω_2 and Ω_4 are contained in $\bar{\Lambda}_4^{4b}(\bar{\nu})$. It only remains to show that

$G_1(\bar{v})$ is transitive on each of these 2 sets and we suggest this as an exercise. □

4.8 The centralizer of an involution

A clique in the Leech graph Γ will be called $\star$-closed if together with an edge it contains the line containing this edge. By (4.7.3) if L is a $\star$-closed clique then its vertices form the set of non-zero vectors of a subspace $V(L)$ in $\bar{\Lambda}$, particularly $|L| = 2^a - 1$ where a is the dimension of $V(L)$. This means that the vertices and $\star$-closed cliques contained in L form the projective geometry of $V(L)$ with respect to the incidence relation defined via inclusion.

Lemma 4.8.1 *If L is a maximal clique in Γ, then L is $\star$-closed, it contains 15 vertices and $G[L]$ induces on L a transitive action. There are 2 orbits $\mathcal{L}_v$ and $\mathcal{L}_t$ of G on the set of maximal cliques in Γ and $|\mathcal{L}_v| = 3 \cdot |\mathcal{L}_t|$.*

Proof. By (4.7.2) a maximal clique L containing $\bar{\lambda}_0$ is of the form $\{\bar{\lambda}_0\} \cup (\varrho^{-1}(K))$ where K is one of the maximal cliques in the sextet graph described in (3.3.8). Thus the maximal cliques in Γ containing $\bar{\lambda}_0$ are in two G_1-orbits bijectively corresponding to the classes $\mathcal{K}_v$ and $\mathcal{K}_t$ of maximal cliques in the sextet graph. Since $|\mathcal{K}_v| = 3 \cdot |\mathcal{K}_t|$ and G acts transitively on the vertex set of Γ, there are two G-orbits, $\mathcal{L}_v$ and $\mathcal{L}_t$ on the set of maximal cliques in Γ so that if $\bar{\lambda}_0 \in L \in \mathcal{L}_\alpha$ then $\varrho(L \setminus \{\bar{\lambda}_0\}) \in \mathcal{K}_\alpha$ for $\alpha = v$ or t. Also because of the transitivity of G on the vertex set of Γ, $G[L]$ is transitive on the set of vertices in L. By the above, whenever L contains an edge incident to $\bar{\lambda}_0$, it contains the line determined by this edge. Hence the transitivity of $G[L]$ on L implies that L is $\star$-closed. □

It is clear that a $\star$-closed clique of size 3 is a line. A $\star$-closed clique of size 7 is the intersection of a clique from $\mathcal{L}_v$ and a clique from $\mathcal{L}_t$; such a clique containing $\bar{\lambda}_0$ is of the form $\bar{\lambda}_0 \cup \varrho^{-1}(M)$ where M is (the set of sextets incident to) an element of type 2 in $\mathcal{G}(Mat_{24})$.

Lemma 4.8.2 *For $i = 1, 2$ and 3 let U_i be the subspace in $\bar{\Lambda}$ generated by $\bar{\Lambda}_4^8$, $\bar{\Lambda}_4^8 \cup \bar{\Lambda}_4^{4a}$ and $\bar{\Lambda}_4^8 \cup \bar{\Lambda}_4^6 \cup \bar{\Lambda}_4^{4a} \cup \bar{\Lambda}_4^{4b} \cup \bar{\Lambda}_4^{4c}$, respectively. Then*

$$1 < U_1 < U_2 < U_3 < \bar{\Lambda}$$

is the only composition series of $\bar{\Lambda}$ as a module for G_1 and the following assertions hold:

(i) $U_1 = C_{\bar{\Lambda}}(Q_1)$ *and* $U_3 = [\bar{\Lambda}, Q_1]$,
(ii) U_2/U_1 *is the centralizer of* Q_1 *in* $\bar{\Lambda}/U_1$,
(iii) $U_2/U_1 \cong \bar{\mathscr{C}}_{11}$ *and* $U_3/U_2 \cong \mathscr{C}_{11}$,
(iv) $U_1^{\perp} = U_3$, $U_2^{\perp} = U_2$.

Proof. For $i = 1$ and 2 let M_i be a $\star$-closed clique in the Leech graph of size $2^{i+1} - 1$ containing $\bar{\lambda}_0$. Then $V(M_i)$ is an $(i+1)$-space in $\bar{\Lambda}$ containing U_1 so that $V(M_i)/U_1$ is an i-space in U_2/U_1. Furthermore $\rho(M_i \setminus \{\bar{\lambda}_0\})$ is a sextet (an element of type 1 in $\mathscr{G}(Mat_{24})$) if $i = 1$ and it is a $\star$-closed clique of size 3 in the sextet graph (an element of type 2 in $\mathscr{G}(Mat_{24})$) if $i = 2$. This means that U_2/U_1 supports a natural representation of $\mathscr{G}(Mat_{24})$ invariant under Mat_{24}. Hence $U_2/U_1 \cong \bar{\mathscr{C}}_{11}$ by (3.3.10). Since G_1 preserves the quadratic form θ on $\bar{\Lambda}$ as in (4.5.6), the remaining assertions are straightforward from (4.4.1) and (4.6.1), noticing that $U_1^{\perp} = U_3$ and $U_2^{\perp} = U_2$, as stated in (iv). □

The following result will play an essential rôle in Section 5.6.

Lemma 4.8.3 *Let* $E_2 = \{v \in \Lambda_2^4, |X(v)| = 1\}$, $E_4 = \{v \in \Lambda_4^{4a}, X(v) = \emptyset\}$, $E_3 = \{v \in \Lambda_2^3, X(v) = P_3(v)\}$ *and* $E = \{0\} \cup E_2 \cup E_4$. *Then*

(i) $\widehat{L}_1 \cong Mat_{24}$ *stabilizes* E_i *setwise for* $i = 2$, 3 *and* 4,
(ii) $L_1 \cong Mat_{24}$ *acts on* $\bar{E}_i$ *as it acts on the set of ordered pairs of element subsets from* $\mathscr{P}$, *on the* 1*-element subsets of* $\mathscr{P}$ *and on the set of sextets, for* $i = 2$, 3 *and* 4, *respectively*,
(iii) $\bar{E}$ *is a complement to* U_1 *in* U_2,
(iv) $\bar{E}_3 = \bar{E}^{\perp} \cap \bar{\Lambda}_2^3$.

Proof. The assertions (i) and (ii) are immediate. A Leech vector and its negative have the same image in $\bar{\Lambda}$; if $v_1, v_2 \in E_4$ and $P_4(v_1)$, $P_4(v_2)$ are tetrads from the same sextet then $\bar{v}_1 = \bar{v}_2$; by the proof of (3.7.3) in any two sextets there are tetrads intersecting in at least two elements. Using these facts it is easy to see that $\bar{E}$ is closed under addition which implies (iii) since $|\bar{E}| = 2^{11}$. Finally (iv) follows directly from the definition of the invariant quadratic form θ. □

Lemma 4.8.4 *Let* B *be an octad,* $\widehat{\Psi}$ *be the set of sextets which refine* B, Ψ *be the preimage of* $\widehat{\Psi}$ *in* $\bar{\Lambda}_4^{4a}$ *and* $W = O_2(L_1[B])$. *Then*

(i) W *is the kernel of the action of* $L_1[B]$ *on* Ψ,
(ii) $[Q_1, W]$ *is the kernel of the action of* Q_1 *on* Ψ.

Proof. By (2.14.1) W is the kernel of $L_1[B]$ on $\widehat{\Psi}$ and by (4.7.5) we obtain (i). If B_1 is an octad such that $|B \cap B_1| \in \{0,8\}$, then B_1 has even intersection with the tetrads in every sextet from $\widehat{\Psi}$ and $\tau(B_1, 1)$ is in the kernel of Q_1 on Ψ by (4.7.6). The elements $\tau(B_1, 1)$ as above generate the commutator subgroup $[Q_1, W]$ of order 2^5. Since the action of Q_1 on Ψ is non-trivial and by (3.8.4) $Q_1/[Q_1, W]$ is irreducible as a module for $L_1[B]$, (ii) follows. □

By (2.15.1) G_1, acting on $Q_1^{\#}$ by conjugation, has two orbits with lengths 759 and 1288 containing elements $\tau(Y, 1)$ where Y is an octad and dodecad, respectively (recall that $\tau(Y, 1) = \tau(\mathscr{P} \setminus Y, 1)$). By (4.6.2) if Y is a dodecad then $\bar{\lambda}_0$ is the only element $\bar{\mu} \in \bar{\Lambda}_4$ such that $\tau(Y, 1) \in Q(\bar{\mu}) := O_2(G(\bar{\mu}))$.

For an octad B and $\delta = \tau(B, 1)$ put

$$\Phi = \Phi(\delta) = \{\bar{\mu} \mid \bar{\mu} \in \bar{\Lambda}_4, \delta \in Q(\bar{\mu})\}$$

and let D be the setwise stabilizer of Φ in G. Then by the above paragraph and (4.6.2) we have the following.

Lemma 4.8.5 *The action of D on Φ is transitive and $D = C_G(\delta)$. The set Φ consists of $\bar{\lambda}_0$, the 35 Q_1-orbits on $\bar{\Lambda}_4^{4a}$ corresponding to the sextets which refine B and the Q_1-orbit on $\bar{\Lambda}_4^6$ which corresponds to B in the sense of (4.6.2 (ii)). In particular*

$$|\Phi| = 1 + 2 \cdot 35 + 2^6 = 135$$

and $\Phi \cap \Gamma(\bar{\lambda}_0) = \Psi$ in the notation of (4.8.4). □

Let R be the kernel of the action of D on Φ and $\bar{D} = D/R$, V_1 be the subspace in $\bar{\Lambda}$ generated by Φ and $V_2 = R/Z(R)$. We continue to follow notation introduced in (4.8.4).

Lemma 4.8.6 *The following assertions hold:*

(i) $\bar{D} \cong \Omega_8^+(2)$,
(ii) $R = [Q_1, W] : W$, $Z(R) = [R, R] = \langle \tau(B, 1) \rangle$ *and* $R \cong 2_+^{1+8}$,
(iii) $V_1 = C_{\bar{\Lambda}}(R)$ *and V_1 is the natural module for $\bar{D}$,*
(iv) $V_1^{\#}$ *consists of Φ, 64 elements from $\bar{\Lambda}_2^2$ and 56 elements from $\bar{\Lambda}_2^4$,*
(v) $V_2 \cong V_1^{\alpha}$ *where α is a diagram automorphism of $\bar{D}$.*

Proof. Let $\mathscr{D}$ be an incidence system whose elements of type 1, 2, 3 and 4 are the vertices, the lines, the cliques from $\mathscr{L}_t$ and the cliques from $\mathscr{L}_v$, respectively, contained in Φ. A clique from $\mathscr{L}_t$ and a clique from $\mathcal{L}_v$ are incident if their intersection is (a $\star$-closed clique) of size 7 and the remaining incidences are via inclusion. Let $\{x_1, x_2, x_3, x_4\}$ be a maximal flag in $\mathcal{D}$ where x_i is of type i and $x_1 = \bar{\lambda}_0$. Let U be the subspace in $\bar{\Lambda}$ generated by Ψ. Then by (4.8.2) U/U_1 is the centralizer of W in $U_1/U_2 \cong \mathscr{C}_{11}$ and by (3.8.2) $U/U_1 \cong \bigwedge^2 W$. This shows that the $\star$-closed cliques of size $2^i - 1$ containing $\bar{\lambda}_0$ and contained in Φ are in a bijection with the totally singular subspaces of dimension $i - 1$ in U/U_1 with respect to the unique quadratic form of plus type on U/U_1 preserved by $L_1[B]/W \cong L_4(2) \cong \Omega_6^+(2)$. This shows that $\mathrm{res}_{\mathcal{D}}(x_1)$ is the projective rank 3 geometry over $GF(2)$ and also that for a $\star$-closed clique N of size 7 such that $\bar{\lambda}_0 \in N \subseteq \Phi$ there are a unique element of type 3 and a unique element of type 4 in $\mathcal{D}$ which contain N. By the latter observation, for $j = 3$ and 4 the residue $\mathrm{res}_{\mathcal{D}}(x_j)$ is isomorphic to the rank 3 geometry of proper subspaces of $V(x_j)$. Hence $\mathcal{D}$ is a Tits geometry with the following diagram:

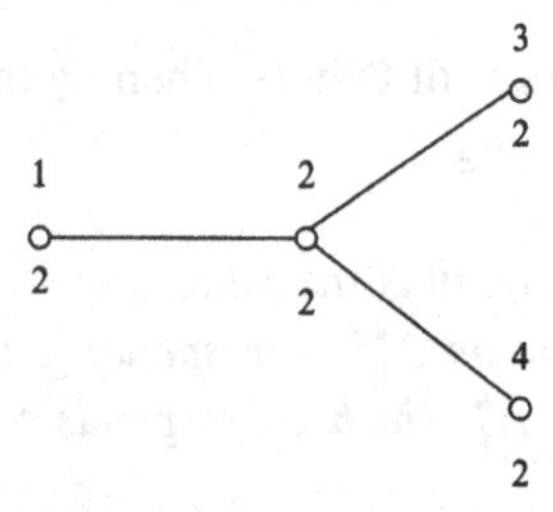

By (1.6.3) $\mathcal{D}$ is the parabolic geometry of $\Omega_8^+(2)$ and by (1.6.5) the latter group is contained in $\bar{D}$. Since $\Omega_8^+(2)$ is the automorphism group of its parabolic geometry (i) follows. By basic properties of $\mathcal{D}$ the stabilizer of $\bar{\lambda}_0$ in $\bar{D}$ acts faithfully on Ψ. This gives the first equality in (ii) because of (4.8.4). By (3.8.2) $\langle \tau(B, 1) \rangle$ is the centralizer of W in $[Q_1, W]$ and also the commutator subgroup of $[Q_1, W]$ and W. Hence noticing that $[Q_1, W]$ is elementary abelian of order 2^5 we obtain the remaining equalities in (ii). By the construction V_1 supports a natural representation of $\mathcal{D}$ and hence it is the 8-dimensional natural module for $\bar{D}$. From (4.4.1) and the table therein it is easy to see that R, besides the elements in Φ, stabilizes 56 elements in $\bar{\Lambda}_2^4$ which are images of vectors from Λ_2^4 with $P_4 \subseteq B$ and 64 elements in $\bar{\Lambda}_2^2$ which are images of vectors from Λ_2^2 with $P_2 = B$. Thus (iii) and (iv) follow. Since $V_2 = R/Z(R)$ is 8-dimensional it is isomorphic either to V_1 or to its image under a diagram automorphism of $\bar{D}$. Since $[Q_1, W]/Z(R)$ is a 4-dimensional subspace in V_2 normalized by the stabilizer $D \cap G_1$ of the vector $\bar{\lambda}_0$ from V_1 we have the latter possibility as stated in (v). □

We will use the same letter Φ to denote the subgraph in Γ induced by Φ. Let κ be the orthogonal form (of plus type) on V_1 preserved by D. Then $\Phi = V_1 \cap \bar{\Lambda}_4$ is the set of isotropic and $V_1 \cap \bar{\Lambda}_2$ is the set of non-isotropic vectors in V_1 with respect to κ and we have the following.

Lemma 4.8.7 *The following assertions hold:*

(i) *if* $\lambda \in \Lambda$ *and* $\bar{\lambda} \in V_1$ *then* $\kappa(\bar{\lambda}) = 0$ *if* $\lambda^2 = 0 \bmod 8$ *and* $\kappa(\bar{\lambda}) = 1$ *if* $\lambda^2 = 4 \bmod 8$,

(ii) $\bar{D}$ *acts transitively on* $V_1 \cap \bar{\Lambda}_2$ *with stabilizer isomorphic to* $Sp_6(2)$. □

By (4.8.7) Φ is a graph on the set of vectors in V_1 isotropic with respect to κ in which two such vectors are adjacent if they are perpendicular. Hence the suborbit diagram of Φ with respect to the action of D is the following:

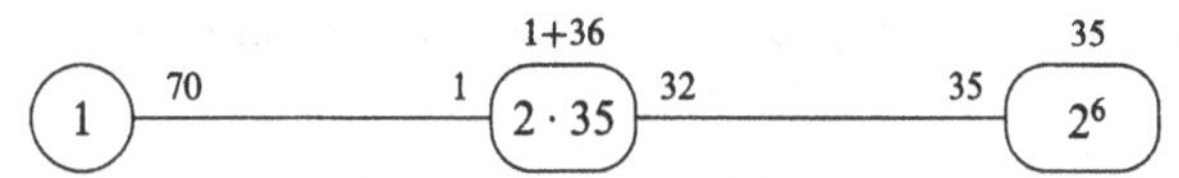

Comparing this diagram with the suborbit diagram of the Leech graph, we observe that whenever $\bar{\nu}, \bar{\mu}$ are non-adjacent vertices in Φ, $\Gamma(\bar{\nu}) \cap \Gamma(\bar{\mu}) \subseteq \Phi$. On the other hand, since $\langle \bar{\nu}, \bar{\mu} \rangle$ is non-singular, $V_1 = \langle \bar{\nu}, \bar{\mu} \rangle \oplus \langle \bar{\nu}, \bar{\mu} \rangle^{\perp}$, where by (4.8.7) the latter summand is $\langle \Gamma(\bar{\nu}) \cap \Gamma(\bar{\mu}) \rangle$. This gives the following.

Lemma 4.8.8 *Let* $\bar{\nu} \in \bar{\Lambda}_4$ *and* $\bar{\mu} \in \bar{\Lambda}_4^6(\bar{\nu})$. *Let* $V_1(\bar{\nu}, \bar{\mu}) = \langle \bar{\nu}, \bar{\mu}, \Gamma(\bar{\nu}) \cap \Gamma(\bar{\mu}) \rangle$ *(a subgroup in* $\bar{\Lambda}$*) and* $\Phi(\bar{\nu}, \bar{\mu})$ *be the subgraph in* Γ *induced by the vertices contained in* $V_1(\bar{\nu}, \bar{\mu})$. *Then*

(i) $V_1(\bar{\nu}, \bar{\mu}) = V_1^g$ *and* $\Phi(\bar{\nu}, \bar{\mu}) = \Phi^g$ *for some* $g \in G$,

(ii) $V_1(\bar{\nu}, \bar{\mu}) = \langle \bar{\nu}, \bar{\mu}, U_2(\bar{\nu}) \cap U_2(\bar{\mu}) \rangle$. □

4.9 Geometries of Co_1 and Co_2

The maximal parabolic geometry of Co_1 can be defined in terms of the Leech graph and some of its subgraphs. To wit, let $\mathscr{H}(Co_1)$ be an incidence system of rank 4, whose elements of type 1, 2, 3 and 4 are the vertices, triangles, maximal cliques from the class $\mathscr{L}_t$ and all images under G of the subgraph Φ defined before (4.8.5), respectively. The incidence relation is via inclusion.

Lemma 4.9.1 *The incidence system $\mathscr{H}(Co_1)$ is a geometry with the diagram*

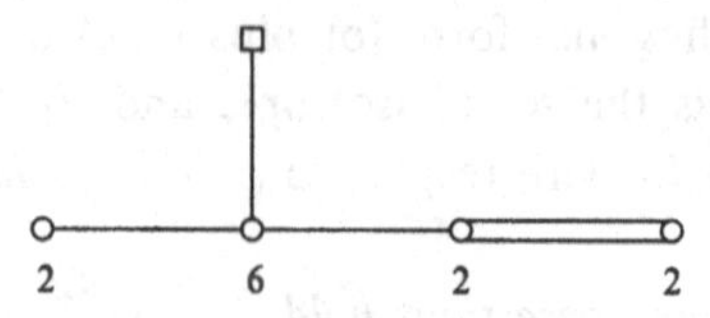

and $G \cong Co_1$ acts on $\mathscr{H}(Co_1)$ flag-transitively.

Proof. Since the incidence relation is via inclusion, it is easy to see that $\mathscr{H}(Co_1)$ is a geometry with a string diagram. By (3.3.8) and (4.8.1) we identify the residue in $\mathscr{H}(Co_1)$ of an element of type 1 (say of $\bar{\lambda}_0$) with the maximal parabolic geometry $\mathscr{H}(Mat_{24})$ of the Mathieu group. Then by (4.8.5) the residue of an element of type 4 (say of Φ) is isomorphic to the geometry of totally isotropic 1-spaces, 2-spaces and one class of 4-spaces in the natural module of $\bar{D} \cong \Omega_8^+(2)$. In other terms this residue is a truncation of the Tits geometry of the latter group and the result follows. □

If $\{H_i \mid 1 \leq i \leq 4\}$ is the amalgam of maximal parabolics associated with the action of G on $\mathscr{H}(Co_1)$, then

$$H_1 \cong G_1 \cong 2^{11} : Mat_{24}, \quad H_2 \cong 2^{4+12}.(3 \cdot Sym_6 \times Sym_3),$$

$$H_3 \cong 2^{2+12}.(L_4(2) \times Sym_3), \quad H_4 \cong D \cong 2_+^{1+8}.\Omega_8^+(2).$$

In order to obtain a diagram without fake nodes, similarly to the case of Mat_{24} we define the minimal parabolic geometry $\mathscr{G}(Co_1)$, whose elements of type 1 and 2 are the same as in $\mathscr{H}(Co_1)$. The elements of type 3 are the $\star$-closed cliques of size 7 in Γ while the elements of type 4 are the cliques from $\mathscr{L}_v$. The incidence relation is via inclusion.

Lemma 4.9.2 *$\mathscr{G}(Co_1)$ is a rank 4 tilde geometry with the diagram*

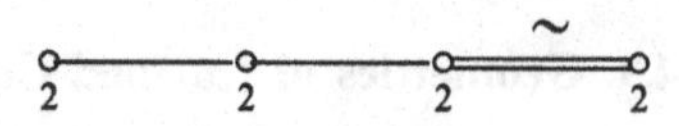

and $G \cong Co_1$ acts on it flag-transitively.

Proof. First notice that a clique from $\mathscr{L}_v$ is contained in a unique subgraph which is an element of type 4 in $\mathscr{H}(Co_1)$ while a clique from $\mathscr{L}_t$ is contained in three such subgraphs. By (3.3.9) the residue in $\mathscr{G}(Co_1)$ of an element of type 1 is the tilde geometry $\mathscr{G}(Mat_{24})$ while the residue of an element of type 4 is the rank 3 projective $GF(2)$-geometry of proper $\star$-closed subgraphs contained in the corresponding maximal clique. □

Let $\{G_i \mid 1 \leq i \leq 4\}$ be the amalgam of maximal parabolics associated with the action of G on $\mathscr{G}(Co_1)$. Then

$$G_1 \cong H_1 \cong 2^{11} : Mat_{24}, \quad G_2 \cong H_2 \cong 2^{4+12}.(3 \cdot Sym_6 \times Sym_3),$$

$$G_3 \cong 2^{2+12+3}.(L_3(2) \times Sym_3), \quad G_4 \cong 2^{1+4+4+6}.L_4(2)$$

(notice that G_4 is the preimage in D of a maximal parabolic subgroup in $\bar{D}$ which is the stabilizer of a maximal totally isotropic subspace in the natural module).

Directly by the definition of the Leech graph we obtain the following.

Lemma 4.9.3 *The Leech graph is the collinearity graph of both $\mathscr{H}(Co_1)$ and $\mathscr{G}(Co_1)$ and each of the geometries possesses a natural representation in $\bar{\Lambda}$.* □

For $\bar{v} \in \bar{\Lambda}$ and $j = 0, 2, 3$ or 4 put

$$\bar{\Lambda}_j(\bar{v}) = \bar{\Lambda}_j + \bar{v} = \{\bar{\lambda} \mid \bar{\lambda} \in \bar{\Lambda}, \bar{v} + \bar{\lambda} \in \bar{\Lambda}_j\},$$

so that $\bar{\Lambda}_j = \bar{\Lambda}_j(\bar{0})$.

Let us fix a vector $\mu_0 \in \Lambda_2$ and let F denote the stabilizer $G(\bar{\mu}_0)$ of $\bar{\mu}_0$ in G, which is isomorphic to Co_2 by (4.5.5) and will be identified with $\widehat{G}(\mu_0)$ in view of the remark before that statement.

Lemma 4.9.4 *For $j = 2$, 3 and 4 the action of F on $\bar{\Lambda}_4 \cap \bar{\Lambda}_j(\bar{\mu}_0)$ is transitive with stabilizers isomorphic to*

$$2^{10} : \mathrm{Aut}\, Mat_{22}, \quad Mat_{23} \text{ and } 2^5 : 2^4 : L_4(2) \cong 2_+^{1+8} : L_4(2),$$

respectively.

Proof. Notice that for $j = 4$, 3 and 2 we have $\bar{\Lambda}_2^j = \bar{\Lambda}_2 \cap \bar{\Lambda}_k(\bar{\lambda}_0)$ for $k = 2$, 3 and 4, respectively. Since G acts transitively on both $\bar{\Lambda}_2$ and $\bar{\Lambda}_4$, in view of this observation the result is immediate from (4.4.1) and the table therein. □

We study the subgraph $\Theta = \Theta(\bar{\mu}_0)$ of the Leech graph Γ induced by $\bar{\Lambda}_4 \cap \bar{\Lambda}_2(\bar{\mu}_0)$. First notice that

$$|\Theta| = |\bar{\Lambda}_4| \cdot |\bar{\Lambda}_2^4| \cdot |\bar{\Lambda}_2|^{-1} = 46\,575.$$

In order to work with the coordinates for Leech vectors introduced in Section 4.4, it is convenient to assume that $\bar{\mu}_0 \in \bar{\Lambda}_2^4$ (so that $\bar{\lambda}_0 \in \Theta$). If in addition we assume that $X(\mu_0) = \emptyset$ and put $\{a, b\} = P_4(\mu_0)$ then an element $\bar{\lambda} \in \bar{\Lambda}_4$ is contained in Θ if and only if $\bar{\lambda}$ is the image of a vector $\lambda \in \Lambda_4$ such that $\lambda(a)+\lambda(b) = 8$. There are exactly two choices for λ when $\bar{\lambda}$ is fixed (in view of the transitivity of F on Θ it is sufficient to check this for $\bar{\lambda} = \bar{\lambda}_0$). Furthermore, $F_1 := F(\bar{\lambda}_0) = G_1(\bar{\mu}_0)$ is the semidirect product $Q_1(\bar{\mu}_0) : L_1(\bar{\mu}_0)$ where $Q_1(\bar{\mu}_0)$ consists of the elements $\tau(Y, 1)$ for Golay sets Y containing $P_4(\mu_0)$ and $L_1(\bar{\mu}_0) = L_1[a, b] \cong \mathrm{Aut}\, Mat_{22}$. By (2.15.3) this shows that $Q_1(\bar{\mu}_0) = O_2(F_1)$ is isomorphic to the 10-dimensional Golay code module.

Lemma 4.9.5 *The group F_1 acting on Θ has five orbits $\Theta \cap \bar{\Lambda}_4^{\alpha}$ for $\alpha =$ 8, 6, 5, 4a and 4b. Their lengths and element stabilizers are given in the table below. The group $F_1/O_2(F_1) \cong \mathrm{Aut}\, Mat_{22}$ acts on the set of $O_2(F_1)$-orbits in $\Theta \cap \bar{\Lambda}_4^{\alpha}$ for $\alpha = 4a, 6, 4b$ and 5 as it acts on the elements in $\mathscr{H}_p$, $\mathscr{H}_h$ and $\mathscr{H}_o$ of the geometry $\mathscr{H}(Mat_{22})$ and on the elements of the Steiner system $S(3, 6, 22)$, respectively.*

Orbits	Lengths	Stabilizers
$\Theta \cap \bar{\Lambda}_4^{8}$	1	$[2^{10}] : \mathrm{Aut}\, Mat_{22}$
$\Theta \cap \bar{\Lambda}_4^{6}$	$2^5 \cdot 77$	$[2^5] : (2^4 : Sym_6)$
$\Theta \cap \bar{\Lambda}_4^{5}$	$2^{10} \cdot 22$	$P\Sigma L_3(4)$
$\Theta \cap \bar{\Lambda}_4^{4a}$	$2 \cdot 231$	$[2^9] : (2^5 : Sym_5)$
$\Theta \cap \bar{\Lambda}_4^{4b}$	$2^6 \cdot 330$	$[2^4].(2 \times 2^3 : L_3(2))$

Proof. We have seen that $\bar{\lambda} \in \Theta$ if and only if $\bar{\lambda}$ is the image of $\lambda \in \Lambda_4$ with $\lambda(a) + \lambda(b) = 8$. This immediately shows that $\Theta \cap \bar{\Lambda}_4^{4c} = \emptyset$, $\Theta \cap \bar{\Lambda}_4^{6}$ consists of the images of $\lambda \in \Lambda_4^6$ with $P_6(\lambda) \subset \{a, b\} \subset P_6(\lambda) \cup P_2(\lambda)$, for $\beta = 4a$ and $4b$ the set $\Theta \cap \bar{\Lambda}_4^{\beta}$ consists of the images of λ with $\lambda(a) = \lambda(b) = 4$ and $\Theta \cap \bar{\Lambda}_4^{5}$ consists of the images of λ with $P_5(\lambda) \subset \{a, b\} \subset P_5(\lambda) \cup P_3(\lambda)$. The transitivity of F_1 on $\Theta \cap \bar{\Lambda}_4^{\alpha}$ can be established in the following way. Choose an appropriate pair $\lambda, \nu \in \Lambda_4^{\alpha}$ such that $\lambda(a) + \lambda(b) = \nu(a) + \nu(b) = 8$ and $\bar{\lambda} = \bar{\nu}$; calculate the stabilizer H of $\{\lambda, \nu\}$ in F_1 and observe that the order of H equals the order of F_1 divided by the size of $\Theta \cap \bar{\Lambda}_4^{\alpha}$ (notice that $F = Co_2$ splits over the centre of $\widehat{G} \cong Co_0$). Along these lines we also obtain the structure of element stabilizers. For this purpose (4.4.1), its proof and the table there are very useful. □

Lemma 4.9.6 *The suborbit diagram of the graph Θ corresponding to the action of F is as given after the proof.*

Proof. It is straightforward to calculate the diagram using (4.9.5), methods and results from Section 4.7, particularly (4.7.4) and the diagrams for $\mathcal{H}(Mat_{22})$ calculated in Section 3.9. □

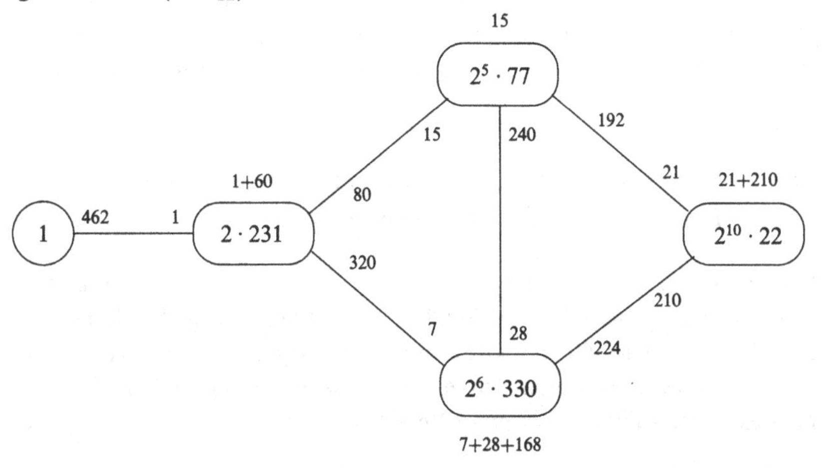

Let $\mathcal{G}(Co_2)$ be the subgeometry in $\mathcal{G}(Co_1)$ formed by the elements ($\star$-closed cliques) contained in Θ.

Lemma 4.9.7 *$\mathcal{G}(Co_2)$ is a rank 4 Petersen geometry with the diagram*

$$\overset{}{\underset{2}{\circ}} \text{———} \underset{2}{\circ} \text{———} \underset{2}{\circ} \overset{P}{\text{———}} \underset{1}{\circ}$$

and $F \cong Co_2$ acts on $\mathcal{G}(Co_2)$ flag-transitively.

Proof. Using (3.4.5) and the remark after its proof, we identify the residue in $\mathcal{G}(Co_2)$ of an element of type 1 with the rank 3 Petersen geometry $\mathcal{G}(Mat_{22})$. For an element of type 4 its residues in $\mathcal{G}(Mat_{22})$ and $\mathcal{G}(Mat_{24})$ are the same. Finally, since F acts transitively on the vertex set of Θ and F_1 acts flag-transitively on the residue of $\bar{\lambda}_0$ in $\mathcal{G}(Co_2)$, we conclude that the action of F on $\mathcal{G}(Co_2)$ is flag-transitive. □

Let $\{F_i \mid 1 \leq i \leq 4\}$ be the amalgam of maximal parabolics associated with the action of $F \cong Co_2$ on $\mathcal{G}(Co_2)$. Then

$$F_1 \cong 2^{10} : \mathrm{Aut}\, Mat_{22}, \qquad F_2 \cong 2^{4+10}.(Sym_5 \times Sym_3),$$

$$F_3 \cong 2^{1+7+6}.(L_3(2) \times 2), \qquad F_4 \cong 2^{1+4+6}.L_4(2).$$

Let B be an octad containing $P_4(\mu_0)$ and $\delta = \tau(B,1)$. By (4.8.7 (ii)) $C_F(\delta) = C_G(\delta) \cap G(\bar{\mu}_0)$ is the preimage in $D = C_G(\delta)$ of the stabilizer of $\bar{\mu}_0$ in $\bar{D}$, so that $C_F(\delta) \cong 2_+^{1+8}.Sp_6(2)$. By (4.8.7 (i)) $\Psi := \Phi(\delta) \cap \Theta$ is the set of isotropic vectors in $V_1 = C_{\bar{\Lambda}}(O_2(D))$ orthogonal to $\bar{\mu}_0$ with respect to the bilinear form associated with κ. Furthermore the elements of type 1, 2 and 3 in $\mathcal{G}(Co_2)$ contained in Ψ correspond to totally singular 1-, 2- and 3-dimensional subspaces in the natural orthogonal module $\bar{\mu}_0^{\perp}$ for $Sp_6(2) \cong O_7(2)$ and hence they form a subgeometry $\mathcal{R}$ in $\mathcal{G}(Co_2)$ isomorphic to the classical $C_3(2)$-geometry $\mathcal{G}(Sp_6(2))$ with the diagram

$$\overset{}{\underset{2}{\circ}}\!\!-\!\!\!-\!\!\!-\!\!\underset{2}{\circ}\!\!=\!\!=\!\!=\!\!\underset{2}{\circ}$$

The residue of $\bar{\lambda}_0$ in $\mathcal{R}$ is the $Sp_4(2)$-subgeometry in $\mathrm{res}_{\mathcal{G}(Co_2)}(\bar{\lambda}_0) \cong \mathcal{G}(Mat_{22})$ as in (3.4.4). This can be summarized as follows.

Lemma 4.9.8 *Let δ be an involution in $F \cong Co_2$ conjugate to an involution from the orbit of length 77 of $F_1/O_2(F_1)$ on $O_2(F_1)$. Then the elements of type 1, 2 and 3 in $\mathcal{G}(Co_2)$ which are pointwise fixed by $O_2(C_F(\delta))$ form a subgeometry $\mathcal{R}$ isomorphic to $\mathcal{G}(Sp_6(2))$ on which $C_F(\delta) \cong 2_+^{1+8}.Sp_6(2)$ induces a flag-transitive action of $Sp_6(2)$.* □

Notice that the orthogonal complement of $\bar{\mu}_0$ in V_1 with respect to the quadratic form κ as in (4.8.7) supports the universal 7-dimensional natural representation of $\mathcal{R}$.

Let $\mathcal{F}$ be $\mathcal{G}(Co_1)$ or $\mathcal{G}(Co_2)$ and let Σ be the collinearity graph of $\mathcal{F}$ (that is Γ or Θ). Any two points in $\mathcal{F}$ are on at most one line and every triangle in Σ is contained in a plane. Hence $\mathcal{F}$ is simply connected if and only if the fundamental group of Σ is generated by triangles. The latter can be established directly using the suborbit diagram of $\mathcal{F}$ and this strategy was realized in [Sh92] for the case of $\mathcal{G}(Co_2)$. Here we follow the strategy developed in [Iv92d] which enables us to deal with a smaller number of cycles. In order to implement the strategy we need some further information about Co_1 and its action on $\bar{\Lambda}$.

4.10 The affine Leech graph

The semidirect product $\bar{\Lambda} : Co_1$ acts naturally on $\bar{\Lambda}$. By (4.5.5) the orbit of a pair $\bar{\lambda}, \bar{\nu} \in \bar{\Lambda}$ under this action is uniquely determined by the value i such that $\bar{\lambda} + \bar{\nu} \in \bar{\Lambda}_i$, where $i = 0, 2, 3$ or 4. Let $\bar{\Lambda}_i$ denote also the corresponding orbital graph on $\bar{\Lambda}$, so that $\bar{\Lambda}_i(\bar{v})$ is the set of elements in $\bar{\Lambda}$ adjacent to $\bar{v}$ in the graph $\bar{\Lambda}_i$. Let p_{ij}^k be the intersection parameter defined by

$$p_{ij}^k = \#\{\bar{\mu} \mid \bar{\mu} \in \bar{\Lambda}_i(\bar{\lambda}), \bar{\mu} \in \bar{\Lambda}_j(\bar{v})\}$$

for $\bar{\lambda} \in \bar{\Lambda}_k(\bar{v})$. Since all the orbitals are self-paired *i.e.*, $\bar{\lambda} \in \bar{\Lambda}_k(\bar{v})$ if and only if $\bar{v} + \bar{\lambda} \in \bar{\Lambda}_k$, we have the following:

$$p_{ij}^k = p_{ji}^k \quad \text{and} \quad |\bar{\Lambda}_k| \cdot p_{ij}^k = |\bar{\Lambda}_j| \cdot p_{ki}^j.$$

We are particularly interested in the graph $\bar{\Lambda}_2$ and call it the *affine Leech graph*. In this section we calculate the suborbit diagram of $\bar{\Lambda}_2$.

As above $\bar{\lambda}_0$ denotes the element in $\bar{\Lambda}_4$ stabilized by $G_1 = Q_1 : L_1$, so that $\{\bar{\lambda}_0\} = \bar{\Lambda}_4^8$. By the tables in (4.4.1) and (4.6.1) and in view of (4.5.6) we obtain the following result which gives the valencies of $\bar{\Lambda}_4$ on the diagram of $\bar{\Lambda}_2$ (as well as on the diagrams of $\bar{\Lambda}_3$ and $\bar{\Lambda}_4$).

Lemma 4.10.1

$$\bar{\Lambda}_2(\bar{\lambda}_0) = \bar{\Lambda}_2^4 \cup \bar{\Lambda}_3^5 \cup \bar{\Lambda}_4^6,$$

$$\bar{\Lambda}_3(\bar{\lambda}_0) = \bar{\Lambda}_2^3 \cup \bar{\Lambda}_3^2 \cup \bar{\Lambda}_3^4 \cup \bar{\Lambda}_4^5,$$

$$\bar{\Lambda}_4(\bar{\lambda}_0) = \bar{\Lambda}_2^2 \cup \bar{\Lambda}_3^3 \cup \bar{\Lambda}_4^{4a} \cup \bar{\Lambda}_4^{4b} \cup \bar{\Lambda}_4^{4c}.$$

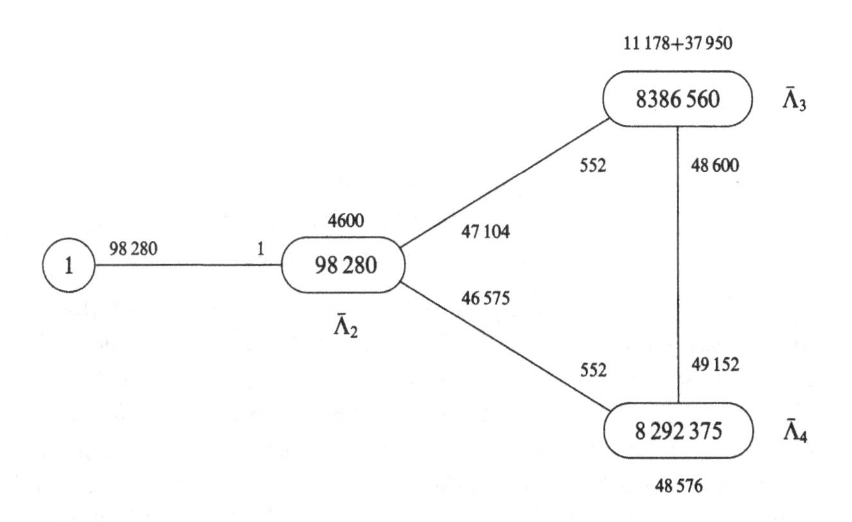

Lemma 4.10.2 *The intersection parameters of $\bar{\Lambda}_2$ are as on the above diagram.*

Proof. By (4.5.6) the graph $\bar{\Lambda}_3$ is the graph on the set of vectors of the 24-dimensional $GF(2)$-space $\bar{\Lambda}$ in which two vectors are adjacent if their sum is non-singular with respect to the non-degenerate quadratic form θ of plus type. This graph is well known to be strongly regular [BCN89], which means that $p_{33}^2 = p_{33}^4$. On the other hand by (4.10.1) we have

$$p_{33}^4 = |\bar{\Lambda}_3^4| + |\bar{\Lambda}_3^2|, \quad p_{34}^2 = |\bar{\Lambda}_2^3| \cdot |\bar{\Lambda}_4| / |\bar{\Lambda}_2|,$$

and hence

$$p_{32}^2 = |\bar{\Lambda}_3| - p_{33}^2 - p_{34}^2 = 47\,104.$$

Now the remaining intersection parameters follow from the obvious identities in which they are involved. □

To complete calculating the diagram there remains to show that $G(\bar{v})$ acts transitively on $\bar{\Lambda}_2(\bar{v}) \cap \bar{\Lambda}_2$ and on $\bar{\Lambda}_2(\bar{v}) \cap \bar{\Lambda}_3$ if $\bar{v} \in \bar{\Lambda}_2$ and that $G(\bar{v})$ has two orbits on $\bar{\Lambda}_2(\bar{v}) \cap \bar{\Lambda}_3$ if $\bar{v} \in \bar{\Lambda}_3$.

Lemma 4.10.3 *The following assertions hold:*

(i) *the subgraph in $\bar{\Lambda}_2$ induced by $\bar{\Lambda}_2^4$ has the following suborbit diagram with respect to the action of G_1*

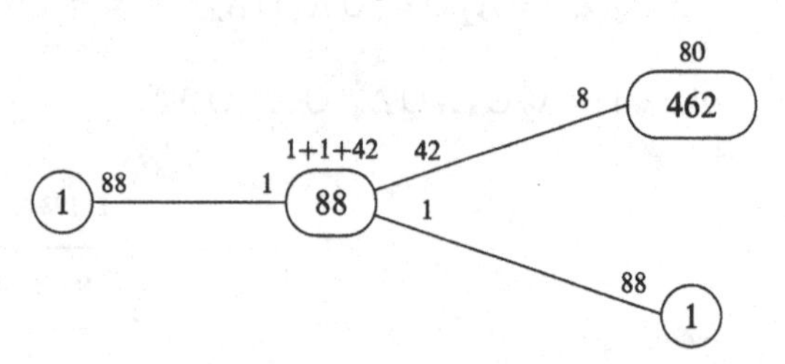

(ii) *G_1 acts on this subgraph vertex- and edge-transitively with the vertexwise stabilizer of an edge being isomorphic to $2^9 : L_3(4)$;*

(iii) *if $\bar{v}_1, \bar{v}_2 \in \bar{\Lambda}_2^4$ and $\bar{v}_2 \in \bar{\Lambda}_2(\bar{v}_1)$, then $\bar{v}_1 + \bar{v}_2 \in \bar{\Lambda}_2^4$.*

Proof. The orbits of Q_1 on $\bar{\Lambda}_2^4$ are all of length 2 and there is an obvious mapping from the set of these Q_1-orbits onto the set of 2-element subsets of $\mathscr{P}$ which commutes with the action of G_1. Two elements from $\bar{\Lambda}_2^4$ are adjacent in $\bar{\Lambda}_2$ if their Q_1-orbits are different and correspond to intersecting pairs from $\mathscr{P}$, which gives (i). Let μ_0 and $\bar{\mu}_0$ be as introduced in the paragraph before (4.9.5), so that $\mu_0 \in \Lambda_2^4$ with $P_4(\mu_0) = \{a, b\}$ and $X(\mu_0) = \emptyset$. Let $c \in \mathscr{P} \setminus \{a, b\}$ and let μ_1, μ_2 be vectors in Λ_2^4 such that $P_4(\mu_1) = \{b, c\}$, $X(\mu_1) = \emptyset$, $P_4(\mu_2) = \{a, c\}$, $X(\mu_2) = \{c\}$. Then $\bar{\mu}_1, \bar{\mu}_2 \in \bar{\Lambda}_2$ and $\mu_0 - \mu_1 = \mu_2$, which means that

$T = \{\bar{\mu}_0, \bar{\mu}_1, \bar{\mu}_2\}$ is a triangle in the graph $\bar{\Lambda}_2$. In particular $\{\bar{\mu}_0, \bar{\mu}_1\}$ is an edge in the subgraph induced by $\bar{\Lambda}_2^4$. It is easy to check that $G(\bar{\lambda}_0, \bar{\mu}_0, \bar{\mu}_1)$ is the semidirect product $(Q_1(\bar{\mu}_0) \cap Q_1(\bar{\mu}_1)) : L_1(a,b,c)$, where $Q_1(\bar{\mu}_0) \cap Q_1(\bar{\mu}_1)$ consists of the elements $\tau(Y,1)$ where Y is a Golay set containing $\{a,b,c\}$ and $L_1(a,b,c) \cong L_3(4)$. In view of (2.15.4) (ii) follows. Now (iii) is straightforward. □

Lemma 4.10.4 *Let $T = \{\bar{\mu}_0, \bar{\mu}_1, \bar{\mu}_2\} \subseteq \bar{\Lambda}_2^4$ and $\bar{\mu}_0 + \bar{\mu}_1 + \bar{\mu}_2 = 0$. For $i = 1$ and 2 put $U(i) = G(\bar{\mu}_0, \bar{\mu}_i)$ and $\Xi(i) = \bar{\Lambda}_2(\bar{\mu}_0) \cap \bar{\Lambda}_2(\mu_i) \cap \bar{\Lambda}_4$. Then*

(i) *$U(i)$ acts transitively on $\Xi(i)$ with stabilizer isomorphic to $2^9 : L_3(4)$,*
(ii) *$U(1) = U(2)$ and $\Xi(1) = \Xi(2)$.*

Proof. By (4.10.3 (ii)) G_1 acts transitively on the set of ordered pairs $(\bar{\omega}, \bar{\eta})$ of elements from $\bar{\Lambda}_2^4$ with $\bar{\omega} \in \bar{\Lambda}_2(\bar{\eta})$ and hence (i) follows. It is clear that $U(1) = G(\bar{\mu}_0, \bar{\mu}_1) = G(\bar{\mu}_0, \bar{\mu}_1 + \bar{\mu}_0) = U(2)$. Hence in view of (i) in order to prove the second equality in (ii) it is sufficient to show that $\Xi(1) \cap \Xi(2) \neq \emptyset$, but since $\bar{\Lambda}_2^4 = \bar{\Lambda}_2 \cap \bar{\Lambda}_2(\bar{\lambda}_0)$, we have $\bar{\lambda}_0 \in \Xi(1) \cap \Xi(2)$. □

In view of the above lemma we put $U = U(i)$ and $\Xi = \Xi(i)$ for $i = 1$ or 2. Taking $i = 1$ we observe that an element $\bar{\lambda} \in \bar{\Lambda}_4$ is contained in Ξ if and only if it is the image of $\lambda \in \Lambda_4$ such that $\lambda(a)+\lambda(b) = \lambda(b)+\lambda(c) = 8$. By (4.4.1) and (4.6.1) we conclude that $\Xi \cap \bar{\Lambda}_4^\alpha$ consists of the images of the λ with $\lambda(a) = \lambda(b) = \lambda(c) = 4$ for $\alpha = 4a$, $\lambda(a) = \lambda(c) = 2$, $\lambda(b) = 6$ for $\alpha = 6$, $\lambda(a) = \lambda(c) = 3$, $\lambda(b) = 5$ for $\alpha = 5$, and $\Xi \cap \bar{\Lambda}_4^\beta$ is empty for $\beta = 4b$ and $4c$. Arguing as in (4.9.6) we obtain the following.

Lemma 4.10.5 *The suborbit diagram of Ξ corresponding to the action of U is the following:*

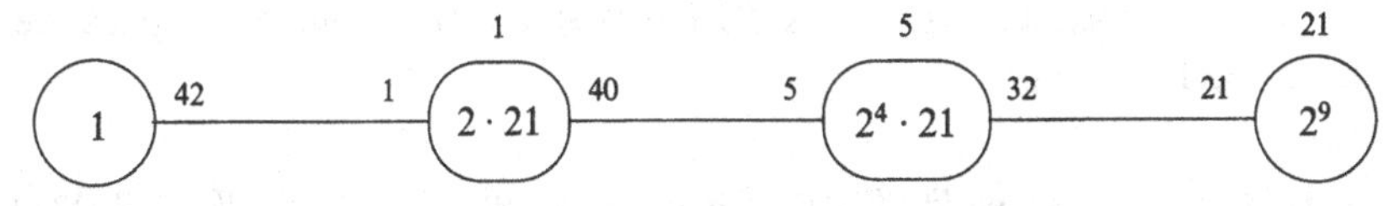

The action of $U_1 = G_1 \cap U \cong 2^9 : L_3(4)$ on Ξ is faithful and transitive on $\Xi \cap \bar{\Lambda}_4^\alpha$ for $\alpha = 4a$, 6 and 5 with $O_2(U_1)$ having orbits of lengths 2, 2^4 and 2^9, respectively. □

Lemma 4.10.6 *The group U is isomorphic to $U_6(2)$ and it acts on Ξ as on the set of maximal totally singular subspaces in the natural 6-dimensional $GF(4)$-module.*

Proof. Let $Y = \{a, b, c\}$. Recall that ϱ is a mapping from $\bar{\Lambda}_4^{4a}$ onto the set of sextets which commutes with the action of G_1. Then $\varrho(\Xi \cap \bar{\Lambda}_4^{4a})$ are the sextets containing Y in a single tetrad and they are identified with elements from the set $\mathscr{P} \setminus Y$ which is the point set of the projective plane $\Pi(Y)$ of order 4 as in (2.9.1). Let B be an octad containing Y, so that B corresponds to a line in $\Pi(Y)$, let $\delta = \tau(B, 1)$ and $D = C_G(\delta)$. Then by (4.8.7 (i)) T is (the non-zero vectors in) a minus 2-space in $V_1 = C_{\bar{\Lambda}}(O_2(D))$ and $\Upsilon := \Xi \cap \Phi(\delta)$ is the set of isotropic vectors orthogonal to T. Hence the vertices and lines of Γ contained in Υ correspond to 1- and 2-dimensional totally singular subspaces in the 6-dimensional orthogonal space $T^{\perp}$ of minus type. By (4.8.7 (ii)) $C_U(\delta)$ contains $O_2(D) \cong 2_+^{1+8}$ and induces on Υ a flag-transitive action of $\Omega_6^-(2) \cong U_4(2)$. This shows that the vertices and lines contained in Ξ together with the images of Υ under U form (with respect to the incidence relation defined via inclusion) a C_3-geometry $\mathscr{F}$ with the diagram

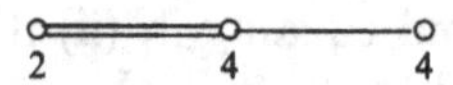

on which U induces a flag-transitive action. By [Ti82] $\mathscr{F} \cong \mathscr{G}(U_6(2))$ and the result follows. □

Since $p_{22}^2 = 4600$ is exactly the index of U in $G(\bar{\mu}_0) \cong Co_2$ we have the following.

Lemma 4.10.7 *If $\bar{v} \in \bar{\Lambda}_2$ then $G(\bar{v})$ acts transitively on $\bar{\Lambda}_2(\bar{v}) \cap \bar{\Lambda}_2$ with stabilizer isomorphic to $U_6(2)$.* □

It is easy to observe that if T is as in (4.10.4) then the setwise stabilizer of T in G_1 is of the form $2^9 : P\Gamma L_3(4)$ and it induces on T the natural action of Sym_3. On the other hand the automorphism group of the C_3-geometry associated with U is $P\Gamma U_6(2) \cong U_6(2) : Sym_3$ and hence we have the following.

Lemma 4.10.8 *Let $\bar{\mu}_0$ be an element from $\bar{\Lambda}_2$, $\bar{\mu}_1$ be an element from $\bar{\Lambda}_2 \cap \bar{\Lambda}_2(\bar{\mu}_0)$ and $\bar{\mu}_2 = \bar{\mu}_0 + \bar{\mu}_1$. Let $T = \{\bar{\mu}_0, \bar{\mu}_1, \bar{\mu}_2\}$, $\widehat{U}$ be the setwise stabilizer of T in G and $\Xi = \bar{\Lambda}_4 \cap \bar{\Lambda}_2(T)$. Then*

(i) *T is a triangle in the affine Leech graph and $T \subseteq \bar{\Lambda}_2$,*
(ii) *$\widehat{U}$ induces on T the natural action of Sym_3,*
(iii) *$\widehat{U}$ is the full stabilizer in G of the subgraph in the Leech graph Γ induced by Ξ,*
(iv) *$\widehat{U} \cong \mathrm{Aut}\,\Xi \cong P\Gamma U_6(2) \sim U_6(2).Sym_3$.* □

Notice that the stabilizer in $F \cong Co_2$ of the subgraph Ξ coincides with the stabilizer of $\bar{\mu}_0$ in U and it is isomorphic to $P\Sigma U_6(2) \sim U_6(2).2$.

Before continuing the calculation of the parameters of the affine Leech graph we would like to mention an important subgeometry of the geometry $\mathcal{G}(U_6(2))$ as in the proof of (4.10.6). Let $K \cong P\Gamma L_3(4)$ be a complement to $O_2(G_1[T])$ in $G_1[T] \cong 2^9 : P\Gamma L_3(4)$ and let ω be an involution in K which realizes the field automorphism. Then ω induces on T an action of order 2. Furthermore, $C_K(\omega) \cong L_3(2) \times 2$ and the centralizer of ω in $O_2(G_2[T])$ is of order 2^6. Hence by (19.9) in [ASei76] $C_{\widehat{U}}(\omega) \cong Sp_6(2) \times 2$. Using (4.10.5) and the paragraph before that lemma it is not difficult to identify the subgraph in Ξ induced by the vertices fixed by ω with the dual polar graph of $Sp_6(2)$ with the following suborbit diagram:

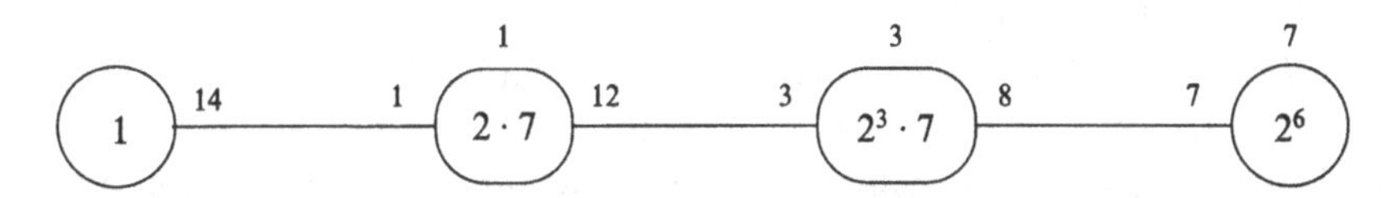

Lemma 4.10.9 *The following assertions hold:*

(i) *up to conjugation $G[T]$ contains a unique involution ω which commutes in $G[T]$ with an element of order 7, and induces on T an action of order 2,*

(ii) $C_{\widehat{U}}(\omega) \cong Sp_6(2) \times 2$,

(iii) *the points, lines and quads in Ξ fixed by ω form a geometry isomorphic to the C_3-geometry $\mathcal{G}(Sp_6(2))$,*

(iv) *ω is conjugated in G to an element $\tau(B,1)$ from Q_1 where B is an octad.*

Proof. By (19.9) in [ASei76] ω is as in the paragraph before the lemma, which gives (i), (ii) and (iii). By (4.10.7) G acts transitively on the triples $\{\bar{v}_0, \bar{v}_1, \bar{v}_2\}$ such that $\bar{v}_i \in \bar{\Lambda}_2 \cap \bar{\Lambda}_2(\bar{v}_j)$ for $0 \le i < j \le 2$ and $\bar{v}_0 + \bar{v}_1 + \bar{v}_2 = 0$. Let $v_0 \in \Lambda_2^3$ with $X(v_0) = P_3(v_0)$, B be an octad disjoint from $P_3(v_0)$ and v_1 be the image of v_0 under $\widehat{\tau}(B,1)$. The $v_2 := v_0 + v_1$ is in Λ_2^2 (with $P_2(v_2) = B$ and $X(v_2) = \emptyset$). By the construction $\widehat{\tau}(B,1)$ permutes v_0 and v_1, stabilizes v_2, and the centralizer of $\widehat{\tau}(B,1)$ in $\widehat{G}(v_0, v_1, v_2)$ is isomorphic to Alt_8 (in particular has order divisible by 7). By (i) ω and $\tau(B,1)$ are conjugate and (iv) follows. □

Let us continue the analysis of the affine Leech graph. For $d \in \mathcal{P}$ let ω_d be a vector in Λ_3^5 such that $P_5(\omega_d) = \{d\}$ and $X(\omega_d) = \emptyset$ and put

$$\Psi_1(d) = \{\mu \in \Lambda_2^4 \mid d \in P_4(\mu), X(\mu) = \emptyset\},$$

$$\Psi_2(d) = \{\mu \in \Lambda_2^3 \mid P_3(\mu) = X(\mu) \neq \{d\}\},$$

$$\Psi_3(d) = \{\mu \in \Lambda_2^2 \mid d \in P_2(\mu), X(\mu) = \emptyset\},$$

$$\Psi_4(d) = \{\mu \in \Lambda_2^3 \mid P_3(\mu) = \{d\}, |X(\mu)| = 7\}.$$

Notice that if $\mu \in \Psi_4(d)$ then $P_3(\mu) \cup X(\mu)$ is an octad. Then $|\Psi_1(d)| = |\Psi_2(d)| = 23$, $|\Psi_3(d)| = |\Psi_4(d)| = 253$ and $\widehat{G}_1(\omega_d) = \widehat{L}_1(d) \cong Mat_{23}$ acts transitively on $\Psi_i(d)$ for $i = 1, 2, 3$ and 4. In addition if $\mu_1 \in \Psi_1(d)$, $\mu_2 \in \Psi_2(d)$ with $P_4(\mu_1) = \{d\} \cup P_3(\mu_2)$ then $\mu_1 + \mu_2 = \omega_d$ and similarly if $\mu_3 \in \Psi_3(d)$, $\mu_4 \in \Psi_4(d)$ with $P_2(\mu_3) = P_3(\mu_4) \cup X(\mu_4)$ then $\mu_3 + \mu_4 = \omega_d$. Hence if $\Psi(d)$ is the union of the $\Psi_i(d)$, then $\bar{\Psi}(d) \subseteq \bar{\Lambda}_2 \cap \bar{\Lambda}_2(\omega_d)$. It is easy to see that different vectors from $\Psi(d)$ have different images in $\bar{\Lambda}$ and since $|\Psi(d)| = 552 = p_{22}^3$ we have the following.

Lemma 4.10.10 $\bar{\Psi}(d) = \bar{\Lambda}_2 \cap \bar{\Lambda}_2(\bar{\omega}_d)$; *the action of* $\widehat{G}_1(\omega_d) = \widehat{L}_1(d) \cong Mat_{23}$ *on* $\Psi_i(d)$ *is similar to its action on* $\mathcal{P} \setminus \{d\}$ *for* $i = 1, 2$ *and to the action on the set of octads containing* d *for* $i = 3, 4$. □

There is an equivalence relation on $\Psi(d)$ with classes of size 2 with respect to which 2 distinct vectors μ, ν are equivalent if $\mu + \nu = \omega_d$. Let Σ be a graph on $\Psi(d)$ such that μ, ν are adjacent if they are not equivalent and $\bar{\mu} + \bar{\nu} \in \bar{\Lambda}_3$. One can check that whenever $\{\mu_1, \mu_2\}$ and $\{\nu_1, \nu_2\}$ are 2 distinct equivalence classes there is a bijection σ of $\{1, 2\}$ onto itself such that μ_i and ν_j are adjacent if and only if $j = \sigma(i)$ (so that Σ is a double cover of the complete graph on 276 vertices). In view of this rule all the adjacencies in Σ are determined by the following 3 conditions: $\Psi_1(d)$ is a coclique; $\mu \in \Psi_1(d)$ and $\nu \in \Psi_3(d)$ are adjacent exactly when $P_4(\mu) \cap P_2(\nu) = \{d\}$; $\mu_1, \mu_2 \in \Psi_3(d)$ are adjacent exactly when $|P_2(\mu_1) \cap P_2(\mu_2)| = 2$. It is straightforward to reconstruct the suborbit diagram of Σ based at $\mu \in \Psi_1(d)$, with respect to $\widehat{G}_1(\omega_d, \mu) \cong Mat_{22}$ (one might find useful the diagrams for $\mathcal{H}(Mat_{24})$ and $\mathcal{H}(Mat_{22})$ from Chapter 3).

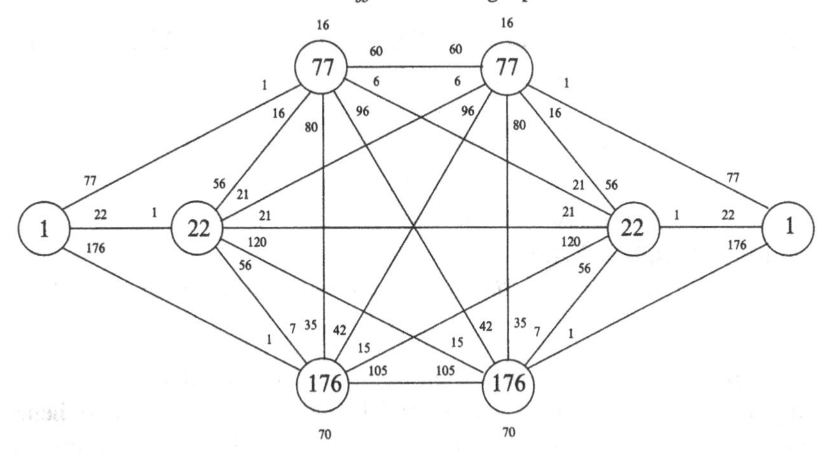

Now for a 3-element subset D in $\mathscr{P}$ let ω_D be a vector from Λ_3^3 such that $P_3(\omega_D) = X(\omega_D) = D$ and put

$$\Phi_1(D) = \{\mu \in \Lambda_2^4 \mid P_4(\mu) \subseteq P_3(\omega_D), X(\mu) = \emptyset\},$$

$$\Phi_2(D) = \{\mu \in \Lambda_2^3 \mid P_3(\mu) = X(\mu) \subseteq P_3(\omega_D)\},$$

$$\Phi_3(D) = \{\mu \in \Lambda_2^2 \mid P_3(\omega_D) \subseteq X(\mu) \subseteq P_2(\mu), |X(\mu)| = 4\},$$

$$\Phi_4(D) = \{\mu \in \Lambda_2^3 \mid P_3(\mu) \not\subseteq P_3(\omega_D), P_3(\omega_D) \subseteq X(\mu), |X(\mu)| = 7\},$$

$$\Phi_5(D) = \{\mu \in \Lambda_2^2 \mid |X(\mu)| = 2, X(\mu) = P_2(\mu) \cap P_3(\omega_D)\},$$

$$\Phi_6(D) = \{\mu \in \Lambda_2^3 \mid P_3(\mu) \subset P_3(\omega_D), |X(\mu)| = 9\}.$$

Let $\Phi(D)$ be the union of the $\Phi_i(D)$. Then $|\Phi_1(D)| = |\Phi_2(D)| = 3$, $|\Phi_3(D)| = |\Phi_4(D)| = 105$, $|\Phi_5(D)| = |\Phi_6(D)| = 168$, $\widehat{G}_1(\omega_D) = \widehat{L}_1[D] \cong P\Gamma L_3(4)$ acts transitively on each of the $\Phi_i(D)$ and we have the following result similar to (4.10.10).

Lemma 4.10.11 $\bar{\Phi}(D) = \bar{\Lambda}_2 \cap \bar{\Lambda}_2(\bar{\omega}_D)$; *the action of* $\widehat{G}_1(\omega_D) = \widehat{L}_1[D] \cong P\Gamma L_3(4)$ *on* $\Phi_i(D)$ *is similar to its action on the elements in* D *for* $i = 1, 2$, *on the maximal flags in* $\Pi(D)$ *for* $i = 3, 4$, *on the hyperovals in* $\Pi(D)$ *for* $i = 5, 6$. □

Recall that by (2.8.2) and (2.9.1), for an octad B intersecting D in two elements, $H = B \setminus D$ is a hyperoval in $\Pi(D)$ and two such hyperovals $H_1 = B_1 \setminus D$ and $H_2 = B_2 \setminus D$ are in the same orbit of $\widehat{L}_1(D) \cong L_3(4)$ if and only if $B_1 \cap D = B_2 \cap D$. This shows that if $\mu \in \Phi_1(D)$ then

$\widehat{G}_1(\omega_D, \mu) \cong P\Sigma L_3(4)$ acting on $\Phi(D)$ has exactly two orbits of length l for $l = 1,\ 2,\ 56,\ 105$ and 112.

Lemma 4.10.12 *Let* $\bar{\omega} \in \bar{\Lambda}_3$ *and* $\bar{\mu}, \bar{\nu} \in \bar{\Lambda}_2 \cap \bar{\Lambda}_2(\bar{\omega})$ *with* $\bar{\mu} - \bar{\nu} \notin \{\bar{0}, \bar{\omega}\}$. *Then*

(i) $G(\bar{\omega})$ *acts transitively on* $\bar{\Lambda}_2 \cap \bar{\Lambda}_2(\bar{\omega})$,
(ii) $G(\bar{\omega}, \bar{\mu}, \bar{\nu}) \cong U_4(3)$,
(iii) $G(\bar{\omega}, \bar{\mu})$ *is the sporadic McLaughlin group* McL *of order* $2^8 \cdot 3^6 \cdot 5^3 \cdot 7 \cdot 11$.

Proof. Since $\widehat{G}$ is transitive on Λ_3 we can assume that $\bar{\omega} = \bar{\omega}_d$. For the same reason there is $g \in \widehat{G}$ which maps ω_d onto ω_D and hence $g\widehat{G}_1(\omega_D)g^{-1}$ is a subgroup of $\widehat{G}(\omega_d)$ which acts on $\Psi(d)$ as $\widehat{G}(\omega_D)$ acts on $\Phi(D)$. Comparing the orbit lengths of $\widehat{G}(\omega_d)$ on $\Psi(d)$ given in (4.10.10) and of $\widehat{G}(\omega_D)$ on $\Phi(D)$ given in (4.10.11), we immediately conclude that $\widehat{G}(\omega_d)$ acts transitively on the 276 equivalence classes of vectors in $\Psi(d)$. Now, comparing the suborbit diagram of Σ and the orbit lengths of $\widehat{G}(\omega_D, \mu)$ on $\Phi(D)$ (for $\mu \in \Phi_1(D)$) given in the paragraph after (4.10.11), in view of the above we conclude that for every $\nu \in \Psi(d)$ the action of $\widehat{G}(\omega_d, \nu)$ on $\Sigma(\nu)$ (which is the set of vertices adjacent to ν in Σ) is transitive. Since Σ is not bipartite we obtain (i).

Let μ be the vector in $\Psi_1(d)$ at which the suborbit diagram of Σ is based, let ν be a vector from $\Psi_2(d)$ which is not equivalent to μ and let $Y = P_4(\mu) \cap P_3(\nu)$. Then one can see from the description of Σ and from its suborbit diagram that the vectors in $\Sigma(\mu) \cap \Sigma(\nu)$ are indexed by the hyperovals constituting two orbits of $\widehat{L}_1(Y)$ on the set of all hyperovals in $\Pi(Y)$; two vertices are adjacent if and only if the corresponding hyperovals have no or three common elements. Hence $\Sigma(\mu) \cap \Sigma(\nu)$ is the point graph of the generalized quadrangle $\mathcal{U}$ as in (2.16.1) and by (2.16.1) it is the classical one associated with the group $U_4(3)$. By the above we have

$$|\widehat{G}(\omega_d, \mu, \nu)| = \frac{|Co_3|}{552 \cdot 275} = 2^8 \cdot 3^6 \cdot 5 \cdot 7.$$

It is easy to see that the action of $\widehat{G}(\omega_d, \mu, \nu)$ on $\Sigma(\mu) \cap \Sigma(\nu)$ is faithful and hence by (1.6.5 (vi)) it contains $U_4(3)$. Since $|\widehat{G}(\omega_d, \mu, \nu)| = |U_4(3)|$ we obtain (ii).

It is an easy combinatorial exercise to check that the subgraph induced by $\Sigma(\mu)$ is a strongly regular graph known as the McLaughlin graph. The group $\widehat{G}(\omega_d, \mu)$ induces on it a rank 3 action corresponding to the following suborbit diagram:

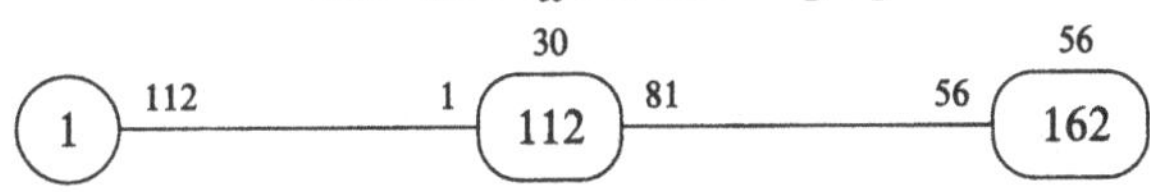

Alternatively one can (as we do) consider (iii) as a definition of the McLaughlin group McL. □

Lemma 4.10.13 *Let Ω be the McLaughlin graph and let $H \cong McL$ act naturally on Ω. Let x, y be a pair of vertices at distance 2 in Ω. Then $H(x,y) \cong L_3(4)$ acting on $\Omega \setminus \{x,y\}$ has three orbits of length 56 and one of length 105.*

Proof. We identify Ω with the subgraph in Σ induced by $\Sigma(\mu)$. Then we can assume that $\mu = \mu_1$, $x = \mu_2 + \omega_D$, $y = \mu_3 + \omega_D$, where $\{\mu_1, \mu_2, \mu_3\} = \Phi_1(D)$. Then $H(x,y)$ contains $G(\omega_D, \mu_1, \mu_2, \mu_3) \cong L_3(4)$ and by (4.10.12) $H(x,y) = G(\omega_D, \mu_1, \mu_2, \mu_3)$ and the orbit lengths follow from (4.10.11). □

The graph Σ represents the two-graph on 276 vertices as in [GS75]. The geometry whose elements are vertices, edges, triangles and maximal cliques (of size 6) in Σ is a doubly extended generalized quadrangle of order $(3,9)$ with the diagram

c c
1 1 3 9

whose full automorphism group is isomorphic to $Co_3 \times 2$; the residue of a vertex is the exceptional EDPS from (1.13.7) associated with McL.

Now there remains to show that $G(\bar{\omega})$ has two orbits on $\bar{\Lambda}_3 \cap \bar{\Lambda}_2(\bar{\omega})$ with lengths given on the diagram.

Lemma 4.10.14 *For $\bar{\eta} \in \bar{\Lambda}_3 \cap \bar{\Lambda}_2(\bar{\omega})$ suppose that*

(i) $\Upsilon = \bar{\Lambda}_2 \cap \bar{\Lambda}_2(\bar{\omega}) \cap \bar{\Lambda}_2(\bar{\eta}) \neq \emptyset$,
(ii) $G[\bar{\omega}, \bar{\eta}] \neq G(\bar{\omega}, \bar{\eta})$.

Let K be the elementwise stabilizer of Υ in G and A be the automorphism group of the subgraph in $\bar{\Lambda}_i$ for $i = 2, 3$ or 4 induced by Υ. Then $|G(\bar{\omega}, \bar{\eta})| \leq \frac{1}{2}|K| \cdot |A|$. □

Proof. It is clear that $G[\bar{\omega}, \bar{\eta}]$ is contained in $G[\Upsilon]$ and the order of the latter group is at most $|K| \cdot |A|$. □

We assume that $\bar{\omega} = \bar{\omega}_d$. If $e \in \mathcal{P} \setminus \{d\}$ then $\omega_d - \omega_e \in \Lambda_2^4$ so that we can put $\bar{\eta} = \bar{\omega}_e$. Clearly there is an element in $\widehat{L}_1$ which permutes ω_d and ω_e. Hence (ii) in (4.10.14) holds and Υ as in (4.10.14 (i)) is the image in $\bar{\Lambda}$ of the union of the $\Upsilon := \Psi_i(d) \cap \Psi_i(e)$ for $i = 1, 2, 3$ and 4. It is straightforward to check that $\Upsilon_4 = \emptyset$, while

$$\Upsilon_1 = \{\mu \in \Lambda_2^4 \mid P_4(\mu) = \{d, e\}\},$$

$$\Upsilon_2 = \{\mu \in \Lambda_2^3 \mid P_3(\mu) = X(\mu) \notin \{d, e\}\},$$

$$\Upsilon_3 = \{\mu \in \Lambda_2^2 \mid \{d, e\} \subseteq P_2(\mu), X(\mu) = \emptyset\},$$

so that $|\Upsilon| = 1 + 22 + 77 = 100$. If we assume that the base vertex μ in the suborbit diagram of Σ is in Υ_1, then besides μ the set Υ contains the 22-orbit of $\widehat{G}_1(\omega, \mu) \cong Mat_{22}$ in the neighbourhood of μ and the 77-orbit at distance 2 from μ. The vectors in Υ_2 are identified with the elements in $\mathscr{P} \setminus \{d, e\}$ while that in Υ_3 with the octads containing $\{d, e\}$; the adjacency relation is via inclusion. Thus (Υ_2, Υ_3) is the residual Steiner system $S(3, 6, 22)$. By (2.9.3) and the remark after its proof the stabilizer of μ in the automorphism group of the subgraph in $\bar{\Lambda}_2$ induced by Υ is isomorphic to $\widehat{L}_1[d, e] \cong \operatorname{Aut} Mat_{22}$. It is not difficult to show that the elementwise stabilizer of Υ in G is trivial and we suggest this as an exercise. Accepting this fact, by (4.10.14) we obtain the following.

Lemma 4.10.15 *If l is the length of the orbit of $\bar{\omega}_e$ under $G(\bar{\omega}_d)$ then*

$$l \geq l_0 := \frac{|Co_3|}{100 \cdot |Mat_{22}|} = 11\,178$$

and the equality holds if and only if $G(\bar{\omega}_d, \bar{\omega}_e)$ is transitive on the set Υ of size 100. □

Let D, ω_D, $\bar{\omega}_D$ be as in (4.10.11) and suppose that $D = \{d, e, f\}$ so that $d \in D$. If $\lambda = \omega_d - \omega_D$ then $N(\lambda) = (8\,4^2 0^{21})$, $P_8(\lambda) = \{d\}$, $P_4(\lambda) = \{e, f\}$, $X(\lambda) = \emptyset$. Hence if $\nu \in \Lambda_2^4$ with $P_4(\nu) = \{d, e\}$ and $X(\nu) = \emptyset$, then $\lambda - 2\nu \in \Lambda_2^4$, which shows that $\bar{\lambda} \in \bar{\Lambda}_2$ and hence $\bar{\omega}_D \in \bar{\Lambda}_3(\bar{\omega}_d)$. Using the description of the sets $\Psi(d)$ and $\Phi(D)$ it is straightforward to check that $\Upsilon := \bar{\Lambda}_2 \cap \bar{\Lambda}(\bar{\omega}_d) \cap \bar{\Lambda}(\bar{\omega}_D)$ is of size 4 (in particular $\bar{\omega}_e$ and $\bar{\omega}_D$ are in different $G(\bar{\omega}_d)$-orbits), consisting of the images of vectors μ_i, $1 \leq i \leq 4$, such that $\mu_1 \in \Lambda_2^4$ with $P_4(\mu_1) = \{d, e\}$, $X(\mu_1) = \emptyset$, $\mu_2 = \omega_d - \mu_1$, $\mu_3 = \omega_D + \mu_1$, $\mu_4 = \omega_d - \mu_3$. In fact $\{\mu_1, \mu_4\} = \Psi_1(d) \cap \Phi_1(D)$, $\{\mu_2, \mu_3\} = \Psi_2(d) \cap \Phi_2(D)$. Since $\bar{\mu}_1 + \bar{\mu}_2 = \bar{\omega}_d$, the elementwise stabilizer K of Υ in G is contained in $G(\bar{\omega}_d)$ and by (4.10.12 (ii)) $K \cong U_4(3)$. The subgraph in $\bar{\Lambda}_2$ induced by Υ has two edges $\{\bar{\mu}_1, \bar{\mu}_4\}$ and $\{\bar{\mu}_2, \bar{\mu}_3\}$, so its automorphism group is D_8. Furthermore, if G contains an element h which induces on Υ an element of order 4 then h permutes $\bar{\omega}_d$ and $\bar{\omega}_D$. By an obvious generalization of (4.10.14) we obtain the following.

Lemma 4.10.16 *If m is the length of the orbit of $\bar{\omega}_D$ under $G(\bar{\omega}_d)$ then*

$$m \geq m_0 := \frac{|Co_3|}{4 \cdot |U_4(3)|} = 37\,950.$$

□

Since $\bar{\omega}_e$ and $\bar{\omega}_D$ are in different $G(\bar{\omega}_d)$-orbits and $p^3_{23} = l_0 + m_0$ where l_0 and m_0 are as in (4.10.15) and (4.10.16), we have the following.

Lemma 4.10.17 *If $\bar{\omega} \in \bar{\Lambda}_3$ then $G(\bar{\omega})$ has two orbits Ω_1 and Ω_2 on $\bar{\Lambda}_3 \cap \bar{\Lambda}_2(\bar{\omega}_d)$ and the following hold, where $\Upsilon = \bar{\Lambda}_2 \cap \bar{\Lambda}_2(\bar{\omega}) \cap \Lambda_2(\bar{\eta})$:*

(i) *if $\bar{\eta} \in \Omega_1$ then $|\Upsilon| = 100$, $G[\bar{\omega}, \bar{\eta}]$ induces the full automorphism group of the subgraph in $\bar{\Lambda}_3$ induced by Υ and $G(\bar{\omega}, \bar{\eta})$ is the Higman–Sims sporadic group HS of order $2^9 \cdot 3^2 \cdot 5^3 \cdot 7 \cdot 11$,*

(ii) *if $\bar{\eta} \in \Omega_2$ then Υ is of size 4 and $G(\bar{\omega}, \bar{\eta}) \sim U_4(3).2^2$.* □

We consider (i) in the above lemma as a definition of the Higman–Sims group, the subgraph on 100 vertices is known as the Higman–Sims graph.

4.11 The diagram of Δ

In this section we study the subgraph Δ in $\bar{\Lambda}_2$ induced by the neighbours of the zero element. In other terms Δ is a graph on $\bar{\Lambda}_2$ in which $\bar{\nu}$ and $\bar{\mu}$ are adjacent exactly when $\bar{\nu} - \bar{\mu} \in \bar{\Lambda}_2$. We call Δ the *shortest vector graph.* We use the suborbit diagram of the affine Leech graph to deduce the suborbit digram of Δ with respect to the action of G, as given below.

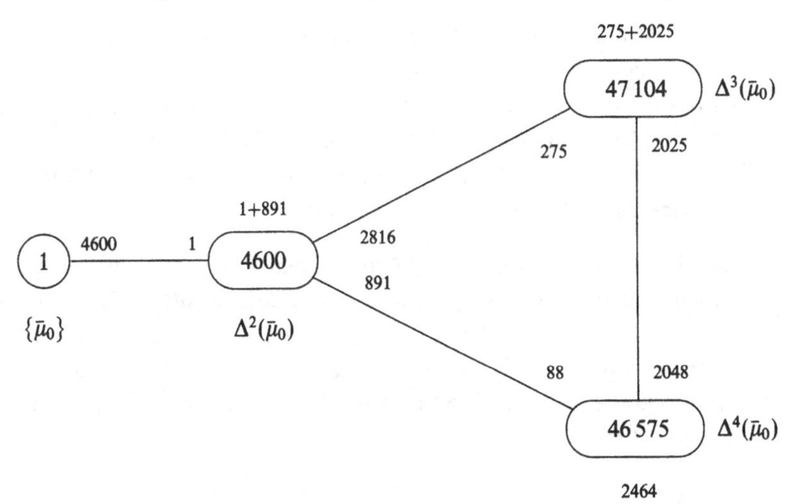

We follow notation introduced before (4.9.5): $\mu_0 \in \Lambda_2^4$, $F = G(\bar{\mu}_0) \cong Co_2$, $X(\mu_0) = \emptyset$, $\{a, b\} = P_4(\mu_0)$. For $\bar{v} \in \Delta = \bar{\Lambda}_2$ we put $\Delta^j(\bar{v}) = \Delta \cap \bar{\Lambda}_j(\bar{v})$.

Directly from (4.10.1), (4.10.12 (iii)), (4.10.6) and (4.10.17) we obtain the following.

Lemma 4.11.1 *For $j = 2$, 3 and 4 the action of F on $\Delta^j(\bar{\mu}_0)$ is transitive with stabilizers isomorphic to $U_6(2)$, McL and $2^{10} : \mathrm{Aut}\, Mat_{22}$, respectively.* □

Let $\bar{\mu}_1 \in \Delta^2(\bar{\mu}_0)$, $\bar{\mu}_2 = \bar{\mu}_0 + \bar{\mu}_1$, $T = \{\bar{\mu}_0, \bar{\mu}_1, \bar{\mu}_2\}$ and $T^{\bullet} = T \cup \{\bar{0}\}$. Then by (4.10.8) T is a triangle in Δ and it will be said to be a *singular triangle.* Thus every edge of Δ is in a unique singular triangle and the set of all singular triangles is clearly preserved by G. In terms introduced before (4.9.5) we have the following.

$$\Delta^4(\bar{\mu}_0) := \bar{\Lambda}_2(\bar{0}) \cup \bar{\Lambda}_4(\bar{\mu}_0) = (\bar{\Lambda}_2(\bar{\mu}_0) \cap \bar{\Lambda}_4(\bar{0})) + \bar{\mu}_0 = \Theta + \bar{\mu}_0.$$

Let $\Xi = \Theta(\bar{\mu}_0) \cap \Theta(\bar{\mu}_1)$. Then by (4.10.4) we have $\Xi = \bar{\Lambda}_4(\bar{0}) \cap \bar{\Lambda}_2(T^{\bullet} \setminus \{\bar{0}\})$ and hence $\Xi + \bar{\mu}_i = \bar{\Lambda}_4(\bar{\mu}_i) \cap \bar{\Lambda}_2(T^{\bullet} \setminus \{\bar{\mu}_i\})$ for $i = 0$, 1 and 2. Thus in view of (4.10.5) and (4.10.8) we have the following

Lemma 4.11.2 *Let $\bar{\mu}_1 \in \Delta^2(\bar{\mu}_0)$. Then*

(i) $\Delta^2(\bar{\mu}_1) \cap \Delta^4(\bar{\mu}_0) = \Xi + \bar{\mu}_0$ *(of size* 891*),*

(ii) $\Delta^2(\bar{\mu}_0) \cap \Delta^2(\bar{\mu}_1) = \{\bar{\mu}_2\} \cup (\Xi + \bar{\mu}_2)$ *where* $\bar{\mu}_2 = \bar{\mu}_0 + \bar{\mu}_1$. □

Lemma 4.11.3 *Let $\bar{\eta} \in \Delta^3(\bar{\mu}_0)$. Then*

(i) *$F(\bar{\eta}) \cong McL$ acts transitively on the set $\Delta^2(\bar{\mu}_0) \cap \Delta^2(\bar{\eta})$ of size* 275 *with stabilizer isomorphic to $U_4(3)$,*

(ii) *the subgraph in Δ induced by $\Delta^2(\bar{\mu}_0) \cap \Delta^2(\bar{\eta})$ is the complement of the McLaughlin graph (which is connected of valency* 162*),*

(iii) *if $\bar{\mu}_1 \in \Delta^2(\bar{\mu}_0)$ then $F(\bar{\mu}_1)$ acts transitively on the set $\Delta^3(\bar{\mu}_0) \cap \Delta^2(\bar{\mu}_1)$ of size* 2816.

Proof. Let $\bar{\omega} \in \bar{\Lambda}_3$, $\bar{\mu}_0, \bar{\eta} \in \bar{\Lambda}_2 \cap \bar{\Lambda}_2(\bar{\omega})$. Then $\bar{\mu}_0 - \bar{\eta} \in \bar{\Lambda}_2$ if and only if $\bar{\eta}$ is at distance 2 from $\bar{\mu}_0$ in the graph Σ defined after (4.10.10). In the latter case $\bar{\omega} + \bar{\mu}_0 \in \Delta^3(\bar{\mu}_0)$, $\bar{\eta} + \bar{\mu}_0 \in \Delta^2(\bar{\mu}_0) \cap \Delta^2(\bar{\omega} + \bar{\mu}_0)$ and (i) is immediate from (4.10.12), while (ii) follows from the definition of the McLaughlin graph, given in the proof of (4.10.12). Finally (iii) is a direct consequence of (i). □

Lemma 4.11.4 *Let $\bar{v} \in \Delta^4(\bar{\mu}_0)$. Then*

(i) *for* $j = 2$, 3 *and* 4 *the subgroup* $F(\bar{\nu}) \cong 2^{10} : \mathrm{Aut}\, Mat_{22}$ *acts transitively on the set* $\Delta^2(\bar{\mu}_0) \cap \Delta^j(\bar{\nu})$ *of size* 88, 2048 *and* 2464 *with stabilizer isomorphic to* $2^9 : L_3(4)$, Mat_{22} *and* $2^5 : 2^4 : Sym_6$, *respectively,*

(ii) *the subgraph induced by* $\Delta^2(\bar{\mu}_0) \cap \Delta^2(\bar{\nu})$ *is connected,*

(iii) *every* $\bar{\mu} \in \Delta^2(\bar{\mu}_0)$ *is adjacent to a vertex from* $\Delta^2(\bar{\mu}_0) \cap \Delta^2(\bar{\nu})$.

Proof. (i) For $j = 2$ the result is by (4.10.3). We assume that $\bar{\nu} = \bar{\lambda}_0 + \bar{\mu}_0$. Then $\Delta^4(\bar{\mu}_0) \cap \Delta^2(\bar{\nu}) = (\Theta \cap \bar{\Lambda}_4^6) + \bar{\mu}_0$ and the case $j = 4$ follows from (4.9.5) and the table therein. With $\bar{\nu}$ as above let $\omega \in \Lambda_2^3$ with $P_3(\omega) \subseteq P_4(\mu_0)$ and $X(\omega) = P_3(\omega)$. Then $\bar{\omega} \in \Delta^2(\bar{\nu}) \cap \Delta^3(\bar{\mu}_0)$ and $F(\bar{\nu}, \bar{\omega}) = L_1(a, b) \cong Mat_{22}$ and the case $j = 3$ follows. The assertion (ii) is immediate from (4.10.3), its proof and the diagram there. If $\bar{\mu}_1 \in \Delta^2(\bar{\mu}_0) \cap \Delta^2(\bar{\nu})$, then by (4.11.2) and (4.10.5) $F(\bar{\nu}, \bar{\mu}_1) \cong 2^9 : L_3(4)$ has three orbits, say Ω_2, Ω_3, Ω_4, on $\Delta^2(\bar{\mu}_0) \cap \Delta^2(\bar{\mu}_1) \setminus \{\bar{\mu}_0 + \bar{\mu}_1\}$ with lengths 42, 512, 336. By the diagram in (4.10.3) and the divisibility condition we have $\Omega_k \subset \Delta^2(\bar{\mu}_0) \cap \Delta^k(\bar{\nu})$ for $k = 2$, 3 and 4, so that (iii) follows. □

Lemma 4.11.5 *Let* $\bar{\eta} \in \Delta^3(\bar{\mu}_0)$. *Then* $F(\bar{\eta}) \cong McL$ *acting on* $\Delta^3(\bar{\mu}_0) \cap \Delta^2(\bar{\eta})$ *has two orbits:* $(\Delta^k(\bar{\mu}_0) \cap \Delta^2(\bar{\eta})) + \bar{\eta}$ *for* $k = 2$ *and* 4 *with lengths* 275 *and* 2025 *and stabilizers isomorphic to* $U_4(3)$ *and* Mat_{22}, *respectively.*

Proof. By (4.11.4) $\Delta^4(\bar{\mu}_0) \cap \Delta^2(\bar{\eta})$ is of size

$$2025 = 2048 \cdot |\Delta^4(\bar{\mu}_0)| \cdot |\Delta^3(\bar{\mu}_0)|^{-1}.$$

If θ is the quadratic form as in (4.5.6) then for $\bar{\alpha} \in \Delta$ we have $\theta(\bar{\mu}_0, \bar{\alpha}) = 1$ if and only if $\bar{\alpha} \in \Delta^3(\bar{\mu}_0)$. Hence every singular triangle which intersects $\Delta^3(\bar{\mu}_0)$ must intersect it in two vertices. Since $275 + 2025 = \frac{1}{2}|\Delta^2(\bar{\eta})|$, the result follows. □

Now by (4.11.1), (4.11.2), (4.11.3), (4.11.4) and (4.11.5) we have the complete suborbit diagram of Δ. In what follows we will make use of the following.

Lemma 4.11.6 *Let* $\bar{\eta} \in \Delta^3(\bar{\mu}_0)$, $\bar{\mu} \in \Delta^2(\bar{\mu}_0) \cap \Delta^2(\bar{\eta})$. *Then* $F(\bar{\eta}, \bar{\mu}) \cong U_3(4)$ *acting on* $\Delta^4(\bar{\mu}_0) \cap \Delta^2(\bar{\eta})$ *has three orbits with lengths* 162, 567, 1296 *and stabilizers isomorphic to* $L_3(4)$, $2^4 : Alt_6$, Alt_7, *respectively.*

Proof. It follows from the paragraph after (4.10.10) that if $\bar{\nu} \in \Delta^4(\bar{\mu}_0) \cap \Delta^2(\bar{\eta})$ then $G(\bar{\eta}, \bar{\mu}_0, \bar{\nu}) \cong Mat_{22}$ has three orbits on $\Delta^2(\bar{\mu}_0) \cap \Delta^2(\bar{\eta})$ with lengths 22, 77 and 176. Hence the result follows from the obvious duality. □

The following useful statement can be checked directly.

Lemma 4.11.7 *Let $\mu \in \Lambda_2$ and $\bar{\mu} \neq \bar{\mu}_0$. Then $\bar{\mu} \in \Delta^j(\bar{\mu}_0)$ for $j = 2$, 3 and 4 if and only if $(\mu, \mu_0) = \pm 2$, ± 1 and 0, respectively.* □

We have deduced the suborbit diagram of Δ from that of the affine Leech graph. In a similar way (using (4.11.2), (4.11.3), (4.11.6) and suggesting the reader fill in some minor details) we obtain the following.

Lemma 4.11.8 *The subgraph $\widehat{\Pi}$ in Δ induced by $\Delta^2(\bar{\mu}_0)$ has the following suborbit diagram with respect to the action of $F = G(\bar{\mu}_0) \cong Co_2$:*

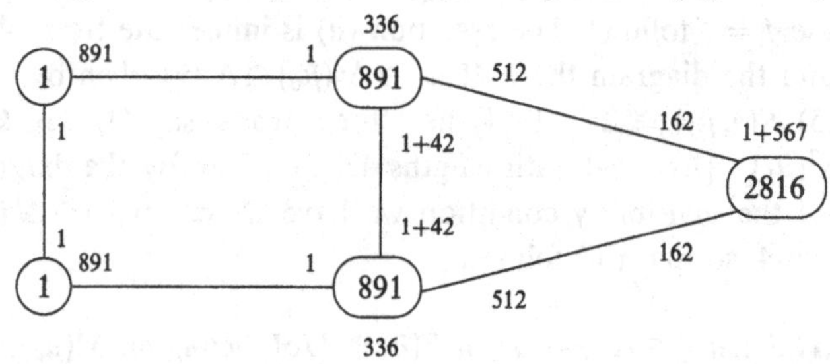

□

There is an equivalence relation on the vertex set of $\widehat{\Pi}$ with classes of size 2 with respect to which two distinct vertices are equivalent if together with $\bar{\mu}_0$ they form a singular triangle. Let Π be the quotient of $\widehat{\Pi}$ with respect to this equivalence relation. Then immediately from (4.11.8) we obtain the following.

Lemma 4.11.9 *The group $F \cong Co_2$ induces on Π a rank 3 action and the distance diagram of Π is the following:*

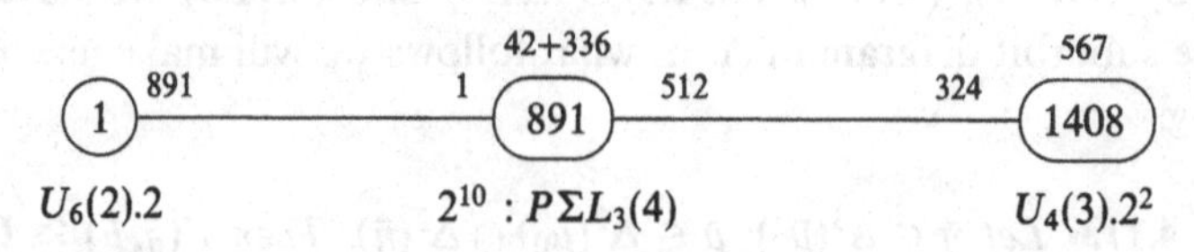

□

Let $\widehat{\Pi}'$ be a graph on $\Delta^2(\bar{\mu}_0)$ in which $\bar{\eta}_1$ and $\bar{\eta}_2$ are adjacent if $\bar{\eta}_2 \in \Delta^4(\bar{\eta}_1)$. Then $\widehat{\Pi}'$ and Π are the point graphs of the exceptional extended dual polar spaces $\mathscr{E}(Co_2 \times 2)$ and $\mathscr{E}(Co_2)$ with the diagram

c

1 2 4 4

from (1.13.7) whose full automorphism groups are isomorphic to $Co_2 \times 2$ and Co_2, respectively. Notice that the residue of an element of type 4 in $\mathcal{E}(Co_2 \times 2)$ is the universal (double) cover of the corresponding residue in $\mathcal{E}(Co_2)$. The following result was established in [Ron81a].

Lemma 4.11.10 *The geometries $\mathcal{E}(Co_2)$ and $\mathcal{E}(Co_2 \times 2)$ are simply connected and $\mathcal{E}(Co_2 \times 2)$ is the universal 2-cover of $\mathcal{E}(Co_2)$.* □

4.12 The simple connectedness of $\mathcal{G}(Co_2)$ and $\mathcal{G}(Co_1)$

In this section we establish the simple connectedness of the P_4-geometry $\mathcal{G}(Co_2)$ and the 2-simple connectedness of the T_4-geometry $\mathcal{G}(Co_1)$. We will see in (7.4.8) that $\mathcal{G}(Co_2)$ is not 2-simply connected.

Let $\varphi : \widetilde{\mathcal{F}} \to \mathcal{F}$ be the universal covering of $\mathcal{F} = \mathcal{G}(Co_2)$, $\widetilde{F}$ be the universal completion of the amalgam of maximal parabolics associated with the action of $F \cong Co_2$ on $\mathcal{F}$ and let $\widetilde{\Theta}$ be the collinearity graph of $\widetilde{\mathcal{F}}$. Then $\widetilde{F}$ acts naturally on $\widetilde{\mathcal{F}}$ and on $\widetilde{\Theta}$. Furthermore (the last paragraph in Section 4.9), φ induces a covering φ_1 of $\widetilde{\Theta}$ onto the collinearity graph $\Theta = \Theta(\bar{\mu}_0)$ of $\mathcal{G}(Co_2)$ and every triangle from Θ is contractible with respect to φ_1.

Let $\mathcal{S}$ be the $\mathcal{G}(Sp_6(2))$-subgeometry of $\mathcal{G}(Co_2)$ as in (4.9.8). By (1.6.4) $\mathcal{S}$ is simply connected and hence every connected component $\widetilde{\mathcal{S}}$ of the preimage of $\mathcal{S}$ in $\widetilde{\mathcal{F}}$ is isomorphic to $\mathcal{S}$, which implies the following.

Lemma 4.12.1 *If $\Theta(\mathcal{S})$ and $\widetilde{\Theta}(\widetilde{\mathcal{S}})$ are the subgraphs in Θ and $\widetilde{\Theta}$ induced by the point-sets of $\mathcal{S}$ and $\widetilde{\mathcal{S}}$, respectively, then φ_1 induces an isomorphism of $\widetilde{\Theta}(\widetilde{\mathcal{S}})$ onto $\Theta(\mathcal{S})$ and $\widetilde{F}[\widetilde{\Theta}(\widetilde{\mathcal{S}})] \cong F[\Theta(\mathcal{S})] \cong 2^{1+8}_+.Sp_6(2)$.* □

Let $\Xi = \Theta(\bar{\mu}_0) \cap \Theta(\bar{\mu}_1)$ be the subgraph in Θ as in (4.10.5) (so that Ξ is isomorphic to the dual polar graph of $U_6(2)$) and assume (the proof of (4.10.6)) that $\Omega := \Xi \cap \Theta(\mathcal{S})$ is a quad in Ξ. This means that the subgraph in Ξ induced by Ω is the point graph of the generalized quadrangle of order (2,4). By (4.10.8) $F[\Xi] \cong U_6(2).2$ (the extension of $U_6(2)$ by a field automorphism). Since the dual polar space of $U_6(2)$ is simply connected by (1.6.4), we have the following.

Lemma 4.12.2 *Let $\widetilde{\Xi}$ be a connected component of the preimage of Ξ in $\widetilde{\Theta}$ such that $\widetilde{\Omega} := \widetilde{\Xi} \cap \widetilde{\Theta}(\widetilde{\mathcal{S}})$ is non-empty. Then*

(i) *φ_1 induces an isomorphism of $\widetilde{\Xi}$ onto Ξ,*

(ii) $\widetilde{\Omega}$ *is a quad in* $\widetilde{\Xi}$,

(iii) $\widetilde{F}[\widetilde{\Xi}] \cong F[\Xi] \cong U_6(2).2$. □

Let x be a vertex of Θ contained in Ω (we can assume that $x = \bar{\lambda}_0$, so that $F(x) = G_1 \cap F \cong 2^{10} : \mathrm{Aut}\, Mat_{22}$). Then by (4.10.5) $O_2(F(x) \cap F[\Xi])$ has index 2 in $O_2(F(x))$, which shows the following.

Lemma 4.12.3 *The graph* Π *as in* (4.11.9) *is a graph on the set of images of* Ξ *under* F *in which two such images* Ξ' *and* Ξ'' *are adjacent if they are distinct and there is* $y \in \Xi' \cap \Xi''$ *such that* Ξ'' *is the image of* Ξ' *under an element from* $O_2(F(y))$. □

Let $\widetilde{\Pi}$ be a graph on the set of images of $\widetilde{\Xi}$ (as in (4.12.2)) under $\widetilde{F}$ in which two such images $\widetilde{\Xi}'$ and $\widetilde{\Xi}''$ are adjacent if they are distinct and there is $\widetilde{y} \in \widetilde{\Xi}' \cap \widetilde{\Xi}''$ such that $\widetilde{\Xi}''$ is the image of $\widetilde{\Xi}'$ under an element from $O_2(\widetilde{F}(\widetilde{y}))$. By (4.12.2) and the paragraph before (4.12.3) the valency of $\widetilde{\Pi}$ is the number of vertices in $\widetilde{\Xi}$ (which is 891) and hence φ induces a covering φ_2 of $\widetilde{\Pi}$ onto Π. We are going to show that φ_2 is an isomorphism.

The definition of Π in (4.12.3) shows that there is a bijection σ from the set of neighbours of Ξ in Π and the vertex set of Ξ such that $\sigma(\Xi') = y$ if and only if Ξ' is the image of Ξ under an element from $O_2(F(y))$. Comparing the suborbit diagram of Ξ in (4.10.5) and the suborbit diagram of Π in (4.11.9), we obtain the following.

Lemma 4.12.4 *Let* Ξ' *and* Ξ'' *be distinct vertices adjacent to* Ξ *in* Π. *Then* Ξ' *and* Ξ'' *are adjacent in* Π *if and only if the distance between* $\sigma(\Xi')$ *and* $\sigma(\Xi'')$ *in* Ξ *is* 1 *or* 2. □

Let $\Theta(\mathcal{S})$ and $\widetilde{\Theta}(\widetilde{\mathcal{S}})$ be as in (4.12.1) while Ω and $\widetilde{\Omega}$ are as in (4.12.2). Let Σ and $\widetilde{\Sigma}$ be the subgraphs of Π and $\widetilde{\Pi}$ induced by the images of Ξ and $\widetilde{\Xi}$ under $F[\Theta(\mathcal{S})]$ and $\widetilde{F}[\widetilde{\Theta}(\widetilde{\mathcal{S}})]$, respectively.

Lemma 4.12.5 *The following assertions hold:*

(i) *both* Σ *and* $\widetilde{\Sigma}$ *are cliques of size* 28,

(ii) *every triangle in* Π *is contractible with respect to* φ_2.

Proof. In terms of (4.8.7) and (4.10.6) $F[\Theta(\mathcal{S})]$ is the stabilizer in $D \cong 2^{1+8}_+.\Omega_8^+(2)$ of a non-isotropic 1-subspace $\langle \bar{\mu}_0 \rangle$ in V_1 while $F[\Xi] \cap F[\Omega]$

is the stabilizer of a minus 2-space $\langle \bar{\mu}_0, \bar{\mu}_1 \rangle$. Hence

$$|\Sigma| = [Sp_6(2) : U_4(2).2] = 28$$

and $F[\Theta(\mathcal{S})]$ acts on Σ doubly transitively. Furthermore, if $x \in \Theta(\mathcal{S})$ then $F[\Theta(\mathcal{S})] \cap F(x)$ contains a Sylow 2-subgroup of $F(x)$ and hence it contains $O_2(F(x))$. This implies that $\sigma^{-1}(x)$ is contained in Σ and the latter is a clique. By (4.12.1) and (4.12.2) $\widetilde{\Sigma}$ is a clique isomorphic to Σ and (i) follows. It is easy to see that the restriction of σ to $\Sigma \setminus \{\Xi\}$ is a bijection onto Ω. In view of this observation (ii) follows from (i) and (4.12.4). □

In the next chapter we will make use of the following result.

Lemma 4.12.6 *Let $S = F[\Sigma] \cong 2^{1+8}_{+}.Sp_6(2)$ (the stabilizer in $F \cong Co_2$ of a $\mathcal{G}(Sp_6(2))$-subgeometry from $\mathcal{G}(Co_2)$). Then*

(i) *S has three orbits: Σ, Σ_1 and Σ_2 on the vertex set of Π with lengths 28, 2016 and 256, respectively, where Σ_1 is the set of vertices at distance 1 from Σ,*

(ii) *$O_2(S)$ acts transitively on Σ_2 with kernel $Z(O_2(S))$ so that the stabilizer is of the form $2.Sp_6(2)$.*

Proof. Clearly Σ is an orbit of S. If $y \in \Sigma$ then the set $\Sigma \setminus \{y\}$ corresponds to the points incident to a plane π in the residue of y in $\mathcal{E}(Co_2)$ isomorphic to $\mathcal{G}(U_6(2))$. Let $u \in \Pi(y) \setminus \Sigma$, so that u corresponds to a point outside π. It is well known and easy to check that the stabilizer of π in $U_6(2).2$ permutes transitively the points outside π. This shows that S is transitive on the set Σ_1 of vertices at distance 1 from Σ. One can see from the intersection diagram of the collinearity graph Ξ of $\mathcal{G}(U_6(2))$ (4.10.5) that there are exactly 1 vertex in π adjacent to u in Ξ and 10 vertices at distance 2. By (4.12.4) this implies that u is adjacent to exactly 12 vertices from Σ and hence $|\Sigma_1| = 2016$. Thus in order to prove (i) it remains to show that S is transitive on the set $\Sigma_2 := \Pi \setminus (\Sigma \cup \Sigma_1)$ of size 256. By (4.11.4) $F_1 \cong 2^{10}.\mathrm{Aut}\, Mat_{22}$ has 3 orbits on Π with lengths 44, 1024 and 1232. If we assume that the element of type 1 stabilized by F_1 is contained in the subgeometry stabilized by S, then $T := F_1 \cap S \sim 2^{10}.2^4.Sym_6$ and it is easy to see (compare the diagram $D_h(Mat_{22})$) that T has 6 orbits on Π with lengths 12, 32, 1024, 16, 256 and 960. In view of the above this immediately shows that Σ_2 is an S-orbit. Since $O_1(S) = \langle [O_2(T), Q], Q \rangle$ where Q is a complement in $O_2(T)$ to $O_2(F_1)$, (ii) is easy to deduce from (4.11.4). □

Below we present the orbit diagram of Π with respect to the action of S.

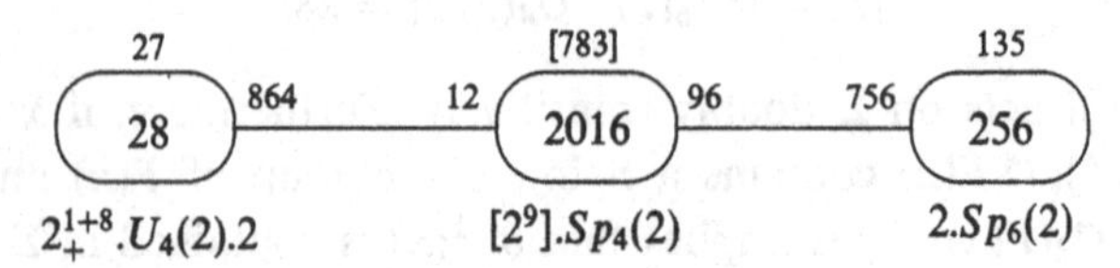

The following result was proved in [Sh92].

Proposition 4.12.7 *The geometry $\mathcal{G}(Co_2)$ is simply connected.*

Proof. By (4.12.5 (ii)) it is sufficient to show that Π is triangulable. Let $\alpha \in \Pi$, $\beta \in \Pi_2(\alpha)$; we put $\Phi = \Pi(\alpha) \cap \Pi(\beta)$ and identify Φ with the subgraph in Π induced by this set. Then by (4.11.9) Φ and $\Pi(\alpha)\backslash\Phi$ are the orbits of $F(\alpha, \beta) \cong U_4(3).2$ with lengths 324 and 567, respectively. Since the diameter of Π is 2, in order to apply (1.14.1) all we have to show is that Φ is connected and that every vertex from $\Pi(\alpha) \setminus \Phi$ is adjacent to a vertex from Φ. Let $\gamma \in \Phi$ and assume that $\gamma = \Xi$. Then by (4.12.4) $\sigma(\alpha)$ and $\sigma(\beta)$ are at distance 3 in Ξ and $\delta \in \Pi(\gamma)$ is contained in Φ if and only if $\sigma(\delta)$ is at distance 1 or 2 from both $\sigma(\alpha)$ and $\sigma(\beta)$. For a pair x, y of vertices at distance k in Ξ let p_{ij}^k denote the number of vertices at distance i from x and at distance j from y. The numbers p_{ij}^k can be calculated from the intersection parameters of Ξ in (4.10.5) [BI84]. In view of the above

$$\Phi(\gamma) = p_{12}^3 + p_{21}^3 + p_{22}^3 = 21 + 21 + 105 = 147.$$

Since the valency of Φ is larger than one third of its size, there are at most two connected components. By (4.11.3 (ii)) and (4.11.8) $O^2(F(\alpha, \beta)) \cong U_4(3)$ (which is the only index 2 subgroup in $F(\alpha, \beta)$) has two orbits, Φ_1 and Φ_2, on Φ of length 162 each and the action on Φ_i is similar to the action of $U_4(3)$ on the set of vertices at distance 2 from a given vertex in the McLaughlin graph. Since the valency of Φ is 147, by (4.10.13) Φ_i can not be a connected component of Φ and hence the latter is connected. Since the valency of Φ is less than the valency of the subgraph induced by $\Pi(\alpha)$ and $F(\alpha, \beta)$ is transitive on $\Pi(\alpha) \setminus \Phi$, every vertex from the latter set is adjacent to a vertex from Φ and the result follows. □

Let us turn to the simple connectedness question for $\mathcal{G} = \mathcal{G}(Co_1)$. Let $\psi : \widetilde{\mathcal{G}} \to \mathcal{G}$ be the universal covering, $\widetilde{G}$ be the universal completion of the amalgam of maximal parabolics corresponding to the action of G on $\mathcal{G}$ and $\widetilde{\Gamma}$ be the collinearity graph of $\widetilde{\mathcal{G}}$. Then by the remark in the last paragraph of Section 4.9 ψ induces a covering ψ_1 of $\widetilde{\Gamma}$ onto the Leech graph Γ and the triangles in Γ are contractible with respect to ψ_1.

Let $\Theta = \Theta(\bar{\mu}_0)$ be the subgraph in Γ defined after the proof of (4.9.4) and let $\Xi = \Theta(\bar{\mu}_0) \cap \Theta(\bar{\mu}_1)$ be as in (4.10.5). Since Θ is the collinearity graph of the geometry $\mathscr{G}(Co_2)$ which is simply connected by (4.12.7), we have the following.

Lemma 4.12.8 *Let $\widetilde{\Theta}$ be a connected component of the preimage of Θ in $\widetilde{\Gamma}$. Then the restriction ξ of ψ_1 to $\widetilde{\Theta}$ is an isomorphism onto Θ and if $\widetilde{\Xi} = \xi^{-1}(\Xi)$ then $\widetilde{G}[\widetilde{\Theta}] \cong G[\Theta] \cong Co_2$ and $\widetilde{G}[\widetilde{\Xi}] \cong G[\Xi] \cong U_6(2).Sym_3$.* □

If $F = G[\Theta]$ then $F[\Xi] \cong U_6(2).2$ has index 3 in $G[\Xi] \cong U_6(2).Sym_3$ and hence we have the following.

Lemma 4.12.9 *The shortest vector graph Δ is the graph on the set of images of Θ under G in which two such images are adjacent if their intersection is an image of Ξ.* □

Let $\widetilde{\Delta}$ be the graph on the set of images of $\widetilde{\Theta}$ under $\widetilde{G}$ in which two images are adjacent if their intersection is an image of $\widetilde{\Xi}$. Then by (4.12.8) and the remark before (4.12.9) the valency of $\widetilde{\Delta}$ is twice the number (which is 2300) of images of Ξ under F. Hence Δ and $\widetilde{\Delta}$ have the same valency 4600 and ψ induces a covering ψ_2 of $\widetilde{\Delta}$ onto Δ.

We assume that $\bar{\lambda}_0 \in \Xi$ and let $\widetilde{\lambda}_0$ be the preimage of $\bar{\lambda}_0$ in $\widetilde{\Xi}$. Let Ψ and $\widetilde{\Psi}$ be the subgraphs in Δ and $\widetilde{\Delta}$ induced by the images of Θ and $\widetilde{\Theta}$ containing $\bar{\lambda}_0$ and $\widetilde{\lambda}_0$, respectively. Since $G[\Theta]$ is transitive on Θ and $\widetilde{G}[\widetilde{\Theta}]$ is transitive on $\widetilde{\Theta}$, Ψ and $\widetilde{\Psi}$ consist of the images of Θ and $\widetilde{\Theta}$ under $G_1 = G(\bar{\lambda}_0)$ and $\widetilde{G}(\widetilde{\lambda}_0)$. This shows that $|\Psi| = |\widetilde{\Psi}| = 552$ and hence ψ_2 induces a bijection of $\widetilde{\Psi}$ onto Ψ. Furthermore, if $\Theta', \Theta'' \in \Psi$ and $\Theta' \cap \Theta''$ is an image of Ξ, then, since Ξ is connected, the preimages of Θ' and Θ'' in $\widetilde{\Psi}$ intersect in an image of $\widetilde{\Xi}$ and hence ψ_2 induces an isomorphism of $\widetilde{\Psi}$ onto Ψ. One can see from the suborbit diagram of Δ that F has two orbits on the set of triangles in Δ (one of the orbits is formed by the singular triangles). On the other hand Ψ is just the subgraph in Δ induced by $\bar{\Lambda}_2^4$ and from its suborbit diagram in (4.10.3) we observe that Ψ contains both singular and non-singular triangles, which gives

Lemma 4.12.10 *Every triangle in Δ is contractible with respect to ψ_2.* □

Proposition 4.12.11 *The geometry $\mathscr{G}(Co_1)$ is 2-simply connected.*

Proof. By (1.6.4) and (3.3.11) every residue in $\mathcal{G}(Co_1)$ is 2-simply connected and hence we only have to show that $\mathcal{G}(Co_1)$ is simply connected. By (4.12.10) in order to prove the simple connectedness of $\mathcal{G}(Co_1)$ it is sufficient to show that Δ is triangulable. We apply (a version of) (1.14.1). If $\bar{\mu}$ is at distance 2 from $\bar{\mu}_0$ in Δ, then the subgraph induced by $\Delta^2(\bar{\mu}_0) \cap \Delta^2(\bar{\mu})$ is connected by (4.11.3 (ii)) if $\bar{\mu} \in \Delta^3(\bar{\mu}_0)$ and by (4.11.4 (ii)) if $\bar{\mu} \in \Delta^4(\bar{\mu}_0)$. Hence the quadrangles in Δ are triangulable. Let $C = \{\bar{\mu}_0, \bar{\mu}_1, \bar{\mu}_2, \bar{\mu}_3, \bar{\mu}_4\}$ be a non-degenerate 5-cycle in Δ. We say that C is of type 1 if $\bar{\mu}_j \in \Delta^4(\bar{\mu}_{j+2})$ for some $0 \leq i \leq 4$ (where addition of indices is modulo 5). Suppose that C is of type 1 and assume without loss of generality that $\bar{\mu}_2 \in \Delta^4(\bar{\mu}_0)$. In this case by (4.11.4 (iii)) there is a vertex $\bar{v} \in \Delta^2(\bar{\mu}_0) \cap \Delta^2(\bar{\mu}_2)$ which is adjacent to $\bar{\mu}_4$. Hence C splits into the triangle $\{\bar{\mu}_0, \bar{v}, \bar{\mu}_4\}$ and the quadrangles $\{\bar{\mu}_0, \bar{\mu}_1, \bar{\mu}_2, \bar{v}\}$, $\{\bar{\mu}_2, \bar{\mu}_3, \bar{\mu}_4, \bar{v}\}$. Thus the 5-cycles of type 1 are triangulable. Suppose now that C is not of type 1 which means that $\bar{\mu}_2, \bar{\mu}_3 \in \Delta^3(\bar{\mu}_0)$. Then by (4.11.5) the third vertex $\bar{\mu}_5 = \bar{\mu}_2 + \bar{\mu}_3$ in the singular triangle containing the edge $\{\bar{\mu}_2, \bar{\mu}_3\}$ is contained in $\Delta^k(\bar{\mu}_0)$ for $k = 2$ or 4. If $k = 2$ then C splits into a triangle and two quadrangles while if $k = 4$ then C splits into the triangle $\{\bar{\mu}_2, \bar{\mu}_3, \bar{\mu}_5\}$ and two 5-cycles of type 1: $\{\bar{\mu}_0, \bar{\mu}_1, \bar{\mu}_2, \bar{\mu}_5, \bar{\mu}_6\}$ and $\{\bar{\mu}_0, \bar{\mu}_4, \bar{\mu}_3, \bar{\mu}_5, \bar{\mu}_6\}$ where $\bar{\mu}_6 \in \Delta^2(\bar{\mu}_0) \cap \Delta^2(\bar{\mu}_5)$. In any case C is triangulable and so is Δ. □

4.13 *McL* geometry

In this section we discuss a geometry related to the Petersen graph and associated with the McLaughlin group. By (4.11.4 (i)) if $\bar{\eta} \in \Delta^3(\bar{\mu}_0)$ then $G(\bar{\mu}_0, \bar{\eta}) \cong McL$ acts transitively on the set $\Delta^2(\bar{\eta}) \cap \Delta^4(\bar{\mu}_0)$ of size 2025 with stabilizer isomorphic to Mat_{22}. We analyse this action in further detail calculating in the Leech lattice, rather than in the Leech lattice modulo 2.

Let μ_0 be as above (*i.e.* $\mu_0 \in \Lambda_2^4$, $X(\mu_0) = \emptyset$, $P_4(\mu_0) = \{a, b\}$), let $\eta \in \Lambda_2^3$ with $P_3(\eta) = X(\eta) = \{a\}$, $\omega = \mu_0 + \eta$, so that $\omega \in \Lambda_3^5$ with $X(\omega) = \emptyset$, $P_5(\omega) = \{b\}$. Put $M^c = \widehat{G}(\mu_0, \eta) = \widehat{G}(\mu_0, \omega)$, $M_1^c = M^c \cap \widehat{G}_1$. Since $\bar{\eta} \in \Delta^3(\bar{\mu}_0)$, by (4.11.1), $M^c \cong G(\bar{\mu}_0, \bar{\eta}) \cong McL$ and it is easy to see that $M_1^c = \widehat{L}_1(a, b) \cong Mat_{22}$. Let $v_0 \in \Lambda_2^4$ with $P_4(v_0) = \{a, b\}$, $X(v_0) = \{a\}$ so that $v_0 = \lambda_0 - \mu_0$ for $\lambda_0 \in \Lambda_4^8$ with $P_8(\lambda_0) = \{b\}$ and $X(\lambda_0) = \emptyset$. Then $\bar{v}_0 \in \Delta^4(\bar{\mu}_0) \cap \Delta^2(\bar{\eta})$ and hence by (4.11.7) the set Ω (of size 2025) of images of v_0 under M^c is the following:

$$\Omega = \{v \mid v \in \Lambda_2, (v, \mu_0) = 0, (v, \eta) = 2\}.$$

Using (4.4.1) and the table therein it is straightforward to check that M_1^c

has the following three orbits on $\Omega \setminus \{v_0\}$:

$$\Omega_1 = \{v \mid v \in \Lambda_2^2, X(v) = P_2(v) \cap \{a,b\} = \emptyset\},$$

$$\Omega_2 = \{v \mid v \in \Lambda_2^2, \{a,b\} \subseteq P_2(v), X(v) = \{a,c\}, c \neq b\},$$

$$\Omega_3 = \{v \mid v \in \Lambda_2^3, v(a) = -1, v(b) = 1, X(v) \subset P_1(v), |X(v)| = 7\}.$$

Then $v \in \Omega_i$ can be identified with the octet $P_2(v)$ if $i = 1$, with the hexad $P_2(v) \setminus \{a,b\}$ having a distinguished element $X(v) \setminus \{a\}$ if $i = 2$ and with the hyperoval $X(v) \setminus \{a\}$ in the residual projective plane $\Pi(P_3(v) \cup \{a,b\})$ if $i = 3$.

Let Ω also denote the graph on this set which is invariant under the action of M^c and such that $\Omega(v_0) = \Omega_1$. In other terms two vectors from Ω are adjacent if they are perpendicular. Then straightforward calculations using the diagrams for $\mathscr{H}(Mat_{22})$ in Section 3.9 show that the suborbit diagram of Ω with respect to the action of M^c is the following:

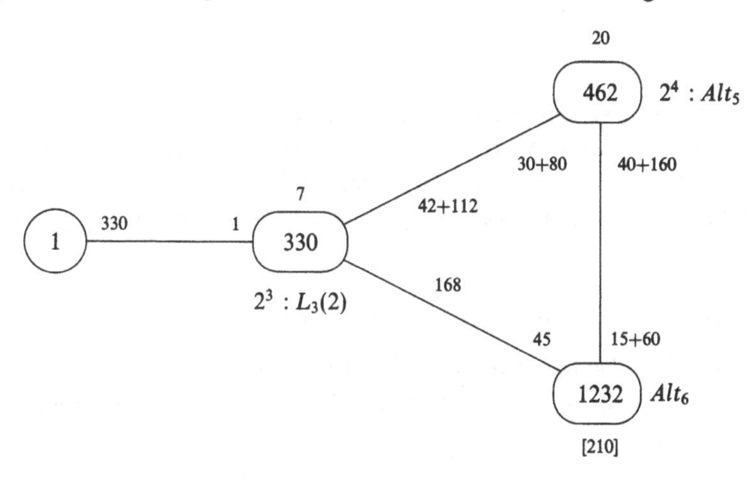

Two vertices in Ω_1 are adjacent if the corresponding octets are disjoint. Hence the subgraph induced by Ω_1 is isomorphic to the derived graph $\Delta(\mathscr{G}(Mat_{22}))$ of the P_3-geometry $\mathscr{G}(Mat_{22})$. Thus the elements of type 1, 2 and 3 in $\mathscr{G}(Mat_{22})$ are identified with the Petersen subgraphs, edges and vertices in the subgraph induced by Ω_1 so that the incidence relation is via inclusion and M_1^c is a flag-transitive automorphism group of $\mathscr{G}(Mat_{22})$. This means that $M_1^c = M^c(v_0)$ acts transitively on the set of Petersen subgraphs in $\Omega_1 = \Omega(v_0)$ and if Θ_1 is such a subgraph then $M_1^c[\Theta_1] \cong 2^4 : Sym_5$. We are going to construct in Ω a family of locally Petersen subgraphs.

Recall that the complement $\bar{T}(n)$ of the triangular graph of order n is a graph on the set of 2-element subsets of an n-element set in which two such subsets are adjacent if they are disjoint. The Petersen graph is $\bar{T}(5)$ while $\bar{T}(6)$ is the point graph of $\mathcal{G}(Sp_4(2))$. It is easy to see that $\bar{T}(n)$ is locally $\bar{T}(n-2)$, in particular $\bar{T}(7)$ is locally Petersen. The incidence system $\mathcal{H}(Alt_7)$ whose elements are the triangles, edges and vertices in $\bar{T}(7)$ with the incidence relation via inclusion is a geometry with the following diagram:

$$\overset{}{\underset{2}{\circ}}\overset{P}{\text{———}}\underset{1}{\circ}\text{———}\underset{1}{\circ}$$

The groups Alt_7 and Sym_7 are flag-transitive automorphism groups of $\mathcal{H}(Alt_7)$ with vertex stabilizers isomorphic to Sym_5 and $Sym_5 \times 2$, respectively.

Consider $\eta' \in \Lambda_2^2$ with $\{a,b\} \cap P_2(\eta') = \{b\}$, $X(\eta') = P_2(\eta')$ and put $\omega' = \mu_0 + \eta'$. Then $\omega' \in \Lambda_3^4$ with $P_4(\omega') = \{a\}$, $b \in P_2(\omega')$ and $X(\omega') = P_2(\omega') \setminus \{b\}$. Let Ω' be a graph on

$$\{v' \mid v' \in \Lambda_2, (v', \mu_0) = 0, (v', \eta') = 2\}$$

in which two vectors are adjacent if they are perpendicular.

It follows from (4.4.1) and its proof that $\widehat{G}_1(\mu_0, \eta')$ is a semidirect product of the subgroup in $\widehat{Q}_1$ formed by the elements $\widehat{\tau}(Y,1)$ for the Golay sets Y disjoint from $P_4(\omega') \cup P_2(\omega')$ and of $\widehat{L}_1(a,b) \cap \widehat{L}_1[P_2(\omega')] \cong Alt_7$.

Lemma 4.13.1 *Let $\Theta' = \{\mu' \mid \mu' \in \Lambda_2^4, X(\mu') = P_4(\mu') \subseteq X(\omega')\}$. Then*

(i) *the subgraph in Ω' induced by Θ' is isomorphic to $\bar{T}(7)$,*

(ii) *$\widehat{G}_1(\mu_0, \eta')$ induces on Θ' an action isomorphic to Alt_7 with kernel 2^4.*

Proof. The mapping $\xi : \mu' \mapsto P_4(\mu')$ establishes a bijection of Θ' onto the set of 2-element subsets of the set $X(\omega')$ of size 7. Furthermore, $(\mu', \mu'') = 0$ if and only if $\xi(\mu') \cap \xi(\mu'') = \emptyset$, which gives (i). The assertion (ii) is immediate by the paragraph before the lemma. □

Since $(\mu_0, \eta') = (\mu_0, \eta) = -1$, by (4.11.7) there is an element $g \in \widehat{G}(\mu_0)$ which maps η' onto η. Clearly such a g maps Ω' onto Ω. Since $M^c = \widehat{G}(\mu_0, \eta)$ acts transitively on the vertex set of Ω, the element g can be chosen in such a way that v_0 is contained in the image Θ of Θ' under g. By (4.13.1) $\Theta \cong \bar{T}(7)$ which is locally Petersen. Hence $M^c[\Theta \cap \Omega_1] \cong 2^4 : Sym_5$ and by (4.13.1) this is exactly the stabilizer in

$\widehat{G}_1(\mu_0, \eta')$ of a vertex of Θ'. This means that $\widehat{G}_1(\mu_0, \eta')^g \cong 2^4 : Alt_7$ is the full stabilizer of Θ in M^c.

Define $\mathscr{G}(McL)$ to be an incidence system of rank 4 whose elements of type 1 are the images of Θ under M^c, elements of type 2, 3 and 4 are the triangles, edges and vertices of Ω and the incidence relation is via inclusion. Then the residue of v_0 is isomorphic to $\mathscr{G}(Mat_{22})$ and the residue of Θ is isomorphic to $\mathscr{G}(Alt_7)$. Since the incidence relation is via inclusion, $\mathscr{G}(McL)$ belongs to a string diagram and we have the following.

Lemma 4.13.2 *$\mathscr{G}(McL)$ is a geometry with the diagram*

$$\overset{}{\underset{2}{\circ}} \!-\!-\!-\! \underset{2}{\circ} \overset{P}{-\!-\!-} \underset{1}{\circ} \!-\!-\!-\! \underset{1}{\circ}$$

and M^c induces on $\mathscr{G}(McL)$ a flag-transitive action. □

The geometry $\mathscr{G}(McL)$ was first constructed in an unpublished work of A. Neumaier (see [Bue85], [ISh88]). It has been shown in [BIP98] that $\mathscr{G}(McL)$ is simply connected and that it possesses a 2-cover associated with a non-split extension $3^{23} \cdot McL$.

Let us define in $\mathscr{G}(McL)$ a subgeometry. By (4.11.6) a subgroup $U_4(3)$ in $M^c \cong McL$ has three orbits on Ω with lengths 162, 567 and 1296. We are interested in the orbit of length 567 and particularly in the subgraph in Ω induced by this orbit. Consider $\mu_1 \in \Lambda_2^2$ with $X(\mu_1) = \emptyset$, $\{a, b\} \subseteq P_2(\mu_1)$ and put $U = \widehat{G}(\omega, \mu_0, \mu_1)$, $U_1 = U \cap \widehat{G}_1$. Since $\bar{\mu}_1 \in \Delta^2(\mu_0) \cap \Delta^2(\eta)$, by (4.10.12 (ii)) we have $U \cong U_4(3)$. It is easy to see that U_1 is the stabilizer in $M_1^c = \widehat{L}_1(a, b) \cong Mat_{22}$ of the hexad $P_2(\mu_1) \setminus \{a, b\}$, so that $U_1 \cong 2^4 : Alt_6$ (which is the stabilizer in M_1^c of the subgeometry $\mathscr{Q}$ as in (3.4.4)), in particular the set Ψ of images of v_0 under U is of size 567 and in fact

$$\Psi = \{v \mid v \in \Omega, (v, \mu_1) = 2\}.$$

We calculate the orbits of U_1 on $\Psi \setminus \{\lambda_0\}$ in the way we have calculated the orbits of M_1^c on $\Omega \setminus \{\lambda_0\}$ and obtain the following four orbits:

$$\Psi_1 = \{v \mid v \in \Omega_1, P_2(v) \cap P_2(\mu_1) = \emptyset\},$$

$$\Psi_2 = \{v \mid v \in \Omega_2, P_2(v) \cap P_2(\mu_1) = \{b\}\},$$

$$\Psi_3 = \{v \mid v \in \Omega_2, |P_2(v) \cap P_2(\mu_1)| = 3, X(v) \subseteq P_2(\mu_1)\},$$

$$\Psi_4 = \{v \mid v \in \Omega_3, P_3(v) \cap P_2(\mu_1) = \{b\}\},$$

of length 30, 96, 120, 320, respectively. Having the suborbits it is straightforward to calculate the suborbit diagram of the subgraph in Ω induced by Ψ:

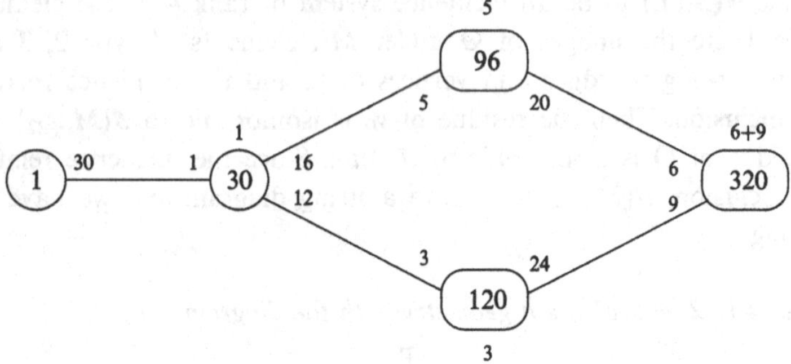

Notice that Ψ_1 is the set of vertices in the derived graph of $\mathcal{G}(Mat_{22})$ incident to the edges (elements of type 2) contained in the $\mathcal{G}(Sp_4(2))$-subgeometry $\mathcal{Q}$ stabilized by U_1.

Let Σ be the sextet used in the definition of the element Θ of type 1 in $\mathcal{G}(McL)$ and assume that $P_2(\mu_1) = S_1 \cup S_2$. Then the subgraph Υ induced by $\Theta \cap \Psi$ is the point graph of the generalized quadrangle $\mathcal{G}(Sp_4(2))$. Let $\mathcal{G}(U_4(3))$ be a geometry whose elements of type 1, 2 and 3 are the vertices of Ψ, the triangles in Ψ and images of Υ under U, respectively; the incidence relation is via inclusion. Then $\mathcal{G}(U_4(3))$ is a GAB, a (*geometry which is almost a building*) described in [Kan81] with the following diagram:

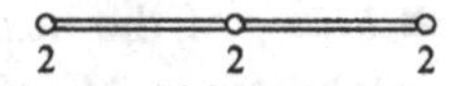

The residue of $\{\lambda_0\}$ (as an element of type 3) can be identified with the subgeometry $\mathcal{Q}$ in $\mathcal{G}(Mat_{22})$ and the residue of Υ (as an element of type 1) is the geometry of vertices and triangles in Υ. The universal cover of $\mathcal{G}(U_4(3))$ is an infinite affine building ([Kan81], [Ti82]). An outer automorphism of $U_4(3)$ performs a diagram automorphism of $\mathcal{G}(U_4(3))$ permuting the sets of elements of type 1 and 3.

There is another geometry associated with Ψ. Let $Q = O_2(U_1)$, Φ_2 be an orbit (of length 16) of Q in Ψ_2, Φ_1 be the set (of size 10) of vertices in Ψ_1 adjacent to vertices in Φ_1 and $\Phi = \{\lambda_0\} \cup \Phi_1 \cup \Phi_2$. Then the subgraph induced by Φ is the Schläfli graph, *i.e.* the line graph of the generalized quadrangle of order (2,4). The geometry $\mathcal{E}(U_4(3))$ whose elements are the images of Φ under U, the triangles and the vertices of Ψ is the extended dual polar space with the diagram

$$\overset{1}{\circ}\overset{c}{\text{———}}\overset{4}{\circ}\text{═══}\overset{2}{\circ}$$

as in (1.13.7) with the automorphism group of the form $U_4(3).2^2$.

4.14 Geometries of $3 \cdot U_4(3)$

Let $\mathscr{H} = \mathscr{H}(Co_1)$ and $\mathscr{G} = \mathscr{G}(Co_1)$ be the maximal parabolic and tilde geometries of $G = Co_1$, respectively. Let $\{y_i \mid 1 \le i \le 4\}$ and $\{x_i \mid 1 \le i \le 4\}$ be maximal flags in $\mathscr{H}$ and $\mathscr{G}$, respectively, such that $x_1 = y_1 = \bar{\lambda}_0$, $x_2 = y_2$, $x_3 \subset y_3$, $x_4 \subset y_4$ (here the y_i and the x_i are considered as vertices, lines and $\star$-closed cliques in the Leech graph). Put $H_i = G(y_i)$ and $G_i = G(x_i)$, $Q_i = O_2(G_i)$ for $1 \le i \le 4$, so that $G_1 = H_1$, $G_2 = H_2$, $G_3 \le H_3$, $G_4 \le H_4$ (Section 4.9).

Lemma 4.14.1 *The group $G_1 \cong 2^{11} : Mat_{24}$ contains two conjugacy classes of subgroups of order 3 with representatives X_s and X_t such that*

(i) $N_{G_1}(X_s) \cong 2^{4+1}.3 \cdot Sym_6$ *and* $N_{G_1}(X_s)O_2(H_1 \cap H_2) = H_1 \cap H_2$,
(ii) $N_{G_1}(X_t) \cong 2^3.(Sym_3 \times L_3(2))$ *and* $N_{G_1}(X_t)O_2(H_1 \cap H_3) = H_1 \cap H_3$,
(iii) X_s *and* X_t *are not conjugate in* G.

Proof. (i) and (ii) follow from (2.12.3), (2.13.3) and (3.8.2) with the remark that X_s and X_t map, respectively, onto $3a$- and $3b$-subgroups in $\bar{G}_1 \cong Mat_{24}$. The elements of order 3 in $3a$- and $3b$-subgroups in the complement $\widehat{L}_1 \cong Mat_{24}$ have different characters in the 24-dimensional real representation of $\widehat{G} \cong Co_0$ in the vector space containing the Leech lattice, which implies (iii). □

By the above lemma and (2.13.1) for $\alpha = s$ and t $N_G(X_\alpha)$ acts transitively on the set $\Phi(X_\alpha)$ of vertices in the Leech graph fixed by X_α.

The parabolic $G_2 \cong 2^{4+12}.(3 \cdot Sym_6 \times Sym_3)$ induces $3 \cdot Sym_6$ on $\mathrm{res}^+_{\mathscr{G}}(x_2) \cong \mathscr{G}(3 \cdot Sym_6)$ and the kernel G_2^- is of order $2^{17} \cdot 3$. Let X be a Sylow 3-subgroup of G_2^-.

Lemma 4.14.2 $N_{G_2}(X)/X \cong 2^{4+1}.3 \cdot Sym_6$.

Proof. By the Frattini argument $N_{G_2}(X)Q_2 = G_2$ and hence all we have to show is that $C_{Q_2}(X)$ is the natural symplectic module for $G_2^+/O_{2,3}(G_2^+) \cong Sp_4(2)$. Since $[Q_1 : Q_1 \cap Q_2] = 2$; Q_1 is the irreducible Golay code module $\mathscr{C}_{11}$ for $\bar{G}_1 \cong Mat_{24}$ and $G_{12}Q_1/Q_1$ is the stabilizer of a sextet in $\bar{G}_1$, we conclude by (3.8.4) that there are three chief factors of $G_2^+/Q_2 \cong 3 \cdot Sym_6 \cong 3 \cdot Sp_4(2)$ inside Q_2; two isomorphic to the hexacode module and one to the natural symplectic module. Since X acts faithfully on Q_2 and XQ_2/Q_2 is in the centre of $O^2(G_2)/Q_2$ the result follows. □

Since X is contained in G_2^-, it fixes $\mathrm{res}_{\mathcal{G}}^+(x_2)$ elementwise. Let u be an element of type j in $\mathrm{res}_{\mathcal{G}}^+(x_2)$ for $j = 3$ or 4. Then $\mathrm{res}_{\mathcal{G}}^-(u)$ is a projective space of rank $j-2$ over $GF(2)$, which contains x_2 and hence there is a unique element $\psi(u, X)$ of type $j-1$ incident to u which X fixes pointwise. Here $\psi(u, X)$ is the centralizer of X in the subgroup of order 2^j which represents u in $\bar{\Lambda}$. Furthermore $\psi(u, X)$ and $\psi(v, X)$ are incident if and only if u and v are incident, and we have the following.

Lemma 4.14.3 *Let $\Psi = \Psi(x_2, X)$ be the subgeometry in $\mathcal{G}$ formed by the elements $\psi(u, X)$ for all $u \in \mathrm{res}_{\mathcal{G}}^+(x_2)$. Then Ψ is isomorphic to $\mathrm{res}_{\mathcal{G}}(x_2)$ (i.e. to the rank 2 tilde geometry $\mathcal{G}(3 \cdot Sym_6)$) and $N_{G_2}(X)$ induces on Ψ its full automorphism group with kernel of order 2^5.* □

If w is an element of type 1 in Ψ, then by the above lemma $G(w) \cap N_{G_2}(X)$ has index 45 in $N_{G_2}(X)$ and since $G(w)$ is a conjugate of G_1, by (4.14.1) we have the following.

Lemma 4.14.4 *There is an element $g \in G$ which conjugates X onto X_s and maps $\Psi(x_2, X)$ onto a subgraph $\Psi(y, X_s)$ contained in $\Phi(X_s)$, isomorphic to the point graph of $\mathcal{G}(3 \cdot Sym_6)$ (here $y = x_2^g$).* □

Let us look more closely at the subgraph $\Phi(X_s)$ in the Leech graph induced by the set of vertices fixed by X_s. We identify X_s with a subgroup of type $3a$ in the complement $L_1 \cong Mat_{24}$ to Q_1 in G_1. Concerning the action of X_s on $\mathcal{P}$ we follow the notation introduced in Section 4.9. Recall that ϱ denotes the mapping of $\Gamma(\bar{\lambda}_0) = \bar{\Lambda}_4^{4a}$ onto the set of sextets which commutes with the action of G_1.

Lemma 4.14.5

(i) *The subgraph $\Phi(X_s)$ is of valency 32; $\Phi(X_s) \cap \bar{\Lambda}_4^{4a} = \Phi_0 \cup \Phi_1$ where $\varrho(\Phi_0) = \Sigma$ and $\varrho(\Phi_1) = \{\Sigma_{ij} \mid 1 \leq i < j \leq 6\}$ (in terms of (2.12.7)),*
(ii) *$N_{G_1}(X_s)$ acts transitively on Φ_1.*

Proof. Directly from (2.12.7) we obtain (i) together with the remark that Φ_j consists of one or two $N_{G_1}(X_s)$-orbits for $j = 0$ and 1. Without loss of generality we assume that the subgraph $\Psi(y, X_s)$ as in (4.14.4) contains $\bar{\lambda}_0$. Since $N_{G(y)}(X_s)$ induces the (flag-transitive) automorphism group of the rank 2 tilde geometry associated with $\Psi(y, X)$, we obtain (ii) together with the remark that $\Psi(y, X) \cap \bar{\Lambda}_4^{4a} \subseteq \Phi_1$. □

By the remark after (4.14.1) $N_G(X_s)$ is transitive on (the vertex set of) $\Phi(X_s)$ and by (4.14.5) it has two orbits, say Ω_0 and Ω_1, on the set of lines contained in $\Phi(X_s)$ with $|\Omega_1| = 15 \cdot |\Omega_0|$. Although $O^2(N_G(X_s)/X_s)$ is still to be identified with $3 \cdot U_4(3)$, let $\mathcal{G}(3 \cdot U_4(3))$ denote the incidence system of rank 3 whose elements of type 1, 2 and 3 are the vertices in $\Phi(X_s)$, the lines from Ω_1 and the images of $\Psi(y, X_s)$ under $N_G(X_s)$, with the incidence relation via inclusion.

Lemma 4.14.6 *$\mathcal{G}(3 \cdot U_4(3))$ is a geometry with the diagram*

$$\underset{2}{\circ}\overset{\sim}{=\!=\!=}\underset{2}{\circ}=\!=\!=\underset{2}{\circ}$$

and $N_{G_1}(X_s)$ induces on it a flag-transitive action with kernel X_s.

Proof. The group $N_{G_1}(X_s)$ acts on the set $\varrho(\Phi_1)$ of lines from Ω_1 incident to $\bar{\lambda}_0$ as $Sym_6 \cong Sp_4(2)$ acts on the point set of $\mathcal{G}(Sp_4(2))$. Then a subset of size 3 in $\varrho(\Phi_1)$ is a line of $\mathcal{G}(Sp_4(2))$ if and only if its stabilizer in $N_{G_1}(X_s)$ has order $2^9 \cdot 3$. On the other hand $\Psi(y, X_s)$ contains three lines from Ω_1 containing $\bar{\lambda}_0$ and by (4.14.3) $N_{G_1 \cap G(y)}(X_s)$ is of order $2^9 \cdot 3$. Since the vertices and the lines (from Ω_1) contained in $\Psi(y, X_s)$ form the rank 2 tilde geometry, the result follows. □

We are going to construct another geometry on $\Phi(X_s)$. Let $B_i = (E \setminus \{p_i\}) \cup T_i$ for some i, $1 \le i \le 6$, so that B_i belongs to the orbit of length 6 of $N_{L_1}(X_s)$ on the set of octads fixed by X_s. Let $\delta = \tau(B_i, 1)$ be the corresponding involution in Q_1 and let $\Phi(\delta)$ be the set of vertices x in the Leech graph such that $\delta \in O_2(G(x))$, $D = C_G(\delta)$, $R = O_2(D)$, $\bar{D} = D/R$ and $V_1 = C_{\bar{\Lambda}}(R)$. Then by (4.8.6) and (4.8.7) $R \cong 2_+^{1+8}$, $\bar{D} \cong \Omega_8^+(2)$ and the subgraph in the Leech graph induced by $\Phi(\delta)$ is the point graph of the parabolic geometry of $\Omega_8^+(2)$.

Lemma 4.14.7 *The subgraph in the Leech graph induced by $\Phi(\delta) \cap \Phi(X_s)$ is isomorphic to the Schläfli graph and $(D \cap N_G(X_s))/X_s \cong 2 \times U_4(2).2$ induces the full automorphism group of the subgraph.*

Proof. Without loss of generality we assume that X_s is contained in a complement $K \cong Alt_8 \cong L_4(2)$ in $C_{G_1}(\delta) \cong 2^{1+4+4+6}.L_4(2)$. Since $C_{G_1}(\delta) \cap N_{G_1}(X_s)$ has index 6 in $N_{G_1}(X_s)$, we conclude that $N_K(X_s) \cong (Sym_5 \times Sym_3)^e$. This means that X_s acts fixed-point freely on the natural module of K and has 4-dimensional centralizer in the exterior square of the natural module. By (4.8.6) this means that $N_R(X_s) = \langle \delta \rangle$ and $N_G(X_s) \cap O_2(C_{G_1}(\delta))$ is of order 2^5. Now the result follows either by direct calculation in the Leech graph or by analysing subgroups of order 3 in $\Omega_8^+(2)$. □

By the above lemma $T := [X_s, V_1]$ is a minus 2-space in V_1 and by the proof of (4.10.6) T is as in (4.10.4). Hence we have the following ((4.10.5) and the diagram therein).

Lemma 4.14.8 *If* $T = [X_s, V_1]$ *with* $T^{\#} = \{\bar{\mu}_0, \bar{\mu}_1, \bar{\mu}_2\}$, *then* $\Phi(\delta) \cap \Phi(Z_s)$ *is a quad in the subgraph* Ξ *of the Leech graph induced by* $\bar{\Lambda}_2(T) \cap \bar{\Lambda}_4$. □

Define $\mathcal{E}(3 \cdot U_4(3))$ to be the incidence system of rank 3 whose elements of type 1 are the images of $\Phi(\delta) \cap \Phi(X_s)$ under $N_G(X_s)$, the elements of type 2 are the lines in the orbit Ω_1 and the elements of type 3 are the vertices in $\Phi(X_s)$, and the incidence is via inclusion.

Lemma 4.14.9

(i) $\mathcal{E}(3 \cdot U_4(3))$ *is an extended dual polar space with the diagram*

c
o———o═══o
1 4 2

on which $N_G(X_s)$ *induces a flag-transitive automorphism group,*

(ii) $N_G(X_s)/X_s \sim 3 \cdot U_4(3).2^2$,

(iii) $\mathcal{E}(3 \cdot U_4(3))$ *possesses a 3-fold covering onto the geometry* $\mathcal{E}(U_4(3))$ *constructed at the end of Section* 4.13.

Proof. The elements of type 1 and 2 incident to $\bar{\lambda}_0$ (considered as an element of type 3) are indexed by the octads B_i, $1 \le i \le 6$, and the sextets Σ_{jk}, $1 \le j < k \le 6$, respectively. Furthermore the element indexed by B_i and the element indexed by Σ_{jk} are incident if and only if $i \in \{j, k\}$. Hence (i) follows directly from (4.14.5) and (4.14.7). The classification of flag-transitive extended dual polar spaces of rank 3 achieved in [DGMP] and [Yos91] (the table after (1.13.7)) implies that $N_G(X_s)/X_s$ must be isomorphic either to $U_3(4).2^2$ or to $3 \cdot U_4(3).2^2$. By (4.14.3) $N_G(X_s)/X_s$ contains a subgroup of the form $2^5.3.Sym_6$ and hence the latter possibility holds, which implies (ii) and (iii). □

Using (4.4.1) it is not difficult to list all the vectors in $\Phi(X_s)$ and to describe the action of $N_{G_1}(X_s)$ on the set of these vectors which gives the suborbit diagram of the action of $N_G(X_s)$ on $\Phi(X_s)$, which earlier appeared in [Yos92], p. 159. We omit the details of the calculations and summarize the result in the following.

Lemma 4.14.10 *The group $N_{G_1}(X_s)/X_s \cong 2^5.Sym_6$ acting on the set $\Phi(X_s)$ of vectors in the Leech graph fixed by X_s has nine orbits with lengths* 1, 2, 30, 60, 96, 192, 320, 360 *and* 640 *which are contained in $\bar{\Lambda}_4^{\alpha}$ for $\alpha = 8, 4a, 4a, 6, 6, 5, 4c, 4b$ and 5, respectively. The suborbit diagram of the graph of valency* 30 *on $\Phi(X_s)$ is as given below.* □

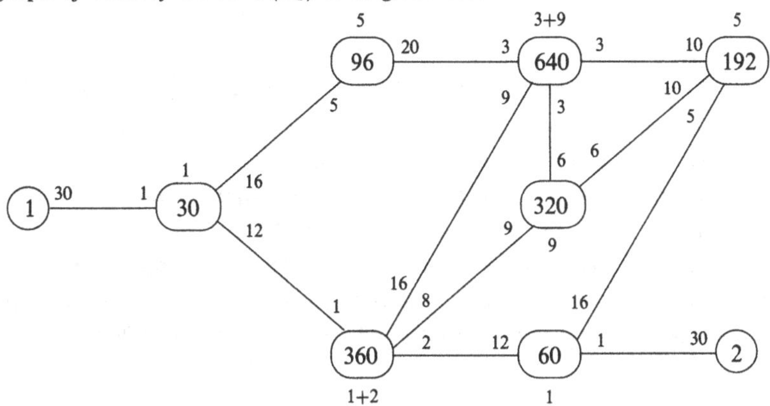

Notice that $O_3(N_G(X_s)/X_s)$ stabilizes every element of type 3 in $\mathcal{G}(3 \cdot U_4(3))$ and the geometry whose elements are the orbits of $O_3(N_G(X_s)/X_s)$ on the element set of $\mathcal{G}(3 \cdot U_4(3))$ with the induced incidence relation and type function is isomorphic to Kantor's GAB $\mathcal{G}(U_4(3))$ discussed at the end of the previous section.

Lemma 4.14.11 *$C_{\bar{\Lambda}}(X_s)$ is a 12-dimensional irreducible module for $N_G(X_s)/X_s$, the restriction to this module of the quadratic form θ is non-zero and $O_3(N_G(X_s)/X_s)$ acts on this module fixed-point freely.*

Proof. By (3.8.2)–(3.8.4) the centralizers of X_s in the irreducible Golay code and Todd modules are 5-dimensional. In view of (4.8.2) this shows that $C_{\bar{\Lambda}}(X_s)$ is 12-dimensional. By (4.14.10) $\Phi(X_s)$ contains $\bar{\lambda}_0$ and some vertices from $\bar{\Lambda}_4^5$, and hence the restriction of θ to $C_{\bar{\Lambda}}(X_s)$ is non-trivial. In order to check the irreducibility one could for instance determine all the vectors in $\bar{\Lambda}$ fixed by X_s and describe the orbits of $N_G(X_s)$ on these vectors. □

Arguing as in the above lemma it is easy to check that $C_{\bar{\Lambda}}(X_t)$ is 8-dimensional.

Let us consider the intersection of $\Phi(X_s)$ and the point set $\Theta = \Theta(\bar{\mu}_0)$ of the subgeometry $\mathcal{G}(Co_2)$ in $\mathcal{G}$. We assume that X_s stabilizes Θ, or, equivalently, that $\{a, b\} := P_4(\mu_0)$ is contained in E. The intersection can be calculated either by removing from $\Phi(X_s)$ the vertices which are not contained in Θ or by determining the vertices in Θ fixed by X_s (where the latter can be considered as a subgroup in F_1).

The result is certainly independent of the particular way of calculating and is summarized in the following.

Lemma 4.14.12 *The following assertions hold:*

(i) *the group* $N_{F_1}(X_s)/X_s \cong 2^4.(Sym_4 \times 2)$ *acting on the set* $\Theta \cap \Phi(X_s)$ *has six orbits with lengths* 1, 2, 12, 24, 32 *and* 64 *contained in* $\bar{\Lambda}_4^{\alpha}$ *for* $\alpha = 8, 6, 4a, 4b, 6$ *and* 5, *respectively,*

(ii) $N_F(X_s)$ *preserves on* $\Theta \cap \Phi(X_s)$ *an imprimitivity system with classes of size* 3 *and the graph obtained by factorizing over this system is the point graph of the generalized quadrangle of order* (2,4),

(iii) $N_F(X_s)/X_s \cong 2 \times U_4(2).2$ *and the suborbit diagram of the subgraph in the Leech graph induced by* $\Theta \cap \Phi(X_s)$ *is as given below.* □

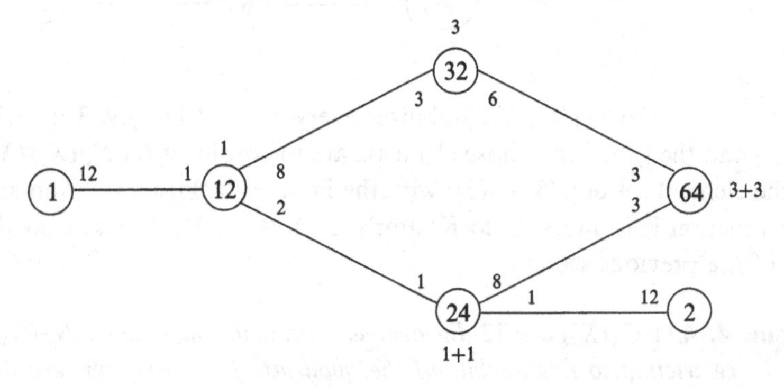

Notice that the non-empty intersections with Θ of the subgraphs realizing the elements of type 3 in $\mathscr{G}(3 \cdot U_4(3))$ are isomorphic to the line graph of the Petersen graph while the non-empty intersections with Θ of the subgraphs realizing elements of type 1 in $\mathscr{E}(3 \cdot U_4(3))$ are isomorphic to the point graph of the generalized quadrangle of order (2,2).

To the end of the section we discuss the fixed points in the Leech graph of a subgroup of order 7. Let S be a subgroup of order 7 in G_1. Then by (2.13.5) and its proof S stabilizes exactly three elements in $\mathscr{P}$, say a, b and c, and exactly three octads forming a trio, say $T = \{B_1, B_2, B_3\}$. Thus without loss of generality we may assume that $S \leq L_1 \cap H_3$. Since S centralizes exactly three non-zero elements in the irreducible Golay code module Q_1, namely the ones corresponding to B_1, B_2 and B_3, we have the following.

Lemma 4.14.13 *$N_{G_1}(S) = N_{G_1 \cap H_3}(S)$ is the direct product of a subgroup isomorphic to Sym_4 and $N_{L_1(a,b,c)}(S) \cong Frob_7^3$.* □

Since a Sylow 3-subgroup X in $N_{L_1(a,b,c)}(S)$ fixes a, b and c, by (4.14.1) we can assume that $X = X_s$.

Using (4.4.1) it is straightforward to find the set Ξ of vertices in the Leech graph fixed by S and to establish the following result applying (2.13.1).

Lemma 4.14.14 *The following assertions hold:*

(i) *the set Ξ is of size 14 consisting of $\bar{\lambda}_0$, three vertices from $\bar{\Lambda}_4^6$, six vertices from $\bar{\Lambda}_4^{4b}$ and 4 vertices from $\bar{\Lambda}_4^5$,*

(ii) *the graph on Ξ in which two vertices $\bar{\nu}, \bar{\mu} \in \Xi$ are adjacent if $\bar{\nu} \in \bar{\Lambda}_4^6(\bar{\mu})$ is isomorphic to the point–line incidence graph of the projective plane of order 2,*

(iii) *$N_G(S)$ induces on Ξ the natural action of $\mathrm{Aut}\, L_3(2)$ with kernel $N_{L_1}(a,b,c)(S) \cong Frob_7^3$, $C_G(S)/S \cong L_3(2)$ has two orbits on Ξ of length 7 each, in particular S is fully normalized in G,*

(iv) *$W := C_{\bar{\Lambda}}(S)$ and W is generated by the vertices from Ξ,*

(v) *W, as a module for $N_G(S)$, is isomorphic to the direct sum of the natural module for $C_G(S)/S \cong L_3(2)$ and its dual; the restriction to W of the quadratic form θ from (4.5.6) is non-trivial,*

(vi) *Ξ is contained in $\Phi(X_s)$ where X_s is a Sylow 3-subgroup of $N_{L_1(a,b,c)}(S)$.* □

5
The Monster

The Monster is the largest among the sporadic simple groups. It was predicted to exist independently by B. Fischer and R. Griess in 1973 and it was constructed by R. Griess in 1980. According to its standard definition the Monster is a simple group M which contains a subgroup Z_1 of order 2 such that

$$G_1 := C_M(Z_1) \sim 2_+^{1+24}.Co_1,$$

which means that $Q_1 := O_2(G_1)$ is an extraspecial group of order 2^{25} (*i.e.* $Z_1 = Z(Q_1)$, and Q_1/Z_1 is elementary abelian of order 2^{24}) and such that $G_1/Q_1 \cong Co_1$ acts on Q_1/Z_1 as it acts on the Leech lattice $\bar{\Lambda}$ modulo 2. Since by (4.9.3) the tilde geometry $\mathcal{G}(Co_1)$ of the Conway group Co_1 possesses a natural representation in $\bar{\Lambda}$, this means that the elements of type i of $\mathcal{G}(Co_1)$ are realized by certain subgroups of order 2^{i+1} in Q_1 which contain Z_1 so that the incidence relation is via inclusion. Let Z_2 and Z_3 realize incident elements of type 1 and 2, respectively, and put $G_i = N_G(Z_i)$, $Q_i = O_2(G_i)$, $\bar{G}_i = G_i/Q_i$, $i = 2, 3$. Then $G_2 \cap G_3$ contains a Sylow 2-subgroup of G_1,

$$\bar{G}_2 \cong Sym_3 \times Mat_{24}, \quad \bar{G}_3 \cong L_3(2) \times 3 \cdot Sym_6$$

and

$$[G_2 : G_{12}] = 3, \quad [G_3 : G_{23}] = [G_3 : G_{13}] = 7, \quad [G_3 : G_{123}] = 21$$

(here as usual $G_{12} = G_1 \cap G_2$ *etc.*) In this chapter we study a group G generated by an amalgam $\mathcal{M} = \{G_1, G_2, G_3\}$ such that the structure of the G_i and the intersection indices are as above. We will call $\mathcal{M}$ *the Monster amalgam.* We will show that G acts flag-transitively on a tilde geometry $\mathcal{G}(M)$ of rank 5. Then we construct a number of subgroups of G associated with certain subgeometries of $\mathcal{G}(M)$. Some of the subgroups involve other sporadic simple groups. We will determine the structure of these subgroups and establish the simple connectedness of the corresponding subgeometries of $\mathcal{G}(M)$. Then we apply the triangulability of a graph on the set of Baby Monster involutions in the Monster group established in [ASeg92] to prove the simple connectedness of $\mathcal{G}(M)$. We start by studying some basic properties of $\mathcal{M}$ and constructing the geometry $\mathcal{G}(M)$.

5.1 Basic properties

We start with a couple of definitions. Let $\mathcal{M} = \{G_1, G_2, G_3\}$ be an amalgam of rank 3, put $Q_i = O_2(G_i)$, $\bar{G}_i = G_i/Q_i$, $1 \leq i \leq 3$, $G_{ij} = G_i \cap G_j$, $T_{ij} = Q_i \cap Q_j$ for $1 \leq i < j \leq 3$, $G_{123} = G_1 \cap G_2 \cap G_3$, $Z_1 = Z(Q_1)$.

Definition 5.1.1 *The amalgam $\mathcal{M}$ is called the* Monster amalgam *if the following hold:*

(i) *Q_1 is an extraspecial group of order 2^{25};*
(ii) *$\bar{G}_1 \cong Co_1$ acts on Q_1/Z_1 as it acts on the Leech lattice $\bar{\Lambda}$ modulo 2;*
(iii) *G_{123} contains a Sylow 2-subgroup of G_1;*
(iv) *$\bar{G}_2 \cong Sym_3 \times Mat_{24}$ and $\bar{G}_3 \cong L_3(2) \times 3 \cdot Sym_6$;*
(v) *$[G_2 : G_{12}] = 3; [G_3 : G_{23}] = [G_3 : G_{13}] = 7; [G_3 : G_{123}] = 21$;*
(vi) *for $1 \leq i < j \leq 3$, we have $Q_i \cap Q_j \neq Q_i$.*

Notice that condition (iv) can be deduced from the other conditions together with certain information on subgroups in Co_1 and Mat_{24} containing Sylow 2-subgroups.

In this chapter $\mathcal{M}$ is a Monster amalgam and G is a faithful completion of $\mathcal{M}$.

Let $\eta : Q_1 \to \bar{\Lambda}$ be the homomorphism commuting with the action of $\bar{G}_1$ whose existence is guaranteed by (5.1.1 (ii)), so that $Z_1 = \ker \eta$. In what follows we identify a subgroup in $\bar{\Lambda}$ with the set of its non-trivial elements. There are a quadratic and the associated bilinear mappings of Q_1/Z_1 onto Z_1 invariant under the action of $\bar{G}_1$ which are defined, respectively, by $p \mapsto p^2$ and $(p,q) \mapsto [p,q]$ for $p,q \in Q_1$. By (4.5.6) there is a unique Co_1-invariant quadratic form on $\bar{\Lambda}$ (denoted by θ) and hence we obtain the following.

Lemma 5.1.2 *Let $p, q \in Q_1 \setminus Z_1$, $\eta(p) \in \bar{\Lambda}_i$, $\eta(q) \in \bar{\Lambda}_j$ and $\eta(pq) \in \bar{\Lambda}_k$. Then*

(i) *p is an involution if $i = 2$ or 4 and it is an element of order 4 if $i = 3$,*

(ii) *p and q commute if and only if $i + j + k$ is even.* □

For a subspace $\bar{\Delta}$ in $\bar{\Lambda}$ let $\bar{\Delta}^{\perp}$ denote the orthogonal complement of $\bar{\Delta}$ with respect to θ. For $S \leq Q_1$ put

$$S^{\perp} = \eta^{-1}(\eta(S)^{\perp}).$$

Since θ is non-singular, when $Z_1 < S$ we have $\frac{1}{2}|S| \cdot |S^{\perp}| = |Q_1|$. By (5.1.2) we have the following.

Lemma 5.1.3 *Let q_1 and q_2 be different elements of $Q_1 \setminus Z_1$ having the same image under η (i.e. $S := \langle q_1, q_2 \rangle = \eta^{-1}(\bar{\lambda})$ for some $\bar{\lambda} \in \bar{\Lambda}$), and let $q \in Q_1$. Then for $i = 1$ and 2 we have $q^{-1} q_i q = q_i$ if $q \in S^{\perp}$ and $q^{-1} q_i q = q_{3-i}$ otherwise.* □

Recall that an element of type i in the T-geometry $\mathscr{G}(Co_1)$ of the Conway group Co_1 is an i-dimensional subspace $\bar{\Delta}$ in $\bar{\Lambda}$ such that $\bar{\Delta} \subset \bar{\Lambda}_4$ and $\bar{\Delta} \leq \bar{\Delta}^{\perp}$ (compare the definition of $\mathscr{G}(Co_1)$ given before (4.9.2)). Hence $\eta^{-1}(\bar{\Delta})$ is elementary abelian of order 2^{i+1}. We will identify the elements of $\mathscr{G}(Co_1)$ with their preimages under η.

Lemma 5.1.4 *G_{12} is the normalizer in G_1 of an element Z_2 of type 1 in $\mathscr{G}(Co_1)$.*

Proof. Since $[G_2 : G_{12}] = 3$, we have $Q_2 \leq G_{12}$ and by (5.1.1 (iv)) $G_{12}/Q_2 \cong 2 \times Mat_{24}$. Since G_{12} contains a Sylow 2-subgroup of G_1 it contains Q_1. On the other hand $Q_1 Q_2 / Q_2$ is non-trivial by (5.1.1 (vi)). Then the structure of G_{12}/Q_2 implies that $Q_1 Q_2 / Q_2$ has order 2 and hence T_{12} is of index 2 in Q_1. Since Q_1 is extraspecial, T_{12} contains Z_1. Since G_{12} normalizes T_{12}, it also normalizes $Z_2 := T_{12}^{\perp}$, where Z_2 has order 4 and $Z_1 < Z_2$. We know that G_{12} contains a Sylow 2-subgroup of G_1 but by (4.5.5) $\bar{\Lambda}_4$ is the only orbit of Co_1 on $\bar{\Lambda}^{\#}$ of odd length and hence $\eta(Z_2) \in \bar{\Lambda}_4$ (i.e. Z_2 is an element of type 1 in $\mathscr{G}(Co_1)$). Since G_{12} involves a chief factor isomorphic to Mat_{24} we have $G_{12} = N_{G_1}(Z_2)$ and the result follows. □

Without loss of generality from now on we assume that $\eta(Z_2) = \bar{\lambda}_0 = \bar{\Lambda}_4^8$.

Lemma 5.1.5 *G_{13} is the normalizer in G_1 of an element Z_3 of type 2 in $\mathcal{G}(Co_1)$ incident to Z_2.*

Proof. Since $[G_3 : G_{13}] = 7$, we have $Q_3 \leq G_{13}$ and $X := G_{13}/Q_3 \cong Sym_4 \times 3 \cdot Sym_6$. Since Q_1 is not contained in Q_3, Q_1Q_3/Q_3 is a non-trivial normal 2-subgroup in X and the structure of X immediately shows that Q_1Q_3/Q_3 is of order 2^2. Since G_{13} contains a Sylow 2-subgroup of G_1, it contains Q_1 and hence T_{13} has index 2^2 in Q_1 and it contains Z_1, since Q_1 is extraspecial. Thus G_{13} normalizes $Z_3 := T_{13}^{\perp}$ which is a subgroup of order 2^3 in Q_1 containing Z_1. Furthermore, Z_3 is normalized by a Sylow 2-subgroup of G_1 which normalizes Z_2. By (4.5.5), for $i = 2$ and 3 the size of $\bar{\Lambda}_i$ is divisible by 8 and hence $\bar{\lambda}_0 = \eta(Z_2) \leq \eta(Z_3) \subseteq \bar{\Lambda}_4$. Finally, by (4.6.1) $\bar{\Lambda}_4^{4a}$ is the only non-trivial suborbit of Co_1 on $\bar{\Lambda}_4$ whose length is not divisible by 4 and hence $\eta(Z_3)$ is a triangle in the Leech graph. Since G_{13} contains a Sylow 2-subgroup of G_1 and has a factor group isomorphic to $Sym_3 \times 3 \cdot Sym_6$, we have $G_{13} = N_{G_1}(Z_3)$ and the result follows. □

By (4.5.5) and (5.1.4) we have $G_{12}/Q_1 \cong 2^{11} : Mat_{24}$ (the semidirect product of the irreducible Golay code module $\mathcal{C}_{11}$ and the Mathieu group Mat_{24}). By (4.8.2) Q_1/Z_1, as a module for G_{12}/Q_1, is uniserial containing a chief factor isomorphic to $\mathcal{C}_{11}$, a chief factor isomorphic to $\bar{\mathcal{C}}_{11}$ and two 1-dimensional chief factors, namely Z_2/Z_1 and $Q_1/Z_2^{\perp}$. This implies that $Z_2 = Z(Q_2)$ and in particular Z_2 is normal in G_2. In a similar way one can see that $Z_3 = Z(Q_3)$. Furthermore, since G_{13} contains Q_1, by (5.1.3) G_3 does not centralize Z_3 and the structure of G_3 immediately implies that $G_3/C_{G_3}(Z_3) \cong L_3(2)$.

Lemma 5.1.6 *The following assertions hold:*

(i) $G_2/C_{G_2}(Z_2) \cong Sym_3$; $G_3/C_{G_3}(Z_3) \cong L_3(2)$;
(ii) $G_{13} = N_{G_3}(Z_1) = N_{G_1}(Z_3)$; $G_{23} = N_{G_3}(Z_2) = N_{G_2}(Z_3)$;
(iii) *Z_2 and Z_3 are the normal closures of Z_1 in G_2 and G_3, respectively;*
(iv) *if $\{i, j, k\} = \{1, 2, 3\}$ then $G_i = \langle G_{ij}, G_{ik} \rangle$.*

Proof. Since both G_{13} and G_{23} have index 7 in G_3 we obtain (i), (ii) and (iii) by the arguments before the lemma. In (iv) for $i = 1$ the result follows from (5.1.4) and (5.1.5) since the Leech graph is connected; the remaining cases are immediate from (i) and (ii). □

Let G_2^+ and G_2^- be the preimages in G_2 of the direct factors Mat_{24} and Sym_3 of $\bar{G}_2$, respectively (notice that by (5.1.6) we have $G_2^+ = C_{G_2}(Z_2)$).

We consider the chief factors of G_2^+ inside Q_2. By (4.8.7)

$$R_2 := \eta^{-1}(\bar{\Lambda}_4^8 \cup \bar{\Lambda}_4^{4a})$$

is the unique subgroup in Q_1 containing Z_2 and normal in G_2^+ such that R_2/Z_2 is the irreducible Todd module for G_2^+/Q_2. The quotient Q_2/R_2 involves two irreducible chief factors, both isomorphic to the irreducible Golay code module. Since $G_2^+ \trianglelefteq G_2$, this shows that $R_2 \trianglelefteq G_2$. Let X be a Sylow 3-subgroup of G_2^-. By (5.1.6) X acts on Z_2 fixed-point freely and by the Frattini argument $N_{G_2}(X)Q_2 = G_2$. By (4.8.7) $R_2^{\perp} = R_2$ and hence R_2 is elementary abelian. Since G_2/Q_2 acts irreducibly on R_2/Z_2, since the dimension of R_2/Z_2 is 11 which is an odd number and since $X/Q_2 \trianglelefteq G_2/Q_2$, X centralizes R_2/Z_2 and we have the following.

Lemma 5.1.7 *If X is a Sylow 3-subgroup of G_2^-, then, as a module for $C_{G_2}(X)$, the subgroup R_2 possesses the decomposition $R_2 = Z_2 \oplus C_{R_2}(X)$ where $C_{R_2}(X) \cong \bar{\mathscr{C}}_{11}$.* □

Let us analyse the structure of Q_2/R_2. Since Q_1 is extraspecial, the commutator subgroup of $Z_2^{\perp} = T_{12}$ is exactly Z_1. Since X does not normalize Z_1, it does not normalize $Z_2^{\perp}$. If d is a generator of X then $Z_2^{\perp}/R_2$ and $(Z_2^{\perp})^d/R_2$ are two abelian normal subgroups in Q_2/R_2 which have trivial intersection and factorize Q_2/R_2. Thus we have the following.

Lemma 5.1.8 *Q_2/R_2 is an elementary abelian 2-group of rank 22 and as a module for $\bar{G}_2 \cong Sym_3 \times Mat_{24}$ it is isomorphic to $Z_2 \otimes \bar{\mathscr{C}}_{11}$. In particular $C_{R_2}(X) = C_{Q_2}(X)$.* □

The structure of G_2 specified above can be expressed by writing

$$G_2 \sim 2^{2+11+22}.(Sym_3 \times Mat_{24}).$$

Lemma 5.1.9 *Let Y be a subgroup of order 2 in Z_3, such that $Y \neq Z_1$, and suppose that $g \in G_3$ conjugates Z_1 to Y. Then*

(i) *Q_1^g is independent of the particular choice of g and $Z(Q_1^g) = Y$,*
(ii) *if $Y \leq Z_2$ then $Q_1^g \leq G_2$ and $Q_1 \cap Q_1^g = R_2$.*

Proof. (i) follows directly from (5.1.6 (i)). If $Y \leq Z_2$ then by (5.1.6 (i)) the element g can be chosen from G_2 in which case $Q_1^g \leq G_2$. We know that on the one hand G_2 normalizes R_2 and on the other hand Q_1 and $Q_1(Y)$ are two extraspecial groups of order 2^{25} with different commutator subgroups Z_1 and Y. Hence their intersection is an elementary abelian 2-group of rank at most 13, which is exactly the rank of R_2. Since $R_2 \leq Q_1$ and R_2 is normalized by G_2, (ii) follows. □

Let G_3^+ and G_3^- be the preimages in G_3 of the direct factors $3 \cdot Sym_6$ and $L_3(2)$ of $\bar{G}_3$, respectively. Then by (5.1.6) we have $G_3^+ = C_{G_3}(Z_3)$ and hence $G_3^+ \leq G_2^+$. In the next lemma we analyse the chief factors of G_3 inside Q_3.

Lemma 5.1.10 *Let $R_3 = core_{G_3}(R_2)$ and T_3 be the normal closure of $R_3^\perp$ in G_3. Then*

$$1 < Z_3 < R_3 < T_3 < Q_3$$

is the only chief series of G_3 inside Q_3. Furthermore, $Q_3/T_3 \cong Z_3 \otimes H$ where H is the hexacode module for $G_3^+/Q_3 \cong 3 \cdot Sym_6$, $T_3/R_3 \cong Z_3^* \otimes V$ where V is the natural symplectic module of $G_3^+/O_{2,3}(G_3^+) \cong Sp_4(2)$ and $R_3/Z_3 \cong H^*$ (here U^* denotes the dual of U).

Proof. Put $S = G_2^+ \cap G_3^+$, so that $S/Q_2 \cong 2^6 : 3 \cdot Sym_6$ is the stabilizer of a sextet in $G_2^+/Q_2 \cong Mat_{24}$. We call $H := O_2(S/Q_2)$ the hexacode module for $S/O_2(S)$ to distinguish it from its dual. By (4.8.7), (3.8.2), (3.8.3), (3.8.4) there are seven non-trivial chief factors of S inside $O_2(S)$, one isomorphic to H^*, three isomorphic to H and three isomorphic to the natural symplectic module V of $S/O_{2,3}(S) \cong Sp_4(2)$. In particular $C_{R_2/Z_3}(O_2(S))$ is the only chief factor isomorphic to H^*, which shows that the preimage R_3 of $C_{R_2/Z_3}(O_2(S))$ in Q_3 is normal in G_3. By (3.8.4) R_3 does not split over Z_3. Since R_3/Z_3 is the only chief factor of S isomorphic to H^*, and $R_3 \leq R_2$, R_3 is contained in every conjugate of R_2 in G_3. On the other hand R_3 is a maximal G_{123}-submodule in R_2 and in view of (5.1.6 (iv)) R_2 cannot be normalized by G_{13}, since it is already normalized by G_{12}, and $\bar{G}_1$ is irreducible on Q_1. Hence $R_3 = core_{G_3}(R_2)$.

By (5.1.3) $Z_3^\perp = Q_1 \cap G_3^+$ and by (4.8.7) $Z_3^\perp/R_3^\perp$ is isomorphic to H (which is dual to R_3/Z_3) as a G_3^+-module. Hence by (4.8.7) and (3.8.4) $R_3^\perp/R_3$ involves exactly two chief factors of G_3^+: R_2/R_3 and $R_3^\perp/R_2$, both isomorphic to V. Let T_3 be the normal closure of $R_3^\perp$ in G_3. Then all chief factors of G_3^+ in T_3/R_3 are isomorphic to V. Let E be a Sylow 7-subgroup in G_3. Then $G_3 = \langle G_{23}, E \rangle$ and hence E does not normalize R_2. This shows that E acts non-trivially on T_3/R_3 and on the other hand it centralizes the action of $G_3^+/O_{2,3}(G_3^+)$ on T_3/R_3. We have noticed that Q_3 involves only three chief factors of G_3^+ isomorphic to V. This shows that T_3/R_3 is elementary abelian and that as a module for G_3 it is isomorphic either to $Z_3 \otimes V$ or to $Z_3^* \otimes V$. Since G_{23} normalizes R_2 as well as the subgroup Z_2 of order 4 in Z_3 the latter possibility holds. In a similar way we identify the structure of Q_3/T_3 (suggesting the reader fill in the details). □

Thus the structure of G_3 as specified above can be expressed in the following way:

$$G_3 \sim 2^{3+6+12+18}.(L_3(2) \times 3 \cdot Sym_6).$$

5.2 The tilde geometry of the Monster

We follow the notation introduced in the previous section. Let Δ be a graph (called the *first Monster graph*) on the set of right cosets of G_1 in G in which two such cosets are adjacent if their intersection is a coset of G_{12}. The group G acts on Δ by right translations; let v_0 be the coset containing the identity (*i.e.* G_1 itself). Then $G(v_0) = G_1$ and for $v = G_1 g$ we have $G(v) = g^{-1}G_1 g$. Put $Q_v = O_2(G(v))$, $Z_v = Z(G(v))$. Since $[G_2 : G_{12}] = 3$, for every coset of G_2 in G there are exactly three vertices in Δ which intersect it in a coset of G_{12}; furthermore these three vertices form a triangle called a *line*. The action of G_1 on the set $L(v_0)$ of lines containing v_0 is similar to its action on the set of cosets of G_{12} in G_1. By (5.1.4) this means that Q_1 is the kernel of the action and $G_1/Q_1 \cong Co_1$ acts on $L(v_0)$ as it acts on $\bar{\Lambda}_4$. Let

$$l = \{v_0, v_1, v_2\}$$

be the line formed by the vertices intersecting G_2 in cosets of G_{12}. Then by (5.1.6) G_2 induces on l the natural action of Sym_3 with kernel G_3^+ and $Z_{v_0} = Z_1$, Z_{v_1}, Z_{v_2} are the subgroups of order 2 in Z_2. This gives the following (where as usual $\Delta(v_0)$ denotes the set of vertices of Δ adjacent to v_0).

Lemma 5.2.1 *Let $\mathscr{X}$ be the set of subgroups X of order 2 in Q_1 such that $\eta(X) \in \bar{\Lambda}_4$. Then*

(i) *the mapping $\varphi : v \mapsto Z_v$ establishes a bijection of $\Delta(v_0)$ onto $\mathscr{X}$,*
(ii) *a triangle $\{u, v, w\}$ in Δ is a line if and only if $\langle Z_u, Z_v, Z_w\rangle$ is of order 4.* □

By the above lemma the orbits of Q_1 on $\Delta(v_0)$ are of length 2 and such an orbit together with v_0 forms a line. Thus if $l = \{v_0, v_1, v_2\}$ is as above, then $\{v_1, v_2\}$ is an orbit of Q_1 and the kernel $Q_1(v_1, v_2)$ is $T_{12} = Z_2^{\perp}$, so we have the following.

Lemma 5.2.2 *If $\{u_1, u_2\}$ is an orbit of Q_1 on $\Delta(v_0)$ then $Q_1(u_1, u_2) = \langle \varphi(u_1), \varphi(u_2)\rangle^{\perp}$; in particular, kernels at different orbits are different.* □

Lemma 5.2.3 *Let $u, v \in \Delta(v_0)$. Then u and v are adjacent in Δ if and only if $\eta\varphi(u)$ and $\eta\varphi(v)$ are either equal or adjacent vertices of the Leech graph.*

Proof. Let Σ be the subgraph in Δ induced by the images of v_0 under G_3. Then by (5.1.6) $|\Sigma| = 7$, Σ contains the line l and G_3 induces on Σ a doubly transitive action of $L_3(2)$. Hence Σ is a clique. Furthermore, by (5.1.6 (iii)) and (5.2.1 (i)) $\{\varphi(v) \mid v \in \Sigma\}$ is the set of subgroups of order 2 in Z_3 and by (5.1.5) $\eta(Z_3)$ is a line in the Leech graph. Since $\bar{G}_1$ acts transitively on the vertices and lines of the Leech graph this proves the "if" part of the statement. Thus $\Delta(v_0) \cap \Delta(v_1)$ contains $\varphi^{-1}(R_2 \cap \mathscr{X})$ where R_2 is defined after (5.1.6) and $\mathscr{X}$, φ are as in (5.2.1). By (5.2.1 (i)) if $u \in \Delta(v_0) \cap \Delta(v_1)$ then $Z_u \in Q_1 \cap Q_{v_1}$ and by (5.1.9 (ii)) the latter intersection is exactly R_2 which proves the "only if" part of the statement. □

Recall that a clique N in Δ is said to be $\star$-closed if together with every edge it contains the unique line containing the edge. By (5.2.1) and (5.2.3) we have the following.

Lemma 5.2.4 *Let N be a clique in Δ containing v_0. Then*

(i) *$\eta\varphi(N \setminus \{v_0\})$ is a clique in the Leech graph,*
(ii) *N is $\star$-closed if and only if $N = \varphi^{-1}\eta^{-1}(L) \cup \{v_0\}$ for a $\star$-closed clique L in the Leech graph.* □

Notice that if N is a $\star$-closed clique of size $2^l - 1$ in Δ containing v_0 then Q_1 normalizes the subgroup $Z(N)$ of order 2^l in Q_1 generated by $\eta^{-1}(N)$ and induces on $Z(N)$ an elementary abelian group of order 2^{l-1}. By the above we have the following result analogous to (4.8.1).

Lemma 5.2.5 *If N is a maximal clique in Δ, then N is $\star$-closed of size 31 and $\{\varphi(u) \mid u \in N\}$ is the set of subgroups of order 2 in an elementary abelian group $Z(N)$ of order 2^5 contained in Q_w for every $w \in N$. Furthermore $G[N]$ acts on N as $GL(Z(N)) \cong L_5(2)$ acts on the set of subgroups of order 2 in $Z(N)$. There are two orbits, $\mathscr{N}_v$ and $\mathscr{N}_t$, of G on the set of maximal cliques in Δ with $|\mathscr{N}_v| = 3 \cdot |\mathscr{N}_t|$ and for $\alpha = v$ or t whenever N*

is a clique from $\mathcal{N}_\alpha$ *containing* v_0 *we have* $\eta\varphi(N \setminus \{v_0\}) \in \mathcal{L}_\alpha$, *where* $\mathcal{L}_\alpha$ *is as in* (4.8.1). □

Define $\mathcal{G}(M)$ to be an incidence system of rank 5 whose elements of type 1, 2, 3, 4 and 5 are the vertices, lines, ⋆-closed cliques of size 7, ⋆-closed cliques of size 15 and the maximal cliques from the orbit $\mathcal{N}_v$, respectively, and the incidence relation is via inclusion. Then v_0, l and Σ (as in the proof of (5.2.3)) are pairwise incident elements in $\mathcal{G}(M)$ stabilized by G_1, G_2 and G_3, respectively. The residue of v_0 is isomorphic to the tilde geometry $\mathcal{G}(Co_1)$ on which $\bar{G} \cong Co_1$ induces a flag-transitive action. If N is an element of type 5 in $\mathcal{G}(M)$, so that $N \in \mathcal{N}_v$, then the residue of N is the projective geometry of the proper subspaces in $Z(N)$ and by (5.2.5) $G[N]$ induces the full automorphism group of this residue. Finally, since the incidence relation is via inclusion, it is easy to see that $\mathcal{G}(M)$ is a geometry with a string diagram and we have the following.

Proposition 5.2.6 *The geometry* $\mathcal{G}(M)$ *is a rank* 5 *tilde geometry with the diagram*

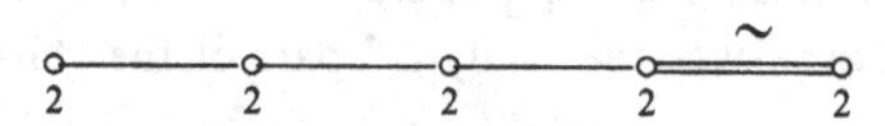

the group G *induces on* $\mathcal{G}(M)$ *a flag-transitive action and* Δ *is the collinearity graph of* $\mathcal{G}(M)$. □

5.3 The maximal parabolic geometry

In this section we construct the maximal parabolic geometry $\mathcal{H}(M)$ for the group G. A crucial step is to show that G contains a subgroup D of the form $2^{10+16}.\Omega_{10}^+(2)$ containing Q_1 such that $(D \cap G_1)/Q_1 = C_{\bar{G}_1}(\bar{\delta}) \cong 2_+^{1+8}.\Omega_8^+(2)$ where $\bar{\delta}$ is a central involution in $\bar{G}_1 \cong Co_1$ as in Section 4.8.

As above, Δ is the first Monster graph (*i.e.* the collinearity graph of $\mathcal{G}(M)$), v_0 is a vertex of Δ such that $G(v_0) = G_1$, $\varphi : v \mapsto Z_v = Z(G(v))$ is a bijection of $\Delta(v_0)$ onto a set $\mathcal{X}$ of subgroups of order 2 in Q_1 as in (5.2.1), η is a mapping of $\mathcal{X}$ onto the vertex set $\bar{\Lambda}_4$ of the Leech graph Γ, commuting with the action of G_1, such that the fibres of η are the orbits of Q_1. Then $u, v \in \Delta(v_0)$ are adjacent if and only if $\eta\varphi(u)$ and $\eta\varphi(v)$ are either equal or adjacent in Γ. Furthermore, we assume that $\{v_0, v_1, v_2\}$ is a line stabilized by G_2 and that $\eta\varphi(v_1) = \eta\varphi(v_2) = \bar{\lambda}_0$.

Let u be a vertex from $\Delta(v_0)$ such that $\eta\varphi(u) \in \bar{\Lambda}_4^\alpha$ for $\alpha = 6, 4b, 4c$ or 5. Then by (5.2.3) u is at distance 2 from v_1. By (5.1.4) $G_{12}/Q_1 = \bar{G}_1(\bar{\lambda}_0) \cong 2^{11}.Mat_{24}$ and by (5.2.2) $Q_1(v_1) \neq Q_1(u)$. Hence $G(v_0, v_1) = G_2^+$

acts transitively on $\varphi^{-1}\eta^{-1}(\bar{\Lambda}_4^\alpha)$ for every α as above. Let $\Delta_2^\alpha(v_1)$ be the orbit of u under $G(v_1)$. By the above this orbit is independent of the particular choice of u. For an arbitrary $w \in \Delta$ we put $\Delta_2^\alpha(w)$ to be the image of $\Delta_2^\alpha(v_1)$ under an element of G which maps v_1 onto w.

Lemma 5.3.1 *Let $u \in \Delta(v_0)$, $\bar{v} = \eta\varphi(u)$ and suppose that $\bar{v} \in \bar{\Lambda}_4^\alpha$ for $\alpha = 6, 4b, 4c$ or 5. Then*

(i) $[Z_u, Z_{v_1}] = Z_1$ *if $\alpha = 5$ and $[Z_u, Z_{v_1}] = 1$ otherwise,*
(ii) *the orbit of u under Q_{v_1} has length 2^8, 2^9, 2^{13} or 2^{13}, respectively,*
(iii) *there are exactly four orbits of G_1 on $\Delta_2(v_0)$, namely the orbits $\Delta_2^\alpha(v_0)$ for $\alpha = 6, 4b, 4c$ and 5,*
(iv) *if $w \in \Delta_2(v_0)$ then $G_1 \cap G(w)$ acts transitively on $\Delta(v_0) \cap \Delta(w)$,*
(v) *if $w \in \Delta_2^5(v_0)$ then there is a unique vertex v adjacent to both v_0 and w and $[Z_w, Z_1] = Z_v$,*
(vi) *if $w \in \Delta_2^\alpha(v_0)$ for $\alpha = 6, 4b$ or $4c$ then $Z_w \le G_1$ and $Z_w \not\le Q_1$.*

Proof. Part (i) follows directly from (4.8.2 (iv)) and (5.1.3). Let $\bar{\Xi}$ be the orbit of $\bar{v}$ under $Q_{v_1} \cap G_1$. Since $O_2(G_{12}) = (Q_{v_1} \cap G_1)(Q_1 \cap G_{v_1})$, by (4.6.2) we have $|\bar{\Xi}| = 2^6$, 2^7, 2^{11} and 2^{11} for $\alpha = 6$, $4b$, $4c$ and 5, respectively. Let $\{u, u_1\} = \varphi^{-1}\eta^{-1}(\bar{v})$. We claim that $Q_{v_1} \cap Q_1$ contains an element q which maps u onto u_1. By (5.1.3) and (5.1.9 (ii)) an element q from Q_1 possesses these properties if $\eta(q) \in \bar{\Lambda}_4^{4a} \cap \bar{\Lambda}_4^5(\bar{v})$. Since the orbitals of the action of Co_1 on the Leech graph are self-paired, such a q exists if and only if $\bar{\Lambda}_4^{4a}(\bar{v}) \cap \bar{\Lambda}_4^5 \neq \emptyset$ or, equivalently, if there is a vertex in $\bar{\Lambda}_4^5$ adjacent to $\bar{v}$ in Γ. One can see from the suborbit diagram of Γ in Section 4.7 that such a vertex exists for every α under consideration. Hence the orbit Ξ of u under $Q_{v_1} \cap G_1$ is twice as long as $\bar{\Xi}$. Since u is adjacent to v_0 and not adjacent to v_2 whenever g is an element from $Q_{v_1} \setminus G_1$ (which maps v_0 onto v_2) we have $\Xi \cap \Xi^g = \emptyset$ and hence the length of the orbit of u under Q_{v_1} is four times the size of $\bar{\Xi}$ and (ii) follows. Now (iii), (iv), (v) and (vi) follow immediately from (i), (ii) and their proof. □

For $u \in \Delta(v_0) \cap \Delta_2^6(v_1)$ put

$$A = A(v_1, u) = \langle Z_w \mid w \in \{v_1, u, \Delta(v_1) \cap \Delta(u)\}\rangle.$$

By (4.8.8) the vertices $\bar{\lambda}_0 = \eta\varphi(v_1)$ and $\bar{v} := \eta\varphi(u)$ determine a subgraph $\Phi = \Phi(\bar{\lambda}_0, \bar{v})$ in the Leech graph which is induced by the vertices fixed by $O_2(C_{\bar{G}_1}(\bar{\delta}))$ for a central involution $\bar{\delta}$ in $\bar{G}_1 \cong Co_1$ and Φ generates in $\bar{\Lambda}$ a subspace $V_1 = V_1(\bar{\lambda}_0, \bar{v})$ of dimension 8.

Lemma 5.3.2 *The following assertions hold:*

(i) $A \leq \langle Z_{v_1}, Z_u, Q_{v_1} \cap Q_u \rangle$;
(ii) *A is an elementary abelian 2-group and the index of $A \cap Q_{v_1}$ in A is at most 2;*
(iii) $A \cap Q_1 = \langle \eta^{-1}(\bar{\mu}) \mid \bar{\mu} \in \Phi \rangle$ *and* $|A \cap Q_1| = 2^9$;
(iv) *whenever $u, v \in \Delta(v_0)$ are such that $\eta\varphi(u)$ and $\eta\varphi(v)$ are distinct vertices in Φ at distance 2 from each other, we have $A = A(u, v)$;*
(v) $A = A(v_0, v)$ *for some* $v \in \Delta_2^6(v_0)$;
(vi) $|A| = 2^{10}$.

Proof. By (5.2.1) if $w \in \Delta(v_1) \cap \Delta(u)$ then $Z_w \leq Q_{v_1} \cap Q_u$ and (i) follows. By (i) the subgroups Z_w taken for all $w \in \Delta(v_1) \cap \Delta(u)$ generate in A an elementary abelian subgroup centralized by Z_{v_1} and Z_u. By (5.1.3) and (4.8.2) we have $[Z_{v_1}, Z_u] = 1$ and (ii) follows. By (5.1.9) we have $Q_1 \cap Q_{v_1} = R_2$ while in terms of (4.8.2) we have $R_2 = \eta^{-1}(U_2(\bar{\lambda}_0))$. Hence (iii) and (iv) follow from (4.8.8 (ii)) and (i). Since the graph which is the complement of Φ is connected (compare the diagram before (4.8.8)), by (iv) $A = A(w, v)$ for some $w \in \Delta(v_1) \cap \Delta(u)$. By the definition of A it is normalized by $G(v_1, u)$ and by (5.3.1 (iv)) $G(v_1, u)$ acts transitively on the set $\Delta(v_1) \cap \Delta(u)$ which contains v_0. Hence (v) follows. By (v), (ii) and (iii) either $|A| = 2^{10}$ or $A \leq Q_1$. In the latter case by the obvious symmetry A must be contained in Q_{v_1} which is impossible since Z_u is not in Q_{v_1}. Hence (vi) follows. □

Let Ψ be the connected component containing v_0 of the subgraph induced by the vertices $u \in \Delta$ such that $A = A(u, v)$ for some $v \in \Delta_2^6(u)$ and put $D = G[\Psi]$. Then D is vertex-transitive on Ψ and by (5.3.2 (iii)) $\Psi(v_0) = \varphi^{-1}\eta^{-1}(\Phi)$ (in particular Ψ is of valency $270 = 2 \cdot 135$). Let $\mathscr{F}$ be a geometry whose elements are the vertices, the lines and the $\star$-closed cliques of size 7 and 31 contained in Ψ. A clique $N_v \in \mathscr{N}_v$ and a clique $N_t \in \mathscr{N}_t$ have different type and they are incident if and only if $N_v \cap N_t$ is a $\star$-closed clique of size 15; the remaining incidences are via inclusion.

Lemma 5.3.3 *The following assertions hold:*

(i) *$\mathscr{F}$ is the natural parabolic D_5-geometry of $\Omega_{10}^+(2)$;*
(ii) *D induces the full automorphism group of $\mathscr{F}$;*
(iii) $D \sim 2^{10+16}.\Omega_{10}^+(2)$ *and* $A = Z(O_2(D))$.

Proof. By (5.3.2 (iii)) the residue of v_0 in $\mathscr{F}$ is the parabolic geometry of $\Omega_8^+(2)$ as in the proof of (4.8.6) what particularly implies that for a $\star$-closed clique J of size 15 and $\alpha \in \{v, t\}$ there is a unique $N_\alpha \in \mathscr{N}_\alpha$

such that $J \subset N_\alpha \subset \Psi$. Hence the residue in $\mathcal{F}$ of a maximal clique N is isomorphic to the rank 4 projective $GF(2)$-geometry of the $\star$-closed cliques contained in N. This shows that $\mathcal{F}$ is a Tits geometry of type $D_5(2)$ and (i) follows from (1.6.3). By (5.3.2 (iii), (iv)) $D \cap G_1$ is the full preimage in G_1 of $\bar{G}_1[\Phi]$ which acts flag-transitively on $\mathrm{res}_{\mathcal{F}}(v_0)$. In view of the vertex-transitivity of D on Ψ this shows that the action of D on $\mathcal{F}$ is flag-transitive and (ii) follows from (1.6.5). Let K be the kernel of the action of D on $\mathcal{F}$. Then K is contained in $D(v_0)$ and the latter is of the form $2_+^{1+24}.2_+^{1+8}.\Omega_8^+(2)$. It is easy to see that A is contained in K and that A is the natural module for $D/K \cong \Omega_{10}^+(2)$. Hence K/A is of order 2^{16} and it involves two irreducible chief factors for $D(v_0)/O_2(D(v_0)) \cong \Omega_8^+(2)$. Since one of these factors, namely $(K \cap Q_1)A/A$, is not normalized by D we obtain (iii) with the remark that K/A is the spin module for D/K. □

By the above lemma Ψ is the graph on the non-zero isotropic vectors in the natural module of $\Omega_{10}^+(2)$ in which two vectors are adjacent if their sum is an isotropic vector. The intersection diagram of Ψ with respect to the action of $D/K \cong \Omega_{10}^+(2)$ is the following:

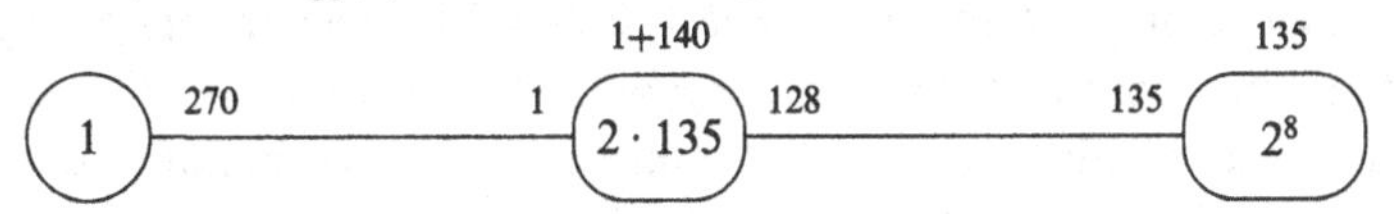

Comparing the diagram and (5.3.1 (ii)) we observe that $\Psi \cap \Delta_2(v_0)$ is an orbit of Q_1 on $\Delta_2^6(v_0)$.

Let $\mathcal{H}(M)$ be an incidence system of rank 5 whose elements of type 1, 2, 3, 4 and 5 are the vertices, lines, $\star$-closed cliques of size 7, $\star$-closed cliques of size 31 from the orbit $\mathcal{N}_t$ and the images under G of the subgraph Ψ stabilized by D; the incidence relation is via inclusion. Then by (4.9.1) and (5.3.3) we have the following.

Lemma 5.3.4 *The incidence system $\mathcal{H}(M)$ is a geometry with the diagram*

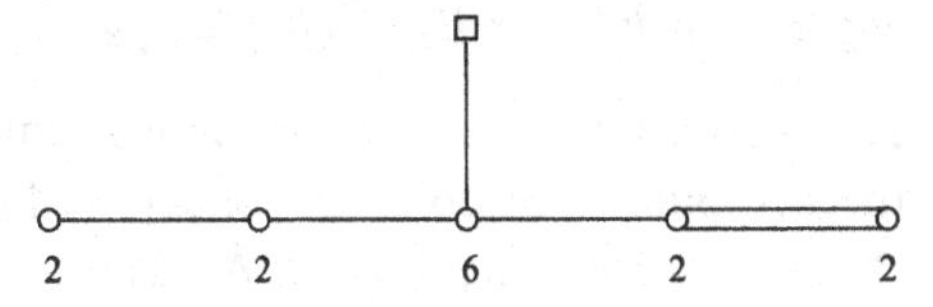

and the group G acts on $\mathcal{H}(M)$ flag-transitively with $D \sim 2^{10+16}.\Omega_{10}^+(2)$ as the stabilizer of an element of type 5. □

Notice that for $i = 1$, 2 and 3 the set of elements of type i in $\mathcal{G}(M)$ coincides with the set of elements of type i in $\mathcal{H}(M)$, so that G_1, G_2

and G_3 are also maximal parabolics of the action of G on $\mathcal{H}(M)$. A clique $N_v \in \mathcal{N}_v$ (an element of type 5 in $\mathcal{G}(M)$) is contained in a unique element of type 5 in $\mathcal{H}(M)$ while a clique $N_t \in \mathcal{N}_t$ (an element of type 4 in $\mathcal{H}(M)$) is in three such elements. If we put $G_5 = G[N_v]$ and $H_4 = G[N_t]$ then

$$G_5 \sim 2^{5+5+10+5+1+10}.L_5(2), \quad H_4 \sim 2^{5+10+20}.(L_5(2) \times Sym_3).$$

Finally the stabilizer G_4 of an element of type 4 in $\mathcal{G}(M)$ has index 31 in H_4 and

$$G_4 \sim 2^{4+1+2+12+8+8+4}.(L_4(2) \times Sym_3).$$

5.4 Towards the Baby Monster

As in the previous section let $A = Z(O_2(D))$ where $D = H_5$ is the stabilizer in G of the subgraph Ψ which is an element of type 5 in $\mathcal{H} = \mathcal{H}(M)$. Then an element $\Pi \in \mathrm{res}_{\mathcal{H}}(\Psi)$ can be identified with the subgroup Z_Π in A generated by the Z_u taken for all $u \in \Pi$. In this way the elements v_0, l and Σ of type 1, 2 and 3 in the corresponding maximal flag are identified with Z_1, Z_2 and Z_3, respectively. For a subgroup in A the terms "isotropic", "non-isotropic", "orthogonal complement" *etc.* are with respect to the orthogonal form of plus type preserved by $D/O_2(D) \cong \Omega^+_{10}(2)$. Let Y_1 be a non-isotropic subgroup of order 2 in A contained in the orthogonal complement of Z_3. We will study the centralizer of Y_1 in the group G; the quotient of this centralizer over Y_1 will eventually be identified with the sporadic simple group known as the *Baby Monster.*

In view of (5.3.3 (iii)) and the standard properties of the orthogonal groups [Tay92] we have the following.

Lemma 5.4.1 *Let* $\widetilde{B}_o = C_D(Y_1)$ *and* $B_o = \widetilde{B}_o/Y_1$. *Then* $B_o \cong 2^{9+16}.Sp_8(2)$ *and* A/Y_1 *is the orthogonal module for* $B_o/O_2(B_o) \cong Sp_8(2) \cong \Omega_9(2)$. □

Since $Z_1 \leq Z_3$ the subgroup Y_1 is in the orthogonal complement of Z_1. By (5.3.2 (iii)) the orthogonal complement of Z_1 is $A \cap Q_1$. Hence $Y_1 \leq Q_1$ and since Y_1 is non-isotropic, $\eta(Y_1) \in \bar{\Lambda}_2$. Hence without loss of generality we can assume that $\eta(Y_1) = \bar{\mu}_0$ where μ_0 is as introduced before (4.9.5), *i.e.* $\mu_0 \in \Lambda^4_2$, $X(\mu_0) = \emptyset$ and $P_4(\mu_0) = \{a, b\}$.

Lemma 5.4.2 *Let* $\widetilde{B}_1 = C_{G_1}(Y_1)$ *and* $B_1 = \widetilde{B}_1/Y_1$. *Then*

(i) $B_1 \cong 2^{1+22}_+.Co_2$ *and* $O_2(\widetilde{B}_1) = \eta^{-1}(\bar{\mu}_0^\perp)$,

(ii) $\widetilde{B}_1$ *acting on* $\Delta(v_0)$ *has three orbits,* $\Omega_j := \eta^{-1}(\bar{\Lambda}_4 \cap \bar{\Lambda}_j(\bar{\mu}_0))$ *for* $j = 2, 3$ *and* 4,

(iii) *if* $u \in \Omega_j$ *then* $[Y_1, Z_u] = 1$ *(equivalently* $Y_1 \leq G(u)$*) if and only if* $j = 2$ *or* 4,

(iv) *for* $j = 2, 3$ *and* 4 *if* $u \in \Omega_j$ *then* $\widetilde{B}_1(u)Q_1/Q_1 \cong 2^{10} : \mathrm{Aut}\, Mat_{22}$, Mat_{23} *and* $[2^5] : (2^4 : L_4(2))$, *respectively,*

(v) *if* $u \in \Omega_4$ *then* $[Q_u : C_{Q_u}(Y_1)] \geq 2^6$,

(vi) v_0 *and* Ω_4 *are in different orbits of* $C_G(Y_1)$ *on* Δ.

Proof. Part (i) follows directly from (4.5.5) and (5.1.3) and (ii), (iii), (iv) from (4.4.1) and the table therein. Let $u \in \Omega_4$ and let g be an element in G_1 such that $u^g = v_1$ (recall that $\eta(v_1) = \bar{\lambda}_0$). Then $\bar{\mu}_0^g \in \bar{\Lambda}_2^2$. By (5.1.9) $Q_1 \cap Q_{v_1}$ is of order 2^{13} and hence $Q_{v_1}Q_1/Q_1 = O_2(\bar{G}_1(\bar{\lambda}_0))$ (where $\bar{G}_1(\bar{\lambda}_0) \cong 2^{11} : Mat_{24}$). We can see from the table in (4.4.1) that the orbits of $O_2(\bar{G}_1(\bar{\lambda}_0))$ on $\bar{\Lambda}_2^2$ are of length 2^6 and we obtain (v). Since $[Q_1 : C_{Q_1}(Y_1)] = 2$ by (i), we have (vi) by (v). □

Let $\widetilde{B}_2 = C_{G_2}(Y_1)$. In terms of (5.4.1) Z_2Y_1/Y_1 is a 2-dimensional isotropic subspace in the 9-dimensional orthogonal space A/Y_1. Since $B_o/O_2(B_o) \cong Sp_8(2)$, the normalizer of Z_2 in $\widetilde{B}_o$ permutes transitively the three subgroups of order 2 in Z_2. This means that $\widetilde{B}_2$ contains an element which maps v_0 onto v_1. Let $\widetilde{B} = \langle \widetilde{B}_1, \widetilde{B}_2 \rangle$ and let Θ be the subgraph in Δ induced by the images of v_0 under $\widetilde{B}$. We call $B := \widetilde{B}/Y_1$ the *Baby Monster* group and denote it also by BM. Since $\widetilde{B} \leq C_G(Y_1)$ and $\widetilde{B}_1 = C_{G_1}(Y_1)$ we have $\widetilde{B}(v_0) = \widetilde{B}_1$. By (5.4.2 (ii)–(vi)) we have the following.

Lemma 5.4.3 $\Theta(v_0) = \Omega_2$ *and* $\eta(\Omega_2)$ *is the point set of a* $\mathcal{G}(Co_2)$-*subgeometry in* $\mathcal{G}(Co_1) \cong \mathrm{res}_{\mathcal{G}(M)}(v_0)$. □

Let $\mathcal{G}(BM)$ denote the subgeometry in $\mathcal{G}(M)$ formed by the elements contained in Θ with respect to the induced incidence relation and type function.

Lemma 5.4.4 $\mathcal{G}(BM)$ *is a rank* 5 *Petersen geometry with the diagram*

$$\overset{}{\underset{2}{\circ}}\!\!-\!\!-\!\!-\!\!\underset{2}{\circ}\!\!-\!\!-\!\!-\!\!\underset{2}{\circ}\!\!-\!\!-\!\!-\!\!\underset{2}{\circ}\overset{P}{-\!\!-\!\!-}\underset{1}{\circ}$$

and BM *acts on* $\mathcal{G}(BM)$ *flag-transitively.*

Proof. By (5.4.3) the residue of v_0 in $\mathcal{G}(BM)$ is isomorphic to $\mathcal{G}(Co_2)$ and by (5.2.4) an element N of type 5 in $\mathcal{G}(BM)$ is a maximal clique in Θ.

Since $\widetilde{B}$ is 1-arc-transitive on Θ, the stabilizer of N in $\widetilde{B}$ acts transitively on its vertex set. Furthermore the stabilizer of $\eta(N)$ in $C_{G_1}(Y_1)Q_1/Q_1$ induces $L_4(2)$ on the element set of $\eta(N)$ and since $N \subseteq (\eta(Y_1))^{\perp}$, $Q_1 \cap C_{G_1}(Y_1)$ induces on N an elementary abelian group of order 2^4. Hence $\widetilde{B}_1[N]/B_1(N) \cong 2^4.L_4(2)$ and so $\widetilde{B}[N]/\widetilde{B}(N) \cong L_5(2)$. Since the incidence relation is via inclusion, $\mathcal{G}(BM)$ belongs to a string diagram and the result follows. □

Let $\{B_i \mid 1 \leq i \leq 5\}$ be the amalgam of maximal parabolic subgroups associated with the action of BM on $\mathcal{G}(BM)$. Then from the structure of the maximal parabolics in the group G we can deduce the following:

$$B_1 \sim 2_+^{1+22}.Co_2, \quad B_2 \sim 2^{2+10+20}.(Sym_3 \times \mathrm{Aut}\, Mat_{22}),$$

$$B_3 \sim 2^3.[2^{32}].(L_3(2) \times Sym_5), \quad B_4 \sim 2^4.[2^{30}].(L_4(2) \times Sym_2),$$

$$B_5 \sim 2^{5+10+10+5}.L_5(2).$$

Notice that because of the choice of the maximal flag in $\mathcal{G}(M)$ we have $B_i = C_{G_i}(Y_1)/Y_1$ for $1 \leq i \leq 5$.

Let Ψ be the subgraph of Δ introduced before (5.3.3) and isomorphic to the point graph of the parabolic geometry of $\Omega_{10}^+(2)$. Then $\Psi \cap \Theta$ consists of the points contained in the orthogonal complement of Y_1 in A. This complement is clearly an 8-dimensional non-singular symplectic space and we have the following.

Lemma 5.4.5 *The elements in $\mathcal{G}(BM)$ of type* 1, 2, 3 *and* 4 *which are contained in $\Psi \cap \Theta$ form a $C_4(2)$-subgeometry $\mathcal{G}(Sp_8(2))$ whose stabilizer in B is the subgroup $B_o \cong 2^{9+16}.Sp_8(2)$ as in* (5.4.1). *The residue of v_0 in this subgeometry is the $\mathcal{G}(Sp_6(2))$-subgeometry in $\mathcal{G}(Co_2)$ as in* (4.9.8). □

Since $\widetilde{B}_1$ acts transitively on the set of lines in $\mathcal{G}(BM)$ incident to v_0, we have the following.

Lemma 5.4.6 *Let x_2 be an element of type* 2 *in $\mathcal{G}(M)$ which contains v_0. Then $C_{G(x_2)}(Y_1) \not\leq C_{G_1}(Y_1)$ if and only if $\eta\varphi(x_2) \in \bar{\Lambda}_2(\bar{\mu}_0) \cap \bar{\Lambda}_4 = \eta(\Omega_2)$.* □

5.5 $^2E_6(2)$-subgeometry

Let Y_2 be a subgroup of order 4 in A generated by Y_1 and a non-isotropic subgroup Y_1' which is not perpendicular to Y_1 (so that Y_2 is a minus 2-space). We assume that Y_2 is contained in the orthogonal complement of Z_3 and that $\eta(Y_2) = \langle \bar{\mu}_0, \bar{\mu}_1, \bar{\mu}_2 \rangle$ where $\mu_1 \in \Lambda_2^4$ such that $\bar{\mu}_2 := \bar{\mu}_0 + \bar{\mu}_1$ is contained in $\bar{\Lambda}_2$ (say $P_4(\mu_1) = \{b, c\}$, $c \neq a$ and $X(\mu_1) = \emptyset$).

In view of the choice of Y_2, by (5.3.3) and standard properties of the orthogonal groups we have the following.

Lemma 5.5.1 *Let $\widetilde{E}_4 = N_D(Y_2)$, $\bar{E}_4 = \widetilde{E}_4/Y_2$ and Υ be the subgraph of Ψ induced by the vertices y such that Z_y is in the orthogonal complement of Y_2 in A. Then $\bar{E}_4 \cong 2^{8+16}.(\Omega_8^-(2) \times 3).2$ and Υ is isomorphic to the point graph of the natural parabolic geometry of $\Omega_8^-(2)$.* □

The suborbit diagram of Υ with respect to the action of $\bar{E}_4/O_{2,3}(\bar{E}_4) \cong \Omega_8^-(2).2$ is the following:

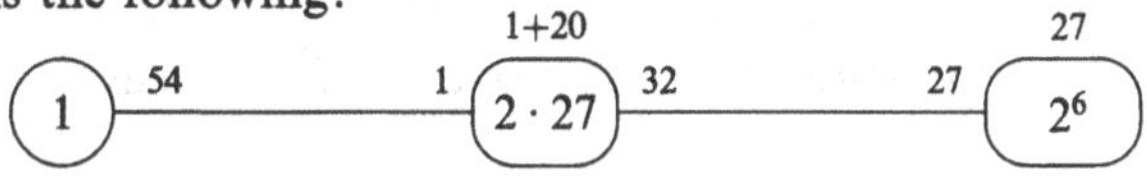

1 —54— 1 [$2 \cdot 27$] (1+20) 32 —27— [2^6] (27)

Directly from (4.10.8) we obtain the following.

Lemma 5.5.2 *Let $\widetilde{E}_1 = N_{G_1}(Y_2)$ and $\bar{E}_1 = \widetilde{E}_1/Y_2$. Then $O_2(\widetilde{E}_1) = \eta^{-1}(T^\perp)$ (where T is as in (4.10.4)) and $\bar{E}_1 \cong 2_+^{1+20}.U_6(2).Sym_3$.* □

Let $\widetilde{E}_2 = N_{G_2}(Y_2)$. Since Z_2 is in the orthogonal complement of Y_2, the normalizer of Z_2 in $\widetilde{E}_4$ permutes transitively the three subgroups of order 2 in Z_2. This means that $\widetilde{E}_2$ contains an element which maps v_0 onto v_1. Put $\widetilde{E} = \langle \widetilde{E}_1, \widetilde{E}_2 \rangle$, $\bar{E} = \widetilde{E}/Y_2$ and let Ξ be the subgraph in Δ induced by the images of v_0 under $\widetilde{E}$.

Lemma 5.5.3 *$\Xi(v_0) = \eta^{-1}(\bar{\Lambda}_4 \cap \bar{\Lambda}_2(\bar{\mu}_0) \cap \bar{\Lambda}_2(\bar{\mu}_1))$ and the action of $\widetilde{E}_1$ on this set is transitive.*

Proof. By the paragraph before the lemma $\Xi(v_0)$ contains v_1 and hence it is non-empty. By (5.5.2) (or rather by (4.10.8)) $\widetilde{E}_1/C_{G_1}(Y_2) \cong Sym_3 \cong \mathrm{Aut}\, Y_2$. This means that Ξ is also the orbit of v_0 under $\widetilde{E} \cap C_G(Y_2)$. Since $C_G(Y_2) = C_G(Y_1) \cap C_G(Y_1')$, by (5.4.2 (ii), (iv)) we conclude that $\Xi(v_0)$ is contained in $\eta^{-1}(\bar{\Lambda}_4 \cap \bar{\Lambda}_2(\bar{\mu}_0) \cap \bar{\Lambda}_2(\bar{\mu}_1))$. Finally by (4.10.5) the action of $\widetilde{E}_1$ on the latter set is transitive and the result follows. □

Define $\mathcal{G}(^2E_6(2))$ (just a name so far) to be the incidence system of rank 4 such that for $1 \leq i \leq 3$ the elements of type i are the elements of type i in $\mathcal{G}(M)$ contained in Ξ and the elements of type 4 are the images under $\widetilde{E}$ of the subgraph Υ, the incidence relation being is via inclusion.

Proposition 5.5.4 *The following assertions hold:*

(i) $\mathcal{G}(^2E_6(2))$ *is a Tits geometry with the diagram*

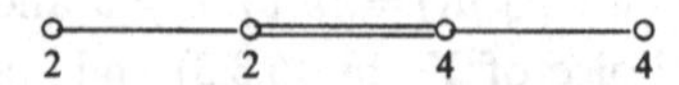

and $\bar{E}$ *induces a flag-transitive automorphism group of* $\mathcal{G}(^2E_6(2))$;

(ii) $\bar{E} \cong {}^2E_6(2).Sym_3$.

Proof. Since the incidence relation is via inclusion, $\mathcal{G}(^2E_6(2))$ belongs to a string diagram. By the proof of (4.10.6) the residue of v_0 is the C_3-geometry of $U_6(2)$ on which $\widetilde{E}_1$ induces a flag-transitive action while the residue of Υ is the C_3-geometry of $\Omega_8^-(2)$ on which $\widetilde{E}_4$ acts flag-transitively. Hence (i) follows. Since the C_3-residues in $\mathcal{G}(^2E_6(2))$ are buildings, by [Ti82] the geometry itself is a building of $^2E_6(2)$. By (1.6.5) and (5.5.2) $\bar{E}$ is the full automorphism group of $\mathcal{G}(^2E_6(2))$ and (ii) follows. □

Below we present the suborbit diagram of Ξ with respect to the action of $\bar{E}$. Notice that Υ contains v_0, 54 vertices from $\Xi(v_0)$ and 64 vertices from $\Xi_2^6(v_0)$. Furthermore, by (4.10.5) it is easy to observe that in the notation of (5.3.1) $\Xi_2^i(v_0)$ is contained in $\Delta_2^i(v_0)$ for $i = 5$ and 6.

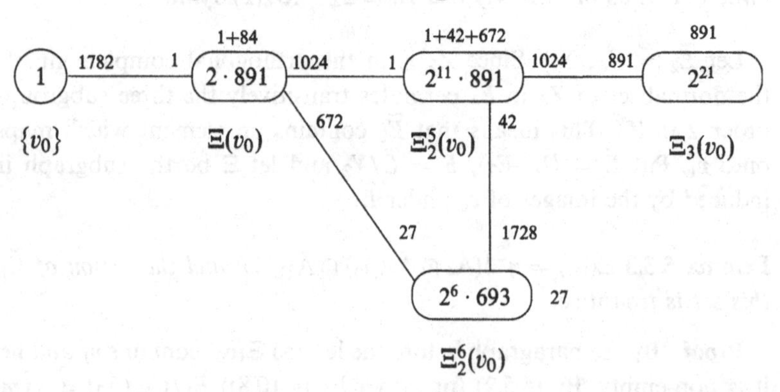

Let $\{\bar{E}_i \mid 1 \le i \le 4\}$ be the amalgam of maximal parabolics associated with the action of $\bar{E}$ on $\mathcal{G}(^2E_6(2))$. Then $\bar{E}_1$ and $\bar{E}_4$ are as above and we have

$$\bar{E}_1 \cong 2_+^{1+20}.U_6(2).Sym_3, \quad \bar{E}_2 \cong 2^{2+9+18}.(P\Gamma L_3(4) \times Sym_3),$$

$$\bar{E}_3 \cong 2^{3+4+12+12}.(L_3(2) \times 3.Sym_5), \quad \bar{E}_4 \cong 2^{8+16}.(\Omega_8^-(2) \times 3).2$$

and because of the choice of the maximal flag in $\mathcal{H}(M)$, $\bar{E}_i$ is the quotient over Y_2 of the normalizer of Y_2 in $H_1 = G_1$, $H_2 = G_2$, $H_3 = G_3$ and $H_5 = D$ for $i = 1, 2, 3$ and 4, respectively.

There is an involution $\omega \in \bar{E} \setminus O^2(\bar{E})$ such that $C_{\bar{E}}(\omega) \cong F_4(2) \times 2$ [ASei76]. The element ω induces on Y_2 an action of order 2 so that $\langle Y_2, \omega \rangle \cong D_8$. The vertices in Ξ fixed by ω form a subgraph which is the collinearity graph of the natural parabolic geometry of $F_4(2)$. The suborbit diagram of this subgraph is presented below.

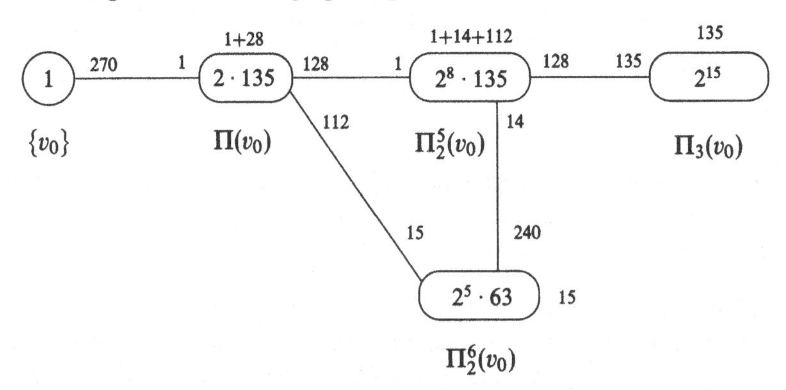

5.6 Towards the Fischer group $M(24)$

Let X_s be a Sylow 3-subgroup in $O_{2,3}(G_3)$. Then X_s is contained in G_3^+ and in G_1 and it maps onto a conjugate of the subgroup in $\bar{G}_1$ introduced under the same name in (4.14.1). We are going to study the normalizer of X_s in the group G and the connected component $\Delta(X_s)$ of the subgraph in Δ induced by the vertices which are fixed by X_s. Notice that if Π is the neighbourhood of v_0 in $\Delta(X_s)$ then in terms of (4.14.10) we have $\eta(\Pi) = \Phi(X_s)$. Let $\widehat{F}$ denote the setwise stabilizer of $\Delta(X_s)$ in $N_G(X_s)$. We are going to identify $F = \widehat{F}/X_s$ with the largest Fischer 3-transposition group $M(24)$. Our first result follows directly from (4.14.9) and (4.14.11).

Lemma 5.6.1 $N_{G_1}(X_s)/X_s \cong 2_+^{1+12} : 3 \cdot U_4(3).2^2$. □

Let X be a Sylow 3-subgroup in G_2^-. Then by (3.8.2) and the Frattini argument we obtain the next result.

Lemma 5.6.2 *Let $N = N_{G_2}(X)$. Then $N \cong 2^{11}.(Sym_3 \times Mat_{24})$ and $O_2(N)$ is isomorphic to the irreducible Todd module $\bar{\mathcal{C}}_{11}$.* □

Since X is contained in G_2^-, it fixes $\mathrm{res}^+_{\mathcal{G}}(l) \cong \mathcal{G}(Mat_{24})$ elementwise (where l is the element of type 2 in $\mathcal{G} = \mathcal{G}(M)$ stabilized by G_2). Arguing as in the paragraph before (4.14.3) we observe that for every element u of type i in $\mathrm{res}^+_{\mathcal{G}}(l)$ there is a unique element $\psi(u, X)$ of type $i-2$ incident to

u which X fixes pointwise (where $3 \leq i \leq 5$) and we obtain the following analogue of (4.14.3).

Lemma 5.6.3 *The subgeometry $\Psi(l, X)$ in $\mathcal{G}$ formed by the elements $\psi(u, X)$ taken for all $u \in \mathrm{res}^+_{\mathcal{G}}(l)$ is isomorphic to $\mathrm{res}^+_{\mathcal{G}}(l) \cong \mathcal{G}(Mat_{24})$ and $N_{G_2}(X)$ induces on $\Psi(l, X)$ the natural action of Mat_{24}.* □

Recall that the sextet graph is the collinearity graph of both $\mathcal{G}(Mat_{24})$ and $\mathcal{H}(Mat_{24})$.

Lemma 5.6.4 *There is an element $g \in G$ which conjugates X onto X_s and maps the set of vertices in $\Psi(l, X)$ onto a subgraph $\Psi(l^g, X_s)$ of $\Delta(X_s)$ isomorphic to the sextet graph.*

Proof. Let u be an element of type 4 in $\mathrm{res}^+_{\mathcal{G}}(l)$ and $v = \psi(u, X)$. Then v is of type 2 and X fixes $\mathrm{res}^-_{\mathcal{G}}(v)$ which means that X is conjugate to a subgroup in G_{12}. By (5.6.2) and (3.3.4) the order of $N_G(X) \cap G(v)$ is $2^{22} \cdot 3^3$. Hence the result follows from (4.14.1), (4.14.11) and the remark after the proof of that lemma. □

By the paragraph before (4.14.6) $N_{G_1}(X_s)$ has 2 orbits on the set of planes ($\star$-closed cliques of size 7) contained in $\Delta(X_s)$ and containing v_0, furthermore one of the orbits is 15 times as long as the other one. This means that F has 2 orbits on the set of planes contained in $\Delta(X_s)$. Let $\mathcal{G}(M(24))$ be an incidence system of rank 4 whose elements of type 1 and 2 are the vertices and lines of $\mathcal{G}(M)$ contained in $\Delta(X_s)$, the elements of type 3 are the planes from the long orbit of F on the set of planes contained in $\Delta(X_s)$, the elements of type 4 are the images under F of the subgraph $\Psi(l^g, X_s)$ as in (5.6.4) and the incidence relation is via inclusion.

Lemma 5.6.5 *$\mathcal{G}(M(24))$ is a geometry with the diagram*

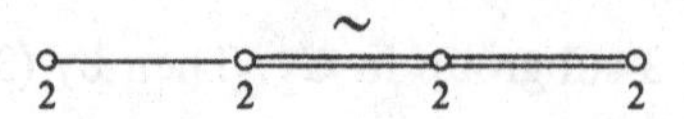

and F induces a flag-transitive automorphism group of $\mathcal{G}(M(24))$.

Proof. Since the incidence relation is via inclusion, $\mathcal{G}(M(24))$ belongs to a string diagram. By (5.6.3) and the paragraph before that lemma the residue of an element of type 4 is isomorphic to the tilde geometry $\mathcal{G}(Mat_{24})$. The residue of an element of type 1 is isomorphic to the geometry $\mathcal{G}(3 \cdot U_4(3))$ by (4.14.6) and the paragraph before that lemma. The flag-transitivity is straightforward. □

Let $\{K_i \mid 1 \leq i \leq 4\}$ be the amalgam of maximal parabolics associated with the action of F on $\mathcal{G}(M(24))$. Then under a suitable choice of the maximal flag of $\mathcal{G}(M)$ we have $K_i = N_{G_i}(X_s)/X_s$ for $1 \leq i \leq 3$ and $K_4 \cong N_{G_2}(X)/X$ where X is a Sylow 3-subgroup in $O_{2,3}(G_2)$ and we have the following:

$$K_1 \cong 2_+^{1+12} : 3 \cdot U_4(3).2^2, \quad K_2 \cong 2^{2+1+4+8+2}.(Sym_3 \times Sym_6),$$

$$K_3 \cong 2^{8+6+4}.(L_3(2) \times Sym_3), \quad K_4 \cong 2^{11+1}.Mat_{24}.$$

The subgraph $\Psi(l^g, X_s)$ in $\Delta(X_s)$ which realizes an element of type 4 in $\mathcal{G}(M(24))$ is the collinearity graph of $\mathcal{H}(Mat_{24})$. An element of type 2 in $\mathcal{H}(Mat_{24})$ (a trio) is realized by a 7-vertex complete subgraph Θ_2 and one can see that Θ_2 is a plane of $\mathcal{G}(M)$ contained in the short orbit of F on the set of planes in $\Delta(X_s)$. An element of type 3 in $\mathcal{H}(Mat_{24})$ (an octad) is realized by a 35-vertex subgraph Θ_3 whose vertices are indexed by the 2-dimensional subspaces in a 4-dimensional $GF(2)$-vector–space with two subspaces being adjacent if their intersection is 1-dimensional. Let $\mathcal{H}(M(24))$ be an incidence system of rank 4 whose elements of type 1 and 4 are as in $\mathcal{G}(M(24))$, whose elements of type 2 and 3 are the images under F of Θ_2 and Θ_3, respectively, and the incidence relation is via inclusion. Then $\mathcal{H}(M(24))$ is a geometry with the following diagram:

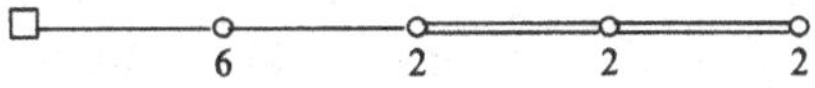

□———o———o═══o═══o
  6 2 2 2

The action of F on $\mathcal{H}(M(24))$ is flag-transitive and if $\{F_i \mid 1 \leq i \leq 4\}$ is the amalgam of maximal parabolic subgroups corresponding to the action of F on $\mathcal{H}(M(24))$ then $F_1 = K_1$, $F_4 = K_4$ while

$$F_2 \cong 2^{3+12}.(L_3(2) \times Sym_6), \quad F_3 \cong 2^{7+8}(L_4(2) \times Sym_3).$$

Let us consider the intersection of $\Delta(X_s)$ and the subgraph Ψ introduced before (5.3.3) which realizes an element of type 5 in $\mathcal{H}(M)$. Since $\Psi(v_0) = \varphi^{-1}\eta^{-1}(\Phi)$ and we can choose Φ to be equal to $\Phi(\delta)$ as in the paragraph before (4.14.7), we can assume that $\eta(\Psi(v_0) \cap \Delta(X_s) \cap \Delta(v_0))$ realizes an element of type 1 in $\mathcal{E}(3 \cdot U_4(3))$ as (4.14.9). By (4.14.8), in this case $[A, X_s]$ is a minus 2-subspace in A which can be identified with Y_2 as in the first paragraph of Section 5.5. Hence $N_D(X_s)$ is contained in $\widetilde{E}_4 = N_D(Y_2)$, $N_D(X_s)/X_s \cong 2^8.\Omega_8^-(2).2$ and $\Delta(X_s) \cap \Psi$ coincides with the subgraph Υ as in (5.5.1).

Define $\mathcal{E}(M(24))$ to be an incidence system of rank 4 whose elements of type 1 are the images under F of the subgraph Υ as above, the elements of type 2 are the planes from the longer orbit of F on the set of planes contained in $\Delta(X_s)$, the elements of type 3 and 4 are the lines and points of $\mathcal{G}(M)$ contained in $\Delta(X_s)$ and the incidence relation is via inclusion. Then by the preceding paragraph and (4.14.9) we obtain the following

Lemma 5.6.6 *$\mathcal{E}(M(24))$ is an extended dual polar space with the diagram*

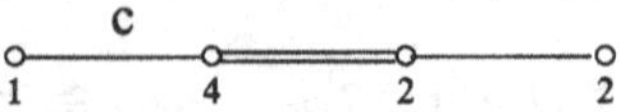

$$\overset{c}{\underset{1}{\circ}\!\!-\!\!-\!\!-\!\!\underset{4}{\circ}}\!\!=\!\!=\!\!=\!\!\underset{2}{\circ}\!\!-\!\!-\!\!-\!\!\underset{2}{\circ}$$

on which F induces a flag-transitive action. □

By the construction $\{N_D(X_s)/X_s, K_3, K_2, K_1\}$ is the amalgam of maximal parabolics associated with the action of F on $\mathcal{E}(M(24))$.

In the next lemma we specify the structure of $O_2(F_4)$.

Lemma 5.6.7 *Let $Q = O_2(F_4)$. Then Q is elementary abelian and as a module for $F_4/Q \cong Mat_{24}$ it is isomorphic to the 12-dimensional Todd module.*

Proof. Recall that $F_4 = K_4 = N/X$ where $N = N_{G_2}(X)$ and X is a Sylow 3-subgroup in $G_2^- \cong 2^{2+11+22}.Sym_3$. Let $\widehat{P} = O_2(N)$, $P = \widehat{P}X/X$ and $\widehat{Q}$ be a Sylow 2-subgroup in the preimage of Q in N. Then $\widehat{Q} \cong Q$, P is a subgroup of index 2 in Q and by (5.6.2) P is the irreducible Todd module for F_4/Q. Since P is an irreducible submodule of codimension 1 in Q, it is easy to see that Q is abelian. Furthermore, P is the only faithful submodule of F_4/Q in Q. Hence in order to prove the lemma it is sufficient to show that F_4/Q has an orbit of length 24 in $Q \setminus P$ on which it acts as on the base set $\mathcal{P}$ of the Golay code (2.15.1). We consider N as a subgroup in G_2. Since $N_{G_2}(X)/N_{G_2^-}(X) \cong Mat_{24}$ has no normal 2-subgroups, $\widehat{Q}$ is contained in one of the three Sylow 2-subgroups of G_2^-. Since $G_{12} \sim 2_+^{1+24}.2^{11}.Mat_{24}$ and $O_2(G_{12})$ is a Sylow 2-subgroup in G_2^- we can assume without loss of generality that $\widehat{Q} \leq O_2(G_{12})$. Moreover, since $\widehat{P} \leq R_2 \leq Q_1$, $O_2(G_{12})(N \cap G_{12}) = G_{12}$ and $G_{12}/O_2(G_{12}) \cong Mat_{24}$ acts irreducibly on $O_2(G_{12})/Q_1$, we conclude that $\widehat{Q} \leq Q_1$. Now given an element from Q_1 we have to decide in what case it normalizes and does not centralize X. Recall that $R_2 = \eta^{-1}(\bar{\Lambda}_4^8 \cup \bar{\Lambda}_4^{4a})$ and $\widehat{P} = C_{R_2}(X)$ is a complement to Z_2 in R_2. This means that $\langle R_2, X\rangle = \langle Z_2, X\rangle \times \widehat{P} \cong Alt_4 \times \widehat{P}$. Hence if Y is a Sylow 3-subgroup in G_2^- then $\widehat{P} = C_{R_2}(Y)$ if and only if Y is one of the four Sylow 3-subgroups in $\langle Z_2, X\rangle$. On the other hand G_2^- contains $2^{24} = [G_2^- : N]$ Sylow 3-subgroups and R_2 contains 2^{22} complements to Z_2. Hence for each complement to Z_2 in R_2 there are exactly four Sylow 3-subgroups Y in G_2^- such that $\widehat{P} = C_{R_2}(Y)$ and Z_2 acts transitively on the set of these four subgroups by conjugation. In particular $[N_{G_2}(\widehat{P}) : N] = 4$ and an element q from Q_1 normalizes $\widehat{P}$ if and only if $X^q \leq \langle Z_2, X\rangle$ and in this case q normalizes and does not centralize a subgroup of order 3 in $\langle Z_2, X\rangle$ if and only if it does not centralize Z_2, *i.e.* if $q \notin Z_2^\perp$. Hence without loss of generality we can assume that $\eta(\widehat{P})$ coincides with the complement $\bar{E}$ to $U_1 = \eta(Z_2)$ in $U_2 = \eta(R_2)$ as in (4.8.3). In terms of (4.8.3) let $V = \eta^{-1}(\bar{E}_3)$. Then by (4.8.3) $N_{G_{12}}(\widehat{P})$ stabilizes V setwise. Since $\bar{E}_3 \subseteq \bar{\Lambda}_2^3$ we have $V \notin Z_2^\perp$ and by (4.8.3 (iv)) $V \subseteq \widehat{P}^\perp$. By the Frattini argument $N_{G_{12}}(\widehat{P})/N_{G_2^-}(\widehat{P}) \cong Mat_{24}$ and by (4.8.3 (ii)) the latter group induces on $\bar{E}_3$ the natural action of degree 24. Finally, since the stabilizer of an element in this action (isomorphic to Mat_{23}) has no subgroups of index 8 or less we conclude that N/X has an orbit of length 24 on VX/X and the result follows. □

Corollary 5.6.8 *A subgroup X of order 3 in G_2^- can be chosen in such a way that for*

$$\bar{E}_3 = \{\bar{v} \mid v \in \Lambda_2^3, X(v) = P_3(v)\}$$

the set $\eta^{-1}(\bar{E}_3)$ is contained in $N_{G_2}(X) \setminus C_{G_2}(X)$ and it maps onto an orbit of length 24 of $N_{G_2}(X)/X$ on $O_2(N_{G_2}(X)/X)$. □

5.7 Identifying $M(24)$

In this section we study the geometry $\mathscr{H} = \mathscr{H}(M(24))$ and the action of F (still to be identified with $M(24)$) on $\mathscr{H}$. Let $\{y_i \mid 1 \leq i \leq 4\}$ be a maximal flag in $\mathscr{H}$ so that $\{F_i = F(y_i) \mid 1 \leq i \leq 4\}$ is the amalgam of maximal parabolics associated with the action of F on $\mathscr{H}$. Put $R_i = O_2(F_i)$, $F_{ij} = F_i \cap F_j$, $R_{ij} = O_2(F_{ij})$ for $1 \leq i, j \leq 4$.

Lemma 5.7.1 *For $i = 1$, 2 and 3 we have $R_{i4} = R_iR_4$.*

Proof. Comparing the shapes of F_i and F_4 we obtain the following:

$$F_{14} \sim 2_+^{1+12}.2^5.3 \cdot Sym_6 \sim 2^{12}.2^6.3 \cdot Sym_6;$$

$$F_{24} \sim 2^{3+12}.(L_3(2) \times (2 \times Sym_4)) \sim 2^{12}.2^6.(L_3(2) \times Sym_3);$$

$$F_{34} \sim 2^{7+8}.(L_4(2) \times 2) \sim 2^{12}.2^4.L_4(2).$$

By (2.10.1), (2.10.2) and (2.10.3) the action of F_{i4}/R_{i4} on R_{i4}/R_4 is irreducible and the result follows. □

Let $\Gamma = \Gamma(M(24))$ be a graph on the set of elements of type 4 in $\mathscr{H}$ in which two such elements are adjacent if they are incident in $\mathscr{H}$ to a common element of type 3. For $z \in \mathscr{H}$ of type 1, 2 and 3 let $\Gamma[z]$ denote the subgraph induced by the vertices incident to z in $\mathscr{H}$. (Recall that for $z \in \Gamma$ of type 4 $\Gamma(z)$ is the neighbourhood of z in the graph Γ.) When talking about octads, trios and sextets we mean those from $\mathrm{res}_{\mathscr{H}}(y_4) \cong \mathscr{H}(Mat_{24})$.

Lemma 5.7.2 *The following assertions hold:*

(i) *F acts on Γ vertex- and edge-transitively;*

(ii) *there is a mapping κ from $\Gamma(y_4)$ onto the set of octads which commutes with the action of F_4;*

(iii) *if B is the octad which corresponds to y_3, then $\kappa^{-1}(B)$ is an orbit of length 2 of R_4 and $\Gamma[y_3] = \{y_4, \kappa^{-1}(B)\}$ is a triangle in Γ;*

(iv) *$\Gamma[y_2]$ is the point graph of $\mathscr{G}(Sp_4(2)) \cong \mathrm{res}^+_{\mathscr{H}}(y_2)$ and $\kappa(\Gamma[y_2] \cap \Gamma(y_4))$ is the set of octads contained in the trio which corresponds to y_2;*

(v) *$\Gamma[y_1]$ is the point graph of (the dual of) $\mathscr{G}(U_4(3)) \cong \mathrm{res}^+_{\mathscr{H}}(y_1)$ and $\kappa(\Gamma[y_1] \cap \Gamma(y_4))$ is the set of octads refined by the sextet which corresponds to y_1;*

(vi) *the valency of Γ is $2 \cdot 759$.*

Proof. Part (i) follows from the flag-transitivity of the action of F on $\mathscr{H}$. By (5.7.1) the subgroup R_4 induces on $\Gamma[y_2]$ an action of order 8. By the basic properties of the generalized quadrangle of order (2,2) this implies that R_4 induces on $\Gamma(y_4) \cap \Gamma[y_2]$ an action of order 4. By (5.6.7) R_4 is the 12-dimensional Todd module which is indecomposable. Hence R_4 acts faithfully on $\Gamma(y_4)$ with orbits of length 2 and the kernels at these orbits correspond to one of the two orbits of length 759 of F_4/R_4 on the dual of R_4 (which is the Golay code module). This implies (ii) and (iii). For $i = 1$ and 2 there is a unique orbit of F_{i4}/R_i on $\Gamma[y_i]$ (with length 30 and 6, respectively) on which R_{i4}/R_i acts with orbits of length 2. Hence (iv) and (v) follow from (iii). Finally (vi) is a direct consequence of (ii) and (iii). □

In our further considerations a crucial rôle is played by the observation that the geometry $\mathscr{E}(U_4(3))$ and the geometry dual to $\mathscr{G}(U_4(3))$ have the same sets of elements of type 2 and 3 (Section 6.13). By this observation $\Gamma[y_1]$ contains the Schläfli graph as a subgraph. More specifically the following holds. Let $\Sigma = \{S_1, ..., S_6\}$ be the sextet which corresponds to y_1, so that $\{S_i \cup S_j \mid 1 \le i < j \le 6\}$ is the image of $\Gamma(y_4) \cap \Gamma[y_1]$ under

κ. Then for every k, $1 \leq k \leq 6$ there is a unique Schläfli subgraph Θ in $\Gamma[y_1]$ containing y_4 which is an element of type 1 in $\mathcal{E}(U_4(3))$ such that $\kappa(\Theta(y_4)) = \{S_k \cup S_i \mid 1 \leq i \leq 6, i \neq k\}$. Notice that Σ (and hence y_1 as well) is uniquely determined by S_k.

Lemma 5.7.3 *The graph $\Gamma = \Gamma(M(24))$ contains a family $\mathcal{S}$ of Schläfli subgraphs with the following properties:*

(i) *for every 4-element subset S of the base set $\mathcal{P}$ of the Golay code associated with y_4 there is a unique $\Theta \in \mathcal{S}$ which contains y_4, such that $\kappa(\Theta(y_4))$ is the set of (five) octads containing S;*

(ii) *the group F acts transitively on $\mathcal{S}$ and*

$$F[\Theta] \sim 2_+^{1+12}.3 : (U_4(2).2 \times 2);$$

(iii) *$F_{14}/O_{2,3}(F_{14}) \cong Sym_6$ has a unique orbit Ω of length 6 on the involutions in $O_{2,3}(F_{14})/O_{2,3}(F_1) \cong 2^5$;*

(iv) *Θ is (the connected component containing y_4 of) the subgraph of $\Gamma[y_1]$ induced by the vertices fixed by an involution from Ω.*

Proof. The assertions (i) and (ii) follow from the paragraph before the lemma. Since $O_{2,3}(F_{14})/O_{2,3}(F_1)$ is of order 2^5 while by the basic properties of the Schläfli graph we have $F_4[\Theta]/F(\Theta) \cong 2^4.Sym_5$ (4.14.7) we obtain (iv). Since $O_{2,3}(F_{14})/O_{2,3}(F_1)$ involves the natural symplectic module of $Sym_6 \cong Sp_4(2)$ we obtain the uniqueness of Ω stated in (iii).□

Recall that the Todd module $\bar{\mathcal{C}}_{12} \cong R_4$ is the quotient of the power set $2^{\mathcal{P}}$ over the Golay code $\mathcal{C}_{12}$ and that it is generated by the images of the subsets of size at most 4 (2.3.3). By (5.7.1) and (2.15.1 (iii)) we have the following.

Lemma 5.7.4 *The unique non-identity element s in $Z(R_1)$ is contained in R_4 and it is the image in $R_4 \cong \bar{\mathcal{C}}_{12}$ of a tetrad from the sextet which corresponds to y_1 and $F_{1i} = C_{F_i}(s)$ for $i = 2, 3$ and 4.* □

For a subset Y of $\mathcal{P}$ let $\bar{\mathcal{C}}_{12}(Y)$ and $\bar{\mathcal{C}}_{12}[Y]$ denote the subspaces in $\bar{\mathcal{C}}_{12}$ generated by the images of the subsets contained in Y and of the subset having even intersection with Y, respectively.

Lemma 5.7.5 *Let $z \in \Gamma(y_4)$ and $\kappa(z) = B$ (an octad). Then*

(i) $O_2(F_4 \cap F(z)) = (R_4 \cap F(z))(O_2(F(z)) \cap F_4))$,

(ii) $|R_4 \cap O_2(F(z))| = 2^7$,

(iii) $R_4 \cap O_2(F(z)) = Z(O_2(F_4 \cap F(z)) \cong \bar{\mathcal{C}}_{12}(B)$,
(iv) $R_4 \cap F(z) \cong \bar{\mathcal{C}}_{12}[\mathcal{P} \setminus B]$.

Proof. By (5.7.2 (iii)) we have $F_4 \cap F(z) \sim 2^{11}.2^4.L_4(2)$ and applying (5.7.1) for the case $i = 3$ we obtain (i) which implies (ii). Since both R_4 and $O_2(F(z))$ are abelian $R_4 \cap O_2(F(z))$ is contained in the centre of $O_2(F_4 \cap F(z))$ which it is easy to identify with $\bar{\mathcal{C}}_{12}(B)$, which gives (iii). By (5.7.2 (iii)) $R_4 \cap F(z)$ is a hyperplane in R_4 normalized by $F_4 \cap F(z)$ and by (2.15.1 (i)) this hyperplane is either $\bar{\mathcal{C}}_{12}[B]$ or $\bar{\mathcal{C}}_{12}[\mathcal{P} \setminus B]$. Since the former does not contain $\bar{\mathcal{C}}_{12}(B) \cong R_4 \cap O_2(F(z))$ (iv) follows. □

Lemma 5.7.6 *In terms of* (5.7.5) *suppose that* $z \in \Gamma[y_3]$. *Then*

(i) $R_4 \cap O_2(F(z)) = Z(R_3)$,
(ii) *if* H *is* F_3 *or* $F_4 \cap F(z)$ *then* H *acting on* $Z(R_3))^\#$ *has four orbits with lengths* 8, 28, 56 *and* 35 *consisting of the images in* $R_4 \cong \bar{\mathcal{C}}_{12}$ *of the i-element subsets of* B *for* $i = 1, 2, 3$ *and* 4, *respectively*,
(iii) *the action induced by* H *on its orbit of length* 8 *in* $Z(R_3)^\#$ *is isomorphic to* Alt_8.

Proof. By (5.7.2 (iii)) we have the factorization $F_{34} = (F_4 \cap F(z))R_4$ and (i) follows from (5.7.5 (iii)). Since R_4 is abelian the above factorization also implies that F_3 and $F_4 \cap F(z)$ have the same orbits on $Z(R_3)^\#$ and hence (ii) and (iii) follow. □

Let t_1, t_2, t_3 and t_4 be distinct elements in $\mathcal{P}$ which we identify with the corresponding involutions in the orbit of length 24 of F_4 on R_4. Let Y_i be the union of the t_j for $1 \le j \le i$ and we identify Y_i with the subgroup (of order 2^i) in R_4 generated by the corresponding involutions. Let Θ_i be the connected component containing y_4 of the subgraph in Γ induced by the vertices u such that $Y_i \le O_2(F(u))$.

Lemma 5.7.7 *The following assertions hold:*

(i) *a vertex* $z \in \Gamma(y_4)$ *is contained in* Θ_i *if and only if* Y_i *is contained in the octad* $\kappa(z)$;
(ii) *the subgroup* $F[\Theta_i]$ *is contained in* $N_F(Y_i)$ *and it acts on* Θ_i *vertex- and edge-transitively*;
(iii) *the valency of* Θ_i *is* $2 \cdot 253$, $2 \cdot 77$, $2 \cdot 21$ *and* $2 \cdot 5$ *for* $i = 1, 2, 3$ *and* 4, *respectively*;
(iv) $N_{F_4}(Y_i)/R_4$ *is isomorphic to* Mat_{23}, $\mathrm{Aut}\, Mat_{22}$, $P\Gamma L_3(4)$ *and* $2^6.3.Sym_5$ *for* $i = 1, 2, 3$ *and* 4, *respectively*.

Proof. Let $z \in \Gamma[y_3]$ and $t = t_j$ for some j, $1 \leq j \leq 4$. Then by (5.7.5 (iv)) $t \in F(z)$ if and only if $t \in \kappa(z)$. On the other hand by (5.7.5 (iii)) and (5.7.6 (i)) if $t \in \kappa(z)$ then $t \in Z(R_3) = R_4 \cap O_2(F(z))$, hence (i) and (iii) follow. Since $R_4 \leq N_{F_4}(Y_i)$ and $N_{F_4}(Y_i)$ acts transitively on the set of octads containing Y_i (2.10.1 (iii)) the action of $N_{F_4}(Y_i)$ on $\Theta_i(y_4)$ is transitive. By (5.7.6 (ii)) $N_{F_3}(Y_i)R_3 = F_3$ and since Θ_i is connected (by the definition), we have $F[\Theta_i] = \langle N_{F_4}(Y_i), N_{F_3}(Y_i)\rangle$ and the action of $F[\Theta_i]$ on Θ_i is vertex- and edge-transitive, which gives (ii). Finally (iv) follows directly by (2.15.1 (iii)). □

Lemma 5.7.8 *The subgraph Θ_4 belongs to the family $\mathscr{S}$ of Schläfli subgraphs as in* (5.7.3).

Proof. Let $\Sigma = \{S_1, ..., S_6\}$ be the sextet corresponding to y_1 and assume that $Y_4 = S_1$. Then by (5.7.4) $s := t_1t_2t_3t_4$ is the unique non-identity element in $Z(R_1)$. Since s is also the unique element in Y_4 which is the image in R_4 of a 4-element subset of $\mathscr{P}$, in view of (5.7.4) we have $N_{F_4}(Y_4) \leq C_{F_4}(s) = F_{14}$. By (5.7.6 (ii)) we also have $N_{F_3}(Y_4) \leq C_{F_3}(s) = F_{13}$. Hence we have $F[\Theta_4] \leq F_1$ which implies $\Theta_4 \subseteq \Gamma[y_1]$ in view of (5.7.7 (ii)) and the flag-transitivity of F_1 on $\mathrm{res}_{\mathscr{H}}(y_1)$. Since $F[\Theta_4] = \langle N_{F_4}(Y_4), N_{F_3}(Y_4)\rangle$, the subgroup Y_4 fixes Θ_4 elementwise. Let $v \in \Gamma(y_4) \cap \Gamma[y_1]$. Then $\kappa(v) = S_k \cup S_l$ for some k, l, $1 \leq k < l \leq 6$ and by (5.7.5 (iii)) $t_m \in F(v)$ if and only if $k = 1$ (equivalently if $Y_4 \subseteq \kappa(v)$), independently of the choice of $m \in \{1, 2, 3, 4\}$. Hence Y_4 induces on $\Gamma[y_1] \cap \Gamma(y_4)$ an action of order 2. Therefore Θ_4 is fixed by an involution from an orbit of length 6 of $F_{14}/O_{2,3}(F_{14}) \cong Sym_6$ on $O_{2,3}(F_{14})/O_{2,3}(F_1) \cong 2^5$. By (5.7.3 (iii), (iv)) we obtain the desired inclusion $\Theta_4 \in \mathscr{S}$. □

Let $\mathscr{T} = \mathscr{T}(M(24))$ be an incidence system of rank 6 in which the elements of type i are the images under F of the subgraphs Θ_i for $i = 1, 2, 3$ and 4, the elements of type 5 are the images of $\Gamma[y_3]$ under F, the elements of type 6 are the vertices of Γ and the incidence relation is via inclusion.

Lemma 5.7.9 *The incidence system $\mathscr{T}$ is a geometry with the diagram*

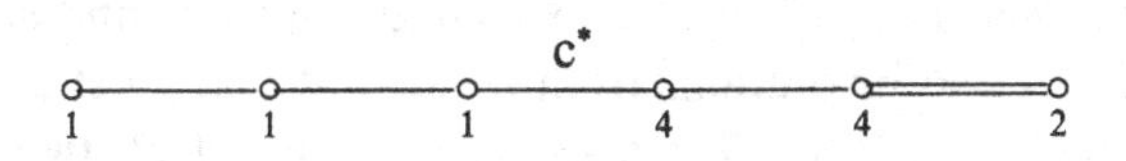

and F induces on it a flag-transitive action.

Proof. Since the incidence relation is via inclusion it is easy to see that $\mathcal{T}$ is a geometry with a string diagram. The elements in $\mathrm{res}^+_{\mathcal{T}}(y_4)$ are the subsets of $\mathcal{P}$ of size at most 4 and the octads. Since the incidence relation is via inclusion, this residue is isomorphic to the geometry $\mathcal{E}(Mat_{24})$ as in (3.1.1). By (5.7.8) $\mathrm{res}^+_{\mathcal{T}}(\Theta_4)$ is isomorphic to the geometry of triangles and vertices of the Schläfli graph (*i.e.* to the generalized quadrangle of order (2,4)), hence the diagram is as given above. The flag-transitivity follows by (5.7.7 (ii)). □

Now we are in a position to apply the geometric characterization of the largest Fischer 3-transposition group achieved in [Mei91].

Proposition 5.7.10

Let $\mathcal{T}_6$ be a geometry with diagram as in (5.7.9) and M_6 be a flag-transitive automorphism group of $\mathcal{T}_6$. Then one of the following holds:

(i) *M_6 is the largest Fischer 3-transposition group $M(24)$ of order*

$$2^{22} \cdot 3^{16} \cdot 5^2 \cdot 7^3 \cdot 11 \cdot 13 \cdot 17 \cdot 23 \cdot 29$$

or the commutator subgroup (of index 2) of $M(24)$;

(ii) *M_6 is the unique non-split extension $3 \cdot M(24)$ or the commutator subgroup of $3 \cdot M(24)$.*

In each of the cases (i) and (ii) the geometry $\mathcal{T}_6$ is uniquely determined up to isomorphism. □

5.8 Fischer groups and their properties

Some intermediate results in the proof of (5.7.10) will play an important rôle in our subsequent exposition and we discuss these steps in the present section. Thus let $\mathcal{T}$ and F be as in the previous section. Let $\{s_i \mid 1 \le i \le 6\}$ be a maximal flag in $\mathcal{T}$, where s_i is of type i. For $3 \le i \le 5$ let $\mathcal{T}_i$ be a geometry whose diagram coincides with that of $\mathrm{res}^+_{\mathcal{T}}(s_{6-i})$; let $\mathcal{T}_6$ be a geometry whose diagram coincides with that of $\mathcal{T}$ and for $3 \le j \le 6$ let M_j be a flag-transitive automorphism group of $\mathcal{T}_j$. The elements from the left to the right on the diagram of $\mathcal{T}_j$ will be called points, lines *etc.* Let Π_j be the collinearity graph of $\mathcal{T}_j$. Recall that a group G is said to be a 3-*transposition* group if it contains a conjugacy class D of involutions which generates G and such that the product of any two involutions from D has order at most 3. In this case the *transposition graph* of G is a graph on D in which two involutions are adjacent if their product is of order 2.

The geometry $\mathcal{T}_3$ is a flag-transitive C_3-geometry in which the residue of a point is the unique generalized quadrangle of order $(2,4)$, which is classical, and the residue of a plane is the unique projective plane of order 4; by [A84] and [Ti82] we have the following.

Lemma 5.8.1 *The geometry $\mathcal{T}_3$ is isomorphic to the natural parabolic geometry of $U_6(2)$, and M_3 contains $U_6(2)$ as a normal subgroup.* □

Since $F(s_6) \cap F(s_3) \cong 2^{12}.P\Gamma L_3(4)$, $F(s_3)$ induces on $\mathrm{res}^+_{\mathcal{T}}(s_3)$ the full automorphism group of the latter geometry isomorphic to $U_6(2).Sym_3$. The graph Π_3 is strongly regular with the following suborbit diagram with respect to the action of $U_6(2).Sym_3$:

1 —180— 1 (180; 3+48) 128 — 45 (512; 135)

$2^{1+8}_+.(U_4(2) \times 3).2$ $\quad 2^{4+8}.[3^3].2^2 \quad (U_4(2) \times 3).2$

The point graph Π_4 of $\mathcal{T}_4$ is locally Π_3 and an important step in the proof of (5.7.10) is the following characterization of locally Π_3-graphs given in [Mei91] (with [BH77] and [DGMP] being credited).

Lemma 5.8.2 *The geometry $\mathcal{T}_4$ is uniquely determined up to isomorphism and M_4 is either the Fischer 3-transposition group $M(22)$ or the extension $M(22).2$ of $M(22)$ by an outer automorphism; Π_4 is the transposition graph of $M(22)$.* □

The suborbit diagram of Π_4 with respect to the action of $M(22).2$ is the following:

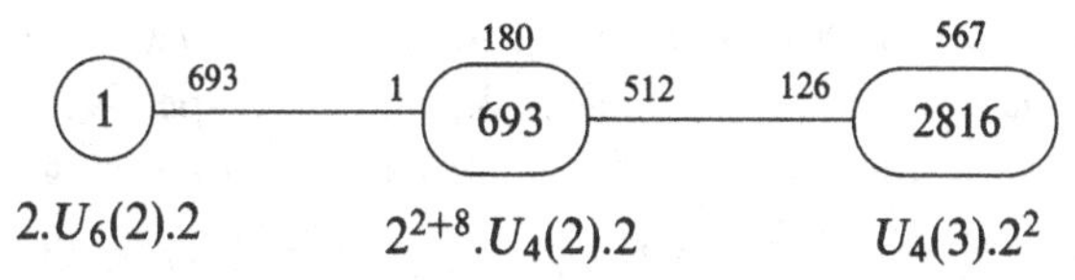

In its turn Π_5 is locally Π_4 and the next step in the proof is the following (Proposition 6.2 in [Mei91]).

Lemma 5.8.3 *The geometry $\mathcal{T}_5$ is uniquely determined up to isomorphism and M_5 is the Fischer 3-transposition group $M(23)$; Π_5 is the transposition graph of $M(23)$.* □

The suborbit diagram of Π_5 with respect to the action of $M(23)$ is the following:

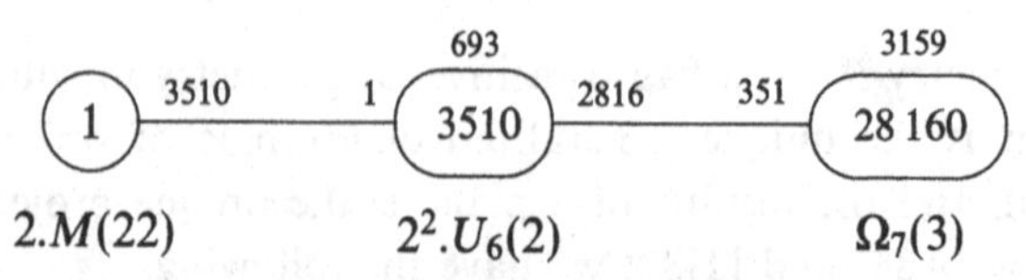

Finally Π_6 is locally Π_5 and (5.7.10) (which is Proposition 6.3 in [Mei91]) completes the picture. In (5.7.10 (i)) Π_6 is the transposition graph of $M(24)$ while in (5.7.10 (ii)) it is a 3-fold antipodal cover of the transposition graph with the following suborbit diagram with respect to the action of $3 \cdot M(24)$ (it is straightforward to deduce from this diagram that of the transposition graph of $M(24)$).

3510 | 28 431 | 3510

1 — 31 671 1 — 31 671 — 28 160 1080 — 825 792 — 2160 28 160 — 63 342 — 1 31 671 — 2

$2 \times M(23)$ $2^2.M(22)$ $\Omega_8^+(3).2$ $2.M(22)$ $M(23)$

It follows from the above result that the geometries $\mathcal{T}_4$, $\mathcal{T}_5$ and $\mathcal{T}_6$ as in (5.7.10) are 2-simply connected, a result, originally established in [Ron81a]. An independent proof of (5.8.2), (5.8.3) and (5.7.10) based on computer calculations can be found in [BW92a] (see also [BW92b]). In [Pase94] it was shown that the flag-transitivity assumptions in (5.8.2), (5.8.3) and (5.7.10) can be removed.

Since a Sylow 2-subgroup of $F(s_6) = F_4 \cong 2^{12}.Mat_{24}$ is of order 2^{22}, by (5.7.10) we have either $F \cong M(24)$ or $F \cong 3 \cdot M(24)$. We are going to show that the former of the possibilities holds by constructing a triple cover of $\mathcal{T} = \mathcal{T}(M(24))$.

As in the beginning of Section 5.6 let $\widehat{F}$ be the setwise stabilizer in $N_G(X_s)$ of the connected component $\Delta(X_s)$ containing v_0 of the subgraph induced by the vertices fixed by X_s, so that $F = \widehat{F}/X_s$. For $1 \le i \le 4$ let $\widehat{K}_i$ and $\widehat{F}_i$ be the preimages in $\widehat{F}$ of K_i and F_i, respectively. Then by the paragraph after the proof of (5.6.5) we have the following:

$\widehat{K}_1 = \widehat{F}_1 \sim 3.2_+^{1+12} : 3 \cdot U_4(3).2^2$; $\widehat{K}_4 = \widehat{F}_4 \sim 2^{11}.(Sym_3 \times Mat_{24})$;

$\widehat{K}_2 \sim [2^{17}].(Sym_3 \times 3 \cdot Sym_6)$; $\widehat{F}_2 \sim 2^{3+12}.(L_3(2) \times 3 \cdot Sym_6)$;

$\widehat{K}_3 \sim 2^{8+6+3}.(L_3(2) \times Sym_3 \times Sym_3)$; $\widehat{F}_3 \sim 2^{6+8}.(L_4(2) \times Sym_3 \times Sym_3)$.

Since $\widehat{F}_2$ is the normalizer in G_3 of a Sylow 3-subgroup from $O_{2,3}(G_3)$, we observe that $\widehat{F}_2$ (and hence $\widehat{F}$ as well) does not split over X_s. Since $\widehat{F}_1 \cap \widehat{F}_2$ contains a section $3 \cdot Sym_6$, $\widehat{F}_1$ does not split over X_s either. On the other hand $\widehat{K}_3$, $\widehat{F}_3$ and $\widehat{K}_4 = \widehat{F}_4$ split over X_s. More precisely we have the following. The number of maximal flags in $\mathcal{G}(M(24))$ is odd and hence there is an element q in the Borel subgroup of the action of

$\widehat{F}$ on $\mathscr{G}(M(24))$ which inverts X_s. It is clear that q is also in the Borel subgroup of the action of $\widehat{F}$ on $\mathscr{H}(M(24))$. Let $\widehat{H}$ be one of $\widehat{K}_3$, $\widehat{F}_3$ and $\widehat{K}_4 = \widehat{F}_4$. Then X_s is a direct factor of $C_{\widehat{H}}(X_s)$ and

$$\widetilde{H} := \langle O^3(C_{\widehat{H}}(X_s)),\ q\rangle$$

is a complement to X_s in $\widehat{H}$. Furthermore, $\widetilde{K}_3 \cap \widetilde{K}_4$ is a complement to X_s in $\widehat{K}_3 \cap \widehat{K}_4$. Put $\mathscr{A} = \{\widehat{K}_1, \widehat{K}_2, \widetilde{K}_3, \widetilde{K}_4\}$ and $\mathscr{B} = \{\widehat{F}_1, \widehat{F}_2, \widetilde{F}_3, \widetilde{F}_4\}$.

Lemma 5.8.4 *Let $\mathscr{G}(3 \cdot M(24)) = \mathscr{G}(\widehat{F}, \mathscr{A})$, $\mathscr{H}(3 \cdot M(24)) = \mathscr{G}(\widehat{F}, \mathscr{B})$. Then*

(i) *$\mathscr{G}(3 \cdot M(24))$ is a geometry with the diagram*

2 2 2 2

on which $\widehat{F}$ acts faithfully and flag-transitively;

(ii) *$\mathscr{H}(3 \cdot M(24))$ is a geometry with the diagram*

6 2 2 2

on which $\widehat{F}$ acts faithfully and flag-transitively.

Proof. Since $\mathscr{G}(M(24))$ and $\mathscr{H}(M(24))$ are geometries on which F acts flag-transitively, it is easy to check that the amalgams $\mathscr{A}$ and $\mathscr{B}$ satisfy the conditions in (1.4.1). Hence $\mathscr{G}(3 \cdot M(24))$ and $\mathscr{H}(3 \cdot M(24))$ are geometries; their diagrams follow from the paragraph before the lemma. □

It is obvious that the natural homomorphism $\varphi : \widehat{F} \to F$ induces 1-coverings of $\widetilde{\mathscr{G}} = \mathscr{G}(3 \cdot M(24))$ onto $\mathscr{G}(M(24))$ and of $\widetilde{\mathscr{H}} = \mathscr{H}(3 \cdot M(24))$ onto $\mathscr{H}(M(24))$. We will denote these 1-coverings by the same letter φ.

Let $\widetilde{\Gamma} = \Gamma(3 \cdot M(24))$ be a graph on the set of elements of type 4 in $\widetilde{\mathscr{H}}$ in which two vertices are adjacent if they are incident to a common element of type 3. Let $\{\widetilde{y}_i \mid 1 \le i \le 4\}$ be a maximal flag in $\widetilde{\mathscr{H}}$ such that $\varphi(\widetilde{y}_i) = y_i$. For $\widetilde{z} \in \widetilde{\mathscr{H}}$ let $\widetilde{\Gamma}[\widetilde{z}]$ denote the subgraph in $\widetilde{\Gamma}$ induced by the vertices incident to $\widetilde{z}$. Since the residue of $\widetilde{y}_4$ in $\widetilde{\mathscr{H}}$ and the residue of y_4 in $\mathscr{H}(M(24))$ are both isomorphic to $\mathscr{H}(Mat_{24})$, we observe that φ induces a covering of $\widetilde{\Gamma}$ onto $\Gamma(M(24))$. Furthermore, $\widetilde{\Gamma}[\widetilde{y}_1]$ is the collinearity graph of the geometry $\mathscr{G}(3 \cdot U_4(3))$ as in (4.14.6) with the suborbit diagram given in (4.14.10). The morphism φ induces also a 1-covering of the residue of $\widetilde{y}_1$ in $\widetilde{\mathscr{H}}$ isomorphic to $\mathscr{G}(3 \cdot U_4(3))$ onto the residue of y_1 in $\mathscr{H}(M(24))$ isomorphic to $\mathscr{G}(U_4(3))$. It is easy to check that the Schläfli graph does not possess connected triple covers which are vertex- and edge-transitive. In view of (4.14.9) this means that φ induces a covering of the geometry

$\mathcal{E}(3 \cdot U_4(3))$ associated with the residue of $\widetilde{y}_1$ in $\widetilde{\mathcal{H}}$ onto the geometry $\mathcal{E}(U_4(3))$ associated with the residue of y_1 in $\mathcal{H}(M(24))$. This gives the following.

Lemma 5.8.5 *The graph $\widetilde{\Gamma}$ contains a family $\widetilde{\mathcal{S}}$ of Schläfli subgraphs and under φ a member of $\widetilde{\mathcal{S}}$ maps isomorphically onto a subgraph from the family $\mathcal{S}$ as in (5.7.3).* □

For $1 \leq i \leq 4$ let $\widehat{Y}_i$ be the subgroup in $\widetilde{F}_4$ such that $\widehat{Y}_i X_s / X_s = Y_i$ (where the Y_i are as in the paragraph before (5.7.7)) and let $\widetilde{\Theta}_i$ be the connected component containing $\widetilde{y}_4$ of the subgraph in $\widetilde{\Gamma}$ induced by the vertices $\widetilde{u}$ such that $\widehat{Y}_i \leq O_2(\widehat{F}(\widetilde{u}))$. Then $\widetilde{\Theta}_i$ maps onto Θ_i and by (5.8.5) $\widetilde{\Theta}_4$ maps onto Θ_4 isomorphically. Let $\widetilde{\mathcal{T}} = \mathcal{T}(3 \cdot M(24))$ be an incidence system whose elements are the vertices of $\widetilde{\Gamma}$ and the images under $\widehat{F}$ of $\widetilde{\Gamma}[\widetilde{y}_3]$, $\widetilde{\Theta}_4$, $\widetilde{\Theta}_3$, $\widetilde{\Theta}_2$ and $\widetilde{\Theta}_1$ with respect to the incidence relation defined via inclusion. Then $\widetilde{\mathcal{T}}$ is a geometry with the diagram as in (5.7.9) and φ induces a covering of $\widetilde{\mathcal{T}}$ onto $\mathcal{T}(M(24))$. Now by (5.7.10) we obtain the following.

Proposition 5.8.6 *The geometry $\mathcal{T}(3 \cdot M(24))$ is the universal 2-cover of $\mathcal{T}(M(24))$;*

$$\widehat{F} \cong 3 \cdot M(24) \quad \text{and} \quad F \cong M(24).$$

By (5.8.1), (5.8.2) and (5.8.3) it is easy to deduce the shapes of the maximal parabolics associated with the action of F on $\mathcal{T}(M(24)$ (or equivalently of $\widehat{F}$ on $\mathcal{T}(3 \cdot M(24)))$:

$$F(s_1) \sim 2 \times M(23), \quad F(s_2) \sim (2 \times 2 \cdot M(22)).2,$$

$$F(s_3) \sim (2 \times 2^2 \cdot U_6(2)).Sym_3, \quad F(s_4) \sim 2^{1+12}.(3 \times U_4(2)).2^2,$$

$$F(s_5) = F_3 \sim 2^{7+8}.(L_4(2) \times Sym_3), \quad F(s_6) = F_4 \sim 2^{12}.Mat_{24}.$$

Proposition 5.8.7 *Let $\mathcal{F}$ be one of the following geometries: $\mathcal{H}(3 \cdot M(24))$, $\mathcal{H}(M(24))$, $\mathcal{G}(3 \cdot M(24))$, $\mathcal{G}(M(24))$ and $\mathcal{E}(M(24))$. Then $\mathcal{F}$ is simply connected.*

Proof. Let $\mathcal{F} = \mathcal{H}(M(24))$ and let $\psi : \widehat{\mathcal{F}} \to \mathcal{F}$ be the universal covering. Let $\widehat{\Gamma}$ be a graph on the set of elements of type 4 in $\widehat{\mathcal{F}}$ in which two elements are adjacent if they are incident to a common element of type 3. For $\widehat{z} \in \widehat{\mathcal{F}}$ let $\widehat{\Gamma}[\widehat{z}]$ be the subgraph in $\widehat{\Gamma}$ induced by the vertices incident to $\widehat{z}$ and let $\{\widehat{y}_i \mid 1 \leq i \leq 4\}$ be a maximal flag in $\widehat{\mathcal{F}}$ such that

$\psi(\widehat{y}_i) = y_i$. Then ψ induces a covering of $\widehat{\Gamma}$ onto $\Gamma = \Gamma(M(24))$ (denoted by the same letter ψ) and the restriction of ψ to $\widehat{\Gamma}[\widehat{y}_1]$ is an isomorphism onto $\Gamma[y_1]$. Since $\Gamma[y_1]$ contains a Schläfli subgraph Θ from the family $\mathcal{S}$ as in (5.7.3), there is a subgraph $\widehat{\Theta}$ in $\widehat{\Gamma}$ which maps isomorphically onto Θ. Let $\widehat{\mathcal{T}}$ be an incidence system whose elements are the vertices of $\widehat{\Gamma}$ and the connected components of the preimages of the subgraphs in Γ which realize elements of $\mathcal{T} = \mathcal{T}(M(24))$; the incidence relation is via inclusion. Then $\widehat{\Theta}$ is an element of type 4 in $\widehat{\mathcal{T}}$, which shows that $\widehat{\mathcal{T}}$ is a geometry and that ψ induces its covering onto $\mathcal{T}$. By (5.8.6) we have either $\widehat{\mathcal{T}} = \mathcal{T}$ or $\widehat{\mathcal{T}} = \widetilde{\mathcal{T}} = \mathcal{T}(3 \cdot M(24))$. In the latter case $\widehat{\Gamma}$ must be $\Gamma(3 \cdot M(24))$ but since $\widetilde{\Gamma}[\widetilde{y}_1]$ is a proper triple cover of $\Gamma[y_1]$ this is impossible. Hence ψ is an isomorphism. Almost the same argument shows that $\mathcal{H}(3 \cdot M(24))$ is simply connected.

By the above paragraph F is the universal completion of $\{F_i \mid 1 \leq i \leq 4\}$ and $\widehat{F}$ is the universal completion of $\{\widehat{F}_1, \widehat{F}_2, \widetilde{F}_3, \widetilde{F}_4\}$. We claim that F is also the universal completion of the amalgam $\{K_i \mid 1 \leq i \leq 4\}$. First of all $K_1 = F_1$ and $K_4 = F_4$. For $i = 2$ or 3 let $P_i = O_2(F_i)$. Then $P_i \leq K_i$ and it is easy to check that F_i is the unique completion of the amalgam $\{N_{F_1}(P_i), K_i, N_{F_4}(P_i)\}$. Hence a completion of $\{K_i \mid 1 \leq i \leq 4\}$ must also be a completion of $\{F_i \mid 1 \leq i \leq 4\}$ and the claim follows. In a similar way one can show that $\widehat{F}$ is also the universal completion of the amalgam $\{\widehat{K}_1, \widehat{K}_2, \widetilde{K}_3, \widetilde{K}_4\}$ and hence both $\mathcal{G}(M(24))$ and $\mathcal{G}(3 \cdot M(24))$ are simply connected. Finally the residue in $\mathcal{G}(M(24))$ of an element of type 4 is isomorphic to $\mathcal{G}(Mat_{24})$ and it is simply connected by (3.3.11). Hence K_4 is the universal completion of $\{K_4 \cap K_i \mid 1 \leq i \leq 3\}$ and F is also the universal completion of $\{K_1, K_2, K_3\}$. Since the latter amalgam is a subamalgam of the amalgam of maximal parabolics associated with the action of F on $\mathcal{E}(M(24))$ (the paragraph after (5.6.6)), this implies that $\mathcal{E}(M(24))$ is simply connected. □

As in the beginning of Section 5.6 let X_s be a Sylow 3-subgroup of $O_{2,3}(G_3)$. Since the subgraph $\Delta(X_s)$ is connected by the definition, (5.8.6) implies the following.

Lemma 5.8.8 *The subgroup of the group G generated by $N_{G_1}(X_s)$ and $N_{G_2}(X_s)$ is isomorphic to $3 \cdot M(24)$.* □

Let $\widehat{Y}_i$ be the preimage of Y_i in $\widetilde{F}_4$ as introduced after (5.8.5), $1 \leq i \leq 4$. Then $\widehat{F}(s_i)$ (which is the preimage of $F(s_i)$ in $\widehat{F}$) is (contained in) the normalizer of $\widehat{Y}_i$ in $\widehat{F}$. It is easy to see that $\widehat{Y}_1$ and $\widehat{Y}_2$ are conjugate in

G to subgroups Y_1 and Y_2 as in Section 5.4 and Section 5.5, respectively, and we have the following.

Lemma 5.8.9 *The Baby Monster group B contains the Fischer group $M(23)$ and $\bar{E} \cong {}^2E_6(2).Sym_3$ contains $M(22).2$.* □

5.9 Geometry of the Held group

In this section we study the normalizer in the group G of a subgroup of order 7. More specifically we analyse the subgroup in G generated by the normalizers in G_1 and G_2 of a subgroup S of order 7 from G_{12}. We start with the following.

Lemma 5.9.1 *Let S be a Sylow 7-subgroup in G_{12} and let Ξ be the subgraph of Δ induced by the vertices fixed by S. Then S is of order 7 and*

(i) *Ξ is of valency* 28,

(ii) *there is a subgroup $T \cong Frob_7^3$ in G_{12} containing S such that $N_{G_i}(S) = N_{G_i}(T)$ for $i = 1$ and 2,*

(iii) *$N_{G_1}(S)/T \cong 2_+^{1+6}.\mathrm{Aut}\, L_3(2)$, S is fully normalized in G_1 and $C_{G_1}(T) \cong 2_+^{1+6}.L_3(2)$,*

(iv) *S is not fully normalized in G_2 and $N_{G_2}(S)/T \cong C_{G_2}(T) \sim [2^8].Sym_3 \times Sym_3$.*

Proof. A vertex $u \in \Xi(v_0)$ is fixed by S if and only if $\eta(u)$ is fixed by the image $\bar{S}$ of S in $\bar{G}_1 \cong Co_1$ and hence (i) follows from (4.14.14 (i)). By (4.14.14 (iv)) $C_{Q_1}(S)$ is generated by the subgroups $\eta^{-1}(\bar{\mu})$ taken for all elements $\bar{\mu} \in \bar{\Lambda}_4$ fixed by $\bar{S}$. From (4.14.14 (v)) it is easy to deduce that $C_{Q_1}(S)$ is extraspecial of order 2^7 and of plus type. This implies in particular that Z_1 is the kernel of the action of $C_{Q_1}(S)$ on $\Xi(v_0)$. Let K be the kernel of the action of $N_{G_1}(S)$ on $\Xi(v_0)$ and put $T = O^2(K)$. Then $T \cong Frob_7^3$ by (4.14.14 (iii)) and $N_{G_1}(S) = N_{G_1}(T)$. Since T fixes l, it is contained in G_2^+ and by the Frattini argument it is centralized by a subgroup of order 3 from G_2^- and hence (ii) follows. Now (iii) is immediate from (4.14.14 (iii)) and (iv) follows from (2.13.5), (5.1.8) and (3.8.4). □

By (4.14.14 (iii)) and since $C_{Q_1}(S)$ induces a non-trivial action on Ξ, we conclude that $N_{G_1}(S)$ acts transitively on $\Xi(v_0)$ while $C_{G_1}(S)$ has two orbits of length 14 each. Let $H = \langle C_{G_1}(T), C_{G_2}(T)\rangle$ and Ξ^c be the subgraph in Δ induced by the images of v_0 under H. By the definition Ξ^c is connected and it is contained in Ξ. Since T is contained in G_1 and normalized by H, it fixes Ξ^c elementwise. Furthermore by (5.9.1) and the above discussions we obtain the following.

Lemma 5.9.2 *The following assertions hold:*

(i) Ξ^c *is of valency* 14 *and it contains the line* l *stabilized by* G_2*;*

(ii) H *acts on* Ξ^c *vertex- and edge-transitively;*

(iii) *there are seven lines incident to* v_0 *and contained in* Ξ^c*;* $C_{G_1}(T)$ *induces on the set of these lines the natural action of* $L_3(2)$. □

Let S_1 be a Sylow 7-subgroup of G_3. Then S_1 is contained in G_3^- and by (5.1.10) we have the following.

Lemma 5.9.3 *There is a unique subgroup* T_1 *in* G_3 *containing* S_1 *such that* $N_{G_3}(S_1) = N_{G_3}(T_1)$*;* S_1 *is not fully normalized in* G_3 *and* $C_{G_3}(T_1) \cong N_{G_3}(S_1)/T_1 \cong 2^6 : 3 \cdot Sym_6$ *where* $O_2(C_{G_3}(T_1))$ *is the hexacode module.* □

Since $S_1 \leq G_3^-$, it fixes $\text{res}^+_{\mathcal{G}}(\Sigma)$ elementwise (where Σ is the element of type 3 in $\mathcal{G} = \mathcal{G}(M)$ stabilized by G_3). Recall that the elements of type i in $\text{res}^+_{\mathcal{G}}(\Sigma)$ are the $\star$-closed cliques in Δ containing Σ of size $2^i - 1$ for $i = 4$ and 5, respectively. Clearly, in such a clique S_1 fixes a vertex if $i = 4$ and a line if $i = 5$. It is easy to see (the paragraph before (5.6.3)) that these vertices and lines form a subgeometry of $\mathcal{G}$ isomorphic to $\text{res}^+_{\mathcal{G}}(\Sigma)$ and we have the following.

Lemma 5.9.4 *Let* Ψ *be the set of vertices of* Δ *fixed by* S_1 *and contained in elements from* $\text{res}^+_{\mathcal{G}}(\Sigma)$. *Then the vertices of* Ψ *together with the lines contained in* Ψ *form the rank* 2 *tilde geometry* $\mathcal{G}(3 \cdot Sym_6)$ *on which* $C_{G_3}(T_1)$ *induces the full automorphism group with kernel* $O_2(C_{G_3}(T_1)) \cong 2^6$. □

By the above lemma S_1 fixes an incident vertex–line pair (v', l') and hence S_1 is a Sylow 7-subgroup of a G-conjugate of G_{12}. Thus there is an element $g \in G$ which conjugates S_1 onto S and maps (v', l') onto (v_0, l). Then Ψ^g is a subgraph in Ξ^c isomorphic to the point graph of $\mathcal{G}(3 \cdot Sym_6)$. Since T_1 fixes Ψ elementwise, it is easy to see that $T_1^g = T$. Hence Ψ^g is stabilized by $C_{G_3^g}(T) \cong 2^6 : 3 \cdot Sym_6$. By (5.9.2 (iii)) there is a structure π of a projective plane of order 2 on the set of lines containing v_0 and contained in Ξ^c which is preserved by $C_{G_1}(T)$. Since $C_{G_3^g}(T) \cap C_{G_1}(T)$ contains a Sylow 2-subgroup of $C_{G_1}(T)$ we see that the lines containing v_0 and contained in Ψ^g form a line in π.

Let $\mathcal{G}(He)$ be the incidence system of rank 3 whose elements of type 1, 2 and 3 are the images of Ψ^g under H, the lines contained in Ξ^c and the vertices of Ξ^c, the incidence relation being via inclusion. Then by (5.9.2 (iii)), (5.9.4) and the above paragraph we have the following.

Lemma 5.9.5 *$\mathcal{G}(He)$ is a rank 3 tilde geometry with the diagram*

$$\circ\overset{}{\underset{2}{}}\!\!-\!\!-\!\!\circ_{2}\overset{\sim}{=\!=}\circ_{2}$$

on which H induces a flag-transitive automorphism group. □

Notice that by (5.9.1 (iii)) an element from $N_{G_1}(T) \setminus C_{G_1}(T)$ induces an outer automorphism of H (this element is not an automorphism of $\mathcal{G}(He)$). Since it is well known and easy to check that Mat_{24} does not possesses outer automorphisms we conclude that $\mathcal{G}(He)$ is not isomorphic to $\mathcal{G}(Mat_{24})$. The flag-transitive rank 3 tilde geometries have been classified in [Hei91], giving the following.

Proposition 5.9.6 *The group H is the sporadic simple Held group He of order*

$$2^{10} \cdot 3^3 \cdot 5^2 \cdot 7^3 \cdot 13$$

and the geometry $\mathcal{G}(He)$ is simply connected. □

By (4.14.14 (vi)) the subgroup X_s as in Section 5.6 is conjugate to a Sylow 3-subgroup in T and we have the following.

Corollary 5.9.7 *The Held group He is a subgroup in the Fischer group $M(24)$.* □

5.10 The Baby Monster graph

We follow notation introduced in Section 5.4 as follows: Y_1 is a non-isotropic subgroup in the orthogonal complement of Z_3 in A; $\widetilde{B} = \langle C_{G_1}(Y_1), C_{G_2}(Y_1)\rangle$; $\Delta(Y_1)$ is the subgraph in Δ induced by the images of v_0 under $\widetilde{B}$ and $B = B/Y_1$ is the action induced by $\widetilde{B}$ on $\Delta(Y_1)$. Let Y_2 be a minus 2-space containing Y_1 and contained in the orthogonal complement of Z_3 in $A \cong 2^{10}$ as in Section 5.5. Put $\widehat{E} = \langle N_{G_1}(Y_2), N_{G_2}(Y_2)\rangle \cap \widetilde{B}$ and $E = \widehat{E}/Y_1$. Then $E \leq B$ and by (5.5.4) $E \cong 2^{\cdot 2}E_6(2).2$. The subgraph Ξ defined before (5.5.3) is the subgraph of $\Delta(Y_1)$ induced by the images of v_0 under E. By (5.5.4) Ξ is isomorphic to the point graph of the natural parabolic geometry of E with the suborbit diagram as given in Section 5.5. By (5.5.3) and (4.10.8) E_1 is the full stabilizer of $\Xi(v_0)$ in B_1 and hence E is the full stabilizer of Ξ in B.

Define the *Baby Monster graph* Ω to be the graph whose vertices are the images of Ξ under B, where two such distinct images Ξ_1 and Ξ_2 are adjacent if there is a Δ-vertex u in $\Xi_1 \cap \Xi_2$ such that Ξ_1 is the image of Ξ_2 under an element from $O_2(B(u))$.

Comparing the structures of $B_1 \cong 2_+^{1+22}.Co_2$ and $E_1 \cong 2^{2+20}.U_6(2).2$ we observe that

$$[O_2(B_1) : (O_2(B_1) \cap E_1)] = 2$$

and hence for every vertex $u \in \Xi$ there is a unique Ω-vertex adjacent to Ξ which is Ξ^g for $g \in O_2(B_1) \setminus E_1$. It is well known (and easy to deduce from the suborbit diagram of Ξ) that the action of E on the vertex set of Ξ is primitive, which gives the following.

Lemma 5.10.1 *The valency of Ω is 3 968 055 which is the number of vertices in Ξ and also the index of E_1 in E.* □

Let z denote Ξ as a vertex of Ω. By the paragraph before (5.10.1) we have a bijection

$$\xi : \Omega(z) \to \Xi$$

which commutes with the action of E_1 (here as usual $\Omega(z)$ denotes the set of neighbours of z in Ω). Let Ξ_2^6 denote the graph on Ξ in which $u, w \in \Xi$ are adjacent if $u \in \Xi_2^6(w)$. The suborbit diagram of Ξ_2^6, as given below, can be deduced from that of Ξ using some standard relations between the parameters of symmetric association schemes ([BI84] and [BCN89]) and these calculations were kindly performed for us by D.V. Pasechnik.

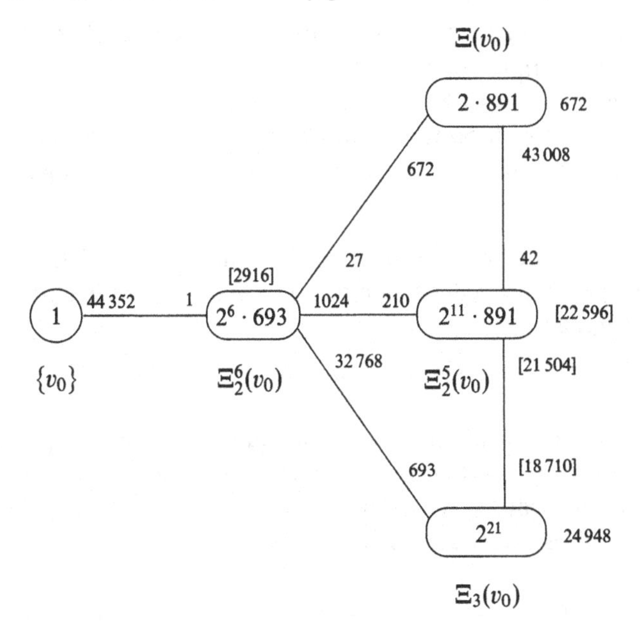

Lemma 5.10.2 *Let a, b be vertices from $\Omega(z)$ such that $\xi(a)$ and $\xi(b)$ are adjacent either in Ξ or in Ξ_2^6. Then a and b are adjacent in Ω.*

Proof. Let Υ be the subgraph of Ξ defined in (5.5.1). Then the stabilizer of Υ in E is of the form $2^{9+16}.\Omega_8^-(2).2$. By the definition Υ is induced by the vertices $y \in \Delta$ such that Z_y is in the orthogonal complement of Y_2 in A. If $y \in \Upsilon$ then since D contains $O_2(G(y))$, D also contains an element which maps z onto $\xi^{-1}(y)$. This shows that $\Upsilon^e = \{z\} \cup \xi^{-1}(\Upsilon)$ is the orbit of z under $B_o := C_D(Y_1)/Y_1 \cong 2^{9+16}.Sp_8(2)$. By the above the action of B_o on Υ^e is similar to the action of $Sp_8(2)$ on the set of cosets of $\Omega_8^-(2).2$. The latter action is doubly transitive (of degree 120) and in view of the paragraph after the proof of (5.5.4) we obtain the result. □

Let $K_z = Y_2/Y_1$. Then K_z is the centre (of order 2) of E and it coincides with the kernel of the action of E on Ξ (equivalently on $\Omega(z)$). For an arbitrary vertex $d \in \Omega$ put $K_d = Z(B(d))$, so that K_d is the kernel of the action of $B(d)$ on $\Omega(d)$.

Lemma 5.10.3 *Let $e \in \Omega(z)$. Then*

(i) $[K_e, K_z] = 1$,
(ii) *if* $u = \xi(e)$ *then* $K_e Z_u = K_z Z_u$,
(iii) *in terms of* (5.10.2) *if* $L = \langle K_v \mid v \in \Upsilon^e \rangle$ *then* $L = Z(O_2(B_o)) \cong 2^9$.

Proof. Since K_z fixes $\Omega(z)$ elementwise, it is contained in $B(e)$ and, since $K_e = Z(B(e))$, (i) follows. The definition of the adjacency relation in Ω implies that $K_e = Y_2^q/Y_1$ for an element $q \in O_2(\widetilde{B}(u)) \leq Q_u$. Since Q_u is extraspecial with centre Z_u, this gives (ii). In order to see (iii) it is sufficient to prove that L is isomorphic to the orthogonal complement of Y_1 in A. □

We need some further properties of the action of $\bar{E} \cong {}^2E_6(2).Sym_3$ on Ξ (the suborbit diagram in Section 5.5). Let $\mathcal{G}(U_6(2))$ be the C_3-geometry with the diagram

$$\underset{2}{\circ} = \underset{4}{\circ} - \underset{4}{\circ}$$

on which $\bar{E}(v_0)/O_2(\bar{E}(v_0))$ induces the full automorphism group. If $u \in \Xi$ then $\bar{E}(u) \cong 2_+^{1+20}.U_6(2).Sym_3$ and the centre of $\bar{E}(u)$ contains a unique non-identity element (a root element) which we denote by $\gamma(u)$. The following result is rather standard.

Lemma 5.10.4 *Let X be an orbit of $\bar{E}(v_0)$ on $\Xi \setminus \{v_0\}$ and $u \in X$. Then* *for X being, respectively, $\Xi(v_0)$, $\Xi_2^6(v_0)$, $\Xi_2^5(v_0)$ and $\Xi_3(v_0)$ the following assertions hold:*

(i) *the product $\gamma(v_0) \cdot \gamma(u)$ has order 2, 2, 4 and 3;*
(ii) *$O_2(\bar{E}(v_0))$ acts on X with orbits of length 2, 2^6, 2^{11} and 2^{21};*
(iii) *$\bar{E}(v_0)/O_2(\bar{E}(v_0))$ acts on the set of orbits of $O_2(\bar{E}(v_0))$ on X as it acts on the set of points, planes, points in $\mathcal{G}(U_6(2))$ and on a 1-element set.* □

We will need some information about involutions in $\bar{E}$.

Lemma 5.10.5 *Let τ be an involution in $\bar{E} \cong {}^2E_6(2).Sym_3$, such that $\tau \cdot \gamma(u)$ is of order 2 or 4 for every $u \in \Xi$. Then*

(i) *$\tau \in \bar{E} \setminus \bar{B}'$ and the conjugacy class of $\bar{E}$ containing τ is uniquely determined,*
(ii) *$C := C_{\bar{E}}(\tau) \cong F_4(2) \times 2$,*
(iii) *C has two orbits $\Xi^{(2)}$ and $\Xi^{(4)}$ on Ξ with lengths 69 615 and 3 898 440 and stabilizers of the shape $[2^{16}].Sp_6(2)$ and $[2^{15}].U_4(2).2$, respectively,*
(iv) *if $u \in \Xi^{(i)}$ for $i = 2$ or 4, then $\tau \cdot \gamma(u)$ is of order i,*
(v) *the subgraph of Ξ induced by $\Xi^{(4)}$ is connected and it is at distance 1 from $\Xi^{(2)}$.*

Proof. Using the computer package [GAP] it is straightforward to calculate the structure constants p_{ij}^k of the group association scheme of $\bar{E}$. The structure constant has the following meaning. If C_k, C_i and C_j are conjugacy classes of $\bar{E}$ then p_{ij}^k is the number of ways in which a given element $a \in C_k$ can be represented as a product $b \cdot c$ where $b \in C_i$ and $c \in C_j$. These calculations give everything except (v) and the structure of the stabilizers. The remaining information is easy to deduce, comparing the suborbit diagrams of the collinearity graphs of $\mathcal{G}({}^2E_6(2))$ and $\mathcal{G}(F_4(2))$ given in Section 5.5. □

The quotient E/K_z is a subgroup in $\bar{E}$ and for $u \in \Xi$ the subgroup in $\bar{E}$ generated by $\gamma(u)$ is the image of $\langle K_z, K_{\xi^{-1}(u)} \rangle$.

Lemma 5.10.6

(i) *Two distinct vertices $a, b \in \Omega(z)$ are adjacent in Ω if and only if $\xi(b)$ and $\xi(a)$ are adjacent either in Ξ or in Ξ_2^6, (equivalently if $\gamma(\xi(a))$ and $\gamma(\xi(b))$ commute);*

(ii) *if* $\pi = (d, e, f)$ *is a 2-path in* Ω *such that* d *and* f *are not adjacent then the B-orbit containing* π *is uniquely determined by the isomorphism type of* $\langle K_d, K_e, K_f \rangle / K_e$, *which is either* $D_6 \cong Sym_3$ *or* D_8;

(iii) $E = B(z)$ *has two orbits,* $\Omega_2^3(z)$ *and* $\Omega_2^4(z)$, *on the set* $\Omega_2(z)$ *of vertices at distance* 2 *from* z, *if* $d \in \Omega_2^i(z)$ *and* $e \in \Omega(z) \cap \Omega(d)$ *then* $\langle K_d, K_z, K_e \rangle / K_e \cong D_{2i}$ *for* $i = 3$ *and* 4;

(iv) *if* $y \in \Omega_2(z)$ *then* $B(z, y)$ *acts transitively on* $\Omega(z) \cap \Omega(y)$.

Proof. Part (i) follows from (5.10.2), (5.10.3 (i)) and (5.10.4). Notice that $\xi^{-1}(\Xi_3(v_0)) \subseteq \Omega_2^3(\xi^{-1}(v_0))$ and $\xi^{-1}(\Xi_2^5(v_0)) \subseteq \Omega_2^4(\xi^{-1}(v_0))$. Since $E(v_0)$ acts transitively on $\Xi_3(v_0)$ and on $\Xi_2^5(v_0)$ we obtain (ii), (iii) and (iv). □

As a corollary of (5.10.6 (i)) and the proof of (5.10.2) we have the following.

Lemma 5.10.7 *Let* $u \in \Xi_2^6(v_0)$. *Then* $\xi^{-1}(u)$ *is contained in a unique* 120-*vertex complete subgraph* Υ^e *which contains* z *and* $\xi^{-1}(v_0)$. *The setwise stabilizer of* Υ^e *in* B *is the stabilizer of a* $\mathcal{G}(Sp_8(2))$-*subgeometry in* $\mathcal{G}(BM)$ *as in* (5.4.5). □

Let $\bar{\Xi}$ be the graph on the vertex set of Ξ in which u and v are adjacent if $u \in \Xi(v) \cup \Xi_2^6(v)$, so that the edge set of $\bar{\Xi}$ is the union of the edge sets of Ξ and Ξ_2^6. Then by (5.10.6 (i)) ξ establishes an isomorphism of $\Omega(z)$ onto $\bar{\Xi}$.

The group $\bar{E}$ contains $M(22).2$ as a subgroup (5.8.9). The following result has been established in [Seg91] (see also Section 5 in [ISa96] and Section 8 in [Iv92c]).

Lemma 5.10.8 *A subgroup in* $\bar{E}$ *isomorphic to* $M(22).2$ *acting on* Ξ *has four orbits* Σ_i, $i = 1, 2, 3, 4$, *with lengths*

$$3510, \quad 142\,155, \quad 694\,980, \quad 3\,127\,410$$

and stabilizers isomorphic, respectively, to

$$2 \cdot U_6(2).2, \quad 2^{10}.\mathrm{Aut}\, Mat_{22}, \quad 2^7.Sp_6(2), \quad 2.(2^9.P\Sigma L_3(4)).$$

Furthermore, if $w \in \Sigma_i$ *for* $i = 2$, 3 *and* 4, *then* w *is adjacent in* $\bar{\Xi}$ *to* 22, 126 *and* $1 + 21$ *vertices from* Σ_1, *respectively.* □

Lemma 5.10.9 *Let* $u \in \Xi_3(v_0)$. *Then the subgraph in* $\bar{\Xi}$ *induced by* $\bar{\Xi}(v_0) \cap \bar{\Xi}(u)$ *is isomorphic to the transposition graph of* $U_6(2)$ *with the suborbit diagram given after* (5.8.1).

Proof. Notice that the action of $\bar{E}$ on the vertex set of $\bar{\Xi}$ is equivalent to its action by conjugation on the set $\{\gamma(w) \mid w \in \bar{\Xi}\}$. Let $H = \bar{E}(v_0) \cap \bar{E}(u)$. Then by (5.10.4 (ii)) H is a (Levi) complement to $O_2(\bar{E}(v_0))$ in $\bar{E}(v_0)$, in particular $H \cong U_6(2).Sym_3$ and by the above remark $H = C_{\bar{E}}(\gamma(v_0)) \cap C_{\bar{E}}(\gamma(u))$. One can readily see from the suborbit diagrams of Ξ and Ξ_2^6 that $\Pi := \bar{\Xi}(v_0) \cap \bar{\Xi}(u) = \Xi_2^6(v_0) \cap \Xi_2^6(u)$ is of size 693. On the other hand by the remark at the beginning of the proof, the set Π is a union of some conjugacy classes of H. Since $|\Pi|$ is odd, at least one of the conjugacy classes must contain central involutions. By [ASei76] the only class of central involutions in H is the class of 3-transpositions of size 693, contained in the simple subgroup $U_6(2)$. Since $w, v \in \bar{\Xi}$ are adjacent if and only if $\gamma(w)$ and $\gamma(v)$ commute, the subgraph induced by Π is exactly the transposition graph of $U_6(2)$. □

Notice that the elements of the C_3-geometry $\mathscr{G}(U_6(2))$ can be realized by the maximal cliques (of size 21), by the 5-vertex cliques contained in more than one maximal clique and by the vertices of the transposition graph of $U_6(2)$.

Lemma 5.10.10 *Let $a, b \in \Omega$, $b \in \Omega_2^3(a)$ and Ψ be the connected component of the subgraph induced by $\Omega(a) \cap \Omega(b)$. Then*

(i) *Ψ is locally the transposition graph of $U_6(2)$,*
(ii) *Ψ is isomorphic to the transposition graph of $M(22)$,*
(iii) *$\Psi = \Omega(a) \cap \Omega(b)$ and $B(a,b) \cong M(22).2$,*
(iv) *every vertex adjacent to a is at distance at most 2 from b,*
(v) *if (d,e,f,h) is a 3-path in Ω such that $\langle K_d, K_e, K_f\rangle / K_e \cong Sym_3$ then the distance from d to h is at most 2.*

Proof. First assume that $z \in \Omega(a) \cap \Omega(b)$ and that $\xi(a) = v_0$. Then by the proof of (5.10.6) $\xi(b) \in \Xi_3(v_0)$ and (i) follows from (5.10.9) and the fact that ξ induces an isomorphism of the subgraph induced by $\Omega(z)$ onto $\bar{\Xi}$ (5.10.6 (i)). By (5.10.6 (iv)) the stabilizer of Ψ in $B(a,b)$ acts transitively on the vertex set of Ψ. Let $\mathscr{Y}$ be a geometry formed by the maximal cliques (of size 22) in Ψ, by the cliques of size 6 contained in more than one maximal clique, by the edges and the vertices of Ψ. Then by the proof of (5.10.9) $B(a,z,b)$ induces on $\Psi(z)$ an action isomorphic to $U_6(2).2$. Hence $\mathscr{Y}$ is flag-transitive and by (5.8.2) we obtain (ii). From (ii) we deduce that $|\Psi| = 3\,510$ and that the stabilizer of Ψ in $B(a,b)$ is isomorphic to $M(22).2$. Now assume that $a = z$. Then by (5.10.8) the image of Ψ in $\bar{\Xi}$ is the orbit Σ_1 and by the last sentence of (5.10.8) every vertex in $\Omega(z)$ is adjacent to a vertex from Ψ, which gives (iii). Since every vertex from $\Omega(a)$ is at distance at most 1 from $\Psi = \Omega(a) \cap \Omega(b)$, it is at distance at most 2 from b and (iv) follows. Finally (v) is a direct consequence of (iv) and (5.10.6 (iii)). □

If $y \in \Omega_2^3(z)$ then by the above lemma there is more than one common neighbour of y and z and in view of (5.10.6) this means that $\langle K_y, K_z \rangle \cong Sym_3$.

Below, Π will denote the graph on 2300 vertices introduced before (4.11.9) on which Co_2 induces a rank 3 action with the suborbit diagram given in (4.11.9).

Let $\Omega(Z_1)$ be the subgraph in Ω induced by the images of z under $B_1 \cong 2_+^{1+22}.Co_2$. If $y \in \Omega(Z_1)$ then the preimage of K_y in $\widetilde{B}$ is a conjugate Y_2^b of Y_2 under an element $b \in \widetilde{B}$. Clearly $Y_2^b \leq Q_1$ and $\eta(Y_2^b)$ is a singular triangle in the shortest vector graph containing $\eta(Y_1)$ (this triangle corresponds to a vertex in the graph Π). On the other hand $Y_2 Z_1$ contains besides Y_2 exactly one image Y_2^b of Y_2 under $b \in \widetilde{B}_1$ (in fact b can be taken from $O_2(\widehat{B}_1)$), so that $Y_2^b/Y_1 = K_{\xi^{-1}(v_0)}$. Hence every vertex of Π corresponds to an edge in $\Omega(Z_1)$ and we obtain the following.

Lemma 5.10.11 *The following assertions hold:*

(i) *the subgraph $\Omega(Z_1)$ contains 4600 vertices;*
(ii) *every orbit of $O_2(B_1)$ on $\Omega(Z_1)$ is of length 2;*
(iii) *there is a mapping δ from the set of $O_2(B_1)$-orbits on $\Omega(Z_1)$ onto the vertex set of Π which commutes with the action of B_1.* □

The subgroups K_y taken for all $y \in \Omega(Z_1)$ generate $O_2(B_1)$, and since the latter is extraspecial of type 2_+^{1+22} and has different kernels at different orbits on $\Omega(Z_1)$, we have the following.

Lemma 5.10.12 *The following assertions hold:*

(i) *the group B_1 acts on $\Omega(Z_1)$ with suborbits 1, 1, $2 \cdot 891$ and $2 \cdot 1408$;*
(ii) *for any two vertices d, f contained in $\Omega(Z_1)$, the subgroup $\langle K_d, K_f \rangle$ is either abelian (of order 2^2) or dihedral of order 8 with centre Z_1.* □

Lemma 5.10.13 *The following assertions hold:*

(i) *$\Omega(z) \cap \Omega(Z_1) = \xi^{-1}(\{v_0\} \cup \Xi(v_0))$;*
(ii) *$u, v \in \Omega(Z_1)$ are adjacent in Ω if and only if the distance in Π between $\delta(u)$ and $\delta(v)$ is 0 or 1;*
(iii) *if $u \in \Omega(Z_1)$ is not adjacent to z then $u \in \Omega_2^4(z)$.*

Proof. Consider the intersection Φ of $\Omega(Z_1)$ and the subgraph Υ^e as in (5.10.7). By (5.10.7) $B(z, \xi^{-1}(v_0))$ acts transitively on the set of images of such a subgraph under B which contain $\varepsilon := \{z, \xi^{-1}(v_0)\}$. Hence this intersection consists of the images of the edge ε under $I = B_1 \cap B[\Upsilon^e]$. By (5.4.5) the group I is the full preimage in B_1 of the stabilizer in $B_1/O_2(B_1) \cong Co_2$ of a $\mathcal{G}(Sp_6(2))$-subgeometry in $\mathcal{G}(Co_2)$ as in (4.9.8). From standard properties of $\mathcal{G}(Sp_6(2))$ it is easy to deduce that ε has 28 images under I and I induces on the set of these images the doubly transitive action of $Sp_6(2)$ on the cosets of $\Omega_6^-(2).2$ and that $\xi(\Phi \setminus \{z\}) \subseteq \{v_0\} \cup \Xi(v_0)$. Since Υ^e is a complete subgraph, in view of (5.10.12 (i)) we obtain (i) and (ii), and the latter implies that the diameter of $\Omega(Z_1)$ is 2. Now (iii) follows from (5.10.6) and (5.10.12 (i)). □

By (5.10.6 (ii)) and (5.10.13 (ii)) every vertex $y \in \Omega_2^4(z)$ is contained in an image of $\Omega(Z_1)$ under an element from $B(z)$.

Lemma 5.10.14 *Let $y \in \Omega(Z_1) \cap \Omega_2^4(z)$. Then*

(i) $B(z, y) \leq B_1(z) = B(z, \xi^{-1}(v_0))$,
(ii) $B(z, y) \cong 2_+^{1+20}.U_4(3).2^2$,
(iii) $\Omega(z) \cap \Omega(y) \subseteq \Omega(Z_1)$ *and* $|\Omega(z) \cap \Omega(y)| = 648$.

Proof. It is clear that $B(z) \cap B_1 = B(z, \xi^{-1}(v_0))$ is the centralizer of Z_1 in $B(z)$. On the other hand by (5.10.12) Z_1 is the centre of $\langle K_z, K_y \rangle$ which implies (i). By (5.10.11 (ii)) and (5.10.13 (i)) $B(z, y) \cap O_2(B_1)$ has index 2 in $O_2(B_1)$ and hence (ii) follows from the suborbit diagram of Π. Finally (iii) is a direct consequence of (i), (ii), (5.10.13 (i)) and the suborbit diagram of Π. □

The following information can be deduced either from the construction of the GAB $\mathcal{G}(U_4(3))$ in [Kan81] or by means of calculations in the Leech lattice. Recall that in $\mathcal{G}(U_6(2))$ a point is incident to 21 planes and a plane is incident to 27 points.

Lemma 5.10.15 *Let $V \cong U_4(3).2^2$ be a subgroup in* $\mathrm{Aut}\, \mathcal{G}(U_6(2)) \cong U_6(2).Sym_3$. *Then*

(i) *V has 2 orbits on the set of* 891 *points of $\mathcal{G}(U_6(2))$ with lengths* 324 *and* 567 *and stabilizers $P\Sigma L_3(4)$ and $2^5.Sym_6$,*
(ii) *V has 2 orbits on the set of* 693 *planes of $\mathcal{G}(U_6(2))$ with lengths* 126 *and* 567 *and stabilizers $U_4(2).2 \times 2$ and $2^5.Sym_6$,*
(iii) *the point–plane incidence graph has the following diagram with respect to the orbits of V:*

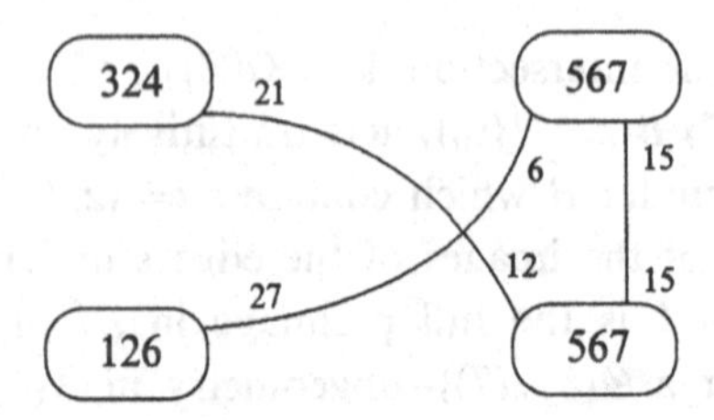

□

Notice that the subgraph in the point–plane incidence graph of $\mathcal{G}(U_6(2))$ induced by the union of the two 567-orbits is the point–plane incidence graph of $\mathcal{G}(U_4(3))$ while the subgraph induced by the 567-orbit on points and the 126-orbit on planes is the point–plane incidence graph of $\mathcal{E}(U_4(3))$.

Lemma 5.10.16 *Let $y \in \Omega_2^4(z) \cap \Omega(Z_1)$. Then $B(z,y) \cong 2_+^{1+20}.U_4(3).2^2$ acting on $\Omega(z)$ has eight orbits $\mathcal{O}_i$, $1 \le i \le 8$, with lengths 1, $2 \cdot 324$, $2 \cdot 567$, $2^6 \cdot 126$, $2^6 \cdot 567$, $2^{11} \cdot 324$, $2^{11} \cdot 567$ and 2^{21} so that $\xi^{-1}(\{v_0\}) = \mathcal{O}_1$, $\xi^{-1}(\Xi(v_0)) = \mathcal{O}_2 \cup \mathcal{O}_3$, $\xi^{-1}(\Xi_2^6(v_0)) = \mathcal{O}_4 \cup \mathcal{O}_5$, $\xi^{-1}(\Xi_2^5(v_0)) = \mathcal{O}_6 \cup \mathcal{O}_7$, $\xi^{-1}(\Xi_3(v_0)) = \mathcal{O}_8$ and $\mathcal{O}_2 = \Omega(z) \cap \Omega(y)$.*

Proof. Since $B(z,y)$ does not contain K_z (which is the kernel of the action of $B(z)$ on $\Omega(z)$), we conclude that $O_2(B(z,y))$ and $O_2(B(z,\xi^{-1}(v_0)))$ have the same orbits on $\Omega(z)$. Hence the result follows from (5.10.4 (ii), (iii)) and (5.10.15). □

Lemma 5.10.17 *In terms of* (5.10.16) *let $u_j \in \mathcal{O}_j$ for $1 \le j \le 8$. Then*

(i) *if $j = 1, 2, 3, 5, 6$ or 7, then u_j is at distance at most* 1 *from $\mathcal{O}_2$,*
(ii) *there is a vertex $x \in \mathcal{O}_2$ such that $\langle K_x, K_{u_8} \rangle \cong Sym_3$,*
(iii) *unless $j = 4$ the orbit $\mathcal{O}_j$ is at distance at most* 2 *from y.*

Proof. For $j = 1$ and 2 the assertion (i) is obvious. For $j = 3$ (i) follows from the fact that the subgraph in Ξ induced by $\Xi(v_0)$ is connected, while for $j = 5$ it follows from the diagram given in (5.10.15). Let $a \in \Xi(v_0)$, $b \in \Xi_2^5(v_0)$ and let $\bar{a}$ and $\bar{b}$ be the orbits under $O_2(\bar{E}(v_0))$ of a and b, respectively. Then in view of (5.10.4 (iii)) $\bar{a}$ and $\bar{b}$ are points of $\mathcal{G}(U_6(2))$. One can see from the suborbit diagram of Ξ that whenever a and b are adjacent in this graph, one has $\bar{a} = \bar{b}$ which implies (i) for $j = 6$. Similarly one can see from the suborbit diagram of Ξ_2^6 that whenever a and b are adjacent in Ξ_2^6 the vertices $\bar{a}$ and $\bar{b}$ are adjacent in the collinearity graph of $\mathcal{G}(U_6(2))$. Since $\bar{E}_1(\bar{b}) = O_2(\bar{E}_1)\bar{E}_1(b)$ and the collinearity graph of $\mathcal{G}(U_6(2))$ is connected, we deduce (i) for $j = 7$. Let $c \in \Xi_3(v_0)$. Then by the suborbit diagram of Ξ we have $|\Xi(v_0) \cap \Xi_3(c)| = 891$ which is the number of orbits of $O_2(\bar{E}(v_0))$ on $\Xi(v_0)$. Since $\bar{E}(v_0, c)$ is a complement to

$O_2(\bar{E}(v_0))$ in $\bar{E}(v_0)$, we conclude that $\mathcal{O}_2 \cap \xi^{-1}(\Xi_3(c))$ is non-empty, which implies (ii) in view of (5.10.4 (i)). Finally, since $\mathcal{O}_2 = \Omega(z) \cap \Omega(y)$, (iii) follows from (i), (ii) and (5.10.10 (v)). □

Lemma 5.10.18 *Let* $u \in \Xi_2^6(v_0)$, $b = \xi^{-1}(u)$, $\Psi = \Omega(b) \cap \Omega(Z_1)$, Υ^e *be as in* (5.10.7) *and* $N = \langle K_w \mid w \in \Psi \rangle$. *Then*

(i) $|\Psi| = 56$ *and* $\Psi = \Upsilon^e \cap \Omega(Z_1)$,

(ii) *N is elementary abelian of order 2^7 and it contains Z_1,*

(iii) *the orbit of b under $O_2(B_1)$ has length 2^6.*

Proof. Since Υ^e is a complete subgraph, b is adjacent to each of the 56 vertices in $\Upsilon^e \setminus \xi^{-1}(\Xi_2^6(v_0))$. By (5.10.13 (i)) and (5.10.14 (iii)) Ψ is contained in $\{z\} \cup \xi^{-1}(\{v_0\} \cup \Xi(v_0))$. By the suborbit diagrams of Ξ and Ξ_2^6 we see that in $\bar{\Xi}$ u is adjacent to exactly 54 vertices from $\Xi(v_0)$ and hence (i) follows. By (i) the subgroup N is the orthogonal complement in A of $\langle Z_1, K_b \rangle$ which gives (ii). By (5.10.4 (ii)) the orbit of b under $B(z, \xi^{-1}(v_0))$ is of length 2^6. From (5.10.7) and the suborbit diagram of Ξ we observe that there are 693 images of Υ^e under B which contain $\{z, \xi^{-1}(v_0)\}$ and these images are transitively permuted by $B[z, \xi^{-1}(v_0)]$. Since the latter group is contained in B_1 and contains $O_2(B_1)$, we conclude that $O_2(B_1)$ stabilizes Υ^e and we obtain (iii). □

In terms of the above lemma put $\widetilde{S} = B_1 \cap B[\Upsilon^e]$. Then $\widetilde{S}$ is the full preimage in B_1 of the stabilizer $S \cong 2_+^{1+8}.Sp_6(2)$ of a $\mathcal{G}(Sp_6(2))$-subgeometry from $\mathcal{G}(Co_2))$. Let R denote the action induced by $\widetilde{S}$ on $\Upsilon^e \setminus \Omega(Z_1)$. Then by (5.10.18 (iii)) we have $R \cong 2^6.Sp_6(2)$. Since $O_2(R)$ is the natural symplectic module for $R/O_2(R) \cong Sp_6(2)$, the latter group permutes transitively the 63 non-identity elements in $O_2(R)$. Let ψ denote the natural homomorphism of $O_2(B_1)$ onto $O_2(R)$. By the above observation if $q \in O_2(B_1)$ and $\psi(q) \neq 1$ then the length of the orbit of q under $\widetilde{S}$ is divisible by 63. By (4.12.6 (i)) $\widetilde{S}$ has 3 orbits Σ, Σ_1 and Σ_2 on the vertex set of Π with lengths 28, 2016 and 256, respectively, and the vertices in Σ_2 are at distance 2 from Σ in Π. Let $\widetilde{\Sigma}$, $\widetilde{\Sigma}_1$ and $\widetilde{\Sigma}_2$ be the preimages of these orbits under δ (5.10.11). Then $\widetilde{\Sigma} = \Upsilon^e \cap \Omega(Z_1)$ and by (5.10.13 (ii), (iii)) the vertices from $\widetilde{\Sigma}_2$ are at distance 2 from $\widetilde{\Sigma}$. The latter implies that $\widetilde{\Sigma}_2 \subseteq \Omega_2^4(z)$. Furthermore since 256 is not divisible by 63 we have $\psi(K_y) = 1$ for $y \in \widetilde{\Sigma}_2$. This means that for every $b \in \Upsilon^e \setminus \Omega(Z_1)$ the subgroups K_y and K_b commute. In terms of (5.10.16) $b \notin \mathcal{O}_2$ and hence b and y are at distance at least 2. On the other hand b is adjacent to z and z is at distance 2 from y, hence the distance between b and y in Ω is at most 3. Since $[K_y, K_b] = 1$ the distance must be 3 by (5.10.6 (iii)) and by (5.10.17 (iii)) $b \in \mathcal{O}_4$.

Lemma 5.10.19 *The group $B(z)$ acts transitively on the set $\Omega_3(z)$ of vertices at distance 3 from z. Furthermore*

(i) *a vertex $y \in \Omega_2^4(z)$ is adjacent to exactly $8064 = 2^6 \cdot 126$ vertices in $\Omega_3(z)$ transitively permuted by $B(z, y)$,*
(ii) *if $u \in \Omega_3(z)$ then $[K_z, K_u] = 1$,*
(iii) *if $v \in \Omega(Z_1) \cap \Omega_2^4(z)$ and $u \in \Omega_3(z) \cap \Omega(v)$, then*

$$B(z) \cap B(u) \cap B_1 \sim [2^{17}].Sp_6(2).$$

Proof. Parts (i) and (ii) follow from the paragraph before the lemma. In terms of that paragraph in order to prove (iii) we have to analyse the structure of $B_1 \cap B(b) \cap B(y)$. We know that the latter is contained in $\widetilde{S} \sim 2_+^{1+22}.2_+^{1+8}.Sp_6(2)$. By (5.10.18 (iii)) we have $[O_2(B_1) : O_2(B_1) \cap B(b)] = 2^6$ and it is easy to see that $\widetilde{\Sigma}$ is the set of vertices in $\Omega(Z_1)$ fixed by $O_2(B_1) \cap B(b)$. In view of (4.12.6 (ii)) this means that $B(b) \cap O_2(\widetilde{S})$ induces on $\widetilde{\Sigma}_2$ a regular action of degree 2^9 and (iii) follows. □

Lemma 5.10.20 *The diameter of Ω is* 3.

Proof. We claim that whenever $(x_0, x_1, x_2 = z, x_3, x_4)$ is a 4-path in Ω, the distance between x_0 and x_4 is at most 3. Clearly we can assume that $x_0, x_4 \in \Omega_2^3(z) \cup \Omega_2^4(z)$. If $x_0 \in \Omega_2^3(z)$ then $z \in \Omega_2^3(x_0)$ and by (5.10.10 (iv)) x_3 is at distance at most 2 from x_0 and the claim follows. Hence we assume that $x_0, x_4 \in \Omega_2^4(z)$ and also that $\xi(x_1) = v_0$. Then by (5.10.17 (iii)) unless $\Delta := \Omega(x_4) \cap \Omega(z)$ is contained in $\mathcal{O}_4$, there is a vertex in Δ which is at distance at most 2 from x_0. We show that the inclusion $\Delta \subseteq \mathcal{O}_4$ is not possible. Since $x_4 \in \Omega_2^4(z)$, by (5.10.13 (i)) and (5.10.14 (iii)) Δ is of size 648 and contained in $\xi^{-1}(\Xi(w))$ for a vertex $w \in \Xi$. We are going to show that for every $w \in \Xi$ the intersection $\Xi(w) \cap \xi(\mathcal{O}_4)$ is of size less than 648. One can see directly from the suborbit diagram of Ξ given in Section 5.5 that $\Xi(w) \cap \Xi_2^6(v_0)$ is of size less than 648 unless $w \in \Xi(v_0)$ (recall that by (5.10.16) $\xi(\mathcal{O}_4) \subset \Xi_2^6(v_0)$). If $w \in \Xi(v_0)$ then by (5.10.15 (iii)) and the diagram therein we observe that $w \in \Xi(v_0)$ is adjacent in Ξ to 0 or $192 = 6 \cdot 32$ vertices in $\xi(\mathcal{O}_4)$. This completes the proof of the claim and also of the lemma. □

Lemma 5.10.21 *Let $u \in \Omega_3(z)$ and τ be the involution generating K_u. Then*

(i) *u is adjacent to 69 615 vertices in $\Omega_3(z)$ and to 3 898 440 vertices in $\Omega_2^4(z)$,*
(ii) *$B(u) \cap B(z) = C_{B(z)}(\tau) \cong F_4(2) \times 2^2$.*

Proof. By (5.10.19 (ii)) $\tau \in B(z)$. By (5.10.19), (5.10.20) and the obvious duality we have

$$\Omega(z) \subseteq \Omega_3(u) \cup \Omega_2^4(u)$$

and hence (compare (5.10.6 (iii)) if γ is the generator of K_v for $v \in \Omega(z)$, then the product $\tau \cdot \gamma$ is of order 2 if $v \in \Omega_3(u)$ and of order 4 if $v \in \Omega_2^4(u)$. By (5.10.5 (ii)) we have $C_{B(z)}(\tau) \cong F_4(2) \times 2^2$ and from (5.10.5 (iii), (iv)) we deduce the sizes of $\Omega(z) \cap \Omega_2^4(u)$ and $\Omega(z) \cap \Omega^3(u)$, which gives (i). Now straightforward calculations give the equality

$$|\Omega_3(z)| = [B(z) : C_{B(z)}(\tau)] = 23\,113\,728.$$

Since $B(u) \cap B(z)$ is obviously contained in $C_{B(z)}(\tau)$ the equality proves (ii). □

By (5.10.6 (i)), (5.10.10), (5.10.14), (5.10.19), (5.10.20) and (5.10.21) we obtain the main result of the section.

Proposition 5.10.22 *The group B is the Baby Monster sporadic simple group BM of order*

$$2^{41} \cdot 3^{13} \cdot 5^6 \cdot 7^2 \cdot 11 \cdot 13 \cdot 17 \cdot 19 \cdot 23 \cdot 31 \cdot 47$$

and the suborbit diagram of Ω with respect to the action of B is as given below, in particular Ω has 13 571 955 000 vertices.

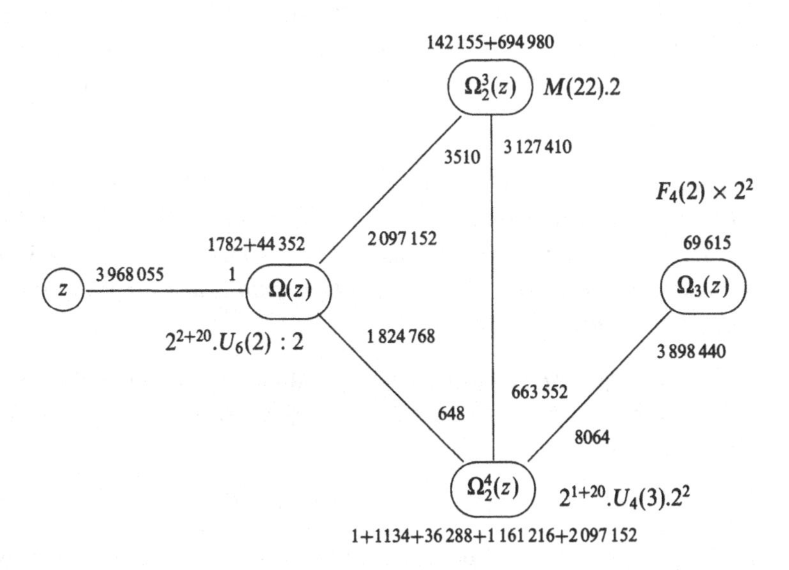

□

5.11 The simple connectedness of $\mathcal{G}(BM)$

The simple connectedness of $\mathcal{G}(BM)$ can be proved following the strategy developed in Section 4.12 using the properties of $\mathcal{G}(BM)$ and its subgeometries established in Sections 5.4, 5.5 and 5.10. In this section we outline the proof.

Let $\mathcal{F} = \mathcal{G}(BM)$ be the P-geometry of the Baby Monster, $\Phi = \{x_1, ..., x_5\}$ be a maximal flag in $\mathcal{F}$ and $\mathcal{B} = \{B_i \mid 1 \leq i \leq 5\}$ be the amalgam of maximal parabolics associated with the action of $B \cong BM$ on $\mathcal{F}$, where $B_i = B(x_i)$ ((5.4.4) and the paragraph after its proof), and let Θ be the collinearity graph of $\mathcal{F}$. Let $\varphi : \widehat{\mathcal{F}} \to \mathcal{F}$ be the universal covering, $\widehat{B}$ be the universal completion of the amalgam $\mathcal{B}$ which by (1.5.2) is a flag-transitive automorphism group of $\widehat{\mathcal{F}}$. Let $\widehat{\Phi} = \{\widehat{x}_1, ..., \widehat{x}_5\}$ be a maximal flag in $\widehat{\mathcal{F}}$ such that $\varphi(\widehat{\Phi}) = \Phi$ and let $\widehat{\mathcal{B}} = \{\widehat{B}_i \mid 1 \leq i \leq 5\}$ be the amalgam of maximal parabolics associated with the action of the group $\widehat{B}$ on $\widehat{\mathcal{F}}$. Then φ induces a homomorphism $\psi : \widehat{B} \to B$ whose restriction to $\widehat{\mathcal{B}}$ is an isomorphism onto $\mathcal{B}$, and also a covering $\chi : \widehat{\Theta} \to \Theta$ of graphs where $\widehat{\Theta}$ is the collinearity graph of $\widehat{\mathcal{F}}$.

Let $\mathcal{S}$ be the $\mathcal{G}(Sp_8(2))$-subgeometry in $\mathcal{F}$ containing x_4, $S \sim 2^{9+16}.Sp_8(2)$ be the stabilizer of $\mathcal{S}$ in B and Σ be the subgraph in Θ induced by the images of x_1 under S. Then $\mathcal{C} = \{S \cap B_i \mid 1 \leq i \leq 4\}$ is the amalgam of maximal parabolics associated with the action of S on $\mathcal{S}$ (the kernel of the action is $O_2(S)$) and Σ is isomorphic to the collinearity graph of $\mathcal{S}$. By (1.6.4) $\mathcal{S}$ is simply connected and hence S is the universal (and in fact the only) completion of $\mathcal{C}$, which gives the following.

Lemma 5.11.1 *Let $\widehat{\mathcal{C}}$ be the subamalgam in $\widehat{\mathcal{B}}$ which maps isomorphically onto the subamalgam $\mathcal{C}$ in $\mathcal{B}$, let $\widehat{\mathcal{S}}$ be the subgroup in $\widehat{B}$ generated by $\widehat{\mathcal{C}}$ and let $\widehat{\Sigma}$ be the subgraph in $\widehat{\Theta}$ induced by the images of $\widehat{x}_1$ under $\widehat{S}$. Then the restriction of ψ to $\widehat{S}$ is an isomorphism onto S and the restriction of χ to $\widehat{\Sigma}$ is an isomorphism onto Σ.* □

Considering $Z(O_2(S))$ as the orthogonal module for $S/O_2(S) \cong Sp_8(2)$, let σ be a non-isotropic involution in $Z(O_2(S))$ which is perpendicular to $Z(O_2(B_4))$. Put

$$\mathcal{D} = \{C_{B_1}(\sigma), C_{B_2}(\sigma), C_{B_3}(\sigma), C_S(\sigma)\},$$

let E be the subgroup in B generated by $\mathcal{D}$ and let Ξ be the subgraph in Θ induced by the images of x_1 under E. Then by (5.5.4) $E \cong 2 \cdot {}^2E_6(2).2$,

$\mathscr{D}$ is the amalgam of maximal parabolics associated with the natural action of E on its Tits geometry $\mathscr{E} = \mathcal{G}(^2E_6(2))$ and Ξ is the collinearity graph of $\mathscr{E}$ with the suborbit diagram given after the proof of (5.5.4). By (1.6.4) $\mathscr{E}$ is simply connected and we have the following.

Lemma 5.11.2 *Let $\widehat{\mathscr{D}}$ be the subamalgam in $\widehat{\mathscr{B}} \cup \widehat{S}$ which maps isomorphically onto the subamalgam $\mathscr{D}$ in $\mathscr{B} \cup S$, let $\widehat{E}$ be the subgroup in $\widehat{B}$ generated by $\widehat{\mathscr{D}}$ and $\widehat{\Xi}$ be the subgraph in $\widehat{\Theta}$ induced by the images of $\widehat{x}_1$ under $\widehat{E}$. Then the restriction of ψ to $\widehat{E}$ is an isomorphism onto E and the restriction of χ to $\widehat{\Xi}$ is an isomorphism onto Ξ.* □

Let $\widehat{\Omega}$ be the graph whose vertices are the images of $\widehat{\Xi}$ under $\widehat{B}$ in which two such distinct images $\widehat{\Xi}_1$ and $\widehat{\Xi}_2$ are adjacent if there is a $\widehat{\Theta}$-vertex $\widehat{u}$ in their intersection such that $\widehat{\Xi}_2$ is the image of $\widehat{\Xi}_1$ under an element from $O_2(\widehat{B}(\widehat{u}))$. Then since the restriction of ψ to $\widehat{B}_1$ is an isomorphism onto B_1 and by (5.11.2) we obtain the following.

Lemma 5.11.3 *The covering φ induces a covering $\eta : \widehat{\Omega} \to \Omega$ of graphs.* □

Arguing as in the proof of (5.10.2) one can easily show that the images of $\widehat{\Xi}$ (considered as a vertex of $\widehat{\Omega}$) under $\widehat{S}$ induce in $\widehat{\Omega}$ a complete subgraph on 120 vertices and by (5.10.6 (i)) we have the following.

Lemma 5.11.4 *Every triangle in Ω is contractible with respect to the covering $\eta : \widehat{\Omega} \to \Omega$.* □

Proposition 5.11.5 *The following assertions hold:*

(i) *the Baby Monster graph is triangulable;*
(ii) *the geometry $\mathcal{G}(BM)$ is simply connected.*

Proof. It is easy to check the conditions in (1.14.1) using (5.10.10), (5.10.17) and (5.10.21) in view of (5.10.5 (v)). By (i) and (5.11.4) η is an isomorphism which forces φ to be an isomorphism and proves (ii). □

In the remainder of the section we study the action on the Baby Monster graph Ω of the subgroup $B_1 \cong 2^{1+22}.Co_2$ of the Baby Monster.

One can see immediately from the suborbit diagram of the Baby Monster graph that whenever u and v are distinct vertices of Ω, $K_u \neq K_v$. This enables us to identify $u \in \Omega$ with the non-identity element in K_u. In this case $B(u) = C_B(u)$ and $u, v \in \Omega$ are adjacent if and only if the product uv is a conjugate of the involution in the centre of B_1.

Lemma 5.11.6 *The subgroup $B_1 \cong 2_+^{1+22}.Co_2$ acting on the vertex set of the Baby Monster graph Ω has five orbits denoted by*

$$\Omega(2a),\ \ \Omega(2d),\ \ \Omega(4a),\ \ \Omega(4d),\ \ \Omega(6a)$$

with lengths

$$2300 \cdot 2,\ \ 56\,925 \cdot 2^6,\ \ 46\,575 \cdot 2^{12},\ \ 56\,925 \cdot 2^{16},\ \ 2300 \cdot 2^{22}$$

and stabilizers of shape

$$2^{2+20}.U_6(2).2,\ \ [2^{26}].Sp_6(2),\ \ [2^{22}].\mathrm{Aut}\,Mat_{22},\ \ [2^{16}].Sp_6(2),\ \ 2 \cdot U_6(2).2,$$

respectively. Here when writing $n \cdot 2^m$ for the length of an orbit $\Omega(i\alpha)$ we mean that $Q_1 = O_2(B_1)$ acting on $\Omega(i\alpha)$ has n orbits of length 2^m each.

Proof. If we put $\Omega(2a) = \Omega(Z_1)$, then by (5.10.11) the properties of $\Omega(2a)$ are as stated in the lemma. Consider the B_1-orbits at distance 1 from $\Omega(2a)$. If $u = \xi^{-1}(v_0)$ then $\{z, u\}$ is the orbit of z under Q_1 and since $B_1 = C_B(zu)$, we have $B(z,u) = B(z) \cap B_1$ and hence by (5.10.4) the orbits of $B(z) \cap B_1$ on $\Omega(z) \setminus \Omega(2a)$ are the $\xi^{-1}(X)$ for $X = \Xi_2^6(v_0)$, $\Xi_2^5(v_0)$ and $\Xi_3(v_0)$. Let $\Omega(2d)$, $\Omega(4d)$ and $\Omega(6a)$ denote the orbits of B_1 containing $\xi^{-1}(X)$ for X as above. By (5.10.2) if $w \in \xi^{-1}(X)$, then w is adjacent to u only if $X = \Xi_2^6(v_0)$. Hence by (5.10.4 (ii)) and (5.10.18 (iii)) the orbits of Q_1 on $\Omega(2d)$, $\Omega(4d)$ and $\Omega(6a)$ are of length 2^6, 2^{12} and 2^{22}, respectively. In particular these 3 orbits are different and we have the following property. Whenever w is at distance 1 from $\Omega(2a)$, the subgroup $B_1(w)$ acts transitively on the set $\Pi(w) := \Omega(2a) \cap \Omega(w)$. By (5.10.18), if $w \in \xi^{-1}(\Xi_2^6(v_0))$ then $\Pi(w)$ has size 56, consists of 28 Q_1-orbits and $B_1(w)$ induces on the set of these orbits the doubly transitive action of $Sp_6(2)$ on the cosets of $U_4(2).2$. If $w \in \xi^{-1}(\Xi_2^5(v_0))$ then one can see from the suborbit diagrams of Ξ and Ξ_2^6 that $\Pi(w)$ is of size 44 and $B_1(w)$ preserves on this set an imprimitivity system with classes of size 2. Using (5.10.4 (iii)) it is not difficult to see that $B_1(w)$ induces on the set of imprimitivity blocks the natural 3-fold transitive action of $\mathrm{Aut}\,Mat_{22}$. Finally, by the suborbit diagrams of Ξ and Ξ_2^6 we observe that z is the only vertex in $\Omega(2a)$ adjacent to a vertex $w \in \xi^{-1}(\Xi_3(v_0))$, which completes the description of the B_1-orbits at distance 1 from $\Omega(2a)$ and leaves us with $56\,925 \cdot 2^{16}$ vertices whose distance from $\Omega(2a)$ is at least 2. We are going to show that the remaining vertices form a single B_1-orbit.

By (5.10.21 (i)) a vertex $w \in \Omega_3(z)$ is adjacent to $3\,898\,440$ vertices in $\Omega_2^4(z)$ while every vertex from $\Omega_2^4(z)$ is contained in a unique image $\Omega(Z_1')$ of $\Omega(Z_1) = \Omega(2a)$ under an element from $B(z)$. By the paragraph before

(5.10.19), if w is at distance 1 from $\Omega(Z_1')$, then it is adjacent to exactly 56 vertices in $\Omega(Z_1')$. Hence w is at distance 1 from $69\,615 = 3\,898\,440/56$ images of $\Omega(2a)$ under $B(z)$. Since there is a natural bijection between the images of $\Omega(2a)$ under $B(z)$ and the edges incident to z, one can see from (5.10.21 (i)) that w is at distance more than 1 from $\Omega(2a)$ if and only if $u \in \Omega_2^4(w)$. This in particular shows that B_1 acts transitively on the set of pairs (z, w) where $z \in \Omega(2a)$, w is at distance 3 from z and at distance more than 1 from $\Omega(2a)$. Moreover by (5.10.5 (iii)) the stabilizer of such a pair is of the form $[2^{16}].U_4(2).2$ (recall that $B(z, w) \cong F_4(2) \times 2^2$). Hence in order to calculate the length of the orbit of w under B_1 it is sufficient to calculate the number of vertices in $\Omega(2a)$ at distance 3 from w. Comparing the suborbit diagrams of Ξ and its subgraph induced by $\xi(\Omega(z) \cap \Omega_3(w))$ given in Section 5.5, we easily calculate that the set Σ of vertices in $\xi(\Omega(z) \cap \Omega_3(w))$ adjacent to $\xi(u)$ in Ξ is of size 27. A more detailed analysis shows that $\xi^{-1}(\Sigma) \cup \{z\}$ contains all the vertices from $\Omega(2a)$ which are at distance 3 from w. Alternatively one can calculate the structure constants of the group association scheme of BM as in Section 3 in [ISh93a]. It can be checked that $B_1(w)$ induces on $\xi^{-1}(\Sigma) \cup \{z\}$ the doubly transitive action of $Sp_6(2)$ on the cosets of $U_4(2).2$. Since the length of the B_1-orbit containing w turns out to be exactly $56\,925 \cdot 2^{16}$ this completes the proof. □

The notation in the above lemma has the following interpretation. If $\tau = zu$ is the involution in the centre of B_1 and $v \in \Omega(i\alpha)$ then the product τv is contained in the conjugacy class $i\alpha$ of BM as in [CCNPW]. The fact that the product is of order i is easily seen from the above proof.

We sketch the proof of the following result (see [ISh93a] for the details).

Lemma 5.11.7 *In terms of* (5.11.6) *let* $u \in \Omega(i\alpha)$ *and let* $B(u)' \cong 2 \cdot {}^2E_6(2)$ *be the commutator subgroup of* $B(u)$. *Then* $Q_1(u)$ *is contained in* $B(u)'$ *if and only if* $i\alpha = 2d$ *or* $4d$.

Proof. It is not difficult to see that $B_1(u)$ is not contained in $B(u)'$. Then the result follows from the following facts. If $i\alpha = 2d$ or $4d$ then the quotient $B_1(u)/Q_1(u)$ does not contain subgroups of index 2 while if $i\alpha = 2a$, $4a$ or $6a$ then the quotient $B_1(u)/O_2(B_1(u))$ is not a section in $B(u)'$. □

5.12 The second Monster graph

Let us turn back to the group G which is a faithful completion of a Monster amalgam, its tilde $\mathcal{G}(M)$ and maximal parabolic $\mathcal{H}(M)$ geometries and the first Monster graph Δ which is the collinearity graph of both $\mathcal{G}(M)$ and $\mathcal{H}(M)$. Recall that Θ is the subgraph of Δ induced by the images of v_0 under $\widetilde{B}$. By (5.10.22) $\widetilde{B}$ is an extension of the Baby Monster sporadic simple group BM by a group of order 2. As in Section 5.5 let Ξ be the subgraph of Δ induced by the images of v_0 under $\widetilde{E} \cong 2^{2\cdot 2}E_6(2).Sym_3$. Notice that Ξ is contained in Θ and that the intersection of $\widetilde{E}$ with $\widetilde{B}$ is of index 3 in $\widetilde{E}$. Let $\Gamma = \Gamma(G)$ be the *second Monster graph* defined as follows. The vertices of Γ are the images of Θ under the elements of G; two such images are adjacent if their intersection is an image of Ξ. By the above there are exactly three images of Θ under $\widetilde{E}$, which gives the following.

Lemma 5.12.1 *The valency of the second Monster graph Γ is twice the number of vertices in the Baby Monster graph Ω.* □

Let t denote Θ considered as a vertex of Γ. The triple $T = \{t = t_1, t_2, t_3\}$ of images of t under $\widetilde{E}$ forms a triangle. This triangle and its images under G will be called *lines*. Thus there is a natural bijection between the lines containing t and the vertices of the Baby Monster graph. The group G acts naturally on Γ with $\widetilde{B}$ being the stabilizer of t. For a vertex x of Γ put $L_x = Z(G(x))$ (a subgroup of order 2) and for a subset X of vertices in Γ put $L_X = \langle L_x \mid x \in X\rangle$. Then in terms of Section 5.4 $L_t = Y_1$ and $L_T = Y_2$ (elementary abelian of order 2^2). This gives the following.

Lemma 5.12.2 *Let $\{x, y\}$ be an edge of Γ. Then*

(i) *there is a unique line $X = \{x, y, z\}$ containing $\{x, y\}$,*
(ii) *L_x, L_y and L_z are the subgroups of order 2 of the elementary abelian subgroup L_X of order 2^2,*
(iii) *there is a mapping $\sigma : \Gamma(t) \to \Omega$ which commutes with the action of $\widetilde{B}$ and for $u \in \Omega$ the set $\sigma^{-1}(u) \cup \{t\}$ is a line,*
(iv) *L_t fixes $\Gamma(t)$ elementwise.* □

Let $x, y \in \Gamma(t)$, X and Y be the lines containing $\{t, x\}$ and $\{t, y\}$, respectively, and suppose that $X \neq Y$. Let $u = \sigma(x)$ and $v = \sigma(y)$. Since $K_z = Y_2/Y_1$ (see the paragraph before (5.10.3)) we have $K_u = L_X/L_t$ and $K_v = L_Y/L_t$, which shows that x and y could be adjacent only if either $v \in \Omega(u)$ or $v \in \Omega_3(u)$. We will see below that in the latter case x and y are *not* adjacent.

Let $D \sim 2^{10+16}.\Omega_{10}^{+}(2)$ be as in Section 5.3 and let Φ be the subgraph of Γ induced by the images of t under D. Then by (5.4.1) and the definition of Y_2 we observe that Φ is a graph on the set of 496 non-isotropic vectors in $A = Z(O_2(D))$ (which is the natural module for $D/O_2(D)$) in which 2 vectors are adjacent if they are not perpendicular. Using some standard properties of strongly regular graphs associated with classical groups [BvL84] or by means of straightforward calculations one can see the following.

Lemma 5.12.3 *The following assertions hold:*

(i) *Φ is a strongly regular graph with the intersection diagram*

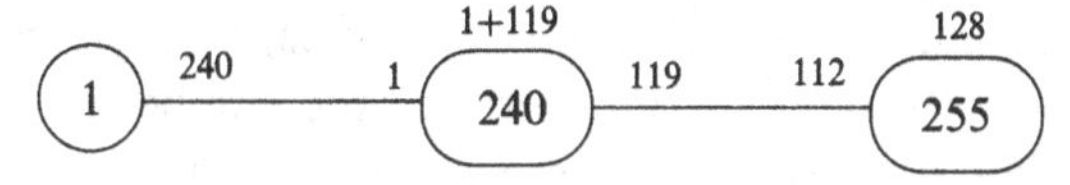

(ii) *D induces on Φ a rank 3 action of $\Omega_{10}^{+}(2)$ on the cosets of $Sp_8(2)$;*

(iii) *the subgraph induced by $\Phi(t)$ is a double antipodal cover of the complete graph on 120 vertices with the suborbit diagram*

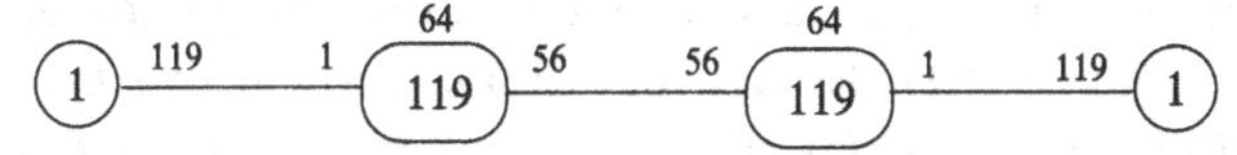

together with a matching which joins the antipodal vertices;

(iv) *$a, b \in \Phi(t)$ are antipodal vertices of the graph in (iii) if and only if $t = a + b$, equivalently if $\{t, a, b\}$ is a line in Γ.* □

Define $\widetilde{\Omega}$ to be a graph on $\Gamma(t)$ in which two vertices x and y are adjacent if they are adjacent in Γ and $\sigma(y) \in \Omega(\sigma(x))$ (notice that the latter inclusion implies that $\sigma(x) \neq \sigma(y)$).

Lemma 5.12.4 *The mapping σ induces a covering of graphs $\widetilde{\Omega} \to \Omega$ (which we denote by the same letter σ). If $X = \{z, u, v\}$ is a triangle in Ω where $u = \xi^{-1}(v_0)$, then X is contractible with respect to σ if and only if $v \in \xi^{-1}(\Xi_2^6(v_0))$.*

Proof. It follows basically by the definition that $\sigma(\Phi(t))$ is the complete 120-vertex subgraph Υ^e in the Baby Monster graph as in the proof of (5.10.2). Since $\Upsilon^e \cap \xi^{-1}(\Xi(v_0))$ and $\Upsilon^e \cap \xi^{-1}(\Xi_2^6(v_0))$ are of size 56 and 64, respectively, the claim follows directly from the intersection diagram in (5.12.3 (iii)). □

Next we are intesested in the suborbit diagram of $\widetilde{\Omega}$ with respect to the action of $G(t) \sim 2 \cdot BM$. Let $\{\widetilde{z}, \widetilde{z}'\} = \sigma^{-1}(z)$ so that $\widehat{E} := G(t, \widetilde{z}) \cong 2^{2.2}E_6(2)$ is a subgroup of index 6 in $\widetilde{E}$. The action of $G(t)$ on $\Gamma(t)$ is equivalent to its action by conjugation on the set $\{L_y \mid y \in \Gamma(t)\}$.

Alternatively this action is similar to that of $B = G(t)/L_t$ on the cosets of $B(z)' = \widehat{E}/L_t \cong 2^{\cdot 2}E_6(2)$. Let $\Omega_\alpha(z)$ be an orbit of $B(z)$ on $\Omega \setminus \{z\}$ and $\widetilde{\Omega}_\alpha(z)$ be the preimage of $\Omega_\alpha(z)$ in $\widetilde{\Omega}$.

Lemma 5.12.5 *If $\Omega_\alpha(z) = \Omega(z)$ or $\Omega_2^3(z)$ then $\widetilde{\Omega}_\alpha(z)$ consists of two $\widehat{E}$-orbits; if $\Omega_\alpha(z) = \Omega_2^4(z)$ or $\Omega_3(z)$ then $\widehat{E}$ acts transitively on $\widetilde{\Omega}_\alpha(z)$.*

Proof. Let $y \in \Omega_\alpha(z)$ and $\{\widetilde{y}, \widetilde{y}'\} = \sigma^{-1}(y)$. First of all it is easy to see from the suborbit diagram of the Baby Monster graph in (5.10.22) that in no cases is $B(z, y)$ contained in the commutator subgroup $B(z)'$ of $B(z)$ which means that $\widehat{E}$ acts transitively on $\Omega_\alpha(z)$ and hence there are at most two $\widehat{E}$-orbits on $\widetilde{\Omega}_\alpha(z)$. Furthermore the number of such orbits is 1 or 2 depending on whether the index n_{zy} of $B(z)' \cap B(y)'$ in $B(z, y)$ is 2 or 4, respectively. Again one can see from the structure of stabilizers given on the suborbit diagram of the Baby Monster graph in (5.10.22) that if $y \in \Omega(z)$ or $y \in \Omega_2^3(z)$ then $B(z, y) \cong 2^{2+20}.U_6(2).2$ or $B(z, y) \cong M(22).2$, respectively, and $n_{zy} = 2$ since there are no subgroups of index 4 in $B(z, y)$. Alternatively the fact that $\widetilde{\Omega}(z)$ consists of two $\widehat{E}$-orbits follows from (5.12.4). If $y \in \Omega_3(z)$, then $B(z, y) \cong K_z \times K_y \times F$ where K_y is not in $B(z)'$, K_z is not in $B(y)'$ and $F \cong F_4(2)$ (5.10.5 (i)) which shows that $n_{zy} = 4$. Similarly one can show that $n_{zy} = 4$ if $y \in \Omega_2^4(z)$ ((5.12.8) below). □

In view of (5.12.5) and (5.10.22) in order to complete the suborbit diagram of $\widetilde{\Omega}$ with respect to the action of $B(t) \cong 2 \cdot BM$ it remains to determine the number of vertices in $\widetilde{\Omega}_2^3(\widetilde{z}')$ adjacent to a given vertex from $\widetilde{\Omega}_2^3(\widetilde{z})$. In terms of (5.10.8) in the graph $\bar{\Xi}$ a vertex from Σ_2 is adjacent to 891 vertices from Σ_1 and a vertex from Σ_3 is adjacent to 24 948 such vertices. Comparing these numbers with the suborbit diagrams of Ξ and Ξ_2^6 we conclude that in $\widetilde{\Omega}$ a vertex from $\widetilde{\Omega}_2^3(\widetilde{z})$ is adjacent to 694980 vertices from $\widetilde{\Omega}_2^3(\widetilde{z})$ and to 142 155 vertices from $\widetilde{\Omega}_2^3(\widetilde{z}')$. Thus we obtain the diagram as given below.

Lemma 5.12.6 *If $y \in \Omega_3(z)$ then a vertex $\widetilde{u} \in \{\widetilde{z}, \widetilde{z}'\}$ and a vertex $\widetilde{v} \in \{\widetilde{y}, \widetilde{y}'\}$ are not adjacent in Γ.*

Proof. Let $\widehat{F}$ be the stabilizer of y in $\widehat{E}$ so that $\widehat{F} \cong L_T \times F$ where $T = \{t, \widetilde{z}, \widetilde{z}'\}$ and $F \cong F_4(2)$. By (5.12.5) $\widehat{F}$ acts transitively on $\{\widetilde{y}, \widetilde{y}'\}$ and hence an element from $\widehat{F}$ conjugates $L_{\widetilde{y}}$ onto $L_{\widetilde{y}'}$. Since F is non-abelian simple and L_t commutes with $L_{\widetilde{v}}$, we conclude that $L_{\widetilde{u}}$ performes the conjugation which implies that $\langle L_{\widetilde{u}}, L_{\widetilde{v}} \rangle \cong D_8$ and the result follows from (5.12.2 (ii)). □

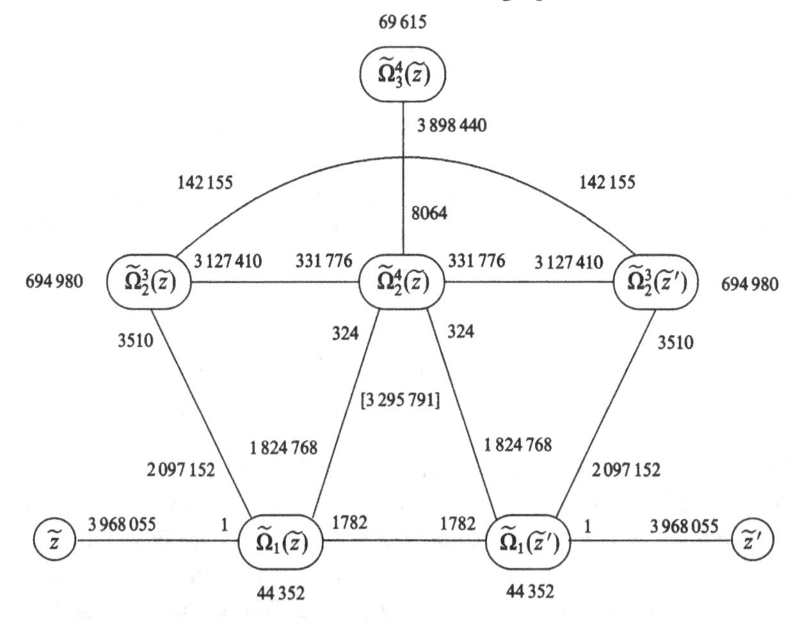

By the paragraph after (5.12.2) the above lemma shows that two vertices from $\Gamma(t)$ are adjacent in $\widetilde{\Omega}$ if and only if they are adjacent in Γ and not on a line containing t.

For $i\alpha = 2a$, $2b$, $3a$, $4a$, $4b$, $6a$ let $\Gamma^{i\alpha}(\widetilde{z})$ denote the set of images under $G(\widetilde{z})$ of the vertices from $\widetilde{\Omega}(\widetilde{z}) \cup \{\widetilde{z}\}$, $\widetilde{\Omega}(\widetilde{z}')$, $\widetilde{\Omega}_2^3(\widetilde{z})$, $\widetilde{\Omega}_2^4(\widetilde{z})$, $\widetilde{\Omega}_3(\widetilde{z})$, $\widetilde{\Omega}_2^3(\widetilde{z}')$, respectively. For $g \in G$ let $\Gamma^{i\alpha}(\widetilde{z}_1^g)$ denote the image of $\Gamma^{i\alpha}(\widetilde{z}_1)$ under g. Notice that in this case $\Gamma^{2a}(t) = \Gamma(t)$ (the meaning of the notation will be explained after (5.14.1)).

Lemma 5.12.7 *Let $u \in \Gamma^{i\alpha}(t)$, $N = \langle L_t, L_u \rangle$ and $v \in \Gamma(t) \cap \Gamma(u)$. Then*

(i) *if $i\alpha = 2a$ then $N \cong 2^2$ and all subgroups of order 2 in N are conjugate to L_t,*
(ii) *if $i\alpha = 2b$ then $N \cong 2^2$ and N contains a conjugate of Z_1,*
(iii) *if $i\alpha = 3a$ then $N \cong D_6$,*
(iv) *if $i\alpha = 4a$ then $N \cong D_8$ and $Z(N)$ is a conjugate of Z_1,*
(v) *if $i\alpha = 4b$ then $N \cong D_8$ and $Z(N) = L_v$, in particular v is uniquely determined,*
(vi) *if $i\alpha = 6a$ then $N \cong D_{12}$ and $Z(N) = L_v$, in particular v is uniquely determined.*

In particular $\Gamma^{2b}(t)$, $\Gamma^{3a}(t)$, $\Gamma^{4a}(t)$, $\Gamma^{4b}(t)$ and $\Gamma^{6a}(t)$ are the orbits of $G(t)$ on $\Gamma_2(t)$.

Proof. Since $L_{\tilde{z}}L_t = L_{\tilde{z}'}L_t$, and in view of (5.12.2), (5.12.5) the result follows from (5.10.3), (5.10.6) and (5.10.19). □

Let Σ be the subgraph of Γ induced by the images of t under G_1. Then the mapping $u \mapsto L_u$ establishes a bijection between the vertex set of Σ and the set of subgroups in Q_1 conjugate to Y_1 in G_1. One may observe that G_1 acting on Σ preserves an equivalence relation with classes of size 2 with respect to which $u, v \in \Sigma$ are equivalent if and only if

$$L_u Z_1 = L_v Z_1.$$

The set of equivalence classes is in a correspondence with the set of vectors in $\bar{\Lambda}_2$ and using (5.12.7) we obtain the following.

Lemma 5.12.8 *Let Σ be the subgraph of Γ induced by the images of t under G_1. Then there is a mapping $\varphi : \Sigma \to \bar{\Lambda}_2$ which commutes with the action of G_1, the fibres of φ are of size 2 and the quotient of Σ over these fibres is the shortest vector graph Δ as in Section 4.11. Furthermore, if $\bar{\mu}_0 = \varphi(t)$ and $\{t, t'\} = \varphi^{-1}(\bar{\mu}_0)$, then $\{t'\} \cup \varphi^{-1}(\Delta^4(\bar{\mu}_0)) \subseteq \Gamma^{2b}(t)$, $\varphi^{-1}(\Delta^3(\bar{\mu}_0)) \subseteq \Gamma^{4a}(t)$; if $s \in \{t, t'\}$ or $s \in \varphi^{-1}(\Delta^3(\bar{\mu}_0))$ then $G(t, s) = G(t) \cap G_1 \cong 2^{2+22}.Co_2$ or $2^{1+22}.McL$, respectively.* □

Let $u \in \Gamma^{3a}$ and $X_s = O_3(\langle L_t, L_u \rangle)$. Then by the paragraph before (5.8.9) one can see that X_s is as in (5.8.8). Let $\widehat{F} = \langle N_{G_1}(X_s), N_{G_2}(X_s) \rangle \cong 3 \cdot M(24)$. Then using the results established in Section 5.8 together with (5.12.7) it is not difficult to deduce the following.

Lemma 5.12.9 *Let Π be the subgraph of Γ induced by the images of t under $\widehat{F} \cong 3 \cdot M(24)$. Then Π is the antipodal triple cover of the transposition graph of $M(24)$. Furthermore, $\Pi_2(t) \cup \Pi_4(t) \subseteq \Gamma^{3a}(t)$ and $\Pi_3(t) \subseteq \Gamma^{6a}(t)$. If $s \in \Pi_3(t)$ or $s \in \Pi_4(t)$ then $G(t, s) = G(t) \cap \widehat{F} \cong 2 \cdot M(22)$ or $M(23)$, respectively.* □

Now by (5.12.7), (5.12.8) and (5.12.9) we obtain the following.

Lemma 5.12.10 *Let $u \in \Gamma^{i\alpha}(t)$. Then*

(i) *if $i\alpha = 2a$ then u is adjacent to t and $G(t, u) \cong 2^{2 \cdot 2}E_6(2)$,*
(ii) *if $i\alpha = 2b$ then $|\Gamma(u) \cap \Gamma(t)| = 9200$ and $G(t, u) \cong 2^{2+22}.Co_2$,*
(iii) *if $i\alpha = 3a$ then $|\Gamma(u) \cap \Gamma(t)| = 31\,671$ and $G(t, u) \cong M(23)$,*
(iv) *if $i\alpha = 4a$ then $|\Gamma(u) \cap \Gamma(t)| = 550$ and $G(t, u) \cong 2^{1+22}.McL$,*
(v) *if $i\alpha = 4b$ then $|\Gamma(u) \cap \Gamma(t)| = 1$ and $G(t, u) \cong 2.F_4(2)$,*
(vi) *if $i\alpha = 6a$ then $|\Gamma(u) \cap \Gamma(t)| = 1$ and $G(t, u) \cong 2 \cdot M(22)$.* □

5.13 Uniqueness of the Monster amalgam

In this section we show that all Monster amalgams are isomorphic. The following result was established in [GMS89], Proposition (2.6).

Lemma 5.13.1 *There exist exactly two isomorphism types of groups* $G_1 \sim 2_+^{1+24}.Co_1$ *satisfying* (5.1.1 (i), (ii)), *where* $Q_1 = O_2(G_1)$.

Proof. We assume that the reader is familiar with standard properties of extraspecial groups and their automorphism groups [Gri73]. First we identify the image $\widehat{G}_1$ of G_1 in the automorphism group A of Q_1 isomorphic to $2^{24} \cdot O_{24}^+(2)$. By (4.5.6) $\bar{G}_1 \cong Co_1$ preserves on $\bar{\Lambda} \cong Q_1/Z_1$ a unique quadratic form θ, which shows that the subgroup $\widehat{G}_1/O_2(A)$ in $A/O_2(A)$ is uniquely determined up to conjugation and hence $\widehat{G}_1$ is uniquely determined in A (up to conjugation). Let V be a 2^{12}-dimensional vector space over the complex numbers, $L = GL(V)$ and $\varphi : Q_1 \to L$ be the unique faithful irreducible representation of Q_1. Then $N_L(\varphi(Q_1))$ realizes all automorphisms of $\varphi(Q_1) \cong Q_1$ and hence up to conjugation L contains a unique subgroup $G_1^{(1)}$ which satisfies the hypotheses of the lemma.

Now it is clear that G_1 is a perfect central extension of $\widehat{G}_1$ by a group of order 2. Let $\widetilde{G}_1$ be the largest perfect central extension of $\widehat{G}_1$ by an elementary abelian 2-group Y. We claim that Y is of order 4. Let Y_3 be the commutator subgroup of $\widetilde{Q}_1 := O_2(\widetilde{G}_1)$. Then the mapping $(q_1, q_2) \mapsto [q_1, q_2]$ for $q_1, q_2 \in \widetilde{Q}_1$ defines a bilinear map from $\bar{\Lambda} \cong \widetilde{Q}_1/Y$ onto Y_3 invariant under the action of Co_1. By (4.5.6) there is a unique such non-zero form, which shows that Y_3 is of order at most 2. Next we observe that $\widetilde{Q}_1/Y_3$ is a $GF(2)$-module for Co_1 which is an extension of $\bar{\Lambda}$ by some trivial submodules. Since $\bar{\Lambda}$ is self-dual and the first cohomology group of $\bar{\Lambda}$ was proved to be trivial in [Gri82], lemma (2.11), we conclude that $\widetilde{Q}_1/Y_3$ contains a submodule which maps isomorphically onto $\bar{\Lambda}$. Factorizing over this submodule we obtain a perfect central extension of Co_1 by Y/Y_3. It is well known that Co_0 is the universal perfect central extension of Co_1, which shows that Y is of order at most 4. Let φ_1 and φ_2 be homomorphisms of Co_0 and $G_1^{(1)}$ onto Co_1 and $G_1^{(0)}$ be the subgroup in the direct product $Co_0 \times G_1^{(1)}$ consisting of the pairs (a, b) with $a \in Co_0$, $b \in G_1^{(1)}$ and $\varphi_1(a) = \varphi_2(b)$. Then it is easy to see that $G_1^{(0)}$ is a perfect central extension of $\widehat{G}_1$ by a group Y of order 4. Let Y_1, Y_2, Y_3 be the subgroups of order 2 in Y, where Y_3 is as above and $Y_1 = Z(Co_0)$. Then $G_1^{(0)}/Y_1 \cong G_1^{(1)}$, $G_1^{(2)} := G_1^{(0)}/Y_2$ is the second group satisfying the hypotheses of the lemma and $G_1^{(3)} := G_1^{(0)}/Y_3$ is a (non-split) extension of Co_1 by an elementary abelian group of order 2^{25}. □

Consider the action of $\bar{H} \cong Co_1$ on its tilde geometry $\mathcal{G}(Co_1)$. Let $\bar{H}_1 \cong 2^{11}.Mat_{24}$ and $\bar{H}_2 \cong 2^{4+12}.(3 \cdot Sym_6 \times Sym_3)$ be the stabilizers of a point and a line incident to the point, respectively. Let $\bar{J}$ be a Sylow 3-subgroup in $O_{2,3}(\bar{H}_2)$ and $\bar{U} = C_{O_2(\bar{H}_2)}(\bar{J})$. Since $O_2(\bar{H}_1)$ is the irreducible Golay code module for $\bar{H}_1/O_2(\bar{H}_1) \cong Mat_{24}$ and $O_2(\bar{H}_1) \cap \bar{H}_2$ has index 2 in $O_2(\bar{H}_1)$, it follows from (3.8.5) that $\bar{U}$ is the natural 4-dimensional symplectic module for $\bar{S} := \bar{H}_2/\bar{R} \cong Sp_4(2)$, where $\bar{R}$ is the largest solvable normal subgroup in $\bar{H}_2$.

We follow notation introduced in the proof of (5.13.1) and in addition put $G_1^{(4)} = Co_0$. For $0 \leq i \leq 4$ let $H_2^{(i)}$ be the preimage of $\bar{H}_2$ in $G_1^{(i)}$ (with respect to the natural homomorphism), $J^{(i)}$ be a Sylow 3-subgroup in $O_{2,3}(H_2^{(i)})$ and $U^{(i)} = C_{O_2(H_2^{(i)})}(J^{(i)})$. We assume that the natural homomorphism sends $H_2^{(i)}$, $J^{(i)}$ and $U^{(i)}$ onto $\bar{H}_2$, $\bar{J}$ and $\bar{U}$, respectively.

Lemma 5.13.2 *For exactly one $i \in \{1, 2\}$ we have that $U^{(i)}$ is a 5-dimensional indecomposable module for $\bar{S}$.*

Proof. Notice that in terms of Section 4.14 $\bar{J}$ is a conjugate of $O_3(N_{\bar{H}}(X_s))$ and by (4.14.11) $\bar{J}$ acts on $\bar{\Lambda}$ fixed-point freely. This shows that $U^{(0)}$ is an extension of $\bar{U}$ by the subgroup Y which is the centre of $G_1^{(0)}$. This shows that $U^{(0)}$ is centralized by the preimage of $\bar{R}$ in $G_1^{(0)}$. Furthermore, since $\bar{S}$ does not preserve a non-zero quadratic form on $\bar{U}$, we conclude that $U_1^{(0)}$ is abelian and hence can be considered as a $GF(2)$-module for $\bar{S}$. We assume that $U^{(i)}$ is a quotient of $U^{(0)}$ for $1 \leq i \leq 3$. The first cohomology group of $\bar{U}$ is 1-dimensional [JP76] and since $\bar{U}$ is self-dual, this means that $U^{(0)}$ is decomposable. We claim that $U^{(3)}$ is indecomposable. Since $\bar{J}$ acts fixed-point freely on $\bar{\Lambda}$, the definition of $G_1^{(3)}$ implies that $U^{(3)}$ is isomorphic to $U^{(4)}$. Let $H_1^{(4)} \cong 2^{12}.Mat_{24}$ be the preimage of $\bar{H}_1$ in $G^{(4)}$. Then $O_2(H_1^{(4)})$ is the Golay code module and by a straightforward calculation we see that $U^{(4)}$ is indecomposable. Hence one of $U^{(1)}$ and $U^{(2)}$ is decomposable and the other one is not. □

Lemma 5.13.3 *In a Monster amalgam $\mathcal{M} = \{G_1, G_2, G_2\}$ the isomorphism type of G_1 is uniquely determined.*

Proof. Let H_2 be the preimage of $\bar{H}_2$ in G_1, J be a Sylow 3-subgroup in $O_{2,3}(H_2)$ and $U = C_{O_2(H_2)}(J)$. By (5.13.1) and (5.13.2) it is sufficient to show that U is indecomposable with respect to the action of $\bar{S}$. We adapt our notation so that J contains the subgroup X of order 3 as in (5.6.2). Then $A := C_{G_2}(X)/X \cong 2^{11}.Mat_{24}$ and $O_2(A)$ is the irreducible Todd module for $A/O_2(A) \cong Mat_{24}$. Let $\widehat{J}$ be the image of J in A. Then $N_A(\widehat{J})/O_2(A) \cong 3 \cdot Sp_4(2)$ and by (3.8.5) $C_{O_2(A)}(\widehat{J})$ is an indecomposable 5-dimensional $GF(2)$-module for $Sp_4(2)$. □

In the proof of the next lemma we make use of some argumentation in [Th79].

Lemma 5.13.4 *The subamalgam $\{G_1, G_2\}$ of the Monster amalgam is uniquely determined up to isomorphism.*

Proof. By (5.13.3) the isomorphism type of G_1 is uniquely determined. It is easy to deduce from the proof of (5.1.4) that up to conjugation in G_1 there is a unique subgroup Z_2 in Q_1 whose normalizer in G_1 contains a Sylow 2-subgroup of the latter and by the statement of (5.1.4) we have $G_{12} = N_{G_1}(Z_2)$. Thus G_{12} (as a subgroup in G_1) is uniquely determined up to conjugation. Comparing (5.6.2) and the structure of G_2 as given after (5.1.8) we conclude that G_2 is generated by G_{12} and a subgroup X of order 3 which normalizes $G_2^+ = C_{G_1}(Z_2) \sim 2^{2+11+22}.Mat_{24}$, which is inverted by every element from $G_2 \setminus G_2^+$, and by (5.6.2) X can be chosen so that $C_{G_1}(X) \sim 2^{11}.Mat_{24}$. Thus to prove the lemma it is sufficient to show that the image of X in the outer automorphism group of G_2^+ is uniquely determined. Suppose that there are two possibilities $X_1 = \langle x_1 \rangle$ and $X_2 = \langle x_2 \rangle$ for X. By (5.1.8) there are exactly three elementary abelian subgroups of order 2^{11} in Q_2/R_2 which are normal in G_2^+/R_2 (one of these subgroups is $(Q_1 \cap G_2)/R_2$). Moreover, these three subgroups are transitively permuted by X_1 and X_2. Hence without loss of generality we can assume that $x_1 x_2^{-1}$ normalizes each of these three subgroups. For $i = 1$ or 2 put $N_i = C_{G_2^+}(X_i)$. Then by (5.6.2) and our assumption $N_i \sim 2^{11}.Mat_{24}$ and $O_2(N_i)$ is the irreducible Todd module for $N_i/O_2(N_i) \cong Mat_{24}$. Hence $O_2(N_i) \leq R_2$. It is well known that the first cohomology group of the irreducible Golay code module is trivial, which enables us to assume that $R_2 N_1 = R_2 N_2$. Since N_i is perfect we have

$$N_1 = (R_2 N_1)' = (R_2 N_2)' = N_2$$

and hence $x_1 x_2^{-1}$ centralizes $N_1 = N_2$. Since $N_1 R_2/R_2 \cong Mat_{24}$ acts irreducibly on $(Q_1 \cap G_2)/R_2$, we conclude that $x_1 x_2^{-1}$ centralizes $(Q_1 \cap G_2)/R_2$ as well as its images under X_i. Finally Z_1 is the commutator subgroup of $Q_1 \cap G_2$ and hence $x_1 x_2^{-1}$ centralizes Z_1. Since $x_1 x_2^{-1}$ also normalizes Z_2 we conclude that the element $x_1x_2^{-1}$ induces the trivial automorphism of G_2^+ and the result follows. □

Proposition 5.13.5 *All Monster amalgams are isomorphic.*

Proof. Let $\mathscr{M} = \{G_1, G_2, G_3\}$ be a Monster amalgam. Then by (5.13.4) the isomorphism type of the subamalgam $\{G_1, G_2\}$ is uniquely determined. By (5.1.5) the subgroup Z_3 is uniquely determined up to conjugation in G_{12} and by (5.1.6) we have $G_{13} = N_{G_1}(Z_3)$ and $G_{23} = N_{G_2}(Z_3)$. Hence the subamalgam $\mathscr{B} = \{G_{13}, G_{23}\}$ in $\{G_1, G_2\}$ is determined uniquely up to conjugation and clearly G_3 is generated by this subamalgam. Hence in order to specify the isomorphism type of $\mathscr{M}$ it is sufficient to indicate the kernel K of the homomorphism onto G_3 of the universal completion U of the amalgam $\mathscr{B}$. Let $G_3^+ = C_{G_{12}}(Z_3)$. Then $G_3^+ \sim 2^{3+6+12+18}.3 \cdot Sym_6$, G_3^+ is normal in both G_{13} and G_{23} (hence it is normal in U). Furthermore $G_{13}/G_3^+ \cong G_{23}/G_3^+ \cong Sym_4$, $G_{123}/G_3^+ \cong D_8$ and Z_3 is the centre of G_3^+. It is clear that $K \cap G_3^+ = 1$, which means that $K \leq C_U(G_3^+)$ and K is a complement to Z_3 in $C_U(G_3^+)$. We claim that there is at most one such complement. Suppose to the contrary that there are two different complements, say K_1 and K_2. Then by the homomorphism theorem $C_U(G_3^+)/Z_3(K_1 \cap K_2) \cong Z_3$, which means that if we put $L = \langle G_3^+, K_1 \cap K_2 \rangle$ then the quotient $\bar{U} := U/L$ is isomorphic to the semidirect product of Z_3 and $L_3(2)$, the latter being the subgroup of the automorphism group of Z_3 (in fact the whole automorphism group) generated by the images of G_{13} and G_{23}. So we have that $\bar{U}$ is generated by its subamalgam $\{G_{13}/L, G_{23}/L\}$. In [Sh88] a very nice lemma was proved asserting that the semidirect product $2^3 : L_3(2)$ is never generated by a subamalgam $\{P_1, P_2\}$ with $P_1 \cong P_2 \cong Sym_4$ and $P_1 \cap P_2 \cong D_8$. This contradicts our assumption on existence of two complements. Hence K is uniquely determined and the result follows. □

5.14 On existence and uniqueness of the Monster

As we have already mentioned, the Monster group M was predicted to exist in 1973 independently by B. Fischer and R.L. Griess. During the 70's many properties of the hypothetical group were established. In particular it was shown that M involves many sporadic groups known by that time and that the centralizers in M of certain elements of order 2, 3 and 5 involve new sporadic simple groups. These sporadic groups (now known as the Fischer Baby Monster BM, the Thompson group Th and the Harada–Norton group HN) were constructed before the Monster itself was proved to exist.

In [Gri76] the number 196 883 was proved to be a lower bound on the dimension of a faithful complex representation of M. In [Th79] J. Thompson proved the uniqueness of the Monster under certain additional assumptions, the most crucial one being existence of a faithful $\mathbb{C}$-module V of dimension 196 883. The proof consists of two principal steps. In the first step effectively it was shown that the subamalgam $\{G_1, G_2\}$ in the Monster amalgam is uniquely determined up to isomorphism. In the second step it was shown that this amalgam possesses (up to conjugation) at most one isomorphism into $GL(V)$.

The Monster was constructed by R.L. Griess in [Gri82] as a subgroup of $GL(V)$ where V is a 196 883-dimensional vector space over the complex numbers. He started with a rather explicit description of the action of G_1 on V which could be the restriction to G_1 of the action of M on V, in other words he realized $G_1 \sim 2^{1+24}.Co_1$ as a subgroup in $GL(V)$. He then found an additional element $\sigma \in GL(V)$ which normalizes $Z_2 \leq G_1$ and together with $G_{12} = N_{G_1}(Z_2)$ generates in $GL(V)$ a subgroup $G_2 \sim 2^{2+11+22}.(Sym_3 \times Mat_{24})$ containing G_{12} with index 3. In constructing the element σ as well as in the identification step a crucial rôle was played by a non-associative algebra B (the *Griess algebra*) which is preserved by G_1 and G_2. The existence of this algebra was earlier pointed out by S.P. Norton who had calculated the values of the (hypothetical at that time) character of degree 196 883. In the stage of identification, the reductions of V over primes $p \geq 5$ were considered. By studying these reductions it was shown that G_1 is the full centralizer of Z_1 in the subgroup M of $GL(V)$ generated by G_1 and G_2. Finally, application of results on characterization of groups by their involution centralizers completed the identification of M with the Monster.

Later, Griess' construction was modified in different directions. In [Con85] the subgroup G_2 was taken as a starting point. This group was explicitly constructed in terms of a so-called Parker loop. This description turned out to be convenient enough to define the action of G_2 on V. It was shown that $C_{G_2}(Z_2)$ possesses inside $GL(V)$ three different extensions to G_1 and in a certain sense these extensions correspond to the involutions in Z_2. The mutual consistency of these extensions was shown and again an isomorphism of the amalgam $\{G_1, G_2\}$ into $GL(V)$ was constructed. In addition a vector from V (called a transposition vector) was pointed out, which has only finitely many images under $M = \langle G_1, G_2 \rangle$. The action of M on the set of these images was proved to be faithful and this implies the finiteness of M. In [Ti85] it was proved that the full automorphism group of the Griess algebra is finite and that G_1 is the full centralizer of Z_1 in the automorphism group of this algebra. An exposition of the construction of the Monster can be found in Chapter 10 of [A94].

In [Iv93b] it was shown that the embedding of the amalgam $\{G_1, G_2\}$ into $GL(V)$ can be easily extended to an embedding of the whole Monster amalgam $\mathcal{M}$. That is, it was shown that the subgroup in $GL(V)$ generated by $N_{G_1}(Z_3)$ and $N_{G_2}(Z_3)$ is of the form $G_3 \sim 2^{3+6+12+18}.(L_3(2) \times 3 \cdot Sym_6)$.

Thus a Monster group which possesses a 196 883-dimensional representation was constructed in [Gri82] but at that time it was not known that *every* Monster group possesses such a representation. This was proved by S.P. Norton in [Nor85]. He considered the action of a Monster group M on the conjugacy class of its involutions ($2a$-involutions) called the *Baby Monster involutions* with centralizers $2 \cdot BM$ and established the following.

Proposition 5.14.1 *Let Γ be the set of Baby Monster involutions in M. Then*

$$|\Gamma| = 97\,239\,461\,142\,009\,186\,000$$

and for $t \in \Gamma$ the subgroup $M(t) = C_M(t) \cong 2 \cdot BM$ acting on $\Gamma \setminus \{t\}$ has eight orbits $\Gamma^{i\alpha}(t)$ where $i\alpha = 2a$, $2b$, $3a$, $3c$, $4a$, $4b$, $5a$ or $6a$ with stabilizers of the form $2^{2.2}E_6(2)$, $2^{2+22}.Co_2$, $M(23)$, Th, $2^{1+22}.McL$, $2.F_4(2)$, HN or $2 \cdot M(22)$, respectively. The centralizer algebra corresponding to the action of M on Γ has a primitive idempotent of dimension 196 883. □

Notation in (5.14.1) has the following meaning: if $u \in \Gamma^{i\alpha}(t)$ then (in terms of [CCNPW]) the product $t \cdot u$ belongs to the conjugacy class $i\alpha$ of the Monster; this notation is consistent with that in (5.12.7). It follows from the general theory of centralizer algebras that the primitive idempotent in (5.14.1) is a $\mathbb{C}$-module for M.

Let $\Gamma = \Gamma(M)$ be the graph on the set Γ as in (5.14.1) in which t is adjacent to the vertices in $\Gamma^{2a}(t)$. Then $\Gamma(M)$ is the second Monster graph as defined in Section 5.12. In [Nor85] S.P. Norton determined the orbits of M on the set of all triples $\{x, y, z\}$ of vertices of the second Monster graph such that x and y are adjacent. The lengths of these orbits enabled him to calculate the structure constants of the centralizer algebra of the action of M on Γ and these constants in turn provide the ranks of the primitive idempotents of the centralizer algebra. The paper [Nor85] contains extremely important information on the structure of the second Monster graph but most of the information is given without proof.

An independent uniqueness proof for the Monster M was given in [GMS89]. The assumption was that M is a finite group containing involutions with centralizers of the form $2_+^{1+24}.Co_1$ and $2 \cdot BM$. A graph on the set of involutions of the Baby Monster type was defined which corresponds to the second Monster graph. The conditions on the centralizers turned out to be strong enough to reconstruct the structure of the graph. The use of the information on the second Monster graph established in [GMS89] was essential in [ASeg92] to prove the following.

Proposition 5.14.2 *The second Monster graph $\Gamma(M)$ is triangulable.* □

5.15 The simple connectedness of $\mathscr{G}(M)$

In this section we establish the simple connectedness of the tilde geometry $\mathscr{G} = \mathscr{G}(M)$ of the Monster group. Let $\varphi : \widehat{\mathscr{G}} \to \mathscr{G}$ be the universal covering and $\widehat{M}$ be the universal completion of the amalgam of maximal parabolics associated with the action of M on $\mathscr{G}(M)$. Then $\widehat{M}$ is a flag-transitive automorphism group of $\widehat{G}$ and φ induces a homomorphism $\chi : \widehat{M} \to M$. Then both $\widehat{M}$ and M are faithful completions of the same Monster amalgam so we can define second Monster graphs $\Gamma(M)$ and $\Gamma(\widehat{M})$ as at the beginning of Section 5.12 and observe that χ induces a morphism $\psi : \Gamma(\widehat{M}) \to \Gamma(M)$. By (5.12.4) and (5.12.6) every triangle in $\Gamma(M)$ is contractible with respect to ψ. Since $\Gamma(M)$ is triangulable by (5.14.2), we conclude that ψ (and hence φ as well) is an isomorphism. Notice that by (5.12.10) we know that every quadrangle in $\Gamma(M)$ is also contractible with respect to ψ, so we only need to know that the fundamental group of $\Gamma(M)$ is generated by the cycles of length 3 and 4. In any case since by (5.13.5) all Monster amalgams are isomorphic, we have the following.

Proposition 5.15.1 *Let $\mathscr{M} = \{G_1, G_2, G_3\}$ be a Monster amalgam and G be a faithful completion of $\mathscr{M}$. Then G is the Monster sporadic simple group of order*

$$2^{46} \cdot 3^{30} \cdot 5^9 \cdot 7^6 \cdot 11^2 \cdot 13^3 \cdot 17 \cdot 19 \cdot 23 \cdot 29 \cdot 31 \cdot 41 \cdot 47 \cdot 59 \cdot 71$$

(so that $G_1 = C_G(Z_1)$), in particular the tilde geometry $\mathscr{G}(M)$ of the Monster group M is simply connected. □

6

From C_n- to T_n-geometries

In this chapter we construct an infinite series of flag-transitive tilde geometries possessing morphisms onto C_n-geometries of symplectic groups over $GF(2)$. For every $n \geq 2$ the series contains one T-geometry $\mathcal{J}(n)$ of rank n, whose automorphism group is isomorphic to

$$3^{\left[n \atop 2 \right]_2} \cdot Sp_{2n}(2).$$

We also prove that these geometries are 2-simply connected. In Section 6.3 we review some known facts about the dual polar graphs associated with C_n-geometries of the symplectic groups. In Section 6.4 we consider the semidirect product $W(n) : Sp_{2n}(2)$ where $W(n)$ is the $GF(3)$-module induced from a 1-dimensional non-trivial module of $O_{2n}^-(2) < Sp_{2n}(2)$. Using the technique presented in Section 6.1, we show that up to conjugation in its automorphism group the semidirect product contains two subamalgams isomorphic to the amalgam of minimal parabolics associated with the action of $Sp_{2n}(2)$ on its C_n-geometry. One of these subamalgams generates a complement to $W(n)$ while the other one leads to a T_n-geometry $\mathcal{J}(n)$ constructed in Section 6.5. Section 6.6 is devoted to a detailed analysis of the rank 3 geometry $\mathcal{J}(3)$. This analysis enables us to identify in Section 6.7 the automorphism group $J(n)$ of $\mathcal{J}(n)$. Since $O_3(J(n))$ is a submodule of the induced module $W(n)$, it contains a family of subgroups of order 3 permuted doubly transitively by $J(n)/O_3(J(n)) \cong Sp_{2n}(2)$. Analysing this family in Section 6.9 we show that the geometries $\mathcal{J}(n)$ are 2-simply connected. In Section 6.10 we characterize the geometries $\mathcal{J}(n)$ in terms of special coverings of the dual polar graphs of the C_n-geometries of symplectic groups. Finally, in Section 6.11 we show that there are no flag-transitive T_3-geometries possessing morphisms onto the C_3-geometry $\mathcal{G}(Alt_7)$.

6.1 On induced modules

We start with a more detailed (but still a bit informal) discussion of the strategy for constructing 1-covers of T- and P-geometries implemented in this chapter.

Let $\mathcal{G}$ be a T- or P-geometry of rank n and $\Phi = \{x_1, x_2, ..., x_n\}$ be a maximal flag in $\mathcal{G}$. Let G be a flag-transitive automorphism group of $\mathcal{G}$, $\mathcal{A} = \{P_i \mid 1 \leq i \leq n\}$ be the amalgam of minimal parabolic subgroups in G, where P_i is the stabilizer of the flag $\Phi \setminus \{x_i\}$. Let $B = \bigcap_{i=1}^{n} P_i$ be the Borel subgroup and $R_i = O_2(P_i) \cap B$ for $1 \leq i \leq n$. Notice that $R_i = O_2(P_i)$ unless $i = n$ and $\mathcal{G}$ is a P-geometry. We assume that $\mathcal{G}$ is a *2-local geometry of* G in the sense that B is a non-trivial 2-group and $N_G(R_i) = P_i$ for $1 \leq i \leq n$.

We consider a group $\widehat{G}$ which possesses a homomorphism onto G with kernel W being an elementary abelian 3-group. Moreover, as a $GF(3)$-module for $G \cong \widehat{G}/W$, the kernel W is induced from a 1-dimensional module W_0 of a subgroup X of G, so that the kernel X_0 of W_0 is of index 2 in X. In some cases $\widehat{G}$ splits over W, in some cases it does not.

The crucial step in the construction is to classify in $\widehat{G}$ all subamalgams $\widetilde{\mathcal{A}} = \{\widetilde{P}_i \mid 1 \leq i \leq n\}$ such that the restriction of φ to $\widetilde{\mathcal{A}}$ is an isomorphism onto $\mathcal{A}$ (we assume that $\varphi(\widetilde{P}_i) = P_i$). We classify these subamalgams up to conjugation in the automorphism group of $\widehat{G}$.

Let $\widetilde{\mathcal{A}}$ be such an amalgam and $\widetilde{B} = \bigcap_{i=1}^{n} \widetilde{P}_i$, so that $\widetilde{B}$ maps isomorphically onto B under φ. Since W has odd order, $\widetilde{B}$ is a Sylow 2-subgroup in the full preimage of B in $\widehat{G}$ and the choice of $\widetilde{B}$ is unique up to conjugation.

For every i, $1 \leq i \leq n$, there is a unique subgroup $\widetilde{R}_i$ in $\widetilde{B}$ which maps isomorphically onto R_i. Since $P_i = N_G(R_i)$, it is clear that $\widetilde{P}_i$ is contained in $N_{\widehat{G}}(\widetilde{R}_i)$ and by the Frattini argument we have the following.

Lemma 6.1.1 *$N_{\widehat{G}}(\widetilde{R}_i) \cap W = C_W(R_i)$ and $N_{\widehat{G}}(\widetilde{R}_i)/C_W(R_i) \cong P_i$.* □

Thus $\widetilde{P}_i$ is a complement to $C_W(R_i)$ in $N_{\widehat{G}}(\widetilde{R}_i)$ and if $C_W(R_i) = 0$ then $\widetilde{P}_i = N_{\widehat{G}}(\widetilde{R}_i)$ is uniquely determined.

For each subamalgam $\widetilde{\mathcal{A}}$ with the prescribed properties we consider the subgroup $\widetilde{G}$ in $\widehat{G}$ generated by $\widetilde{\mathcal{A}}$. For $1 \leq i \leq n$ let $\widetilde{G}_i$ denote the subgroup in $\widetilde{G}$ generated by all the $\widetilde{P}_j$ except for $\widetilde{P}_i$. Define $\widetilde{\mathcal{G}}$ to be the geometry whose elements of type i are all the cosets of $\widetilde{G}_i$ in $\widetilde{G}$ and two cosets are incident if they have a non-empty intersection. In the final step of the construction we show that $\widetilde{\mathcal{G}}$ is a geometry which possesses a

1-covering onto $\mathscr{G}$ and specify the diagram of $\widetilde{\mathscr{G}}$. The covering is proper unless $\widetilde{G}$ is isomorphic to G. In the case of such an isomorphism $\widetilde{G}$ is a complement to W in $\widehat{G}$ and this is certainly impossible if $\widehat{G}$ does not split over W.

Thus in order to realize the strategy outlined, we should be able to construct non-split extensions of groups by induced modules and to calculate centralizers of various subgroups in such modules. In the remainder of the section we discuss the necessary machinery.

It is well known that if G is a finite group and U is a $GF(p)$-module for G then the non-split extension $U \cdot G$ exists if and only if the second cohomology group $H^2(G, U)$ is non-trivial. By the Eckmann–Shapiro lemma (Shapiro lemma in [Bro82]) the cohomology group of the induced module is isomorphic to that of the original module. This gives the following.

Lemma 6.1.2 *Let G be a group, X be a subgroup in G. Let W_0 be a 1-dimensional $GF(3)$-module for X and W be the $GF(3)$-module for G induced from W_0. Then a non-split extension $W \cdot G$ exists if and only if there exists a non-split extension $W_0 \cdot X$.* □

In the concrete situations we consider, the kernel X_0 of W_0 in X is a non-abelian simple group and the question about non-split extensions $W_0 \cdot X$ reduces to consideration of the 3-part of the Schur multiplier of X_0.

Let us turn to the calculation of the centralizers. We consider a slightly more general situation. Let G be a group, X, Y be subgroups of G, X_0 be a subgroup of index 2 in X and $\mathbb{F}$ be a field, whose characteristic is not 2. Let W_0 be a 1-dimensional $\mathbb{F}$-space, turned into an X-module by the following rule: the elements from X_0 centralize W_0 and every element from $X \setminus X_0$ inverts W_0. Let W be the module for G induced from W_0. We are interested in the dimension of $C_W(Y)$.

Since W is an induced module, it possesses the direct sum decomposition

$$W = \bigoplus_{i \in \mathscr{T}} W_i,$$

where the W_i are 1-dimensional $\mathbb{F}$-spaces indexed by the cosets from $\mathscr{T} = G/X$. The group G acting on W permutes the subspaces W_i in the way it permutes the cosets in $\mathscr{T}$. If i_0 denotes the coset $X \cdot 1$, then W_{i_0} and W_0 are isomorphic as X-modules. For an arbitrary coset $i = X \cdot g \in \mathscr{T}$

let $X_i = X^g$ be the stabilizer in G of this coset. Then the elements from $X_{0,i} := X_0^g$ centralize W_i while every element from $X_i \setminus X_{0,i}$ inverts W_i.

Let T be an orbit of Y on $\mathcal{T}$ and $W_T = \bigoplus_{i \in T} W_i$.

Lemma 6.1.3 *$C_{W_T}(Y)$ is an $\mathbb{F}$-subspace in W_T, whose dimension is* 1 *if $Y \bigcap X_i \leq X_{0,i}$ for every $i \in T$ and* 0 *otherwise.*

Proof. Let $\Omega = \bigcup_{i \in T} W_i^{\#}$ be the set of all non-zero elements contained in the subspaces W_i. Since W_T is a direct sum of the W_i, for every element $w \in W_T$ there is a uniquely determined subset $\Omega(w) \subseteq \Omega$ such that $|\Omega(w) \cap W_i| \leq 1$ for every $i \in T$ and w is equal to the sum of all elements in $\Omega(w)$. Moreover, $w \in C_{W_T}(Y)$ if and only if Y stabilizes $\Omega(w)$ as a whole. Suppose first that $Y \cap X_i$ is not contained in $X_{0,i}$ for some (and hence for all) $i \in \mathcal{T}$. Then $Y \cap X_i$ contains an element inverting W_i. Hence if $u \in W_i$ then the orbit of u under Y contains $-u$. Thus in this case there are no non-zero elements in W_T centralized by Y. If $Y \cap X_i \leq X_{0,i}$, then $Y \cap X_i$ centralizes W_i and for every non-zero element $u \in W_i$ the sum $\sigma(u)$ of all its images under Y is a non-trivial element from $C_{W_T}(Y)$. Furthermore, for $u, v \in W_i$ and $\lambda, \mu \in \mathbb{F}$ we have

$$\lambda\sigma(u) + \mu\sigma(v) = \sigma(\lambda u + \mu v).$$

Finally, since T is a Y-orbit, every orbit of Y on Ω intersects $W_i^{\#}$ and the result follows. □

Definition 6.1.4 *An orbit T of Y on $\mathcal{T}$ will be called* untwisted *if $Y \cap X_i \leq X_{0,i}$ for every $i \in T$ and* twisted *otherwise.*

By (6.1.3) we have the following.

Lemma 6.1.5 *In the above notation $C_W(Y)$ is an $\mathbb{F}$-subspace in W whose dimension is equal to the number of untwisted orbits of Y on $\mathcal{T} = G/X$.* □

Let us give a reformulation of the above result in terms of complex characters. Notice that the number of untwisted orbits has nothing to do with the field $\mathbb{F}$ and it is uniquely determined by the triple (X, X_0, Y) of subgroups in G.

Lemma 6.1.6 *In the above terms assume that $\mathbb{F} = \mathbb{C}$. Let χ_0 be the character of W_0 and $\chi = \chi_0^G$ be the induced character of W. Then*

$$\dim C_W(Y) = \langle \chi|_Y, 1_Y \rangle.$$

□

When constructing the covers we will always take $\mathbb{F}$ to be $GF(3)$, but the above lemma shows that the calculations of the centralizers can be carried out in the complex number field (provided that the corresponding characters are known). Finally (6.1.5) implies a purely group-theoretical condition for the triviality of $C_W(Y)$.

Lemma 6.1.7 *In the above terms $C_W(Y)$ is trivial if and only if for every $g \in G$ we have $Y^g \cap X \neq Y^g \cap X_0$.* □

6.2 A characterization of $\mathcal{G}(3 \cdot Sp_4(2))$

In this section we apply the strategy outlined in the previous section to construct the rank 2 T-geometry $\mathcal{G}(3 \cdot Sp_4(2))$ as a 1-cover of the generalized quadrangle $\mathcal{G}(Sp_4(2))$ of order (2,2). This construction provides us with a characterization of the rank 2 T-geometry as well as with a background for construction of the infinite series of T-geometries.

Hypothesis 6.2.1 *$\widetilde{G}$ is a group having a normal subgroup Y of order 3 such that $\widetilde{G}/Y \cong Sym_6$. The centralizer C of Y in $\widetilde{G}$ is a perfect group having index 2 in $\widetilde{G}$. This means that C is a non-split central extension of Alt_6 by Y and Y is inverted by the elements of $\widetilde{G}$ which map onto odd permutations of Sym_6.*

We will prove a sequence of lemmas which are of independent interest and imply the following.

Proposition 6.2.2 *There exists a unique (up to isomorphism) group $\widetilde{G}$ which satisfies Hypothesis 6.2.1, and $\widetilde{G}$ is the automorphism group of the rank 2 T-geometry $\mathcal{G}(3 \cdot Sp_4(2))$.*

First observe that by (2.6.1) the automorphism group $\widetilde{G}$ of the rank 2 T-geometry satisfies the conditions of Hypothesis 6.2.1.

We are going to construct a group which contains every group satisfying Hypothesis 6.2.1 as a subgroup. Let $G \cong Sym_6 \cong Sp_4(2))$, X be a subgroup in G isomorphic to Sym_5 and X_0 be the commutator subgroup of X isomorphic to Alt_5. Let $\Omega = \{1,2,3,4,5,6\}$ denote the set of cosets of X in G. Let X_i be the stabilizer in G of the coset i and $X_{0,i}$ be the commutator subgroup of X_i (which is the intersection of X_i with the commutator subgroup of G, isomorphic to Alt_6). We assume that $X_1 = X$.

Let W_0 be a 1-dimensional $GF(3)$-vector–space on which X acts by the following rule: the elements from X_0 centralize W_0 while every element from $X \setminus X_0$ inverts W_0. Let W be the $GF(3)$-module for G induced from W_0. Then W possesses the direct sum decomposition

$$W = \bigoplus_{i \in \Omega} W_i$$

into 1-dimensional submodules permuted naturally by G. Moreover W_1 and W_0 are isomorphic as X-modules. Furthermore, X_i and $X_{0,i}$ are the normalizer and the centralizer of W_i in G, respectively, for $1 \leq i \leq 6$. Notice that whenever an odd element $g \in G$ normalizes a subspace W_i, g inverts W_i.

When restricted to $G' \cong Alt_6$ the module W becomes the permutational module of G' acting on Ω. The centralizer of G' in W is a 1-dimensional submodule Z. A generator z of Z is the sum over the orbit under G' of a non-zero element from W_i for some $i \in \Omega$. It is clear that Z is a submodule for G, inverted by every odd element (*i.e.* by every element from $G \setminus G'$).

Lemma 6.2.3 *Let $\widehat{G}$ be a group which possesses a homomorphism φ onto G whose kernel is elementary abelian of order 3^6 and as a $GF(3)$-module for $G \cong Sym_6$ the kernel is isomorphic to the module W defined above. Then $\widehat{G}$ is isomorphic to the semidirect product $W : G \cong 3^6 : Sym_6$.*

Proof. The 3-part of the Schur multiplier of Alt_5 is trivial and hence by (6.1.2) $\widehat{G}$ splits over the kernel of φ and the result follows. □

Lemma 6.2.4 *Let $\widetilde{G}$ be a group satisfying Hypothesis 6.2.1. Then $\widetilde{G}$ is isomorphic to a subgroup of the group $\widehat{G}$ defined above.*

Proof. By the construction there is an action of G on W. By Hypothesis 6.2.1 there is a surjective homomorphism ψ of $\widetilde{G}$ onto G. Define the action of $\widetilde{G}$ on W so that $\widehat{g} \in \widetilde{G}$ acts as $\psi(\widehat{g}) \in G$ (in particular Y is the kernel of the action). Let $\bar{G}$ be the semidirect product of W and $\widetilde{G}$ with respect to this action. Then Y and Z are normal subgroups of order 3 in $\bar{G}$. Each of these two subgroups is centralized by the commutator subgroup of $\bar{G}$ and is inverted by every element outside the commutator subgroup. Let $U = \langle Y, Z \rangle$. Then U is elementary abelian of order 9 and besides Y and Z it contains two "diagonal" subgroups of order 3, which we denote by U_1 and U_2. Then for $i = 1$ and 2 the subgroup U_i is normal in $\bar{G}$ and $U_i \cap W = U_i \cap \widetilde{G} = 1$. Hence both W and $\widetilde{G}$ map isomorphically onto their images in $\bar{G}_i := \bar{G}/U_i$. It is easy to check that $\bar{G}_i$ satisfies the hypothesis of (6.2.3) and hence $\bar{G}_i \cong \widehat{G}$ which means that $\widehat{G}$ contains a subgroup isomorphic to $\widetilde{G}$. Notice that if σ is the automorphism of $\bar{G}$ which centralizes $\widetilde{G}$ and inverts every element in W, then $\sigma(U_i) = U_{3-i}$ for $i = 1$ and 2. □

Lemma 6.2.5 *Let $\mathscr{A} = \{P_1, P_2\}$ be the amalgam of minimal (which are also maximal) parabolic subgroups corresponding to the action of $G \cong Sym_6$ on $\mathscr{G}(Sp_4(2))$. Then every group $\widetilde{G}$ satisfying Hypothesis 6.2.1 is generated by its subamalgam $\widetilde{\mathscr{A}}$ which maps isomorphically onto $\mathscr{A}$ under the homomorphism of $\widetilde{G}$ onto G.*

Proof. We have $P_1 \cong P_2 \cong Sym_4 \times 2$ and $B = P_1 \cap P_2 \cong D_8 \times 2$ is a Sylow 2-subgroup of G. Let $R_i = O_2(P_i)$. Then R_i is elementary abelian of order 2^3, it contains odd elements of G and $P_i = N_G(R_i)$ for i =1 and 2. Let ψ be a homomorphism of $\widetilde{G}$ onto G and let $\widetilde{B}$ be a Sylow 2-subgroup in $\psi^{-1}(B)$. Then the restriction ψ_1 of ψ to $\widetilde{B}$ is an isomorphism onto B. Let $\widetilde{R}_i = \psi_1^{-1}(R_i)$ for $i = 1$ and 2. Since R_i contains odd elements of G, we have $C_Y(R_i) = 1$. Now the Frattini argument (compare (6.1.1)) implies that the subamalgam $\widetilde{\mathscr{A}} = \{N_{\widetilde{G}}(\widetilde{R}_1), N_{\widetilde{G}}(\widetilde{R}_2)\}$ maps isomorphically onto $\mathscr{A}$. Since $\widetilde{G}$ does not split over Y, this subamalgam generates the whole of $\widetilde{G}$. □

Lemma 6.2.6 *The group $\widehat{G}$ (up to conjugation in its automorphism group) contains at most two subamalgams which map isomorphically under $\varphi : \widehat{G} \to G$ onto the subamalgam $\mathscr{A}$ of minimal parabolic subgroups associated with the action of G on $\mathscr{G}(Sp_4(2))$. One of these subamalgams generates a complement to W in $\widehat{G}$.*

Proof. Let $\mathscr{A} = \{P_1, P_2\}$, $B = P_1 \cap P_2$, $R_i = O_2(P_i)$, so that $P_i = N_G(R_i)$ for $i = 1$ and 2. Let $\widetilde{\mathscr{A}} = \{\widetilde{P}_1, \widetilde{P}_2\}$ be the subamalgam in $\widehat{G}$ which maps isomorphically onto $\mathscr{A}$ under φ. Since $\widehat{G} = W : G$, at least one such subamalgam is contained in the complement to W. Notice that $\widetilde{B} := \widetilde{P}_1 \cap \widetilde{P}_2$ is a Sylow 2-subgroup in the full preimage of B in $\widehat{G}$. Hence $\widetilde{B}$ is uniquely determined up to conjugation and without loss of generality we assume that $\widetilde{B}$ is contained in the complement G to W in $\widehat{G}$. The restriction φ_1 of φ to $\widetilde{B}$ is an isomorphism onto B. For $i = 1$ and 2 let $\widetilde{R}_i = \varphi_1^{-1}(R_i)$ and $\widetilde{N}_i = N_{\widehat{G}}(\widetilde{R}_i)$. Then $\widetilde{R}_i \leq \widetilde{N}_i$ and by Frattini argument $\widetilde{N}_i \cap W = C_W(R_i)$ and $\widetilde{N}_i / C_W(R_i) \cong P_i$. Thus we have to calculate the centralizers in W of R_1 and R_2.

Without loss of generality we assume that P_1 is the stabilizer in G of the partition $\Omega = \{1,2\} \cup \{3,4\} \cup \{5,6\}$ and P_2 is the stabilizer in G of the pair $\{1,2\}$. Then the orbits of R_1 on Ω are the pairs forming the partition. The transposition $(1,2)$ is contained in R_1 and it stabilizes the subspace W_i for $i \geq 3$. Since this transposition is an odd element, it inverts every subspace W_i which it normalizes. Since the transposition $(3,4)$ is also in R_1, it is clear that all orbits of R_1 on Ω are twisted.

The orbits of R_2 on Ω are $\{1,2\}$ and $\{3,4,5,6\}$. Since the transposition $(1,2)$ is contained in R_2 as well, the orbit of length 4 is twisted. On the other hand since every odd element from R_2 switches the elements in the orbit of length 2, this orbit is untwisted.

Thus $\widetilde{P}_1 = \widetilde{N}_1$ is uniquely determined and it is contained in the complement G to W in $\widehat{G}$. But $\widetilde{N}_2$ is an extension of P_2 by the subgroup $C_W(R_2)$ of order 3. Since P_2 contains the transposition $(3,4)$, it induces a non-trivial action on $C_W(R_2)$. This implies that $\widetilde{N}_2/\widetilde{R}_2$ is an elementary abelian group of order 3^2, extended by an involutory fixed-point free automorphism. By the definition $\widetilde{P}_2$ is a subgroup of index 3 in $\widetilde{N}_2$ containing the Sylow 2-subgroup $\widetilde{B}$ of $\widetilde{N}_2$ which maps isomorphically onto P_2 under φ. It is easy to see that there are exactly four subgroups of index 3 in $\widetilde{N}_2$ containing $\widetilde{B}$, say $\widetilde{S}_j$ for $1 \leq j \leq 4$. One of them, say $\widetilde{S}_4$, has non-trivial intersection with W and for this reason cannot map isomorphically onto P_2. Each of the remaining three subgroups does map isomorphically onto P_2 and hence forms together with $\widetilde{P}_1$ a subamalgam which maps isomorphically onto $\mathcal{A}$. We can assume that $\widetilde{S}_3$ is contained in the complement G to W in $\widehat{G}$ and hence $\{\widetilde{P}_1, \widetilde{S}_3\}$ generates this complement. Let σ be the automorphism of $\widehat{G}$ which commutes with the complement G and inverts every element from W. Then it is easy to see that $\sigma(\widetilde{S}_j) = \widetilde{S}_{3-j}$ for $j = 1$ and 2. Hence up to conjugation in the automorphism group of $\widehat{G}$ there are two subamalgams $\widetilde{\mathcal{A}}$ which map isomorphically onto $\mathcal{A}$ and one of them generates a complement to W. □

Now (6.2.2) follows from (6.2.4), (6.2.5) and (6.2.6). As a corollary we observe that the subamalgams $\{\widetilde{P}_1, \widetilde{S}_1\}$ and $\{\widetilde{P}_1, \widetilde{S}_3\}$ are not conjugate in the automorphism group of $\widehat{G}$ since they generate non-isomorphic subgroups.

Notice that $G \cong Sym_6$ contains two conjugacy classes of subgroups isomorphic to Sym_5 fused in the automorphism group of G. So there are two possibilities for the module W which are equivalent with respect to the automorphism group. On the other hand the outer automorphism of

G performs a duality of the geometry $\mathcal{G}(Sp_4(2))$. In what follows we will need some information on the action of the parabolics P_1 and P_2 on the module W irrelevant to the above duality. This information is contained in the following two lemmas. The former one is a reformulation of (6.2.6).

Lemma 6.2.7 *Consider $\mathcal{G}(Sp_4(2))$ as a geometry of* 1- *and* 2-*dimensional totally singular subspaces of a* 4-*dimensional $GF(2)$-space V with respect to a non-singular symplectic form. Let P_i be the stabilizer in $G \cong Sp_4(2) \cong Sym_6$ of a totally singular $(3-i)$-dimensional subspace and let $R_i = O_2(P_i)$ for $i = 1, 2$. Let W be a $GF(3)$-module for G induced from the unique non-trivial* 1-*dimensional module W_0 for a subgroup $X \cong O_4^-(2) \cong Sym_5$ in G which stabilizes in V a quadratic form. Then $C_W(R_1)$ is trivial and $C_W(R_2)$ is* 1-*dimensional.* □

Lemma 6.2.8 *In terms of* (6.2.7) *let $\widehat{G} = W : G$. Then, up to conjugation in the automorphism group of $\widehat{G}$, there are exactly two subamalgams in $\widehat{G}$ which map isomorphically onto the subamalgam $\mathcal{A}$ of minimal parabolic subgroups associated with the action of G on $\mathcal{G}(Sp_4(2))$. Moreover, one of the subamalgams generates a complement to W while the other one generates a subgroup isomorphic to the automorphism group $3 \cdot Sp_4(2)$ of the rank* 2 *T-geometry.*

Proof. By (6.2.6) up to conjugation there are two subamalgams $\widetilde{\mathcal{A}}$ in $\widehat{G}$ with the prescribed properties and one of them generates a complement to W. The automorphism group $\widetilde{G}$ of $\mathcal{G}(3 \cdot Sp_4(2))$ satisfies Hypothesis 6.2.1 and by (6.2.4) $\widetilde{G}$ is a subgroup of $\widehat{G}$. Finally by (6.2.5) $\widetilde{G}$ is generated by a subamalgam $\widetilde{\mathcal{A}}$ which maps isomorphically onto $\mathcal{A}$ and the result follows. □

6.3 Dual polar graphs

In this section we introduce notation to be used till the end of the chapter and discuss some further properties of $C_n(2)$-geometries.

Let V be a $2n$-dimensional $GF(2)$-space, $n \geq 3$, and Ψ be a non-singular symplectic form on V. If $\{v_1^1, ..., v_n^1, v_1^2, ..., v_n^2\}$ is a (symplectic) basis of V then we can take

$$\Psi(v_i^k, v_j^l) = \delta_{i,j}\delta_{k,3-l}.$$

Let $\mathcal{G} = \mathcal{G}(Sp_{2n}(2))$ be the $C_n(2)$-geometry associated with the pair (V, Ψ) (Section 1.8). Since $n \geq 3$, by (1.6.5) $G \cong Sp_{2n}(2)$ is the only flag-transitive automorphism group of $\mathcal{G}$.

Let Γ be the *dual polar graph* of $\mathcal{G}$ whose vertices are the elements of type n in $\mathcal{G}$ and two vertices are adjacent if they are incident to a common element of type $n-1$. In other terms the vertices of Γ are maximal (n-dimensional) totally singular subspaces in V and two such subspaces are adjacent if their intersection is $(n-1)$-dimensional.

The following result can be found in [BCN89], Theorem 9.4.3.

Lemma 6.3.1 *With Γ and G as above the following assertions hold:*

(i) *Γ is distance-transitive of diameter n with the following intersection numbers: $c_i = a_i = \left[{i \atop 1}\right]_2$, $b_i = 2\left(\left[{n \atop 1}\right]_2 - \left[{i \atop 1}\right]_2\right)$ for $0 \le i \le n$;*

(ii) *G acts distance-transitively on Γ;*

(iii) *for $x, y \in \Gamma$ we have $d(x,y) = i$ for $0 \le i \le n$ if and only if $\dim(x \cap y) = n - i$;*

(iv) *every edge of Γ is in a unique triangle and whenever T is a triangle and x is a vertex there is a unique vertex $y \in T$ such that $d(x,T) = d(x,y)$;*

(v) *if $x \in \Gamma$ then the subgraph induced on $\Gamma_n(x)$ is connected.* □

Let U be an element in $\mathcal{G}$ of type $n-i$ where $0 \le i \le n-1$. Define $\Gamma(U)$ to be the subgraph of Γ induced on the vertices which are incident to (which means contain) U. It is easy to see that $\Gamma(U)$ consists of a single vertex if $i = 0$, otherwise it is isomorphic to the dual polar graph of the geometry $\mathcal{G}(Sp_{2i}(2))$ associated with the pair $(U^\perp/U, \Psi')$ where Ψ' is the form induced by Ψ on $U^\perp/U$. The subgraph $\Gamma(U)$ will be called a *geometrical subgraph of type $n-i$* in Γ. The geometrical subgraphs of type $n-2$ are isomorphic to the point graph of the generalized quadrangle of order $(2,2)$ and will be called *quads*. The whole graph Γ can be considered as a geometrical subgraph of type 0. Since $\Gamma(U) \subseteq \Gamma(W)$ if and only if $W \le U$, it is easy to see that the mapping $U \mapsto \Gamma(U)$ establishes an isomorphism of $\mathcal{G}$ onto the geometry whose elements of type j are geometrical subgraphs in Γ of type j, $1 \le j \le n$, and the incidence relation is via inclusion.

Since a geometrical subgraph of type $n-i$ is isomorphic to the dual polar graph associated with $\mathcal{G}(Sp_{2i}(2))$ and since by (6.3.1 (i)) the intersection parameters c_i and a_i are independent of n we have the following.

Lemma 6.3.2 *If $x, y \in \Gamma$ with $d(x,y) = i$ then $\Gamma(x \cap y)$ is the unique geometrical subgraph of type $n-i$ containing x and y. Every geometrical subgraph in Γ is strongly geodetically closed.* □

Proposition 6.3.3 *The geometry $\mathcal{G}(Sp_{2n}(2))$ is 2-simply connected.*

Proof. Let $\widetilde{\mathcal{G}}$ be the incidence system of rank n which is the universal 2-cover of $\mathcal{G} = \mathcal{G}(Sp_{2n}(2))$, let $\varphi : \widetilde{\mathcal{G}} \to \mathcal{G}$ be the universal 2-covering and let $\widetilde{\Gamma}$ be the graph on the set of elements of type n in $\widetilde{\mathcal{G}}$ in which two such elements are adjacent if they are incident to a common element of type $n-1$. Let $\{V_1, ..., V_n\}$ be a maximal flag in $\mathcal{G}$ where V_i is of type i. Since $\mathrm{res}_{\mathcal{G}}(V_n)$ is a projective geometry, it is 2-simply connected. In addition any two elements of type n in $\mathcal{G}$ are incident to at most one common element of type $n-1$. From this it is easy to conclude that φ induces a covering of $\widetilde{\Gamma}$ onto Γ (which we denote by the same letter φ). Since φ is a 2-covering, the subgraph in $\widetilde{\Gamma}$ induced by the vertices incident to an element of type $n-2$ maps isomorphically onto a quad in Γ. In view of (6.3.2) this implies that the cycles in Γ of length 3, 4 and 5 are contractible with respect to φ. Thus to prove the proposition it is sufficient to show that the fundamental group of Γ is generated by its cycles of length 3 and 4. The latter is equivalent to the statement that every non-degenerate cycle in Γ can be decomposed into triangles and quadrangles. We proceed by induction and assume that the statement is true for all geometries under consideration of rank less than n. By (6.3.2) every non-degenerate cycle of length $2i$ or $2i+1$ is contained in a geometrical subgraph of type $n-i$ and hence unless $i = n$ it is decomposable by the induction hypothesis. Let $C = (y_0, y_1, ..., y_{n-1}, y_n, y_{n+1}, ..., y_{2n} = y_0)$ be a non-degenerate cycle of length $2n$, which means that $y_n \in \Gamma_n(y_0)$, $y_{n-1}, y_{n+1} \in \Gamma_{n-1}(y_0)$. We claim that there is a vertex $z \in \Gamma_{n-2}(y_0)$ which is adjacent to both y_{n-1} and y_{n+1}. In fact, considering the vertices in C as subspaces in V which are maximal isotropic with respect to Ψ, we can put

$$z = \langle y_{n-1} \cap y_{n+1}, (y_{n-1} \cap y_{n+1})^{\perp} \cap y_0 \rangle$$

and it is straightforward to check that z possesses the required properties. Hence C can be decomposed into a quadrangle and two cycles of length $2n-2$. Now let $D = (y_0, ..., y_{2n+1})$ be a non-degenerate cycle of length $2n+1$. Then by (6.3.1 (iv)) the unique vertex z adjacent to both y_n and y_{n+1} is contained in $\Gamma_{n-1}(y_0)$, which shows that D is decomposable into a triangle and two cycles of length $2n$. □

Certainly the above proposition is nothing but a special case of (1.6.4) and we present a proof for the sake of completeness and to illustrate on an easy example the technique of decomposing cycles.

As above let $\Phi = \{V_1, V_2, ..., V_n\}$ be a maximal flag in $\mathcal{G}$. Let $\mathcal{A} = \{P_i \mid 1 \le i \le n\}$ be the amalgam of minimal parabolic subgroups in G

associated with Φ, $\mathscr{B} = \{G_i \mid 1 \le i \le n\}$ be the amalgam of maximal parabolics and $\mathscr{C} = \{P_{ij} \mid 1 \le i < j \le n\}$ be the amalgam of rank 2 parabolics. Let $B = \bigcap_{i=1}^{n} P_i$ be the Borel subgroup. For $1 \le k \le n$ let $\mathscr{C}^{(k)} = \{P_{ij} \mid 1 \le i < j \le n, k \notin \{i,j\}\}$. By (6.3.3) $\mathscr{G}$ is 2-simply connected. Since every rank 2 parabolic P_{ij} is contained in at least one of G_{n-2}, G_{n-1} and G_n, we have the following.

Lemma 6.3.4 *In the above terms $G \cong Sp_{2n}(2)$ is the universal completion of the amalgam $\mathscr{C}$ as well as of the amalgam $\{G_{n-2}, G_{n-1}, G_n\}$. If $k = n$ or $n-1$ then the parabolic G_k is the universal completion of the amalgam $\mathscr{C}^{(k)}$.* □

Let $Q_i = O_2(G_i)$ and $R_i = O_2(P_i)$ for $1 \le i \le n$. Then Q_i is the kernel of the action of G_i on $\mathrm{res}_{\mathscr{G}}(V_i)$ and $G_i/Q_i \cong L_i(2) \times Sp_{2n-2i}(2)$ for $1 \le i \le n$ where $L_1(2)$ and $Sp_0(2)$ are assumed to be the identity groups.

The element $x = V_n$ is a vertex of Γ and $G_n = G(x)$ is its stabilizer in G. We need some more detailed information on the structure of this parabolic. As usual let $G_i(x)$ denote the elementwise stabilizer in G of all the vertices which are at distance at most i from x in Γ.

Lemma 6.3.5 *The following assertions hold:*

(i) *if $y \in \Gamma_n(x)$ then $H := G(x) \cap G(y)$ is isomorphic to $L_n(2)$; x can be considered as the natural module for H, in which case y is the dual of the natural module; $G(x)$ is the semidirect product of Q_n and H;*

(ii) *$G_2(x) = 1$;*

(iii) *Q_n is elementary abelian of order $2^{(n^2+n)/2}$;*

(iv) *Q_n is a quotient of the permutational $GF(2)$-module of H acting on the set of non-zero vectors in x;*

(v) *$Q_n/G_1(x)$ is isomorphic to x as a module for H;*

(vi) *$G_1(x)$ is isomorphic to $\bigwedge^2 x$ as a module for H.*

Proof. The group $G_n = G(x)$ induces the full automorphism group $L_n(2)$ of the projective space $\mathrm{res}_{\mathscr{G}}(x)$. The kernel of the action is contained in the Borel subgroup which is a 2-group. Hence the kernel is exactly Q_n. In terms of the symplectic basis we can put $x = \langle v_1^1, ..., v_n^1 \rangle$ and $y = \langle v_1^2, ..., v_n^2 \rangle$ in which case it is clear that $G(x) \cap G(y) = GL(x) \cong L_n(2)$ is a complement to Q_n in G_n and (i) follows. One can see from the intersection numbers of Γ that any two vertices in Γ have at most three common neighbours. If $g \in G_2(x)$ and $z \in \Gamma_3(x)$ with $z^g \ne z$ then seven vertices in $\Gamma_2(x) \cap \Gamma(z)$ must be common neighbours of z and z^g which is impossible. Hence $G_2(x) = G_3(x)$ and (ii) follows from (9.1.4). Thus Q_n acts faithfully on $\{x\} \cup \Gamma(x) \cup \Gamma_2(x)$ *i.e.* on the union of quads containing x. Since Q_n stabilizes every such quad as a whole, by (2.5.3 (v)) Q_n is elementary abelian. By (i) and (6.3.1 (ii)) Q_n acts regularly on $\Gamma_n(x)$ whose size is $2^{(n^2+n)/2}$ and (iii) follows.

Let $y \in \Gamma_n(x)$. Then there is a bijection ψ between the non-zero vectors in x and the vertices in $\Gamma(y) \cap \Gamma_n(x)$. To wit, for $u \in x^\#$ we have

$$\psi(u) = \langle y \cap u^\perp, u + u_1 \rangle, \quad \text{where} \quad u_1 \in y \setminus (y \cap u^\perp).$$

It is straightforward to check that the transvection $t(u, \langle u \rangle^\perp)$ is contained in Q_n and maps y onto $\psi(u)$. By (6.3.1 (v)) the subgraph induced by $\Gamma_n(x)$ is connected and hence such transvections taken for all non-zero vectors u in x generate a subgroup acting transitively on $\Gamma_n(x)$. This subgroup is clearly Q_n and (iv) follows.

Considering the transvections $t(u, \langle u \rangle^\perp)$, it is easy to observe that $Q_n/G_1(x)$ is non-trivial. Let Π_x be the projective space dual to $\mathrm{res}_{\mathcal{G}}(x)$, so that the points of Π_x are the triangles in Γ containing x. For such a triangle T let $K(T)$ be the elementwise stabilizer of T in Q_n. Then $K(T)$ has index 2 in Q_n. Triangles T_1, T_2, T_3 form a line in Π_x if and only if they are contained in a common quad. By (2.5.3 (iv)), in this case $K(T_1) \cap K(T_2) \leq K(T_3)$. Hence the dual of $Q_n/G_1(x)$ supports a natural representation of Π_x and (v) follows from (1.11.1).

By (ii), (iii) and (v) $G_1(x)$ is non-trivial and it acts faithfully on $\Gamma_2(x)$. By (2.5.3 (vi)) the dual of $G_1(x)$ is generated by subgroups of order 2 indexed by the quads containing x, *i.e.* by the lines of Π_x. By (v) such subgroups corresponding to quads containing a given vertex $z \in \Gamma_1(x)$ generate the natural module of $(G(x) \cap G(z))/O_2(G(x) \cap G(z)) \cong L_{n-1}(2)$, so (vi) follows from (2.4.6). □

Notice that in terms of (2.4.7) Q_n is isomorphic to the quotient $W/\langle W^1, W_3 \rangle$ of the permutational $GF(2)$-module W of H acting on 1-dimensional subspaces of x.

We formulate the following direct consequence of (6.3.5) and (2.4.7 (vi)).

Lemma 6.3.6 *Let Σ be the quad in Γ stabilized by G_{n-2}. Let H be the full preimage in G_{n-2} of the subgroup of index 2 in G_{n-2}/Q_{n-2} isomorphic to $L_{n-2}(2) \times Alt_6$. Then Q_n contains elements which are not contained in H.* □

6.4 Embedding the symplectic amalgam

We follow the notation introduced in the previous section. In addition let Δ be the set of quadratic forms of minus type on V associated with Ψ in the sense that if $f \in \Delta$ then

$$\Psi(v,u) = f(v+u) + f(v) + f(u)$$

for $v, u \in V$ (notice that since the characteristic is 2 there is no difference between plus and minus). The group G acts transitively on Δ; the stabilizer $O(f)$ of $f \in \Delta$ in G is the orthogonal group $O_{2n}^{-}(2)$ containing a subgroup $\Omega(f)$ of index 2 which is the non-abelian simple group $\Omega_{2n}^{-}(2)$. We have

$$|\Delta| = [Sp_{2n}(2) : O_{2n}^{-}(2)] = 2^{n-1}(2^n - 1).$$

The following result (p. xii in [CCNPW]) enables one to distinguish the elements from $\Omega(f)$ and those from $O(f) \setminus \Omega(f)$.

Lemma 6.4.1 *An element $g \in O(f)$ is contained in $\Omega(f)$ if and only if the dimension of $C_V(g)$ is even.* □

Recall that Q_1 is the kernel of the action of G_1 on $\bar{V}_1 = V_1^{\perp}/V_1$. Since V_1 is 1-dimensional, $\bar{V}_1$ is $(2n-2)$-dimensional, Ψ induces on this space a non-singular symplectic form $\bar{\Psi}$ and $G_1/Q_1 \cong Sp_{2n-2}(2)$ acts as the full stabilizer of $\bar{\Psi}$. Let c be the transvection with centre w and axis $\langle w \rangle^{\perp}$, then it is easy to see that c is in the centre of Q_1.

Lemma 6.4.2 *Let a be a non-identity element in Q_1. Then*

$$\dim C_V(a) = \begin{cases} 2n-1 & \text{if } a = c; \\ 2n-2 & \text{otherwise.} \end{cases}$$

Proof. Since a acts trivially on $\bar{V}_1$, we have $v^a - v \in V_1$ for every $v \in V_1^{\perp}$. Suppose first that a acts trivially on $V_1^{\perp}$. Then $v^a - v \in (V_1^{\perp})^{\perp} = V_1$ for every $v \in V$. Hence $a = c$ and $V(a) = V_1^{\perp}$. Now suppose that a acts non-trivially on $V_1^{\perp}$. Then $X = C_V(a) \cap V_1^{\perp}$ has dimension $2n-2$. Suppose that a centralizes a vector $v \in V \setminus V_1^{\perp}$. Then for every $u \in V_1^{\perp} \setminus X$ we have $\Psi(v^a, u^a) = \Psi(v, u + w) \neq \Psi(v, u)$ where w is the non-zero element from V_1. This contradiction shows that $C_V(a) = X$. □

Let $f_0 \in \Delta$ and $O(f_0) \cong O_{2n}^{-}(2)$ be the stabilizer of f_0 in G. Let W_0 be a 1-dimensional $GF(3)$-space on which $O(f_0)$ acts by the following rule: all elements from $\Omega(f_0)$ centralize W_0 and every element from

$O(f_0) \setminus \Omega(f_0)$ inverts W_0. Then W_0 becomes the unique non-trivial 1-dimensional $GF(3)$-module for $O(f_0)$. Let $W = W(n)$ be the $GF(3)$-module for G induced from W_0. Then W possesses the direct sum decomposition

$$W = \bigoplus_{f \in \Delta} W_f,$$

where W_f is a 1-dimensional subspace whose normalizer and centralizer in G are $O(f)$ and $\Omega(f)$, respectively.

Let $\widehat{G} = W : G$ be the semidirect product of $W = W(n)$ and G with respect to the natural action and $\varphi : \widehat{G} \to G$ be the canonical homomorphism. We are going to classify the subamalgams in $\widehat{G}$ which map isomorphically onto $\mathscr{A}$ under φ. By Section 6.1 in order to classify these subamalgams we have to describe the centralizers in W of the subgroups R_i for $1 \leq i \leq n$. We proceed by induction on n noticing that $W(2)$ is the module for $Sp_4(2) \cong Sym_6$ as in Section 6.2 and we assume that $n \geq 3$.

Our nearest goal is to describe $C_W(Q_1)$. Since G_1 is the stabilizer in G of a 1-dimensional subspace in V, by Witt's theorem (Theorem 7.4 in [Tay92]), G_1 has exactly two orbits on Δ. One of these orbits, say Δ_1, contains all forms from Δ vanishing on V_1 and the other orbit, say Δ_2, contains the remaining forms. We claim that $c \in O(f)$ for every $f \in \Delta_2$. In fact, c acts trivially on $V_1^\perp$ and for $v \in V \setminus V_1^\perp$ we have $f(v^c) = f(v + w) = f(v) + f(w) + \Psi(v, w)$. Since both $f(w)$ and $\Psi(v, w)$ are equal to 1, $f(x^c) = f(x)$ and the claim follows. By (6.4.1) and (6.4.2) $c \notin \Omega(f)$ and hence we have the following.

Lemma 6.4.3 *Every Q_1-orbit in Δ_2 is twisted.* □

Now let $f \in \Delta_1$. If $a \in Q_1$ then f^a coincides with f on $V_1^\perp$. But one can easily check that there is a unique form $g \in \Delta_1$ distinct from f which coincides with f on $V_1^\perp$ and the element c switches f and g. This and (6.4.2) imply the following.

Lemma 6.4.4 *Every Q_1-orbit on Δ_1 is untwisted and of length* 2. □

Every form $f \in \Delta_1$ induces a quadratic form $\bar{f}$ of minus type on $\bar{V}_1$ and $\bar{g} = \bar{f}$ if and only if $g = f$ or $g = f^c$. Hence $\bar{G}_1 = G_1/Q_1$ acts on the set of pairs $\{f, f^c\}$ as it acts on the set of minus forms defined on $\bar{V}_1$ and associated with $\bar{\Psi}$. Hence $C_W(Q_1)$ is a module for $\bar{G}_1$ induced from a 1-dimensional $GF(3)$-module for its subgroup $O(\bar{f}) \cong O_{2n-2}^-(2)$.

Since $G_1 \cap O(f)$ is a maximal parabolic subgroup in $O(f)$ (the stabilizer of the 1-dimensional subspace V_1 isotropic with respect to f), it contains a Sylow 2-subgroup of $O(f)$. This means that Δ_1 is a twisted orbit for G_1 and the above mentioned 1-dimensional $GF(3)$-module for $O(\bar{f})$ is non-trivial. This gives the following.

Lemma 6.4.5 *As a $GF(3)$-module for $\bar{G}_1 = G_1/Q_1 \cong Sp_{2n-2}(2)$ the centralizer $C_W(Q_1)$ is isomorphic to the induced module $W(n-1)$. In particular* $\dim C_W(Q_1) = 2^{n-2}(2^{n-1}-1)$. □

For $1 \le j \le n-1$ let $N_j = \bigcap_{i=1}^{j} G_i$ so that $G_1 = N_1 > N_2 > \dots > N_{n-1} = P_n$. It is easy to see that $O_2(N_{i+1}) \ge O_2(N_i)$ and the image of N_{i+1} in $N_i/O_2(N_i) \cong Sp_{2n-2i}(2)$ is a maximal parabolic subgroup which is the stabilizer of a 1-dimensional subspace in the $(2n-2i)$-dimensional space $V_i^{\perp}/V_i$ for $1 \le i \le n-2$. By (6.4.5) this enables us to calculate the dimensions of the $C_W(O_2(N_i))$ inductively and we arrive at the following lemma (recall that $R_i = O_2(P_i)$).

Lemma 6.4.6 $\dim C_W(O_2(N_i)) = 2^{n-i-1}(2^{n-i}-1)$ *and in particular* $\dim C_W(R_n) = 1$. □

Notice that $C_W(O_2(N_{n-2}))$ is the 6-dimensional module $W(2)$ associated with $N_{n-2}/O_2(N_{n-2}) \cong Sp_4(2)$ and hence the latter equality in the above lemma is consistent with (6.2.7).

An important rôle in the subsequent construction will be played by the following.

Lemma 6.4.7 *The parabolic G_{n-1} normalizes but does not centralize $C = C_W(R_n)$.* □

Proof. Let $U = V_{n-1}$, Σ be the set of forms in Δ vanishing on U and

$$E = \bigoplus_{f \in \Sigma} W_f.$$

Then by the arguments before (6.4.6) $C_W(R_n) \le E$. Since the forms in Σ are of minus type, U is a maximal totally singular subspace to each of them. We claim that Q_{n-1} acts transitively on Σ. Recall that an element from G is contained in Q_{n-1} if and only if it acts trivially both on U and on $U^{\perp}/U$. Let $f, g \in \Sigma$. Then both f and g vanish on U and equal 1 on $U^{\perp} \setminus U$. There is a linear transformation a of V which maps f onto g. By Witt's theorem there is a linear transformation b of V which preserves

g and realizes $(a|_{U^\perp})^{-1}$ on $(U^\perp)^a$. Then ab maps f onto g (and hence preserves the form Ψ) and acts trivially on $U^\perp$. Hence $ab \in Q_{n-1}$ and the claim follows.

Since Σ is a Q_{n-1}-orbit, dim $C_E(Q_{n-1}) \leq 1$. On the other hand $C \leq C_E(Q_{n-1})$ since $Q_{n-1} \leq R_n$. Hence $C = C_E(Q_{n-1})$. Since E is invariant under G_{n-1} and Q_{n-1} is normal in G_{n-1}, we conclude that C is invariant under G_{n-1}. Finally by the arguments as before (6.4.6) the action of G_{n-1} on C is non-trivial. □

Lemma 6.4.8 $C_W(Q_n) = 0$.

Proof. Since V_n is maximal totally isotropic with respect to Ψ, for any two forms f and g from Δ their restrictions to V_n have radicals of codimension 1. Hence these restrictions are equivalent in the sense that there is a linear transformation of V_n which maps one restriction onto the other. Now arguments similar to those from the proof of the previous lemma show that there is an element in G_n which maps f onto g. Hence G_n is transitive on Δ and the orbits of Q_n on Δ are either all twisted or all untwisted. On the other hand there are twisted orbits since the element c from the centre of Q_1 acts trivially on V_n and hence $c \in Q_n$. □

Now we are ready to prove the main result of the section.

Proposition 6.4.9 *The group $\widehat{G} \cong W : G \cong W(n) : Sp_{2n}(2)$ contains (up to conjugation in its automorphism group) at most two subamalgams $\widetilde{\mathscr{A}}$ which map isomorphically onto $\mathscr{A}$ under φ. One of the subamalgams is contained in the complement G to W.*

Proof. Let $\widetilde{\mathscr{A}} = \{\widetilde{P}_i \mid 1 \leq i \leq n\}$ be as stated and $\widetilde{B}$ be the Borel subgroup of $\widetilde{\mathscr{A}}$. Then $\widetilde{B}$ is uniquely determined up to conjugation since it is a Sylow 2-subgroup of $\widehat{G}$. We assume that $\widetilde{B}$ is contained in the complement G to W. Let $\widetilde{R}_i$ be the preimage of R_i in $\widetilde{B}$ for $1 \leq i \leq n$. For $1 \leq i \leq n-1$ R_i contains Q_n whose centralizer in W is trivial by (6.4.8) and hence $\widetilde{P}_i = N_{\widehat{G}}(\widetilde{R}_i)$ is uniquely determined. By (6.4.6) dim $C_W(R_n) = 1$ and arguing as in the proof of (6.2.6) we see that there are three candidates for $\widetilde{P}_n$. One of then, say $\widetilde{S}_1$, is contained in the complement to W while two others, say $\widetilde{S}_2$ and $\widetilde{S}_3$, are permuted by the automorphism of $\widehat{G}$ which commutes with G and inverts W. □

6.5 Constructing T-geometries

We use the notation as in the previous section. Let $\widetilde{\mathscr{A}} = \{\widetilde{P}_i \mid 1 \leq i \leq n\}$ denote the subamalgam in $\widehat{G}$ which maps isomorphically onto $\mathscr{G}$ under the canonical homomorphism φ and such that, in terms of that of the proof of (6.4.9) $\widetilde{P}_n = \widetilde{S}_2$. For $k = n$ and $n-1$ put

$$J_k = \langle \widetilde{P}_i \mid 1 \le i \le n,\ i \ne k \rangle.$$

Lemma 6.5.1 *Under φ the subamalgam $\widetilde{\mathcal{D}} = \{J_n, J_{n-1}\}$ maps isomorphically onto the subamalgam $\mathcal{D} = \{G_n, G_{n-1}\}$.*

Proof. The preimage $\widetilde{Q}_n$ of Q_n in $\widetilde{B}$ is contained and normal in $\widetilde{P}_i$ for every $1 \le i \le n-1$. Hence $J_n \le N_{\widehat{G}}(\widetilde{Q}_n)$ and by (6.4.8) the restriction of φ to J_n is an isomorphism onto G_n. This implies that $\langle \widetilde{P}_i \mid 1 \le i \le n-2 \rangle$ (which is in $J_n \cap J_{n-1}$) maps isomorphically onto $G_n \cap G_{n-1}$. This means that all we have to show is that J_{n-1} maps isomorphically onto G_{n-1}. Let $C = C_W(R_n)$ and $F = \langle \widetilde{G}_{n-1}, C \rangle$. By (6.4.7) $F \cong C : G_{n-1}$ and since $Q_{n-1} \le R_n$, C is centralized by Q_{n-1}. Since C is centralized by R_n and normalized by G_{n-1}, it is easy to see that C is centralized by G_{n-1}^{+} and hence the latter is normal in F. Then F/G_{n-1}^{+} is an elementary abelian group of order 9 extended by a fixed-point free involution. Now it is easy to see that J_{n-1} and $\widetilde{P}_n$ have the same image in F/G_{n-1}^{+} isomorphic to Sym_3. This shows that J_{n-1} maps isomorphically onto G_{n-1} and the result follows. □

Lemma 6.5.2 *Let*

$$S_{n,n-1} = \langle \widetilde{P}_n, \widetilde{P}_{n-1} \rangle.$$

Then the quotient $S_{n,n-1}/O_2(S_{n,n-1})$ is isomorphic to the automorphism group of the geometry $\mathcal{G}(3 \cdot Sp_4(2))$.

Proof. Let $\widetilde{Q}_1$ be the preimage of Q_1 in $\widetilde{B}$. Then by (6.4.5)

$$N_{\widehat{G}}(\widetilde{Q}_1)/\widetilde{Q}_1 \cong C_W(Q_1) : (G_1/Q_1) \cong W(n-1) : Sp_{2n-2}(2).$$

Clearly $\widetilde{P}_n$ and $\widetilde{P}_{n-1}$ are contained in $N_{\widehat{G}}(\widetilde{Q}_1)$. Thus proceeding by induction on n, we reduce the calculations into the group $W(2) : Sp_4(2)$ and apply (6.2.8). □

As above let Γ be the dual polar graph of $\mathcal{G}$. Every element $U \in \mathcal{G}$ can be identified with the geometrical subgraph $\Gamma(U)$ so that the incidence relation is via inclusion. Let $\Gamma^{(i)} = \Gamma(V_i)$. Then $\Gamma^{(n)} = \{x\}$ is a vertex, $\Gamma^{(n-1)}$ is a triangle, $\Gamma^{(n-2)}$ is the point graph of the generalized quadrangle of order $(2,2)$ and for $1 \le i \le n-3$ $\Gamma^{(i)}$ is the dual polar graph of $\mathrm{res}^{+}_{\mathcal{G}}(V_i)$. Every edge of Γ is in a unique triangle which represents an element of type $n-1$ incident to both ends of the edge. The parabolic G_n is the stabilizer of x in G and it induces the natural doubly transitive action of $L_n(2)$ on the set of triangles containing x. For $1 \leq i \leq n-2$ the subgraph $\Gamma^{(i)}$ can also be defined as the one induced on all the images of x under G_i and since $\mathcal{G}$ is connected we have $G_i = \langle G_n \cap G_i, G_{n-1} \cap G_i \rangle$.

Let $J = J(n)$ be the subgroup of $\widehat{G} = \widehat{G}(n)$ generated by $\widetilde{\mathcal{D}} = \{J_n, J_{n-1}\}$. By (6.5.1) the restriction φ_1 of the canonical homomorphism φ to $\widetilde{\mathcal{D}}$ is an isomorphism onto $\mathcal{D} = \{G_n, G_{n-1}\}$. Define J_i to be the subgroup in J generated by $\varphi_1^{-1}(G_n \cap G_i)$ and $\varphi_1^{-1}(G_{n-1} \cap G_i)$ for $1 \leq i \leq n-2$. Let $\widetilde{\Gamma}$ be the graph whose vertices are the cosets of J_n in J and two vertices are adjacent if they intersect in a common coset of J_{n-1}. Then by (6.5.1) the restriction of φ to J induces a covering ψ of $\widetilde{\Gamma}$ onto Γ. Let $\widetilde{\Gamma}^{(n)} = \{\widetilde{x}\}$ be the vertex $J_n \cdot 1$ and $\widetilde{\Gamma}^{(n-1)}$ be the triangle formed by the vertices intersecting $J_{n-1} \cdot 1$. Then obviously $\psi(\widetilde{x}) = x$ and $\psi(\widetilde{\Gamma}^{(n-1)}) = \Gamma^{(n-1)}$. Let $\widetilde{\Gamma}^{(i)}$ be the subgraph in $\widetilde{\Gamma}$ induced by all the images of $\widetilde{x}$ under J_i. Because of (6.5.1) $\widetilde{\Gamma}^{(i)}$ can equivalently be defined as the connected component containing $\widetilde{x}$ of $\psi^{-1}(\Gamma^{(i)})$.

Let $\mathcal{J} = \mathcal{J}(n)$ be the geometry whose elements are all the vertices and triangles of $\widetilde{\Gamma}$ as well as all the images under J of the subgraphs $\widetilde{\Gamma}^{(i)}$ for $1 \leq i \leq n-2$. The incidence relation is via inclusion and the type function is inherited from $\mathcal{G}$.

Proposition 6.5.3 *$\mathcal{J} = \mathcal{J}(n)$ is a T-geometry of rank n, possessing a 1-covering onto $\mathcal{G}$, and $J = J(n)$ acts on $\mathcal{J}$ faithfully and flag-transitively.*

Proof. The only claim we still have to prove is that $\mathcal{J}$ is a T-geometry. For this we have to show that $\widetilde{\Gamma}^{(n-2)}$ is the point graph of $\mathcal{G}(3 \cdot Sp_4(2))$. Let L be the kernel of the action of G_{n-2} on $\mathrm{res}^+_{\mathcal{G}}(V_{n-2})$. Then L is the largest subgroup of $G_n \cap G_{n-1}$ which is normal in G_{n-2} and $G_{n-2}/L \cong G^+_{n-2} \cong Sp_4(2)$. Let $\widetilde{L} = \varphi_1^{-1}(L)$. Then $\widetilde{L}$ is the largest subgroup of $J_n \cap J_{n-1}$ normal in J_{n-2}. Then $J_{n-2}/\widetilde{L} \cong S_{n,n-1}\widetilde{L}/\widetilde{L}$ and $S_{n,n-1} \cap L = O_2(S_{n,n-1})$. Hence by (6.5.2) we have $J_{n-2}/\widetilde{L} \cong 3 \cdot Sp_4(2)$. □

It follows from the construction that $\{J_i \mid 1 \leq i \leq n\}$ is the amalgam of maximal parabolics which corresponds to the action of J on $\mathcal{J}$.

6.6 The rank 3 case

Let $\mathcal{J} = \mathcal{J}(n)$ and $J = J(n)$ be as in (6.5.3). In this and the next two sections we specify the structure of J and show that $\mathcal{J}$ is 2-simply connected.

Recall that, if $\mathcal{H}$ is a geometry, we write $\mathcal{H}^i$ for the set of elements of type i in $\mathcal{H}$ and, if Θ is a flag in $\mathcal{H}$, then $\mathrm{res}_{\mathcal{H}}(\Theta)^i$ denotes the set of elements of type i incident to Θ. Let K be the kernel of the homomorphism of J onto $G \cong Sp_{2n}(2)$, which is the restriction of φ. Then $K = J \cap W$, in particular K is an elementary abelian 3-group. It is easy to deduce from the proof of (6.5.3) that $\widetilde{G}_{n-2} \cap K$ is of order 3. In view of the flag-transitivity this implies that for every element x of type $n-2$ in $\mathcal{J}$ the intersection K_x of K and the stabilizer of x in J is a subgroup of order 3.

Lemma 6.6.1 $K = \langle K_x \mid x \in \mathcal{J}^{n-2}\rangle$.

Proof. It is clear that $L = \langle K_x \mid x \in \mathcal{J}^{n-2}\rangle$ is normal in J and the image in J/L of the amalgam $\{J_{n-2}, J_{n-1}, J_n\}$ is isomorphic to the subamalgam $\{G_{n-2}, G_{n-1}, G_n\}$ in G. By (6.3.4) $J/L \cong G \cong Sp_{2n}(2)$ and the result follows. □

We are intending show that in order to generate K it is sufficient to take the subgroups K_x for all elements x of type $n-2$ incident to a fixed element of type n. For this purpose we use the following result established in [Hei91] and independently in [ISh89b] (in both cases computer calculations were used).

Proposition 6.6.2 *Let $\mathcal{F}$ be a T-geometry of rank 3 and F be a flag-transitive automorphism group of $\mathcal{F}$. Suppose that $\mathcal{F}$ possesses a 1-covering onto $\mathcal{G}(Sp_6(2))$ which commutes with the action of F. Then F is isomorphic to a non-split extension $3^7 \cdot Sp_6(2)$, $O_3(F)$ is isomorphic to the E_7-lattice taken modulo 3 and $F/O_3(F) \cong Sp_6(2)$ acts irreducibly on $O_3(F)$ as a subgroup of $Cox(E_7) \cong Sp_6(2) \times 2$.* □

This immediately gives

Corollary 6.6.3 *The geometry $\mathcal{J}(3)$ is simply connected and $J(3)$ is isomorphic to the group F from* (6.6.2). □

In the remainder of the section we deal with the case $n = 3$ only. Let y be an element of type 3 in $\mathcal{J}$ stabilized by J_3 in J. Since J_3 maps isomorphically onto a maximal parabolic in $Sp_6(2)$ and $J_3 \cong 2^6 : L_3(2)$ by (6.3.5), where $O_2(J_3)$ is a direct summand of the permutational module of $J_3/O_2(J_3) \cong L_3(2)$ acting naturally and doubly transitively on the elements of type 1 incident to y. For $x \in \mathcal{J}^1$ let K_x be the subgroup of order 3 in K as in (6.6.1). Then J_3 stabilizes as a whole the set

$$B_y = \{K_x \mid x \in \mathrm{res}_{\mathcal{J}}(y)^1\}$$

of seven such subgroups indexed by the elements of type 1 incident to y. The subgroup $O_2(J_3)$ normalizes every $K_x \in B_y$. Moreover, $q \in O_2(J_3)$ inverts K_x if and only if q projects onto an odd element in $J(x)/O_{2,3}(J(x)) \cong Sym_6$ and by (6.3.6) there are such elements in $O_2(J_3)$. Since $J_3/O_2(J_3)$ permutes the subgroups in B_y doubly transitively, this shows that different subgroups in B_y have different centralizers in $O_2(J_3)$. Hence the subgroups in B_y are linearly independent and hence they generate the whole of K.

Lemma 6.6.4 *Let* $n = 3$ *and* $y \in \mathcal{J}^3$. *Then the subgroups in the set* $B_y = \{K_x \mid x \in \mathrm{res}_{\mathcal{J}}(y)^1\}$ *are linearly independent and generate* K. □

The set B_y of subgroups from K as in (6.6.4) will be called *the special basis of* K *associated with* y. Of course, in order to obtain a basis in the ordinary sense one should choose a non-trivial vector from each of the subspaces in B_y, but most of our arguments are independent of such a choice (a similar convention will be assumed for $n \geq 4$). In particular for every cyclic subgroup $M \in K$ there is a well defined *support of* M *with respect to* B_y which is the set of elements from $x \in \mathrm{res}_{\mathcal{J}}(y)^1$ such that a vector from K_x has a non-zero component in a decomposition of $m \in M^{\#}$ in the basis consisting of vectors from the K_x. By (6.6.2) there is a quadratic form f on K which is preserved by J/K and one may observe that in the basis B_y the form f can be written as a sum of squares.

Let x and z be elements of type 1 in $\mathcal{J}$ which are in the same K-orbit. Then, since K is abelian, we have $K_x = K_z$. On the other hand K-orbits on $\mathcal{J}^1$ are indexed by the elements of $\mathcal{G}^1$. Since $G \cong Sp_6(2)$ acts primitively on the set $\mathcal{G}^1$ of size 63, we conclude that the set

$$B = \{K_x \mid x \in \mathcal{J}^1\}$$

contains 63 subgroups and we know that 7 of them are in B_y.

The structure of $O_2(J_3)$ implies that for any subset in B_y of even size there is exactly one element in $O_2(J_3)$ which inverts every subspace from the subset and centralizes the remaining subspaces. In addition there is a unique projective plane structure $\pi(y)$ on B_y which is preserved by $J_3/O_2(J_3)$. This specifies K as a module for J_3 and enables one to calculate the orbits of J_3 on the set of 1-dimensional subspaces in K. Clearly $B \setminus B_y$ is a union of some of these orbits. The calculations (which are quite elementary) show that only one union of orbits has size 56, which is an orbit itself. To wit, for a subgroup from this orbit its support with respect to B_y is a complement of a line in $\pi(y)$. For further reference we state this fact explicitly in the following.

Lemma 6.6.5 *Let $n = 3$, $y \in \mathscr{J}^3$ and $z \in \mathscr{J}^1$ so that $K_z \notin B_y$. Then the support of K_z with respect to B_y is the complement of a line in $\pi(y)$.* □

This lemma provides us with a rule for rewriting one special basis in terms of another one. Let $z \in \mathscr{J}^2$ and y, y', y'' be the elements of type 3 incident to z. Then $l = \{K_x \mid x \in \mathrm{res}_{\mathscr{J}}(z)^1\}$ is of size 3. It is clear that $l = B_y \cap B_{y'} \cap B_{y''}$ and that l is a line in $\pi(y)$, $\pi(y')$ and $\pi(y'')$. For $i = 1$ and 2 let v_i be distinct elements of type 1 incident to y' but not to z and a_i be non-trivial elements from K_{v_i}. Since K_{v_i} is orthogonal to every subspace in l, by (6.6.5) the support of a_i in the basis B_y is the complement of l in $\pi(y)$. Since a_1 and a_2 are orthogonal, a_1a_2 has two non-zero coordinates in the basis B_y. Finally, the set D of elements in K whose support is $\pi(y) \setminus l$ has size 2^4 and it is closed under taking inverses. Thus 2^3 subgroups of order 3 in K have support $\pi(y) \setminus l$. On the other hand 2^3 such subgroups are in $(B_{y'} \setminus B_y) \cup (B_{y''} \setminus B_y)$ and hence we have the following.

Lemma 6.6.6 *Let $n = 3$, $z \in \mathscr{J}^2$ and $\{y, y', y''\} = \mathrm{res}_{\mathscr{J}}(z)^3$. Let K_{v_1} and K_{v_2} be distinct subgroups from $B_{y'} \setminus B_y$. Let a_1 and a_2 be non-trivial elements from K_{v_1} and K_{v_2}, respectively. Then for $i = 1$ and 2 the support of a_i with respect to the special basis B_y is the 4-element set $B_y \setminus B_{y'}$; the product a_1a_2 in this basis has exactly two non-zero components. Moreover, every subgroup of order 3 in K whose support is $B_y \setminus B_{y'}$ is contained in $B_{y'} \cup B_{y''}$.* □

6.7 Identification of $J(n)$

We follow notation introduced in Section 6.5 and assume that $n \geq 4$. By (6.6.2) if $u \in \mathscr{J}^{n-3}$ then $\mathrm{res}^+_{\mathscr{J}}(u)$ is isomorphic to $\mathscr{J}(3)$ which is the T-geometry of the group $3^7 \cdot Sp_6(2)$. This implies in particular, that the subgroups in $\{K_x \mid x \in \mathrm{res}_{\mathscr{J}}(u)^{n-2}\}$ generate in K an elementary abelian subgroup of order 3^7. Moreover, by (6.6.4), if y is an element of type n

incident to u then

$$\{K_x \mid x \in \mathrm{res}_{\mathcal{J}}(\{u,y\})^{n-2}\}$$

is a special basis for this subgroup.

For an element y of type n in $\mathcal{J}$ let

$$B_y = \{K_x \mid x \in \mathrm{res}_{\mathcal{J}}(y)^{n-2}\}.$$

Let y and y' be elements of type n which are incident to a common element z of type $n-1$. We are intending to express the non-trivial elements from every subgroup $K_v \in B_{y'}$ in terms of elements contained in the subgroups from B_y. If v is incident to z then $K_v \in B_y$ and the expression is obvious. So suppose that v and z are not incident. Since $\mathrm{res}_{\mathcal{J}}(y')$ is a projective geometry, in this case there is a unique element u of type $n-3$ which is incident to both v and z. Since u is incident to z, it is also incident to y and y'. For pairwise incident elements u, z and y of type $n-3$, $n-1$ and n, respectively, let $B(u,z,y)$ denote the set of subgroups K_x for all elements x of type $n-2$ which are incident to u and y but not to z. It is easy to see from the diagram of $\mathcal{J}$ that $B(u,z,y)$ contains four different subgroups. Now (6.6.6) gives the following.

Proposition 6.7.1 *In the above terms suppose that $n \geq 4$. Let $z \in \mathcal{J}^{n-1}$ and $\{y, y', y''\} = \mathrm{res}_{\mathcal{J}}(z)^n$. For $i = 1$ and 2 let v_i be an element of type $n-2$ incident to y' but not to z, let u_i be the unique element of type $n-3$ incident to both v_i and z (so that $K_{v_i} \in B(u_i, z, y')$) and let a_i be a non-trivial element from K_{v_i}. Then*

(i) *a_1 is a product of four non-trivial elements taken from different subgroups in $B(u_1, z, y)$,*

(ii) *if $u_1 \neq u_2$ then $B(u_1, z, y') \cap B(u_2, z, y') = \emptyset$,*

(iii) *if $u_1 = u_2$ then $a_1 a_2$ is a product of exactly two non-trivial elements taken from different subgroups in $B(u_1, z, y)$,*

(iv) *if d is any product of four non-trivial elements taken from different subgroups in $B(u_1, z, y)$, then $d \in K_x$ for some $K_x \in B(u_1, z, y') \cup B(u_1, z, y'')$.* □

Proposition 6.7.2 *In the above terms if $y \in \mathcal{J}^n$ and $K_y = \langle K_x \mid x \in \mathrm{res}_{\mathcal{J}}(y)^{n-2}\rangle$ then $K = K_y$.*

Proof. By (6.7.1) if y' is incident with y to a common element of type $n-1$ then $K_{y'} \leq K_y$ and hence $K_{y'} = K_y$ because of the obvious symmetry. Now the result immediately follows from the connectedness of $\mathcal{J}$. □

Now we are ready to specify the structure of $J(n)$.

Proposition 6.7.3 *If $n \geq 3$ then $J(n)$ is a non-split extension of $G(n) \cong Sp_{2n}(2)$ by an elementary abelian 3-group $K(n)$ of rank $\sigma(n) = [{n \atop 2}]_2$. The action of $J(n)/K(n)$ on $K(n)$ is irreducible.*

Proof. In view of (6.6.2) we can assume that $n \geq 4$. Let y be an element of type n in $\mathcal{J}$ stabilized by J_n in J. By (6.7.2) K is generated by the order 3 subgroups from the set $B_y = \{K_x \mid x \in \text{res}_{\mathcal{J}}(y)^{n-2}\}$ whose size is exactly $\sigma(n)$. The subgroup J_n maps isomorphically onto the parabolic G_n in G. From the basic properties of the latter group it is easy to deduce (6.3.6) that for every $K_x \in B_y$ there is an element in $O_2(J_n)$ which inverts K_x. On the other hand $J_n/O_2(J_n) \cong L_n(2)$ acts primitively on the set of subgroups in B_y. Hence different subgroups from B_y have different centralizers in $O_2(J_n)$. This implies that the subgroups in B_y are linearly independent and that the action of J_n (and hence also of J/K) on K is irreducible.

There remains to prove that J does not split over K. Let t be an element of type 1 incident to y and stabilized by J_1 and let $Q = O_2(J_1)$. Then by the previous paragraph J_1/Q is some extension of $Sp_{2n-2}(2)$ by an elementary abelian 3-group of rank $\sigma(n-1)$. The latter group is generated by the subgroups K_x for $x \in \text{res}_{\mathcal{J}}(\{y,t\})^{n-2}$. Therefore $\dim C_K(Q) \geq \sigma(n-1)$. On the other hand $N := J_1 \cap J_n$ acting on the set of subgroups from B_y has two orbits consisting of the subgroups $K_x \in B_y$ with x incident and non-incident to y, respectively. Since Q is normal in N and acts faithfully on K, it cannot centralize subspaces in both orbits, so $\dim C_K(Q) = \sigma(n-1)$. Hence $J_1 = N_{J(n)}(Q)$ and if J were split over K, J_1 would split over $O_3(J_1)$. Now the non-splitness follows by induction since for $n = 2$ we have a non-split extension $3 \cdot Sym_6 \cong 3 \cdot Sp_4(2)$. □

6.8 A special class of subgroups in $J(n)$

We start by constructing a family $E = E(n)$ of subgroups of order 3 in $K = K(n)$ such that J/K acts doubly transitively on $\mathcal{E}$ by conjugation with stabilizer isomorphic to $O^-_{2n}(2)$. After that we show that a similar family of subgroups must exist in the automorphism group of the universal 2-cover of $\mathcal{J}$.

Lemma 6.8.1 *There is a non-singular quadratic form on K preserved by J/K.*

Proof. For $y \in \mathcal{J}^n$ let us define a quadratic form χ_y to be the sum of squares in the basis B_y. This form is obviously invariant under the action of the stabilizer of y in J. Since $\mathcal{J}$ is connected, to prove the lemma it is sufficient to show that $\chi_{y'} = \chi_y$ whenever y and y' are incident to a common element of type $n-1$. But the equality easily follows from the rewriting rules given in (6.7.1). □

Corollary 6.8.2 *The subgroup K, considered as a $GF(3)$-module for J/K, is self-dual.* □

Lemma 6.8.3 *There is a family $\mathcal{E}$ of 1-dimensional subspaces in K, such that J/K acting on K by conjugation preserves $\mathcal{E}$ as a whole and induces on it a doubly transitive action of $Sp_{2n}(2)$ on the cosets of $O_{2n}^-(2)$.*

Proof. By (6.8.2) it is sufficient to indicate the required family of subspaces in the module dual to K. Recall that J was constructed as a subgroup of the semidirect product $W : G$ where $G \cong Sp_{2n}(2)$ and W is a $GF(3)$-module for G induced from a non-trivial 1-dimensional module for a subgroup O in G isomorphic to $O_{2n}^-(2)$. So K is a submodule in W. Since W is an induced module, it possesses a direct sum decomposition into 1-dimensional subspaces W_f indexed by the cosets of O in G and G induces on the set of these subspaces a doubly transitive action. Then the desired set of hyperplanes in K is formed by the kernels of the projections of K onto the subspaces W_f. □

Let us calculate the support with respect to the special basis B_y of a subspace $E \in \mathcal{E}$. Let O and J_n be the stabilizers in J of E and y, respectively. Then the orbits of $O \cap J_n$ on B_y are the same as the orbits of $\varphi(O \cap J_n)$ on the set of elements of type $n-2$ incident to the image of y in $\mathcal{G} = \mathcal{G}(Sp_{2n}(2))$. Hence the calculations of the orbits can be carried out in the latter geometry. In terms of Section 6.4 we can assume that the image of y in $\mathcal{G}$ is the maximal totally singular subspace V_n and $\varphi(O)$ is the stabilizer $O(f) \cong O_{2n}^-(2)$ of a quadratic form f of minus type associated with the symplectic form Ψ. Then B_y is in the natural bijection with the set of codimension 2 subspaces in V_n. Let $U = U(f)$ be the unique subspace of codimension 1 in V_n which is totally singular with respect to f. Then by Witt's theorem the orbit under $O(f) \cap G_n$ of a subspace X of codimension 2 in V_n depends on whether or not U contains X. Let

$\xi = \xi(y, E)$ be the unique element from $\text{res}_{\mathcal{J}}(y)^{n-1}$ which maps onto U under the 1-covering of $\mathcal{J}$ onto $\mathcal{G}$. Thus the following two sets are the orbits of $O \cap L$ on B_y:

$$\Theta_1(y, \xi) = \{K_x \mid x \in \text{res}_{\mathcal{J}}(\xi)^{n-2}\}, \quad \Theta_2(y, \xi) = B_y \setminus \Theta_1(y, \xi).$$

Lemma 6.8.4 *In the above terms $\Theta_1(y, \xi)$ is the support of $E \in \mathcal{E}$ with respect to the basis B_y.*

Proof. Since $O \cap J_n$ acts transitively on $\Theta_i(y, \xi)$ for $i = 1$ and 2, the support is either one of these orbits or the whole of B_y. Suppose that $\Theta_2(y, \xi)$ is contained in the support. Then E is not orthogonal (in the sense of (6.8.1)) to any subspace $K_x \in \Theta_2(y, \xi)$. Let $\{y, y', y''\}$ be the set of elements of type n incident to ξ. It is clear that $\Theta_1(y', \xi) = \Theta_1(y, \xi)$, hence the support of E with respect to $B_{y'}$ contains $\Theta_2(y', \xi)$ and E is not orthogonal to any $K_x \in \Theta_2(y', \xi)$ (and similarly for y''). Let u be an element of type $n-3$ incident to ξ and let w_1, w_2, w_3, w_4 be the elements of type $n-2$ incident to u and y but not to ξ. Then by the assumption made, E is generated by a product $e_1e_2e_3e_4f$ where e_i is a non-trivial element from K_{w_i} for $1 \leq i \leq 4$, and f is an element orthogonal to $\langle K_{w_i} \mid 1 \leq i \leq 4 \rangle$. By (6.7.1 (iv)) the subgroup I generated by the product $e_1e_2e_3^{-1}e_4^{-1}$ is contained in $\Theta_2(y', \xi) \cup \Theta_2(y'', \xi)$. Since I is orthogonal to E this is a contradiction. □

6.9 The $\mathcal{J}(n)$ are 2-simply connected

The 2-simple connectedness of $\mathcal{J}(3)$ follows from (6.6.2) so we assume in this section that $n \geq 4$.

Let $\mathcal{S} = \{S_{ij} \mid 1 \leq i < j \leq n\}$ be the amalgam of rank 2 parabolic subgroups associated with the action of J on $\mathcal{J}$. Let $\widehat{J}$ be the universal completion of $\mathcal{S}$. Then there is a homomorphism $\varepsilon : \widehat{J} \to J$ such that the composition δ of ε and the restriction of φ to J is a homomorphism onto $G \cong Sp_{2n}(2)$ which maps $\mathcal{S}$ onto the amalgam $\mathcal{C}$ of rank 2 parabolics associated with the action of G on $\mathcal{G}$.

Lemma 6.9.1 *$\widehat{J}$ is a flag-transitive automorphism group of a rank n T-geometry $\widehat{\mathcal{J}}$ possessing a 2-cover ω onto $\mathcal{J}$. Moreover, ε is an isomorphism if and only if ω is an isomorphism.*

Proof. For $k = n$ and $n-1$ let $\widehat{J}_k$ be the subgroups in $\widehat{J}$ generated by the subamalgams $\mathcal{S}^{(k)} = \{S_{ij} \mid 1 \leq i < j \leq n, k \notin \{i, j\}\}$. The subamalgams

$\mathscr{S}^{(n)}$ and $\mathscr{S}^{(n-1)}$ map under δ isomorphically onto the subamalgams $\mathscr{C}^{(n)}$ and $\mathscr{C}^{(n-1)}$, respectively. Hence by (6.3.4) each mapping in the following sequence is an isomorphism:

$$\{\widehat{J}_n, \widehat{J}_{n-1}\} \to \{J_n, J_{n-1}\} \to \{G_n, G_{n-1}\}.$$

Let $\widehat{\Gamma}$ be a graph whose vertices are the cosets of $\widehat{J}_n$ in $\widehat{J}$ with two vertices being adjacent if they intersect a common coset of $\widehat{J}_{n-1}$. Then the above isomorphisms induce the following sequence of graph coverings:

$$\widehat{\Gamma} \to \widetilde{\Gamma} \to \Gamma.$$

Define $\widehat{\mathscr{J}}$ to be the geometry whose elements are the vertices and the triangles in $\widehat{\Gamma}$ and the connected components of the full preimages of the subgraphs representing the elements of $\mathscr{J}$ in $\widetilde{\Gamma}$ (equivalently of the elements of $\mathscr{G}$ in Γ), where the incidence relation is via inclusion and the type function is the obvious one. Then $\widehat{\mathscr{J}}$ possesses 1-coverings onto $\mathscr{J}$ and $\mathscr{G}$. Now arguments as in the proof of (6.5.3) show that $\widehat{\mathscr{J}}$ is a T-geometry possessing a 2-cover ω onto $\mathscr{J}$. □

Let $\{\widehat{J}_i \mid 1 \le i \le n\}$ be the amalgam of maximal parabolic subgroups corresponding to the action of $\widehat{J}$ on $\widehat{\mathscr{J}}$. We assume that ε and δ map $\widehat{J}_i$ onto J_i and G_i, respectively. The elements of $\widehat{\mathscr{J}}$ will be denoted by letters with hats; the same letter without a hat will denote the image in $\mathscr{J}$. A similar convention will be applied to the elements of $\widehat{J}$.

Let $\widehat{K}$ be the kernel of δ. Then clearly ε maps $\widehat{K}$ onto K. It is easy to see from (6.9.1) that $\widehat{J}_{n-2} \cap \widehat{K}$ is of order 3 and hence with every element $\widehat{x}$ of type $n-2$ in $\widehat{\mathscr{J}}$ we can associate a unique subgroup $\widehat{K}_{\widehat{x}}$ of order 3 in $\widehat{K}$ which stabilizes $\widehat{x}$. Moreover, $\widehat{K}_{\widehat{x}}$ maps onto K_x under ε. Let $\widehat{u}$ be an element of type $n-3$ in $\widehat{\mathscr{J}}$. Then by (6.6.2) both $\text{res}^{+}_{\widehat{\mathscr{J}}}(\widehat{u})$ and $\text{res}^{+}_{\mathscr{J}}(u)$ are isomorphic to the rank 3 T-geometry of the group $3^7 \cdot Sp_6(2)$ and we have the following.

Lemma 6.9.2 *In the above notation the group*

$$\langle \widehat{K}_{\widehat{x}} \mid \widehat{x} \in \text{res}_{\widehat{\mathscr{J}}}(\widehat{u})^{n-2} \rangle$$

is elementary abelian of order 3^7 and it maps isomorphically onto its image in K. In particular $\widehat{K}_{\widehat{x}}$ and $\widehat{K}_{\widehat{z}}$ commute whenever $\widehat{x}$ and $\widehat{z}$ are incident to a common element of type $n-3$. □

For an element $\widehat{y}$ of type n in $\widehat{\mathscr{J}}$ define

$$\widehat{B}_{\widehat{y}} = \{\widehat{K}_{\widehat{x}} \mid \widehat{x} \in \text{res}_{\widehat{\mathscr{J}}}(\widehat{y})^{n-2}\}.$$

In view of (6.9.2), arguing as in (6.7.1) and (6.7.2) we observe that the subgroups in $\widehat{B}_{\widehat{y}}$ generate the whole of $\widehat{K}$. Moreover, if $\widehat{y}'$ is an element of type n incident with $\widehat{y}$ to a common element of type $n-1$ then there is a canonical way to express elements from the subgroups in $\widehat{B}_{\widehat{y}'}$ in terms of elements from the subgroups in $\widehat{B}_{\widehat{y}}$ as in (6.7.1).

Lemma 6.9.3 *The homomorphism $\varepsilon : \widehat{J} \to J$ is an isomorphism if and only if any two subgroups from $\widehat{B}_{\widehat{y}}$ commute.*

Proof. The "only if" part is obvious so suppose that the subgroups in $\widehat{B}_{\widehat{y}}$ pairwise commute. Since $\widehat{K}$ is generated by these subgroups, we conclude that $\widehat{K}$ is elementary abelian of rank at most the number of subgroups in $\widehat{B}_{\widehat{y}}$. Since the latter number is the rank of K, this means that ε restricted to $\widehat{K}$ is an isomorphism onto K. Since $\widehat{J}/\widehat{K} \cong J/K \cong Sp_{2n}(2)$ the result follows. □

Whenever two elements of type $n-2$ in $\widehat{\mathcal{J}}$ are incident to a common element of type n they are always incident to a common element of type $n-4$. Thus by the above lemma the isomorphism of $\widehat{J}$ and J for $n = 4$ would imply the isomorphism for the higher ranks. But we are going to prove the isomorphism uniformly on n by constructing in $\widehat{K}$ a family $\widehat{\mathcal{E}}$ of subgroups of order 3 similar to the family $\mathcal{E}$ in K constructed in Section 6.8.

For an element $\widehat{\xi}$ of type $n-1$ in $\widehat{\mathcal{J}}$ put

$$\widehat{Y}(\widehat{\xi}) = \langle \widehat{K}_{\widehat{x}} \mid \widehat{x} \in \mathrm{res}_{\widehat{\mathcal{J}}}(\widehat{\xi})^{n-2} \rangle.$$

Since $\mathrm{res}_{\widehat{\mathcal{J}}}(\widehat{\xi})$ is a projective geometry, any two elements of type $n-2$ incident to $\widehat{\xi}$ are incident to a common element of type $n-3$. Hence $\widehat{Y}(\widehat{\xi})$ is an elementary abelian 3-group of rank $2^{n-1}-1$ which is the number of elements of type $n-2$ incident to $\widehat{\xi}$ and it maps isomorphically onto its image $Y(\xi)$ in K, where

$$Y(\xi) = \langle K_x \mid x \in \mathrm{res}_{\mathcal{J}}(\xi)^{n-2} \rangle.$$

By (6.8.4) every subgroup $E \in \mathcal{E}$ is contained in $Y(\xi)$ for a unique element $\xi = \xi(y, E)$ of type $n-1$ incident to y. In the above terms let $\widehat{E}(\widehat{y})$ be the unique subgroup in $\widehat{Y}(\widehat{\xi})$ which maps onto E under ε (here $\widehat{\xi}$ is the unique element of type $n-1$ incident to $\widehat{y}$ which maps onto ξ) and let $\widehat{\mathcal{E}}(\widehat{y})$ be the set of all subgroups obtained in this way. Clearly $\widehat{\mathcal{E}}(\widehat{y})$ maps bijectively onto $\mathcal{E}$.

Lemma 6.9.4 *The family $\widehat{\mathscr{E}}(\widehat{y})$ is independent of the particular choice of $\widehat{y}$.*

Proof. Because of the connectedness it is sufficient to show that whenever $\widehat{y'}$ is an element of type n incident with $\widehat{y}$ to a common element $\widehat{z}$ of type $n-1$, we have $\widehat{\mathscr{E}}(\widehat{y'}) = \widehat{\mathscr{E}}(\widehat{y})$. In order to prove the latter equality it is sufficient to show that every subgroup $\widehat{E} \in \widehat{\mathscr{E}}(\widehat{y'})$ is contained in $\widehat{\mathscr{E}}(\widehat{y})$. Thus let $\widehat{E} \in \widehat{\mathscr{E}}(\widehat{y'})$ and $\widehat{\xi}$ be the element of type $n-1$ incident to $\widehat{y'}$ such that $\widehat{E}$ is the unique preimage in $\widehat{Y}(\widehat{\xi})$ of a subgroup $E \in \mathscr{E}$. If $\widehat{z} = \widehat{\xi}$ (equivalently if $\widehat{y}$ is incident to $\widehat{\xi}$), then $\widehat{E} \in \widehat{\mathscr{E}}(\widehat{y})$ by the definition. Thus we assume that $\widehat{z} \neq \widehat{\xi}$. Let $\widehat{e}$ be a non-trivial element from $\widehat{E}$. Then $\widehat{e} = \widehat{d}_1\widehat{d}_2...\widehat{d}_s$, where $s = 2^{n-1} - 1$ and the $\widehat{d}_i$ are non-trivial elements taken from different subgroups in the set

$$\widehat{B}(\widehat{\xi}) = \{\widehat{K}_{\widehat{x}} \mid \widehat{x} \in \mathrm{res}_{\widehat{\mathscr{G}}}(\widehat{\xi})^{n-2}\}.$$

Since the subgroups in $\widehat{B}(\widehat{\xi})$ commute pairwise, the factors $\widehat{d}_i$ can be rearranged in an arbitrary way. Let $\widehat{v}_1$ be the unique element of type $n-2$ incident to both $\widehat{\xi}$ and $\widehat{z}$. Let $\widehat{w}_1, ..., \widehat{w}_t$ be the elements of type $n-3$ incident to $\widehat{v}_1$, where $t = 2^{n-2} - 1$. Then for $1 \leq i \leq t$ the element $\widehat{w}_i$ is incident to three elements of type $n-2$ incident to $\widehat{\xi}$, one of them is $\widehat{v}_1$ and two others we denote by $\widehat{v}_{2i}$ and $\widehat{v}_{2i+1}$. In this case

$$\mathrm{res}_{\widehat{\mathscr{G}}}(\widehat{\xi})^{n-2} = \{\widehat{v}_j \mid 1 \leq j \leq 2^{n-1} - 1\}.$$

Since the subgroups in $\widehat{B}(\widehat{\xi})$ commute pairwise, without loss of generality we can assume that $\widehat{d}_j \in \widehat{K}_{\widehat{v}_j}$ for $1 \leq j \leq s$. Then by the analogue of (6.7.1 (iii)) the product $\widehat{d}_{2i}\widehat{d}_{2i+1}$ is equal to the product $\widehat{f}_{2i}\widehat{f}_{2i+1}$ of two non-trivial elements taken from different subgroups in the set $\widehat{B}_{\widehat{y}}$. Since $\widehat{K}_{\widehat{v}_1}$ belongs to $\widehat{B}_{\widehat{y}}$ as well, we conclude that $\widehat{e} = \widehat{f}_1\widehat{f}_2...\widehat{f}_s$, where the $\widehat{f}_i$ are non-trivial elements taken from different subgroups in $\widehat{B}_{\widehat{y}}$. Then $f_1f_2...f_s$ is the decomposition of e in the basis B_y. By (6.8.4) there is an element v of type $n-1$ incident to y such that the f_i are non-trivial elements taken one from each subgroup from $\{K_x \mid x \in \mathrm{res}_{\mathscr{G}}(v)^{n-2}\}$. Hence $e \in Y(v)$ and $\widehat{e}$ is the unique preimage of e in $\widehat{Y}(\widehat{v})$ where $\widehat{v}$ is the element incident to $\widehat{y}$ which maps onto v. Thus $\widehat{E} \in \widehat{\mathscr{E}}(\widehat{y})$ and the result follows. □

By the above lemma we can denote the family $\widehat{\mathscr{E}}(\widehat{y})$ simply by $\widehat{\mathscr{E}}$.

Lemma 6.9.5 *$\widehat{\mathscr{E}}$ is a conjugacy class of subgroups in $\widehat{J}$; $\widehat{J}$ acting on $\widehat{\mathscr{E}}$ induces the doubly transitive action of $Sp_{2n}(2)$ on the cosets of $O^-_{2n}(2)$ and $\widehat{K}$ is in the kernel of the action.*

Proof. Let $\widehat{\xi}$ be the element of type $n-1$ in $\widehat{\mathcal{J}}$ stabilized by $\widehat{J}_{n-1}$. Since the latter group maps isomorphically onto J_{n-1} and $\widehat{Y}(\widehat{\xi})$ maps isomorphically onto $Y(\xi)$, it is easy to see that $\widehat{J}_{n-1}$ stabilizes as a whole the set of subgroups from $\widehat{\mathcal{E}}$ contained in $\widehat{Y}(\widehat{\xi})$. Let $\widehat{y}$ be the element of type n stabilized by $\widehat{J}_n$. It is easy to see from the above that $\widehat{J}_n$ stabilizes $\widehat{\mathcal{E}} = \widehat{\mathcal{E}}(\widehat{y})$ as a whole. Now the result follows from (6.9.4). □

Since the subgroups in $\widehat{\mathcal{E}}$ have order 3, none of them can be inverted by an element from $\widehat{K}$. Hence by (6.9.5) the subgroups from $\widehat{\mathcal{E}}$ are contained in the centre of $\widehat{K}$. Thus in order to prove that $\widehat{K}$ is abelian, it is sufficient to show that $\widehat{K}$ is generated by the subgroups in $\widehat{\mathcal{E}}$. The subgroup $\widehat{Y}(\widehat{\xi})$ for an element $\widehat{\xi}$ of type $n-1$ contains subgroups from $\widehat{\mathcal{E}}$ as well as subgroups $\widehat{K}_{\widehat{x}}$ which are known to generate $\widehat{K}$. Thus the statement we need is a direct consequence of the following lemma whose proof is very similar to that of (6.7.3).

Lemma 6.9.6 *The stabilizer of $\widehat{\xi}$ in $\widehat{J}$ acts irreducibly on $\widehat{Y}(\widehat{\xi})$.* □

Thus $\widehat{K}$ is abelian, $\widehat{J}$ is isomorphic to J and we have the main result of the section.

Proposition 6.9.7 *The geometries $\mathcal{J}(n)$ are 2-simply connected for all $n \geq 3$.* □

6.10 A characterization of $\mathcal{J}(n)$

We start with an elementary but important result about the point graph of the rank 2 T-geometry. Let Θ and $\widetilde{\Theta}$ be the point graphs of $\mathcal{G}(Sp_4(2))$ and $\mathcal{G}(3 \cdot Sp_4(2))$, respectively, and let $\mu : \widetilde{\Theta} \to \Theta$ be the corresponding covering of graphs. The following result can be deduced directly from the intersection diagram of $\widetilde{\Theta}$.

Lemma 6.10.1 *The subgroup of the fundamental group of Θ associated with μ is generated by the cycles of length 3 and by the non-degenerate cycles of length 5; it does not contain cycles of length 4.* □

Let Γ be the dual polar graph of $\mathcal{G}$ and $\widetilde{\Gamma}$ be the graph on the set of elements of type n in $\mathcal{J}$ and $\psi : \widetilde{\Gamma} \to \Gamma$ be the corresponding covering of graphs. Then Γ is a near n-gon with quads. A quad Θ is the subgraph induced by the vertices incident to a given element of type $n-2$. Every cycle in Γ whose length is 4 or 5 is contained in a unique quad. Every connected component of $\psi^{-1}(\Theta)$ is the point graph $\widetilde{\Theta}$ of the rank 2 T-geometry.

Let F be the fundamental group of Γ. Let $F(4)$ be the subgroups of F generated by the cycles of length 4 and $F(3,5)$ be the subgroups of F generated by the cycles of length 3 and by the non-degenerate cycles of length 5. It follows directly from the 2-simple connectedness of $\mathcal{G}$ that $F(4)$ together with $F(3,5)$ generates the whole of F. By the previous paragraph we have the following.

Lemma 6.10.2 *The subgroups of the fundamental group of Γ associated with ψ contains $F(3,5)$ and it does not contain $F(4)$.* □

Let $\chi : \widehat{\Gamma} \to \Gamma$ be the covering of Γ associated with $F(3,5)$. Let $\widehat{\mathcal{J}}$ be the geometry whose elements are vertices and triangles of $\widehat{\Gamma}$ together with the connected components of full preimages of subgraphs in Γ which represent the elements of $\mathcal{G}$ with respect to the natural incidence relation and type function. It is clear that $F(3,5)$ is normal in F and it is also normalized by the action of G on F. Hence every element $g \in G$ can be lifted to an automorphism of $\widehat{\Gamma}$. It is clear that all these liftings form a flag-transitive automorphism group $\widehat{J}$ of $\widehat{\mathcal{J}}$ which commutes with χ, and the action induced by $\widehat{J}$ on $\mathcal{G}$ coincides with G. By (6.10.2) $F(4)$ is not contained in $F(3,5)$ and in view of flag-transitivity and (6.10.1), every connected component $\widehat{\Theta}$ of $\chi^{-1}(\Theta)$ is the point graph of the rank 2 T-geometry. Since χ is a covering of graphs, this shows that $\widehat{\mathcal{J}}$ is a flag-transitive T-geometry. Also by (6.10.2) we have a covering $\lambda : \widehat{\Gamma} \to \widetilde{\Gamma}$ which induces a 2-covering of $\widehat{\mathcal{J}}$ onto $\mathcal{J}$. By (6.9.7) λ is an isomorphism and we have

Lemma 6.10.3 *The subgroup of the fundamental group of Γ associated with ψ is exactly $F(3,5)$.* □

Let $\mathcal{H}$ be a rank n geometry with the following diagram:

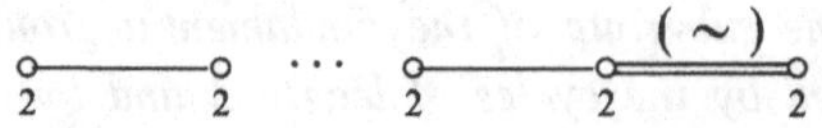

```
                              (~)
o--------o  ...  o--------o========o
2        2       2        2        2
```

where the rightmost edge indicates that for an element z of type $n-2$ the residue $\text{res}^+_{\mathcal{H}}(z)$ is either $\mathcal{G}(Sp_4(2))$ or $\mathcal{G}(3 \cdot Sp_4(2))$, possibly depending on the choice of z. Suppose that $\mathcal{H}$ possesses a 1-covering ν onto $\mathcal{G}$. Let Δ be the graph on the set of elements of type n in $\mathcal{H}$ in which two elements are adjacent if they are incident to a common element of type $n-1$. Then ν induces a covering $\omega : \Delta \to \Gamma$ and every connected component of $\omega^{-1}(\Theta)$ is the point graph of either $\mathcal{G}(Sp_4(2))$ or $\mathcal{G}(3 \cdot Sp_4(2))$. This shows that the subgroup of F associated with ω contains $F(3,5)$ and by

(6.10.3) there is a covering $\delta : \widetilde{\Gamma} \to \Delta$ which induces a covering of $\mathcal{H}$ onto $\mathcal{J}$. This gives

Proposition 6.10.4 *Let $\mathcal{H}$ be a geometry with the above diagram, possessing a 1-cover onto $\mathcal{G}$. Then $\mathcal{H}$ is a quotient of $\mathcal{J}$. In particular every T-geometry of rank n (maybe not flag-transitive) possessing a 1-cover onto $\mathcal{G}(Sp_{2n}(2))$ is a quotient of $\mathcal{J}(n)$.* □

An example of a quotient T-geometry of $\mathcal{J}$ can be constructed as follows. Let L be a non-trivial subgroups of K which intersects K_x trivially for every element x of type $n-2$ in $\mathcal{J}$. Let $\bar{\mathcal{J}}$ be the geometry whose elements are the orbits of L on $\mathcal{J}$ with the type function and incidence relation induced by those in $\mathcal{J}$. Then it is easy to see that $\bar{\mathcal{J}}$ is a T-geometry.

6.11 No tilde analogues of the Alt_7-geometry

In this section we show that there are no geometries $\mathcal{G}$ satisfying the following.

Hypothesis 6.11.1 *$\mathcal{G}$ is a flag-transitive T-geometry of rank 3, G is a flag-transitive automorphism group of $\mathcal{G}$ and there is a 1-covering $\varphi : \mathcal{G} \to \mathcal{G}(Alt_7)$ which commutes with the action of G (i.e. the fibres of φ are unions of G-orbits).*

If $\mathcal{G}$ and G satisfy the above hypothesis then G induces on $\mathcal{G}(Alt_7)$ its unique flag-transitive automorphism group which is Alt_7. Since $3 \cdot Alt_6$ is the only flag-transitive automorphism group of the rank 2 T-geometry which possesses a homomorphism onto Alt_6, it is easy to see that $\mathcal{G}$ and G satisfy the following.

Hypothesis 6.11.2 *$\mathcal{G}$ is a rank 3 T-geometry; G is a flag-transitive automorphism group of $\mathcal{G}$ such that the amalgam $\mathcal{B} = \{G_1, G_2, G_3\}$ of maximal parabolic subgroups satisfies the following:*

$$G_1 \cong 3 \cdot Alt_6; \quad G_2 \cong (Sym_3 \times Sym_4)^e; \quad G_3 \cong L_3(2).$$

(Here G_2 is isomorphic to the setwise stabilizer in Alt_7 of a 3-element subset.)

Thus the non-existence of geometries satisfying Hypothesis 6.11.1 will follow from the non-existence of geometries satisfying Hypothesis 6.11.2.

Lemma 6.11.3 *Let $\mathcal{G}$ and G satisfy Hypothesis* 6.11.1. *Then the isomorphism type of $\mathcal{B}$ is uniquely determined.*

Proof. Since $\mathcal{G}$ is a T-geometry, it is easy to see that $G_{ij} := G_i \cap G_j \cong Sym_4$ for $1 \leq i < j \leq 4$ and that $B := G_1 \cap G_2 \cap G_3 \cong D_8$. First, we observe that all subgroups Sym_4 in G_i are conjugate in the automorphism group of G_i. Notice that G_2 contains three classes of Sym_4-subgroups and the outer automorphism group of G_2 induces Sym_3 on the set of these classes. Since all automorphisms of Sym_4 are inner this shows that the subamalgam $\{G_2, G_3\}$ is uniquely determined. Hence to complete the proof we have to show that there is a unique way to adjoin G_1 to this subamalgam intersecting it in the subamalgam $\{G_{12}, G_{13}\}$. The outer automorphism group of the latter amalgam has order 2 and it is represented by the automorphism which centralizes G_{12} and acts on $G_{13} \setminus G_{12}$ by means of conjugation by the non-trivial element from the centre of B. But this automorphism can be realized inside the normalizer of $\{G_{12}, G_{13}\}$ in the automorphism group of G_1. □

It is easy to see that the unique amalgam $\mathcal{B}$ from (6.11.3) possesses a homomorphism (in the obvious sense) on the amalgam of maximal parabolic subgroups corresponding to the action of Alt_7 on $\mathcal{G}(Alt_7)$. This shows that a simply connected geometry satisfying Hypothesis 6.11.2 possesses a 1-covering onto $\mathcal{G}(Alt_7)$ commuting with the action of its automorphism group (*i.e.* satisfies Hypothesis 6.11.1).

We are going to describe a presentation for the universal completion of the amalgam $\mathcal{B}$ as in (6.11.3), which is due to S.V. Tsaranov (private communication). All the facts claimed below can be easily verified by coset enumeration with a computer.

(a) The Coxeter group of the diagram

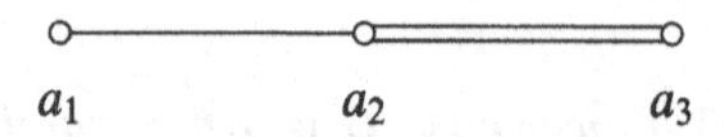

is isomorphic to $Sym_4 \times 2$. To eliminate the centre, one can add the relation $(a_1a_2a_3)^3 = 1$.

(b) The Coxeter group of the diagram

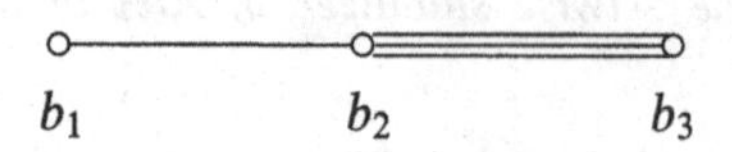

is isomorphic to $Alt_5 \times 2$. To eliminate the centre we put $(b_1b_2b_3)^5 = 1$.

(c) The presentation

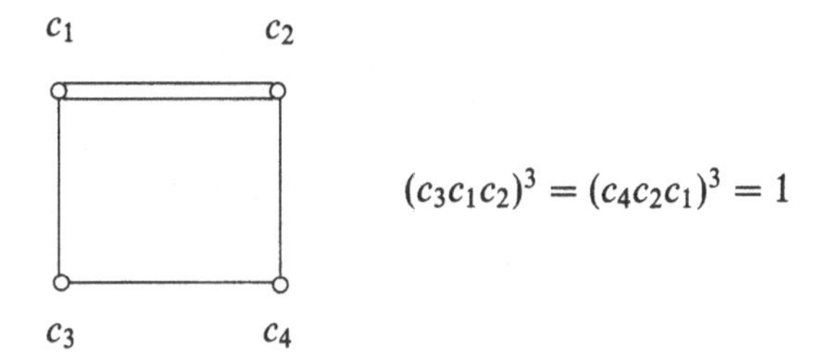

$$(c_3c_1c_2)^3 = (c_4c_2c_1)^3 = 1$$

defines the group $L_3(2)$.

(d) The presentation

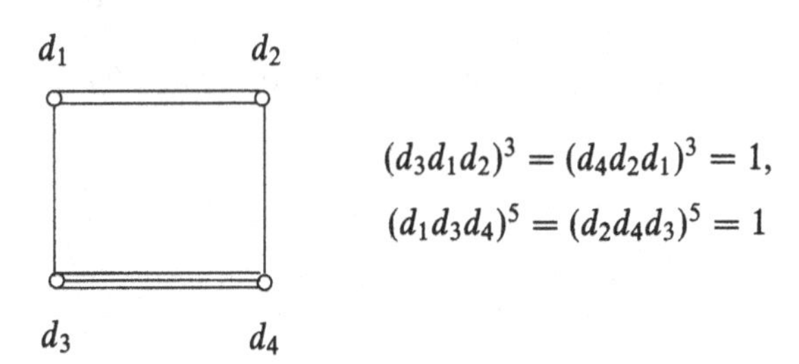

$$(d_3d_1d_2)^3 = (d_4d_2d_1)^3 = 1,$$
$$(d_1d_3d_4)^5 = (d_2d_4d_3)^5 = 1$$

defines the non-split extension $3 \cdot Alt_6$.

(e) The presentation

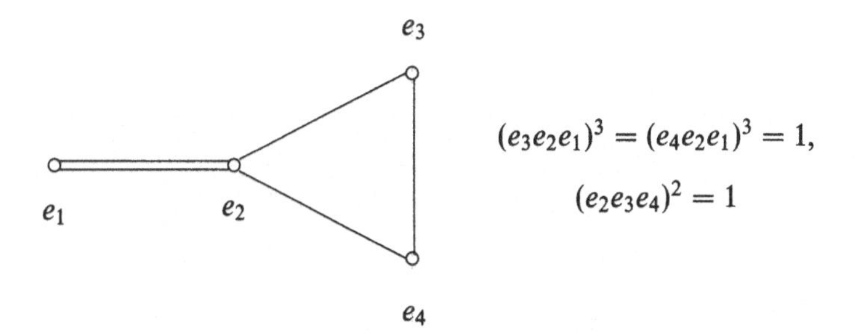

$$(e_3e_2e_1)^3 = (e_4e_2e_1)^3 = 1,$$
$$(e_2e_3e_4)^2 = 1$$

defines the group $(Sym_4 \times Sym_3)^e$. Notice that the relation $(e_2e_3e_4)^3 = 1$ reduces the corresponding Coxeter group $Cox(\widetilde{A}_2) \cong \mathbb{Z}^2 : Sym_3$ to the group $Sym_4 \cong 2^2 : Sym_3$.

Now let us consider the group F defined by the following presentation:

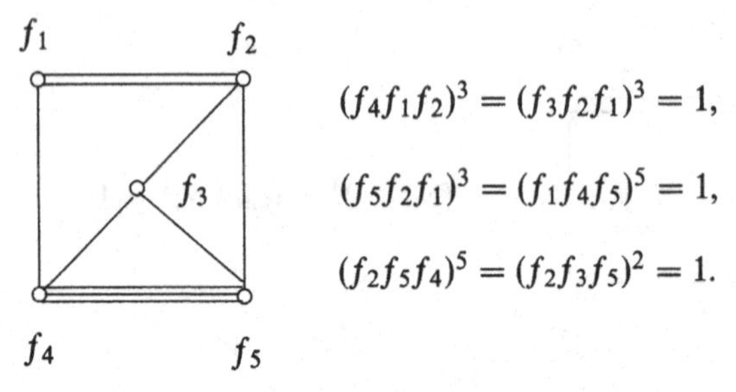

$$(f_4f_1f_2)^3 = (f_3f_2f_1)^3 = 1,$$

$$(f_5f_2f_1)^3 = (f_1f_4f_5)^5 = 1,$$

$$(f_2f_5f_4)^5 = (f_2f_3f_5)^2 = 1.$$

It is easy to see from the above that F is the universal completion of the unique amalgam $\mathscr{B}$ from (6.11.3). On the other hand coset enumeration on a computer implementing the Todd–Coxeter algorithm shows that $F \cong Alt_7$. Since the latter group does not contain $3 \cdot Alt_6$ this shows that $\mathscr{B}$ has no faithful completions and we obtain the following.

Proposition 6.11.4 *There are no geometries* $\mathscr{G}$ *satisfying Hypothesis* 6.11.1 *or* 6.11.2. □

7
2-Covers of P-geometries

In this chapter we construct the univesal 2-covers $\mathcal{G}(3 \cdot Mat_{22})$, $\mathcal{G}(3^{23} \cdot Co_2)$ and $\mathcal{G}(3^{4371} \cdot BM)$ of the P-geometries $\mathcal{G}(Mat_{22})$, $\mathcal{G}(Co_2)$ and $\mathcal{G}(BM)$, respectively. The construction goes as follows. We start with a group G acting flag-transitively on a P-geometry $\mathcal{G}$ and consider an extension $\widehat{G}$ of G by an elementary abelian 3-group W. We show that $\widehat{G}$ contains a subamalgam $\widetilde{\mathcal{A}}$ which maps isomorphically onto the amalgam of rank 2 parabolics associated with the action of G on $\mathcal{G}$. Then the subgroup $\widetilde{G}$ in $\widehat{G}$ generated by $\widetilde{\mathcal{A}}$ is a flag-transitive automorphism group of a geometry $\widetilde{\mathcal{G}}$ which possesses a 2-covering onto $\mathcal{G}$. So in a sense our construction strategy is similar to that in Chapter 6, the difference being that here $\widehat{G}$ does not split over W. We identify the structure (particularly the order) of $\widetilde{G}$ by establishing an upper bound on the size of a flag-transitive 2-cover of $\mathcal{G}$. This bound also enables us to prove the 2-simple connectedness of $\widetilde{\mathcal{G}}$. We start by reviewing some properties of a generic P-geometry and of the geometries $\mathcal{G}(Mat_{22})$, $\mathcal{G}(Co_2)$ and $\mathcal{G}(BM)$.

7.1 On P-geometries

Let $\mathcal{G}$ be a P-geometry of rank $n \geq 3$, *i.e.* a geometry with the diagram

$$P_n: \quad \underset{2}{\circ}\!\!-\!\!\!-\!\!\!-\!\!\underset{2}{\circ}\ \cdots\ \underset{2}{\circ}\!\!-\!\!\!-\!\!\!-\!\!\underset{2}{\circ}\overset{P}{-\!\!\!-\!\!\!-}\underset{1}{\circ}$$

Let $\Phi = \{x_1, x_2, ..., x_n\}$ be a maximal flag in $\mathcal{G}$, where x_i is of type i. Let G be a flag-transitive automorphism group of $\mathcal{G}$. For $1 \leq i \leq n$ let G_i and P_i be the maximal and minimal parabolics of type i associated with the action of G on $\mathcal{G}$, so that $G_i = G(x_i)$ and P_i is the intersection of the G_j for $1 \leq j \leq n$, $j \neq i$. Let

$$B = \bigcap_{i=1}^{n} P_i = \bigcap_{i=1}^{n} G_i$$

be the Borel subgroup and let Q_i denote the kernel of the action of G_i on $\mathrm{res}_{\mathscr{G}}(x_i)$.

Let $\Delta = \Delta(\mathscr{G})$ be the *derived graph* of $\mathscr{G}$, which is a graph on the set of elements of type n in $\mathscr{G}$, in which two such elements are adjacent if they are incident to a common element of type $n-1$. In particular x_n is a vertex of Δ. Let $\pi = \pi(x_n)$ denote the projective space dual to $\mathrm{res}_{\mathscr{G}}(x_n)$. This means that the points of π are the elements of type $n-1$ in the residue of x_n. Then $\bar{G}_n = G_n/Q_n$ is a flag-transitive automorphism group of π and by (1.6.5) either $\bar{G}_n \cong L_n(2)$ or $n = 3$ and $\bar{G}_3 \cong Frob_7^3$. In any case the action of $\bar{G}_n$ on the point set of π is primitive and this immediately gives the following.

Lemma 7.1.1 *If $\{x, y\}$ is an edge of Δ then there is a unique element of type $n-1$ incident to both x and y.* □

The above lemma enables us to identify the elements of type $n-1$ with the edges of Δ so that the vertex–edge incidence in Δ corresponds to the incidence in $\mathscr{G}$. In this way the set $\Delta(x_n)$ of vertices adjacent to x_n is naturally identified with the set of points in π. In fact it is easy to show [Sh85] that the possibility $\bar{G}_3 \cong Frob_7^3$ cannot be realized in a flag-transitive P-geometry, which means that (with respect to the action of G) Δ is a locally projective graph of type $(n, 2)$ (see Section 9.1 for the definitions).

For an element $y \in \mathscr{G}$ let $\Delta[y]$ denote the subgraph in Δ induced by the vertices incident to y. By the above, if y is of type n or $n-1$ then $\Delta[y]$ is a vertex or an edge, respectively. Using the diagram of $\mathscr{G}$ and the fact that in a projective geometry every element is uniquely determined by the set of points it is incident to, it is easy to check the following.

Lemma 7.1.2 *The following assertions hold:*

(i) *if $1 \le i \le n-2$ then $\Delta[x_i]$ is isomorphic to the derived graph of the P-geometry $\mathrm{res}^+_{\mathscr{G}}(x_i)$, in particular $\Delta[x_{n-2}]$ is a Petersen subgraph;*

(ii) *$\Delta[z] \subset \Delta[y]$ if and only if z and y are incident elements of type i and j, respectively, and $1 \le j < i \le n$;*

(iii) *$\Delta[x_i] \cap \Delta(x_n)$ is the point set of a subspace in π of (projective) dimension $n-i-1$, in particular $\Delta[x_i]$ is of valency $2^{n-i}-1$;*

(iv) *if Φ is the point set of a subspace in π of dimension $n-i-1$, for $1 \le i \le n-1$, then there is a unique element y of type i incident to x_n, such that $\Phi = \Delta(x_n) \cap \Delta[y]$;*

(v) *if $u \in \Delta_2(x_n)$ then there is a unique element y of type $n-2$ which is incident to both x_n and u (equivalently such that the Petersen subgraph $\Delta[y]$ contains both x_n and u).* □

In terms of Sections 9.6 and 9.8 the above lemma says that the subgraphs $\Delta[y]$ taken for all $y \in \mathcal{G}$ form a complete family of geometrical subgraphs in the locally projective graph Δ. One can also observe from the lemma that the elements of $\mathcal{G}$ can be identified with the subgraphs $\Delta[y]$ so that the incidence relation is via inclusion and the type of a subgraph is determined by its valency. Since $\mathcal{G}$ is a geometry, the graph Δ and all the subgraphs $\Delta[y]$ are connected and, since G is a flag-transitive automorphism group of $\mathcal{G}$, for every $1 \leq i \leq n-2$ the action of G_i on $\Delta[x_i]$ is 1-arc-transitive, which implies the following.

Lemma 7.1.3 *$G = \langle G_n, G_{n-1}\rangle$ and for every $1 \leq i \leq n-2$ we have $G_i = \langle G_i \cap G_n,\ G_i \cap G_{n-1}\rangle$.* □

Sometimes it is convenient to study 2-coverings of P-geometries in terms of their derived graphs.

Lemma 7.1.4 *Let Δ be the derived graph of a P-geometry $\mathcal{G}$. Let $\varphi : \widetilde{\Delta} \to \Delta$ be a covering of graphs such that for every element y of type $n-2$ in $\mathcal{G}$ every connected component of $\varphi^{-1}(\Delta[y])$ is isomorphic to the Petersen graph. Then $\widetilde{\Delta}$ is the derived graph of a P-geometry $\widetilde{\mathcal{G}}$ which possesses a 2-covering onto $\mathcal{G}$.*

Proof. We define $\widetilde{\mathcal{G}}$ to be the incidence system whose elements are the connected components of the preimages in $\widetilde{\Delta}$ of the subgraphs $\Delta[y]$ taken for all $y \in \mathcal{G}$, the incidence relation is via inclusion and the type function is induced by that in $\mathcal{G}$. We claim that $\widetilde{\mathcal{G}}$ is a P-geometry. Since the incidence relation is via inclusion $\widetilde{\mathcal{G}}$ belongs to a string diagram; since φ is a covering of graphs, it is easy to see that for $\widetilde{x}_n \in \varphi^{-1}(x_n)$ we have $\mathrm{res}_{\widetilde{\mathcal{G}}}(\widetilde{x}_n) \cong \mathrm{res}_{\mathcal{G}}(x_n)$. Finally, by the hypothesis of the lemma, if $\widetilde{y}$ is an element of type $n-2$ in $\widetilde{\mathcal{G}}$ then $\mathrm{res}^{\pm}_{\widetilde{\mathcal{G}}}(\widetilde{y})$ is the Petersen graph geometry and the claim follows. It is clear that φ induces a 2-covering of $\widetilde{\mathcal{G}}$ onto $\mathcal{G}$. □

Let K be the kernel of the action of G_{n-2} on $\mathrm{res}^{+}_{\mathcal{G}}(x_{n-2})$. Then K is the largest subgroup in $G_n \cap G_{n-1}$ which is normal in both $G_{n-2} \cap G_n$ and $G_{n-2} \cap G_{n-1}$ and G_{n-2}/K is a flag-transitive automorphism group of the Petersen graph geometry isomorphic to Sym_5 or Alt_5. The following lemma describes some 2-covers of $\mathcal{G}$ in group-theoretical terms.

Lemma 7.1.5 *Let $\widetilde{G}$ be a group possessing a homomorphism ψ onto G. Suppose that*

(a) *$\widetilde{G}$ contains a rank 2 subamalgam $\widetilde{\mathscr{B}} = \{\widetilde{G}_n, \widetilde{G}_{n-1}\}$ which generates $\widetilde{G}$ and such that the restriction ψ_0 of ψ to $\widetilde{\mathscr{B}}$ is an isomorphism onto the subamalgam $\{G_n, G_{n-1}\}$ in G.*

Put

$$\widetilde{K} = \psi_0^{-1}(K) \quad and \quad \widetilde{G}_{n-2} = \langle N_{\widetilde{G}_n}(\widetilde{K}), N_{\widetilde{G}_{n-1}}(\widetilde{K})\rangle.$$

Suppose also that

(b) *$\widetilde{G}_{n-2}/\widetilde{K} \cong G_{n-2}/K$.*

Then $\widetilde{G}$ is a flag-transitive automorphism group of a P-geometry which possesses a 2-cover onto $\mathscr{G}$.

Proof. Let $\widetilde{\Delta}$ be a graph on the set of (right) cosets of $\widetilde{G}_n$ in $\widetilde{G}$ in which two such cosets are adjacent if they intersect a common coset of $\widetilde{G}_{n-1}$. Then $\widetilde{G}$ acts 1-arc-transitively on $\widetilde{\Delta}$ and by (a) ψ induces a covering $\varphi : \widetilde{\Delta} \to \Delta$ of graphs. Let $\widetilde{x}_n$ denote $\widetilde{G}_n$ considered as a vertex of $\widetilde{\Delta}$ and let Π be the subgraph in $\widetilde{\Delta}$ induced by the images of $\widetilde{x}_n$ under $\widetilde{G}_{n-2}$. Then by (b) the restriction of φ to Π is an isomorphism onto $\Delta[x_{n-2}]$ and the result follows directly from (7.1.4). □

Notice that the vertices of the Petersen graph can be identified with the transpositions in Sym_5 so that two vertices are adjacent if and only if the corresponding transpositions commute.

Lemma 7.1.6 *Suppose that Q_n is finite and non-trivial. Then the following assertions hold:*

(i) *there is $q \in Q_n$ such that $Q_n = \langle q, Q_n \cap Q_{n-2}\rangle$ and acting on the Petersen graph $\text{res}^+_{\mathscr{G}}(x_{n-2})$ the element q induces the transposition from G^+_{n-2} which corresponds to x_n;*
(ii) *G_{n-2} induces Sym_5 on $\text{res}^+_{\mathscr{G}}(x_{n-2})$;*
(iii) *$Q_n = O_2(G_n)$;*
(iv) *Q_{n-1} has index 2 in $O_2(G_{n-1})$ and $Q_n \le Q_{n-1}$.*

Proof. It is easy to see that Q_n is the kernel of the action of G_n on $\Delta(x_n)$. By the definition Q_n stabilizes the Petersen subgraph $\Delta[y]$ whenever y is an element of type $n-2$ incident to x_n. Suppose that Q_n fixes every vertex in such a Petersen subgraph. Then by (7.1.2 (v)) Q_n acts trivially on $\Delta_2(x_n)$ which implies the triviality of Q_n by (9.1.4). Hence Q_n induces on $\Delta[x_{n-2}]$ an action of order 2 generated by the transposition corresponding to x_n (recall that $\Delta[x_{n-2}]$ is isomorphic to the Petersen graph). Hence (i) follows and immmediately implies (ii). We have seen that Q_n induces a 2-group on $\Delta_2(x_n)$. By (9.1.5) Q_n is a 2-group itself. Since $O_2(G_n/Q_n) = 1$ (iii) follows. Since all automorphisms of $L_{n-1}(2)$ are inner, we see from the diagram of $\mathcal{G}$ that $G_{n-1}/Q_{n-1} \cong L_{n-1}(2) \times 2$ and in view of (iii) we obtain (iv). □

Each of the five P-geometries we have constructed so far, namely

$$\mathcal{G}(Mat_{22}),\ \mathcal{G}(3 \cdot Mat_{22}),\ \mathcal{G}(Mat_{23}),\ \mathcal{G}(Co_2) \text{ and } \mathcal{G}(BM),$$

contains a subgeometry which is a C_m- or T_m-geometry, namely

$$\mathcal{G}(Sp_4(2)),\ \mathcal{G}(3 \cdot Sp_4(2)),\ \mathcal{G}(Alt_7),\ \mathcal{G}(Sp_6(2)) \text{ and } \mathcal{G}(Sp_8(2))$$

((3.4.2), (3.4.4), (3.5.8), (4.9.8) and (5.4.5) and the table in Section 1.10). We are going to present a systematic way to construct such subgeometries.

Recall that the edge graph of the Petersen graph is an antipodal distance-transitive graph with the intersection diagram

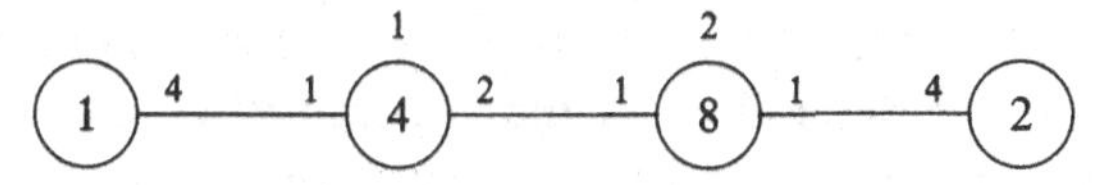

Thus there is an equivalence relation on the set of edges of the Petersen graph with classes of size 3.

With $\mathcal{G}$ and Δ as above let Σ be a graph on the set of edges of Δ in which two edges are adjacent if they are incident to a common element y of type $n-2$ and if they are equivalent in the Petersen subgraph $\Delta[y]$. For any two distinct elements u and v of type $n-2$ the Petersen subgraphs $\Delta[u]$ and $\Delta[v]$ have at most one common edge and hence every edge of Σ is contained in a unique triangle which corresponds to a Petersen subgraph in Δ. Let Σ^c denote the connected component of Σ which contains x_{n-1}. Let $\mathcal{S} = \mathcal{S}(x_{n-1})$ be a geometry whose elements of type i, $1 \leq i \leq n-1$, are the non-empty intersections $\Sigma^c \cap \Delta[y]$, where y is an element of type i in $\mathcal{G}$; the incidence relation is via inclusion.

Lemma 7.1.7 *In the above terms suppose that $n \geq 4$. Then*

(i) *if $\text{res}^+_{\mathcal{G}}(x_{n-3}) \cong \mathcal{G}(Mat_{22})$ then $\mathcal{S}$ is a C_{n-1}-geometry with the diagram*

2 2 ... 2 2 2

(ii) *if* $\mathrm{res}^+_{\mathcal{G}}(x_{n-3}) \cong \mathcal{G}(3 \cdot Mat_{22})$ *then* $\mathcal{S}$ *is a* T_{n-1}*-geometry with the diagram*

$$\underset{2}{\circ}\!\!-\!\!\!-\!\!\!-\!\!\underset{2}{\circ}\ \cdots\ \underset{2}{\circ}\!\!-\!\!\!-\!\!\!-\!\!\underset{2}{\circ}\overset{\sim}{=\!=\!=}\underset{2}{\circ}$$

In both cases the stabilizer of $\mathcal{S}$ *in* G *induces on* $\mathcal{S}$ *a flag-transitive action.*

Proof. Since the incidence relation is via inclusion $\mathcal{S}$ belongs to a string diagram. It is easy to observe that there is a natural isomorphism between the residue of x_{n-1} in $\mathcal{S}$ and $\mathrm{res}^-_{\mathcal{G}}(x_{n-1})$, so that both the residues are projective geometries of rank $n-2$ over $GF(2)$. Finally it is an easy combinatorial exercise to check that in cases (i) and (ii) the geometry formed by $\Sigma^c \cap \Delta[x_{n-3}]$ coincides with the subgeometry as in (3.4.4) and (3.5.8), respectively. □

Recall that by the main result of [Sh85] every flag-transitive P-geometry of rank 3 is isomorphic either to $\mathcal{G}(Mat_{22})$ or to $\mathcal{G}(3 \cdot Mat_{22})$ so that the cases (i) and (ii) in the above lemma cover all the possibilities.

To the end of the section we discuss the properties of some concrete P-geometries. The description of the flag-transitive automorphism groups of P-geometries comes most naturally from the complete classification (Preface) which gives all the pairs $(\mathcal{G}, G)$ where $\mathcal{G}$ is a P-geometry and G is a flag-transitive automorphism group of $\mathcal{G}$. In particular cases the flag-transitive automorphism groups can be determined by *ad hoc* analysis of subamalgams in the amalgam of maximal parabolics in the full automorphism group. We suggest the reader check the details and formulate the final result.

Lemma 7.1.8 *Let* G *be one of* Mat_{22}, Mat_{23}, Co_2 *and* BM, *and* $\mathcal{G}(G)$ *be the* P*-geometry on which* G *acts flag-transitively. Let* H *be a flag-transitive automorphism group of* $\mathcal{G}(G)$. *Then either* $G = H$, *or* $G = Mat_{22}$ *and* $H = \mathrm{Aut}\, Mat_{22}$. □

Lemma 7.1.9 *Let* $\mathcal{G}$ *be isomorphic to* $\mathcal{G}(Mat_{22})$, $\mathcal{G}(Co_2)$ *or* $\mathcal{G}(BM)$ *and* G *be the full automorphism group of* $\mathcal{G}$. *Then*

$$G_i = N_G(Q_i) \ \ for \ \ i = n-1 \ \ and \ \ n.$$

Proof. For $\mathcal{G} = \mathcal{G}(Mat_{22})$ the result is by (3.9.3 (iv)). For $\mathcal{G} = \mathcal{G}(Co_2)$ using (4.9.5) and the table therein it is not difficult to check that the sets of vertices in the collinearity graph Θ fixed by Q_3 and Q_4 are exactly the

$\star$-closed cliques of size 7 and 15 which represent x_3 and x_4, respectively. In the case of the Baby Monster BM we have $G_1 = C_G(Z(Q_1))$ which enables us to identify x_i with $Z(Q_i)$ for $1 \leq i \leq 5$ and immediately gives the result. □

7.2 A sufficient condition

In this section we formulate and prove a sufficient condition for existence of proper 2-covers of a flag-transitive P-geometry in terms of non-split extensions of its automorphism group. We start with a preliminary technical result.

Let $F \cong Sym_5$ act naturally on the Petersen graph Π and $\mathscr{E} = \{E_1, E_2\}$ be a subamalgam in F consisting of the stabilizers in F of a vertex and an edge from Π which are incident. This means that $E_1 \cong 2 \times Sym_3$, $E_2 \cong D_8$ and $E_1 \cap E_2 \cong 2^2$.

Lemma 7.2.1 *Let $\widetilde{F}$ be a group possessing a homomorphism onto $F \cong Sym_5$ whose kernel is an elementary abelian 3-group U. Suppose that U is the centre of the full preimage of $F' \cong Alt_5$ in $\widetilde{F}$. Let $\widetilde{\mathscr{E}}$ be a subamalgam in $\widetilde{F}$ which maps isomorphically onto $\mathscr{E}$. Then $\widetilde{\mathscr{E}}$ generates in $\widetilde{F}$ a subgroup isomorphic to Sym_5.*

Proof. Since the 3-part of the Schur multiplier of Alt_5 is trivial, it is easy to show that $\widetilde{F}$ contains a normal subgroup H isomorphic to Alt_5. Consider the image of $\widetilde{\mathscr{E}}$ in $\widetilde{F}/H$. It is easy to see that this image generates a subgroup of order 2 and the result follows. □

Now we are ready to prove the main result of the section.

Proposition 7.2.2 *Let $\mathscr{G}$ be a P-geometry of rank $n \geq 3$ and G be a flag-transitive automorphism group of $\mathscr{G}$ such that*

$$\text{(a)} \quad G_i = N_G(Q_i) \ \ for \ \ i = n-1 \ \ and \ \ n.$$

Let $\widehat{G}$ be a group which possesses a homomorphism ψ onto G whose kernel W is an elementary abelian 3-group. Suppose further that

- (i) *$\widehat{G}$ does not split over W,*
- (ii) *the preimage of Q_n in $\widehat{G}$ does not centralize non-identity elements in W,*

Then there is a subgroup $\widetilde{G}$ of $\widehat{G}$ such that the restriction of ψ to $\widetilde{G}$ is a non-injective homomorphism onto G. Moreover, $\widetilde{G}$ acts faithfully and flag-transitively on a P-geometry $\widetilde{\mathscr{G}}$ which possesses a 2-covering onto $\mathscr{G}$.

Proof. First notice that the condition (a) implies that Q_n is non-trivial and that

$$\text{(b)} \quad G_n \cap G_{n-1} = N_{G_n}(Q_{n-1}) = N_{G_{n-1}}(Q_n).$$

The minimal parabolic subgroup P_n is a 2-group which contains B with index 2. Let $\widetilde{P}_n$ be a Sylow 2-subgroup in the full preimage of P_n in $\widehat{G}$. Then the restriction ψ_1 to ψ to $\widetilde{P}_n$ is an isomorphism onto P_n. Put $\widetilde{Q}_i = \psi_1^{-1}(Q_i)$ for $i = n,\ n-1,\ n-2$ and $\widetilde{G}_i = N_{\widehat{G}}(\widetilde{Q}_i)$ for $i = n, n-1$. By (ii) and (7.1.6 (iv)) $C_W(\widetilde{Q}_n) = C_W(\widetilde{Q}_{n-1}) = 1$ and hence by the Frattini argument (compare (6.1.1)) $\widetilde{G}_i \cong G_i$ for $i = n, n-1$. Also by (b) $\widetilde{G}_n \cap \widetilde{G}_{n-1} \cong G_n \cap G_{n-1}$ and hence the restriction ψ_0 of ψ to the subamalgam $\widetilde{\mathscr{B}} = \{\widetilde{G}_n, \widetilde{G}_{n-1}\}$ in $\widehat{G}$ is an isomorphism onto the subamalgam $\mathscr{B} = \{G_n, G_{n-1}\}$ in G. Let $\widetilde{G}$ be the subgroup in $\widehat{G}$ generated by $\widetilde{\mathscr{B}}$. We are going to show that $\widetilde{G}$ acts faithfully and flag-transitively on a P-geometry which possesses a 2-covering onto $\mathscr{G}$. By (7.1.3) the restriction of ψ to $\widetilde{G}$ is a homomorphism onto G. This restriction must have a non-trivial kernel, since otherwise $\widetilde{G}$, would be a complement to W in $\widehat{G}$, which is impossible, since by (i) $\widehat{G}$ does not split over W. We are going to show that condition (b) from (7.1.5) holds.

Let $\widetilde{G}_{n-2} = \langle \widetilde{F}_n, \widetilde{F}_{n-1} \rangle$, where $\widetilde{F}_j = \psi_0^{-1}(F_j)$ with $F_j = G_{n-2} \cap G_j$ for $j = n$ and $n-1$ and ψ_2 be the restriction of ψ to $\widetilde{G}_{n-2}$. Then $U := \widetilde{G}_{n-2} \cap W$ is the kernel of ψ_2. Notice that $\widetilde{Q}_{n-2} := \psi_0^{-1}(Q_{n-2})$ is contained and normal in both $\widetilde{F}_n$ and $\widetilde{F}_{n-1}$. Hence $\widetilde{Q}_{n-2}$ is normal in $\widetilde{G}_{n-2}$ and $U \leq C_W(\widetilde{Q}_{n-2})$.

Let K be the kernel of the action of G_{n-2} on $\text{res}^+_{\mathscr{G}}(x_{n-2})$. Then K is the largest subgroup in $F_n \cap F_{n-1}$ normal in both F_n and F_{n-1} and $\mathscr{D} = \{F_n/K, F_{n-1}/K\}$ is the subamalgam in $G_{n-2}/K \cong Sym_5$ (7.1.6 (ii)) consisting of the stabilizers of incident vertex and edge in the Petersen graph $\text{res}^+_{\mathscr{G}}(x_{n-2})$ (*i.e.* $\mathscr{D} \cong \{2 \times Sym_3, D_8\}$). Let $\widetilde{K} = \psi_0^{-1}(K)$. Then $\widetilde{K}$ is the largest subgroup in $\widetilde{F}_n \cap \widetilde{F}_{n-1}$ normal in both $\widetilde{F}_n$ and $\widetilde{F}_{n-1}$ and ψ_2 induces an isomorphism of $\widetilde{\mathscr{D}} = \{\widetilde{F}_n/\widetilde{K}, \widetilde{F}_{n-1}/\widetilde{K}\}$ onto $\mathscr{D}$. We claim that the induced homomorphism

$$\psi_3 : \widetilde{G}_{n-2}/\widetilde{K}W \to G_{n-2}/K$$

is an isomorphism. The kernel of ψ_3 equals the kernel U of ψ_2 and in order to apply (7.2.1) it is sufficient to show that U is centralized by the commutator subgroup of $\widetilde{G}_{n-2}/\widetilde{K}$ (isomorphic to Alt_5).

Let $q \in Q_n$ be as in (7.1.6 (i)) and $\widetilde{q} = \psi_0^{-1}(q)$. Since $C_W(\widetilde{Q}_n) = 1$ and $\widetilde{Q}_n = \langle \widetilde{q}, \widetilde{Q}_n \cap \widetilde{Q}_{n-2} \rangle$, the element $\widetilde{q}$ inverts every element of $C_W(\widetilde{Q}_{n-2})$ and hence of U as well. On the other hand q induces on $\text{res}^+_{\mathscr{G}}(x_{n-2})$ the transposition from $G_{n-2}/K \cong Sym_5$ which corresponds to x_n. Considering other elements of type n incident to x_{n-2}, we conclude that every transposition from G_{n-2}/K inverts U and hence every even permutation centralizes U. Thus (7.2.1) gives $\widetilde{G}_{n-2}/\widetilde{K} \cong Sym_5$, completing the proof. □

The reader might notice that we did not quite follow the strategy outlined in Section 6.1 and instead of embeddings of the amalgam $\mathscr{A} = \{P_i \mid 1 \leq i \leq n\}$ we consider embeddings of the amalgam $\mathscr{B} = \{G_n, G_{n-1}\}$. But in fact these embeddings are related since $\mathscr{B}$ contains $\mathscr{A}$.

7.3 Non-split extensions

In this section we construct some extensions which satisfy the conditions in (7.2.2).

Let $\mathscr{G}$ be one of the geometries $\mathscr{G}(Mat_{22})$, $\mathscr{G}(Co_2)$ and $\mathscr{G}(BM)$, G be the full automorphism group of $\mathscr{G}$ (isomorphic, respectively, to $\mathrm{Aut}\, Mat_{22}$, Co_2 and BM). Let X be a subgroup in G defined as follows. If $G = \mathrm{Aut}\, Mat_{22}$ then X is the stabilizer of an element from $\mathscr{P} \setminus Y_2$ (Section 3.4) and $X \cong P\Sigma L_3(4)$. If $G = Co_2$ then X is the stabilizer of the subgraph Ξ in the collinearity graph of $\mathscr{G}(Co_2)$ isomorphic to the dual polar graph of $U_6(2)$ (the remark after (4.10.8)) and $X \cong P\Sigma U_6(2)$. Equivalently X is the stabilizer in Co_2 of a vertex in the rank 3 graph Π as in (4.11.9). Finally if $G = BM$ then X is the stabilizer of a vertex of the Baby Monster graph (Section 5.10), so that $X \cong 2^{\cdot 2} E_6(2).2$. Let X_0 denote the commutator subgroup (of index 2) in X.

Let W_0 be a 1-dimensional $GF(3)$-module for X whose kernel is X_0. This means that the elements from X_0 centralize and the elements from $X \setminus X_0$ invert every element of W_0. Let W be the $GF(3)$-module for G induced from the module W_0 of X. Notice that the dimension of W is 22, 2300 and 13 571 955 000 (the latter being the number of vertices in the Baby Monster graph) for $G = \mathrm{Aut}\, Mat_{22}$, Co_2 and BM, respectively.

Lemma 7.3.1 *In the above terms we have the following:*

(i) *there exists a group $\widehat{X}$ which is a non-split extension of X by a subgroup of order 3 isomorphic to W_0 as a $GF(3)$-module for X;*
(ii) *there exists a group $\widehat{G}$ which possesses a homomorphism ψ onto G such that* $\ker \psi$ *is an elementary abelian 3-group isomorphic to W as a $GF(3)$-module for X and such that $\widehat{G}$ does not split over* $\ker \psi$.

Proof. Let U be the natural $GF(4)$-module of X, so that $\bar{X} = X/Z(X)$ acts faithfully on the projective geometry of U and the dimension of U

is 3, 6 and 27 for $X \cong P\Sigma L_3(4)$, $P\Sigma U_6(2)$ and $2\cdot^2 E_6(2).2$, respectively. Let $\widetilde{X}$ be the preimage of $\bar{X}$ in the group of semilinear transformations of U. It is a standard fact (proved in (2.7.12) for the case $X \cong P\Sigma L_3(4)$) that $\widetilde{X}$ is a non-split extension of $\bar{X}$ by the multiplicative group of $GF(4)$ which is centralized by X_0 and inverted by the elements from $X \setminus X_0$ (the latter elements act semilinearly but not linearly on U). This gives (i) and implies (ii) by (6.1.2). □

Since the 3-part of the Schur multiplier of X_0 is of order 3 the extension $\widehat{X}$ in (7.3.1 (i)) is unique.

Lemma 7.3.2 *With $\mathcal{G}$, G, X and W as above let G_n be the stabilizer in G of an element of type n in $\mathcal{G}$ and $Q_n = O_2(G_n)$. Then $C_W(Q_n) = 1$.*

Proof. We discuss the cases $G = \mathrm{Aut}\, Mat_{22}$, Co_2 and BM separately.

In the case $G = \mathrm{Aut}\, Mat_{22}$ we apply (6.1.7) by showing that for every $g \in G$ the intersection $X^g \cap Q_3$ is not contained in $X_0^g \cap Q_3$. In the case considered X^g is the stabilizer in G of an element from $\mathcal{P} \setminus Y_2$ and $X_0^g = X^g \cap G'$ where $G' \cong Mat_{22}$ is the commutator subgroup of G. On the other hand $Q_3' := Q_3 \cap G'$ has index 2 in Q_3 and by (3.9.3 (iii)) Q_3' and Q_3 have the same orbits on $\mathcal{P} \setminus Y_2$, which shows that for every $c \in \mathcal{P} \setminus Y_2$ we have $[G(c) \cap Q_3 : G(c) \cap Q_3'] = 2$ and the result follows.

In the case $G = Co_2$ we apply (6.1.6). The induced character χ of G is the permutational character of G on the cosets of $X_0 \cong U_6(2)$ minus the permutational character of G on the cosets of X. The suborbit diagrams of these two permutational actions are given in (4.11.8) and (4.11.9), respectively. Since the former of the actions has rank 5 while the latter has rank 3, we conclude that χ involves two irreducible characters. Looking at the character table of Co_2 in [CCNPW] we deduce that the irreducibles are the ones of degree 23 and 2277. The character table of $G_4 \cong (2^{1+6} \times 2^4).L_4(2)$ can be found in GAP computer package [GAP]. Along with the character table comes the fusion map of the classes of G_4 into the classes of G. This information enables us to calculate the class function on Q_4:

$$(1a)^1,\ (2a)^{15+70},\ (2b)^{1+15+210},\ (2c)^{840},\ (4b)^{336},\ (4c)^{560},$$

where $(2a)^{15+70}$ means that Q_4 contains 85 elements from the Co_2-class $2a$ and that these elements form two G_4-classes of size 15 and 70. Now it is straightforward to check that the restriction of χ to Q_4 is zero, which gives the result. An alternative proof making use of (6.1.7) can be found in [Sh92].

The case $G = BM$ is more complicated technically (but not conceptually). The induced character of the action of BM on the complex analogoue of W has been calculated by D.G. Higman in [Hig76]. The character is the sum of three irreducible characters of degree 4371, 63 532 485 and 13 508 418 144. Thus if the class function on Q_5 were known, to check the claim would be a matter of straightforward calculations. The character table and the fusion pattern of $G_5 \sim 2^{5+10+10+5}.L_5(2)$ are not available in GAP now, but they probably will be in due course. Meanwhile the claim has been checked by two different methods. In [Wil92] and [Wil93] R.A. Wilson has checked the condition in (6.1.7) by explicit calculations with elements of the Baby Monster represented by 4370×4370 matrices over $GF(2)$. A different strategy was implemented in [ISh93b]. It follows directly from (5.11.7) that for $G_1 \cong 2_+^{1+22}.Co_2$ and $Q_1 = O_2(G_1)$ the centralizer $C := C_W(Q_1)$, as a module for $\bar{G}_1 \cong Co_2$, possesses a direct sum decomposition

$$C = C^{(1)} \oplus C^{(2)} \oplus C^{(3)},$$

where $C^{(1)}$ and $C^{(2)}$ are isomorphic to the module W (of dimension 2300) from the case $G = Co_2$ and the module $C^{(3)}$ (of dimension 46 575) is induced from a 1-dimensional non-trivial module of $G_{12}/Q_1 \cong 2^{10}.\mathrm{Aut}\, Mat_{22}$ (the direct summands correspond to the orbits $\Omega(2a)$, $\Omega(6a)$ and $\Omega(4a)$ of G_1 on the vertex set of the Baby Monster graph). We have checked that $O_2(G_{15})$ has trivial centralizer in $C^{(i)}$. For $i = 1$ and 2 this follows from the proof of the lemma in the case $G = Co_2$ and for $i = 3$ the result can be achieved by similar calculations, since the corresponding induced character (computed by GAP) has three irreducible components whose degrees are 23, 2277 and 44 275.

Finally we have shown (Lemma 3.10 in [ISh93b]) that in every module of $\bar{G}_5 \cong L_5(2)$ induced from any 1-dimensional module of any subgroup the subgroup, $O_2(G_{15}/Q_5) \cong 2^4$ has a non-trivial centralizer. This shows that if Q_5 had a non-trivial centralizer in W, then $O_2(G_{15})$ would have a non-trivial centralizer as well, which is not the case. □

Now by (7.1.5), (7.2.2), (7.3.1) and (7.3.2) we obtain the main result of the section.

Corollary 7.3.3 *Let $\mathcal{G}$ be one of the geometries $\mathcal{G}(Mat_{22})$, $\mathcal{G}(Co_2)$ and $\mathcal{G}(BM)$. Let G be the full automorphism group of $\mathcal{G}$ (isomorphic to*

Aut Mat_{22}, Co_2 and BM, respectively). Then there exist a geometry $\widetilde{\mathcal{G}}$ which possesses a 2-covering $\varphi : \widetilde{\mathcal{G}} \to \mathcal{G}$, and a flag-transitive automorphism group $\widetilde{G}$ of $\widetilde{\mathcal{G}}$ which possesses a homomorphism ψ onto G, and the kernel K of ψ is a non-trivial elementary abelian 3-group. Moreover, K as a module for G is a submodule of the module W induced from a non-trivial 1-dimensional module of a subgroup X in G isomorphic to $P\Sigma L_3(4)$, $P\Sigma U_6(2)$ and $2 \cdot {}^2 E_6(2).2$, respectively. □

In (7.3.3) if $\mathcal{G} = \mathcal{G}(Mat_{22})$ then by (3.5.5) the order of K is at most (and hence exactly) 3. This shows that φ in (7.3.3) is a 3-fold covering which is universal by (3.5.7) and hence $\widetilde{\mathcal{G}} \cong \mathcal{G}(3 \cdot Mat_{22})$.

7.4 $\mathcal{G}(3^{23} \cdot Co_2)$

In this section we identify the structure of the group $\widetilde{G}$ as in (7.3.3) in the case $\mathcal{G} = \mathcal{G}(Co_2)$.

Throughout the section $\mathcal{G}$ is a P-geometry of rank 4, G is a flag-transitive automorphism group of $\mathcal{G}$ and it is assumed that $\mathcal{G}$ possesses a non-bijective 2-covering φ onto $\bar{\mathcal{G}} := \mathcal{G}(Co_2)$ and φ commutes with the action of G. This means that the action $\bar{G}$ which G induces on $\bar{\mathcal{G}}$ coincides with the unique flag-transitive automorphism group of $\bar{\mathcal{G}}$ which is Co_2 (7.1.8).

By (4.12.7) $\mathcal{G}(Co_2)$ is simply connected and hence φ is not a covering. Let ψ denote the natural homomorphism of G onto $\bar{G}$ induced by φ and let K be the kernel of ψ.

Let $\Phi = \{x_1, ..., x_4\}$ be a maximal flag in $\mathcal{G}$, where x_i is of type i, and $G_i = G(x_i)$ be the stabilizer of x_i in G. Let $\bar{x}_i$ be the image of x_i in $\bar{\mathcal{G}}$ and $\bar{G}_i$ be the image of G_i in $\bar{G}$. Since $\mathrm{res}_{\bar{\mathcal{G}}}(\bar{x}_i)$ is 2-simply connected for $i = 2$, 3 and 4 and φ is not a covering, we conclude that $K \cap G_i$ is trivial for $i = 2, 3, 4$ and non-trivial for $i = 1$. By (3.5.7) the universal 2-cover of $\mathrm{res}_{\bar{\mathcal{G}}}(\bar{x}_1) \cong \mathcal{G}(Mat_{22})$ is its triple cover $\mathcal{G}(3 \cdot Mat_{22})$. Thus we have the following.

Lemma 7.4.1 $\mathrm{res}_{\mathcal{G}}(x_1) \cong \mathcal{G}(3 \cdot Mat_{22})$ *and* $K \cap G_1$ *is of order* 3. □

By the above lemma, for every element v of type 1 in $\mathcal{G}$ the subgroup $K_v := K \cap G(v)$ is of order 3. Since $\mathcal{G}(Co_2)$ is simply connected, the subgroups K_v taken for all elements v of type 1 in $\mathcal{G}$ generate the whole of K. For an arbitrary element z in $\mathcal{G}$ put

$$B_z = \{K_v \mid v \in \mathrm{res}_{\mathcal{G}}(z)^1\}, \; K_z = \langle K_v \mid v \in \mathrm{res}_{\mathcal{G}}(z)^1 \rangle.$$

Let $\mathcal{S} = \mathcal{S}(x_3)$ and $\bar{\mathcal{S}} = \bar{\mathcal{S}}(\bar{x}_3)$ be the subgeometries of $\mathcal{G}$ and $\bar{\mathcal{G}}$ defined as in (7.1.7) with respect to the elements x_3 and $\bar{x}_3$, respectively. Let S be the stabilizer of $\mathcal{S}$ in G and $\bar{S}$ be the stabilizer of $\bar{\mathcal{S}}$ in $\bar{G}$. By (4.9.8) we have

$$\bar{\mathcal{S}} \cong \mathcal{G}(Sp_6(2)) \text{ and } \bar{S} \cong 2_+^{1+8}.Sp_6(2).$$

By (7.1.7 (ii)) and (7.4.1) $\mathcal{S}$ is a T-geometry of rank 3 possessing a 1-cover onto $\mathcal{G}(Sp_6(2))$ which commutes with the action of S. By (6.10.4)

$$\mathcal{S} \cong \mathcal{J}(3) \text{ and } S/O_2(S) \cong 3^7 \cdot Sp_6(2),$$

in particular $K_{\mathcal{S}} := K \cap S$ is an elementary abelian 3-group of rank 7. By (6.6.4) this gives the following.

Lemma 7.4.2 *In the above terms $K_{\mathcal{S}}$ is elementary abelian of order 3^7 and it coincides with K_{x_3}. In particular, if u and v are elements of type 3 from the same T-subgeometry, then $K_u = K_v$.* □

In what follows, for a subgroup K' in K the statement "rk$(K') = n$" will mean that K' is an elementary abelian group of order 3^n.

Lemma 7.4.3 rk$(K_{x_4}) = 15$.

Proof. The set B_{x_4} has size 15. Since $\text{res}_{\mathcal{G}}(x_4)$ is a projective geometry, every two elements of type 1 incident to x_4 are incident to a common element of type 3. Hence the subgroups in B_{x_4} commute pairwise and it remains to show that they are linearly independent. The parabolic G_4 acts primitively on B_{x_4} and Q_4 is the kernel of the action. It is easy to see that the quotient of Q_4 over the centralizer of K_{x_4} in Q_4 has order more than 2. Hence different subgroups in B_{x_4} have different centralizers in Q_4. In particular they are linearly independent and the result follows. □

Lemma 7.4.4 *Let $\{z_1 = x_3, z_2, z_3\} = \text{res}_{\mathcal{G}}(\{x_2, x_4\})^3$. Then $K_{x_4} = \langle K_{z_i} \mid 1 \leq i \leq 3 \rangle$.*

Proof. The elements z_1, z_2, z_3 are three hyperplanes in $\text{res}_{\mathcal{G}}(x_4)$ containing a common line (which is x_2). Hence every point (an element of type 1 in the residue) is incident to z_i for $i = 1$, 2 or 3. □

Lemma 7.4.5 *Let y_4 be the element of type 4 other than x_4 incident to x_3. Then $K_{x_4} \cap K_{y_4} = K_{x_3}$, so that the quotient of $\langle K_{x_4}, K_{y_4} \rangle$ over its commutator subgroup has rank 23.*

Proof. Let $E = B_{x_4} \setminus B_{y_4}$, so that $B_{x_4} = B_{x_3} \cup E$, and let X be the subgroup in K_{x_4} generated by the subgroups in E. Since G_4 induces the full automorphism group of $\mathrm{res}_{\mathcal{G}}(x_4)$, $G(x_4) \cap G(y_4) = G_{34}$ acts transitively on B_{x_3} and on E. Since different subgroups from B_{x_4} have different centralizers in Q_4, this implies that K_{x_3} and X are the only proper subgroups in K_{x_4} normalized by $G(x_4) \cap G(y_4)$. Hence either $K_{x_4} \cap K_{y_4} = K_{x_3}$ or $K_{x_4} = K_{y_4}$. In the latter case since $\mathcal{G}$ is connected, we immediately obtain $K = K_{x_4}$, in particular K is an elementary abelian 3-group of rank 15. Since the shortest orbit of $\bar{G}_1/\bar{Q}_1 \cong \mathrm{Aut}\, Mat_{22}$ on the set of hyperplanes of $\bar{Q}_1$ has length 22, $\bar{G}$ has no faithful $GF(3)$-representations of dimension less than 22. Since K cannot be centralized by $\bar{G}$ either, the result follows. □

Lemma 7.4.6 *Let U be the subgroup of K generated by the subgroups K_u taken for all elements $u \in \mathrm{res}_{\mathcal{G}}(x_2)^3$. Then $U = \langle K_{x_4}, K_{y_4} \rangle$.*

Proof. By (7.4.4) both K_{x_4} and K_{y_4} are contained in U, so it is sufficient to show that $\langle K_{x_4}, K_{y_4} \rangle$ contains U. With z_1, z_2, z_3 as in (7.4.4) let z_1, z_4 and z_5 be the elements of type 3 incident to both x_2 and y_4. There are exactly five T-subgeometries in $\mathcal{G}$ containing elements of type 3 incident to x_2 and the elements z_i for $1 \le i \le 5$ are in pairwise different such subgeometries. Hence whenever u is an element of type 3 incident to x_2, $\mathcal{S}(u) = \mathcal{S}(z_i)$ for some $1 \le i \le 5$. By (7.4.2) this means that $K_u = K_{z_i}$ and hence $K_u \le \langle K_{z_i} \mid 1 \le i \le 5 \rangle \le \langle K_{x_4}, K_{y_4} \rangle$. □

Lemma 7.4.7 $\mathrm{rk}(K) = 23$.

Proof. Let x, y be elements of type 4 in $\mathcal{G}$ incident to a common element z of type 3 and suppose that z is incident to x_2. Then by (7.4.6) $\langle K_x, K_y \rangle = U = \langle K_{x_4}, K_{y_4} \rangle$. We are going to show that $K = U$. Let w be an element of type 4 in $\mathcal{G}$. Since $\mathcal{G}$ is connected, there is a sequence of vertices $w_0, w_1, ..., w_s = w$ such that $\{w_0, w_1\} = \{x_4, y_4\}$ and for $0 \le i \le s-1$ the elements w_i and w_{i+1} are distinct and incident to a common element of type 3. In this case for every i, $0 \le i \le s-2$, the vertices w_i, w_{i+1} and w_{i+2} are incident to common element of type 2 which means (7.4.6) that $\langle K_{w_i}, K_{w_{i+1}} \rangle = \langle K_{w_{i+1}}, K_{w_{i+2}} \rangle$. This shows that $K_w \le U$. Since this is true for every $w \in \mathcal{G}^4$, $K = U$. By (7.4.3) and (7.4.5) $K_{x_1} \le K_{x_3}$ is in the centre of $U = K$. Since K is generated by the subgroups K_v taken for all elements of type 1, we conclude that K is abelian. □

It is well known and easy to check that Co_2 has a unique faithful

$GF(3)$-representation of dimension less than or equal to 23, which is the unique faithful irreducible section L of the Leech lattice taken modulo 3. In the present context the isomorphism between K and L can be established along the following lines.

Let N be the $GF(3)$-module for $\bar{G} = Co_2$ induced from the unique non-trivial 1-dimensional module of the group $\bar{G}_1 \cong 2^{10}.\mathrm{Aut}\, Mat_{22}$. Then

$$N = \bigoplus_{u \in \bar{\mathcal{G}}^1} N_u,$$

where the N_u are 1-dimensional. Let $\bar{S} \cong 2_+^{1+8}.Sp_6(2)$ be the stabilizer in $\bar{G}$ of the C_3-subgeometry $\bar{\mathcal{S}}$. Put $M = \bigoplus_{u \in \bar{\mathcal{S}}^1} N_u$, so that M is a 63-dimensional $\bar{S}$-submodule in N. Then M possesses a unique homomorphism onto the 7-dimensional $\bar{S}$-module which is the E_7-lattice taken modulo 3. Let M_0 be the kernel of this homomorphism and N_0 be the smallest $\bar{G}$-submodule in N which contains M_0. Then by the arguments almost identical to those for (7.4.2)–(7.4.7) one can show that N/N_0 has dimension at most 23. On the other hand both K and L are quotients of N/N_0 and hence $K \cong L$.

Proposition 7.4.8 *Let $\mathcal{G}$ be a P-geometry of rank 4 and G be a flag-transitive automorphism group of $\mathcal{G}$. Suppose that $\mathcal{G}$ possesses a non-bijective 2-cover onto $\mathcal{G}(Co_2)$ which commutes with the action of G. Then the kernel K of the homomorphism of G onto Co_2 is an elementary abelian 3-group of rank 23 isomorphic to the unique faithful irreducible Co_2-section in the Leech lattice taken modulo 3 and G does not split over K.* □

Thus in terms of (7.3.3) $\tilde{G} \cong 3^{23} \cdot Co_2$ and the corresponding P-geometry will be denoted by $\mathcal{G}(3^{23} \cdot Co_2)$.

7.5 The rank 5 case: bounding the kernel

Let $\mathcal{G}$ be a P-geometry of rank 5, G be a flag-transitive automorphism group of $\mathcal{G}$ and suppose that $\mathcal{G}$ possesses a non-bijective 2-covering φ onto $\bar{\mathcal{G}} \cong \mathcal{G}(BM)$ which commutes with the action of G. In this case by (7.1.8) the action $\bar{G}$ induced by G on $\bar{\mathcal{G}}$ is the only flag-transitive automorphism group of the latter geometry which is the Baby Monster group BM. Let ψ be the natural homomorphism of G onto $\bar{G}$. For $x \in \mathcal{G}$ or $x \in G$ we write $\bar{x}$ to denote $\varphi(x)$ or $\psi(x)$, respectively.

Let $\Phi = \{x_1, ..., x_5\}$ be a maximal flag in $\mathcal{G}$, where x_i is of type i, $G_i = G(x_i)$ be the stabilizer of x_i in G, so that $\mathcal{B} = \{G_i \mid 1 \le i \le 5\}$ is the amalgam of maximal parabolic subgroups associated with the action of G on $\mathcal{G}$. As usual we put $G_{ij} = G_i \cap G_j$ for $1 \le i < j \le 5$. Let K be the kernel of ψ and $K_i = K \cap G_i$ for $1 \le i \le 5$. Then the elements of $\bar{\mathcal{G}}$ can be identified with the orbits of K on $\mathcal{G}$ with respect to the induced incidence relation and type function. Therefore the restriction of φ to $\mathrm{res}_{\mathcal{G}}(x_i)$ associates with every element in the residue its orbit under K_i. In particular the restriction of φ to $\mathrm{res}_{\mathcal{G}}(x_i)$ is an isomorphism if and only if $K_i = 1$. Since $\mathrm{res}_{\mathcal{G}}(x_i)$ is 2-simply connected for $i = 3, 4$ and 5, we have $K_3 = K_4 = K_5 = 1$. On the other hand by (5.11.5 (i)) $\mathcal{G}(BM)$ is simply connected and since φ is non-bijective, some of the K_i must be non-trivial. Since $\mathrm{res}_{\mathcal{G}}(x_i)^-$ is 2-simply connected for every $1 \le i \le 5$, K_1 must be non-trivial. Hence the restriction of φ to $\mathrm{res}_{\mathcal{G}}(x_1)$ is a non-bijective 2-covering onto $\mathrm{res}_{\bar{\mathcal{G}}}(\bar{x}_1) \cong \mathcal{G}(Co_2)$. By (7.4.8), this gives the following:

Lemma 7.5.1

(i) $\mathrm{res}_{\mathcal{G}}(x_1) \cong \mathcal{G}(3^{23} \cdot Co_2)$ *and* K_1 *is elementary abelian of order* 3^{23};
(ii) $\mathrm{res}_{\mathcal{G}}(x_2)^+ \cong \mathcal{G}(3 \cdot Mat_{22})$ *and* K_2 *is of order* 3. □

Lemma 7.5.2 *If v is an element of type 2 in $\mathcal{G}$ then $K_v := K \cap G(v)$ is of order 3. Moreover, the subgroups K_v taken for all elements v of type 2 in $\mathcal{G}$ generate K.*

Proof. By (7.5.1) we only have to prove the statement about the generation. It is clear that $L := \langle K_v \mid v \in \mathcal{G}^2 \rangle$ is normal in G and the image in G/L of the amalgam $\mathcal{B}$ is isomorphic to the amalgam $\bar{\mathcal{B}}$ associated with the action of $\bar{G}$ on $\bar{\mathcal{G}}$. Since $\bar{\mathcal{G}}$ is simply connected, this shows that $G/L \cong \bar{G}$ and hence $L = K$. □

For an arbitrary element z in $\mathcal{G}$ put

$$B_z = \{K_v \mid v \in \mathrm{res}_{\mathcal{G}}(z)^2\}.$$

If L is a subgroup in K then the statement "rk $(L) = n$" will mean that L is an elementary abelian 3-group of order 3^n. In this case a set of n subgroups of order 3 in L which generate L will be called a *basis* of L.

In these terms (7.5.1 (i)) can be reformulated as follows.

Lemma 7.5.3 $\mathrm{rk}(K_{x_1}) = 23$ *and as a module for* $G_1/O_{2,3}(G_1) \cong Co_2$, K_{x_1} *is a section of the Leech lattice modulo* 3. □

Let $\bar{\mathcal{S}} = \bar{\mathcal{S}}(\bar{x}_4)$ and $\mathcal{S} = \mathcal{S}(x_4)$ be the subgeometries of $\bar{\mathcal{G}}$ and $\mathcal{G}$ respectively defined as in (7.1.7). Let $\bar{S}$ and S be the stabilizers of $\bar{\mathcal{S}}$ and $\mathcal{S}$ in $\bar{G}$ and G, respectively. Then by (5.4.5) we have

$$\bar{\mathcal{S}} \cong \mathcal{G}(Sp_8(2)) \quad \text{and} \quad \bar{S} \cong 2^{9+16}.Sp_8(2).$$

By (7.5.1 (ii)) and (7.1.7 (ii)) $\mathcal{S}$ is a rank 4 T-geometry. The restriction of φ to $\mathcal{S}$ is a 1-cover onto $\bar{\mathcal{S}}$. Hence by (6.10.4)

$$\mathcal{S} \cong \mathcal{J}(4) \quad \text{and} \quad S/O_2(S) \cong 3^{35} \cdot Sp_8(2).$$

By (6.7.2) we have the following.

Lemma 7.5.4 *Let $\mathcal{S} = \mathcal{S}(x_4)$ be the rank 4 T-subgeometry in $\mathcal{G}$ corresponding to x_4, S be the stabilizer of $\mathcal{S}$ in G and $K_{\mathcal{S}} = S \cap K$. Then* $\mathrm{rk}(K_{\mathcal{S}}) = 35$ *and B_{x_4} is a basis of $K_{\mathcal{S}}$.* □

Lemma 7.5.5 *Let $i = 3, 4$ or 5. Then* $\mathrm{rk}(K_{x_i}) = \left[{i \atop 2}\right]_2$ *(which is 7, 35 or 155, respectively) and B_{x_i} is a basis of K_{x_i}.*

Proof. If $i = 3$ or 4 then the result follows immediately from (7.5.4), so suppose that $i = 5$. Since $\mathrm{res}_{\mathcal{G}}(x_5)$ is a projective space over $GF(2)$, any two elements, say v_1 and v_2 of type 2 incident to x_5 are incident to a common element of type 4. By (7.5.4) K_{v_1} and K_{v_2} commute. To prove linear independence, consider the action of $Q_5 = O_2(G_5)$ on K_{x_5}. Since Q_5 is the kernel of the action of G_5 on $\mathrm{res}_{\mathcal{G}}(x_5)$, Q_5 normalizes each K_v with $v \in \mathrm{res}_{\mathcal{G}}(x_5)^2$. Let $\mathcal{S} = \mathcal{S}(x_4)$. By (7.5.4), Q_5 normalizes $K_{\mathcal{S}}$. Comparing the orders of Q_5 (which is 2^{30}) and $O_2(S)$ (which is 2^{25}) we observe that Q_5 induces on K_{x_5} an action of order at least 2^5. Since G_5 acts primitively on $\mathrm{res}_{\mathcal{G}}(x_5)^2$, we conclude that B_{x_5} is a basis of K_{x_5}. □

It follows directly from the proof of the above lemma that different subgroups in B_{x_5} have different centralizers in $O_2(G_5)$.

Lemma 7.5.6 *If $3 \le i \le 5$ then $B_{x_1} \cap B_{x_i}$ is a basis of $K_{x_1} \cap K_{x_i}$, in particular* $\mathrm{rk}(K_{x_1} \cap K_{x_i}) = 3$, 7 *and* 15 *for* $i = 3$, 4 *and* 5, *respectively.*

Proof. Let $D_i = B_{x_i} \cap B_{x_1}$, $E_i = B_{x_i} \setminus B_{x_1}$ and let X_i and Y_i be the subgroups in K_{x_i} whose bases are D_i and E_i, respectively. Then $\mathrm{rk}(X_i) = 3$, 7 and 15 while $\mathrm{rk}(Y_i) = 4$, 28 and 140 for $i = 3$, 4 and 5, respectively. Clearly X_i is contained in $K_{x_1} \cap K_{x_i}$ for $3 \le i \le 5$. The parabolic G_i induces the full automorphism group of the projective space $\mathrm{res}_{\mathcal{G}}(x_i)^-$ and the kernel N_i of this action contains $O_2(G_5)$. This means that G_{1i}

acts transitively on D_i and on E_i and in view of the remark after the proof (7.5.5) different subgroups in B_{x_i} have different centralizers in N_i. This shows that X_i and Y_i are the only subgroups in K_{x_i} normalized by G_{1i}. Hence either $K_{x_1} \cap K_{x_i} = X_i$ or K_{x_1} contains K_{x_i}. If $i = 4$ or 5 then the latter is impossible since the rank of K_{x_1} is only 23 by (7.5.3). If K_{x_1} were to contain K_{x_3} then since $Y_3 \leq Y_4$, it would contain the whole K_{x_4}, and we have seen that this is impossible. □

Lemma 7.5.7 *Let y_5 be the element of type 5 incident to x_4 other than x_5. Then $K_{x_5} \cap K_{y_5} = K_{x_4}$.*

Proof. It is clear that K_{x_4} is contained in $K_{x_5} \cap K_{y_5}$. Let Z be the subgroup in K_{x_5} whose basis is $B_{x_5} \setminus B_{x_4}$. Since G_5 induces the full automorphism group of $\mathrm{res}_{\mathcal{G}}(x_5)$, in view of the remark after the proof of (7.5.5) K_{x_4} and Z are the only subgroups in K_{x_5} normalized by G_{45}. Hence if $K_{x_4} \neq K_{x_5} \cap K_{y_5}$ then $K_{x_5} = K_{y_5}$. Since $\mathcal{G}$ is connected, the latter equality implies that $K_{x_5} = K$ (7.5.2), which is impossible since K_{x_1} is not contained in K_{x_5}, by (7.5.6). □

Lemma 7.5.8 *In terms of* (7.5.7), $K_{x_1} \leq \langle K_{x_5}, K_{y_5} \rangle$.

Proof. By (7.5.6) and (7.5.7) $\mathrm{rk}(K_{x_1} \cap K_{x_5}) = \mathrm{rk}(K_{x_1} \cap K_{y_5}) = 15$ and $\mathrm{rk}(K_{x_1} \cap K_{x_5} \cap K_{y_5}) = \mathrm{rk}(K_{x_1} \cap K_{x_4}) = 7$. Hence $\mathrm{rk}(K_{x_1} \cap \langle K_{x_5}, K_{y_5} \rangle) \geq 15 + 15 - 7 = 23$ which is the rank of K_{x_1} by (7.5.3). □

Lemma 7.5.9 *Let y_1 be an element of type 1 other than x_1 incident to x_2. Then the subgroups K_{x_1} and K_{y_1} commute.*

Proof. First we analyse the structure of K_{x_1} as a module for a certain subgroup in $G(x_1) \cap G(y_1)$. By (7.5.3) G_1 acts on K_{x_1} as Co_2 acts on a faithful section of the Leech lattice modulo 3. If L is the kernel of this action then

$$L = K_{x_1} \times O_2(G_1) \cong 3^{23} \times 2_+^{1+22}.$$

By (7.5.1 (ii)) and in view of the structure of the maximal parabolic subgroups associated with the action of Co_2 on $\mathcal{G}(Co_2)$, we have $G_{12}/Q_1 \cong 3.2^{10}.\mathrm{Aut}\, Mat_{22}$. Let H be the preimage in G_{12} of the unique subgroup of index 2 in G_{12}/Q_1, so that $\bar{H} := HL/L \cong 2^{10}.Mat_{22}$. Since there are only three elements of type 1 incident to x_2, H is contained in $G(x_1) \cap G(y_1)$. We claim that $K_{x_1} = K_{x_2} \oplus [K_{x_1}, O_2(\bar{H})]$ and $[K_{x_1}, O_2(\bar{H})]$ is an irreducible 22-dimensional $GF(3)$-module for $\bar{H}$. This can be seen either by restricting to $\bar{H}$ the action of Co_2 on the Leech lattice modulo 3 or directly, since the shortest orbit of $\bar{H}/O_2(\bar{H})$ on the hyperplanes in $O_2(\bar{H})$ has length 22. In view of this decomposition and since K_{x_2} is contained in K_{y_1}, in order to prove the lemma it is sufficient to find an element in $K_{x_1} \setminus K_{x_2}$ which commutes with K_{y_1}.

By (7.5.7) and (7.5.5) if y_5 is an element of type 5 other than x_5 incident to x_4 then K_{x_4} is contained in the centre of $\langle K_{x_5}, K_{y_5}\rangle$ and by (7.5.8) the latter group contains both K_{x_1} and K_{y_1}. Hence $K_{x_1} \cap K_{x_4}$ (which is of rank 7 by (7.5.6)) commutes with K_{y_1} and the result follows. □

Let us introduce some notation. Put

$$U_1(x_5) = \langle K_x \mid x \in \mathrm{res}_{\mathcal{G}}(x_5)^1\rangle,$$

$$U_2(x_5) = \langle K_{x_5}, K_y \mid \{x_5, y\} = \mathrm{res}_{\mathcal{G}}(v)^5 \text{ for some } v \in \mathcal{G}^4\rangle,$$

so that $U_1(x_5)$ and $U_2(x_5)$ are the subgroups in K generated by the subgroups K_x taken for all elements of type 1 incident to x_5 and for all elements of type 5 incident with x_5 to a common element of type 4, respectively.

Since $\mathrm{res}_{\mathcal{G}}(x_5)$ is a projective space, any two elements of type 1 incident to x_5 are incident to a common element of type 2. By (7.5.9) this shows that $U_1(x_5)$ is abelian.

Lemma 7.5.10 *In the above terms* $U_1(x_5) = U_2(x_5)$.

Proof. By (7.5.8) if y_5 is an element of type 5 other than x_5 incident to x_4 then $K_{x_1} \le \langle K_{x_5}, K_{y_5}\rangle$ and hence $U_1(x_5) \le U_2(x_5)$. Let y_5 be as above and a be an element of type 2 incident to y_5. Since $\mathrm{res}_{\mathcal{G}}(y_5)$ is a projective space there is an element x of type 1 in this residue which is incident to both a and x_4 (a line and a hyperplane always have a common point). It is clear that x is incident to x_5. Hence $K_a \le K_x \le U_1(x_5)$. Since K_{y_5} is generated by all such subgroups K_a, we have $U_2(x_5) \le U_1(x_5)$. □

Let Δ be the derived graph of $\mathcal{G}$ whose vertices and edges will be identified with the elements of type 5 and 4 in $\mathcal{G}$, respectively. For a pair x, y of vertices in Δ let $d(x, y)$ be the distance in Δ between these vertices. As usual for a vertex y let $\Delta_i(y)$ denote the set of vertices at distance i from y.

For a vertex y of Δ put

$$V_i(y) = \langle K_z \mid z \in V(\Delta),\ d(z, y) \le i\rangle.$$

Then $V_0(y) = K_y$ and $V_1(y) = U_1(y) = U_2(y)$ by (7.5.10).

Lemma 7.5.11 $\mathrm{rk}(V_1(y)/V_0(y)) \le 248$.

Proof. By (7.5.10), $V_1(y)$ is generated by 31 subgroups K_x for $x \in \mathrm{res}_{\mathcal{G}}(y)^1$, each of them has rank 23 and intersects $V_0(y)$ in a subgroup of rank 15 (7.5.6). Hence $\mathrm{rk}(V_1(y)/V_0(y)) \le 31 \cdot 8 = 248$. □

Lemma 7.5.12 *Let* $z \in \Delta_1(y)$. *Then* $\mathrm{rk}(V_1(z)/\langle V_0(y), V_0(z)\rangle) \le 128$.

Proof. Let v be the unique element of type 4 in $\mathcal{G}$ which is incident to both y and z. The group $V_1(z)$ is generated by the subgroups K_x taken for all $x \in \mathrm{res}_{\mathcal{G}}(z)^1$. If such an x is incident to v, then $K_x \le \langle V_0(y), V_0(z)\rangle$ by (7.5.8). There are exactly 16 elements x which are not adjacent to v, each of them giving contribution of rank at most 8 (indeed, by (7.5.3) the rank of K_x is 23, while the rank of $K_x \cap K_z$ is 15 by (7.5.6)). □

If w is a vertex at distance 2 from y in Δ then K_w is contained in $V_1(z)$ for the unique vertex z adjacent to both w and x. Hence $V_1(y)$ and the subgroups $V_1(z)$ taken for all vertices z adjacent to y generate $V_2(y)$. Since there are exactly 31 such vertices z, by (7.5.12) we obtain the following.

Corollary 7.5.13 *The quotient* $V_2(y)/V_1(y)$ *is generated by at most* $31 \cdot 128 = 3968$ *subgroups of order* 3. □

By the above result if $V_2(2)$ is abelian then $\mathrm{rk}(V_2(y)/V_1(y)) \le 3968$.

Lemma 7.5.14 *Let* y *be a vertex of* Δ, u *be an element of type* 3 *incident to* y. *Let* $\{v_1, v_2, v_3\}$ *be the elements of type* 4 *incident to both* u *and* y; *and for* $i = 1, 2, 3$ *let* z_i *be the element of type* 5 *other than* y *incident to* v_i. *Then*

$$V_1(y) = \langle V_0(y), V_0(z_1), V_0(z_2), V_0(z_3)\rangle.$$

Proof. In the residue of y the elements v_1, v_2 and v_3 are three hyperplanes having the subspace u of codimension 2 in common. Therefore, every element x of type 1 incident to y is incident to z_i for some i, $1 \le i \le 3$. By (7.5.8), $K_x \le \langle V_0(y), V_0(z_i)\rangle$. □

Lemma 7.5.15 $V_3(y) = V_2(y)$.

Proof. Let $z \in \Delta_2(y)$. By (7.1.2 (v)) we can find an element u of type 3 in $\mathcal{G}$ which is incident to both y and z. Let $\{v_1, v_2, v_3\}$ be the elements of type 4 incident to both u and z. For $i = 1, 2$ and 3 let z_i be the element of type 5 other than y incident to v_i. Since $\mathrm{res}_{\mathcal{G}}(u)^+$ is the geometry of edges and vertices of the Petersen graph, and since the Petersen graph has diameter 2, we have

$$V_1(z) \leq \langle V_0(z), V_0(z_1), V_0(z_2), V_0(z_3) \rangle \leq V_2(y),$$

where the first of the inclusions is forced by (7.5.14). □

Since Δ is connected, we immediately obtain the following.

Corollary 7.5.16 *For every $y \in \mathcal{G}^5$ we have $K = V_2(y)$.* □

Lemma 7.5.17 *K is abelian.*

Proof. By (7.5.16), $K = V_2(y)$. Let us show that $V_0(y)$ is in the centre of K. Let z be a vertex at distance at most 2 from y in Δ and a be a vertex which is at distance at most 1 from both y and z. By the remark before (7.5.10), $V_1(a)$ is abelian. Hence $V_0(y)$ and $V_0(z)$ commute. By (7.5.15) the subgroups $V_0(z)$ taken for all vertices z with distance at most 2 from y in Δ generate K. Hence $V_0(y)$ is in the centre of K. Again using the fact that K is generated by the subgroups $V_0(y)$ for all the vertices y of Δ, K is abelian. □

Proposition 7.5.18 *Let $\mathcal{G}$ be a P-geometry of rank 5 and G be a flag-transitive automorphism group of $\mathcal{G}$. Suppose that $\mathcal{G}$ possesses a non-bijective 2-cover onto the geometry $\mathcal{G}(BM)$ which commutes with the action of G and let K be the kernel of the action of G on $\mathcal{G}(BM)$. Then K is an elementary abelian 3-group of rank at most 4371.*

Proof. By (7.5.17), K is abelian. By (7.5.5), (7.5.11) and (7.5.13),

$$\text{rk}(K) \leq 155 + 248 + 3968 = 4371.$$

□

7.6 $\mathcal{G}(3^{4371} \cdot BM)$

Let $\widetilde{\mathcal{G}}$ be the geometry as in (7.3.3) possessing a 2-covering onto $\mathcal{G}(BM)$. Then $\widetilde{\mathcal{G}}$ and $\widehat{G}$ satisfy the hypothesis of (7.5.18) and hence the intersection of $\widetilde{G}$ and W (which is the kernel of the homomorphism of $\widetilde{G}$ onto BM) is of rank at most 4371.

Lemma 7.6.1 *In the above terms let $K = \widetilde{G} \cap W$. Then*

(i) $\text{rk}(K) = 4371$,
(ii) *K is an irreducible $GF(3)$-module for BM,*
(iii) *$\widetilde{G}$ does not split over K.*

Proof. We are going to analyse the structure of K as a module for $\widetilde{G}/K \cong BM$. Let $\widetilde{x}$ be an element of type 1 in $\widetilde{\mathscr{G}}$ and H be the stabilizer of $\widetilde{x}$ in $\widetilde{G}$. Then by (7.5.1) $H \cap K$ is of rank 23. The quotient $\bar{H} := H/(H \cap K)$ can be naturally identified with the stabilizer in BM of an element of type 1 from $\mathscr{G}(BM)$, so that $\bar{H} \cong 2^{1+22}_{+}.Co_2$. By (7.5.3) $Q := O_2(\bar{H})$ is the kernel of the irreducible action of $\bar{H}$ on $K \cap H$. Let Z be the centre of Q which is also the centre of $\bar{H}$. Put

$$K_1 = C_K(Q), \quad K_2 = C_K(Z) \cap [K, Q], \quad K_3 = [K, Z].$$

Then clearly $K = K_1 \oplus K_2 \oplus K_3$ as a module for $\bar{H}$. Since K_1 contains $K \cap H$, by the above $\mathrm{rk}(K_1) \geq 23$. Since $\bar{H}$ acts faithfully on K, K_3 must be non-trivial. Since Q is isomorphic to 2^{1+22}_{+}, it must act faithfully on K_3, the rank of the latter is at least $2048 = 2^{11}$ which is the dimension of the unique faithful irreducible $GF(3)$-representation of Q. From the structure of $\mathscr{G}(BM)$ we know that Z is conjugate in BM to a non-central subgroup Z' in Q which means that $C_K(Z)$ and $C_K(Z')$ have the same rank. By definition K_1 is centralized by both Z and Z'. On the other hand since Z is in the centre of $\bar{H}$ while Z' is not in the centre, we observe that $C_{K_3}(Z)$ is trivial while $C_{K_3}(Z')$ is non-trivial. Since the ranks of $C_K(Z)$ and $C_K(Z')$ are isomorphic, K_2 must be non-trivial. Since Q is normal in $\bar{H}$ and it induces on K_2 an elementary abelian 2-group Q/Z, Clifford's theorem implies that $\mathrm{rk}(K_2)$ is at least the length of the shortest orbit of $Co_2 \cong \bar{H}/Q$ on the set of hyperplanes in Q/Z. Since Q/Z is self-dual, this gives $\mathrm{rk}(K_2) \geq 2300$. Now summing up we obtain

$$\mathrm{rk}(K) = \mathrm{rk}(K_1) + \mathrm{rk}(K_2) + \mathrm{rk}(K_3) \geq 23 + 2300 + 2048 = 4371.$$

Since this lower bound meets the upper bound from (7.5.18), (i) follows.

By the above it is clear that K does not involve trivial composition factors. For a faithful submodule L in K put $L_2 = C_L(Z) \cap [L, Q]$ and $L_3 = [L, Z]$. Then by arguments as in the above paragraph both L_2 and L_3 must be non-trivial of rank at least 2300 and 2048, respectively. Since the sum of these numbers exceeds half of the rank of K, L cannot be proper and we obtain (ii). From the proof of (i) we see that $K_1 = C_K(Q)$. Since $\bar{H} = N_{BM}(Q)$ we conclude that $\bar{N} := N_{\widetilde{G}}(Q)/Q$ is the full automorphism group of the residual geometry $\mathrm{res}_{\widetilde{\mathscr{G}}}(\widetilde{x}) \cong \mathscr{G}(3^{23} \cdot Co_2)$. If $\widetilde{G}$ should split over K, $\bar{N}$ would split over K_1, but we know that this is not the case. □

Thus $\widetilde{G} \cong 3^{4371} \cdot BM$ and the corresponding P-geometry $\widetilde{\mathscr{G}}$ from (7.3.3) will be denoted by $\mathscr{G}(3^{4371} \cdot BM)$.

Proposition 7.6.2 *$\mathcal{G}(3^{4371} \cdot BM)$ is the universal 2-cover of $\mathcal{G}(BM)$.*

Proof. By the construction $\mathcal{G} := \mathcal{G}(3^{4371} \cdot BM)$ possesses a 2-cover onto $\bar{\mathcal{G}} := \mathcal{G}(BM)$. Thus to prove the proposition it is sufficient to show that $\widetilde{G}$ is the universal completion of the amalgam $\mathcal{C}$ of rank 2 parabolic subgroups associated with the action of BM on $\bar{\mathcal{G}}$. Let H be the universal completion of $\mathcal{C}$ and χ be the homomorphism of H onto BM. Since $\mathcal{G}$ possesses a 2-covering onto $\bar{\mathcal{G}}$ which commutes with the action of $\widetilde{G}$, $\mathcal{C}$ is also the amalgam of rank 2 parabolic subgroups associated with the action of $\widetilde{G}$ on $\mathcal{G}$. Hence there is a homomorphism η of H onto $\widetilde{G}$, such that χ is the composition of η and the homomorphism ψ of $\widetilde{G}$ onto BM. All proper residues in $\mathcal{G}$ are 2-simply connected and hence there is a subamalgam $\mathcal{D}$ in H which maps isomorphically under ψ onto the amalgam of maximal parabolic subgroups in $\widetilde{G}$ associated with its action on $\mathcal{G}$. Thus we can construct a P-geometry $\mathcal{H}$ acted on flag-transitively by H and possessing a covering onto $\mathcal{G}$. To wit, the elements of $\mathcal{H}$ are the cosets in H of the subgroups constituting $\mathcal{D}$. By the construction $\mathcal{H}$ possesses a 2-cover onto $\bar{\mathcal{G}}$ which commutes with the action of H. Now by (7.5.18) the kernel of χ has order at most 3^{4371} and ψ must be an isomorphism. □

Let K be as in (7.6.1). Using the technique developed in this chapter (Section 6 in [ISh93b] for details) one can show that K is the unique faithful $GF(3)$-module of BM of dimension 4371 or less. In particular if M_p is the module obtained by taking modulo p the BM-module M of dimension 4371 over the rationals, then $K \cong M_3$. We conclude the section with the following result concerning cohomology of certain representations of BM.

Lemma 7.6.3 *Let G be a group possessing a homomorphism onto BM with kernel isomorphic to the module M_p as above for $p \geq 3$. Then either G splits over M_p or G is the automorphism group of the geometry $\mathcal{G}(3^{4371} \cdot BM)$.*

Proof. One can easily check that for every $p \neq 2$ the centralizer in M_p of Q_5 is trivial. Then the result follows from a straightforward generalization of (7.2.2), for the case when the characteristic of W is more than 3, and (7.6.2). □

Let $\widehat{G}$ be the non-split extension of BM by the $GF(3)$-module W induced from the non-trivial 1-dimensional module of $X \cong 2{\cdot}{}^2E_6(2).2$, so that $\widehat{G}$ contains the automorphism group $\widetilde{G}$ of $\mathcal{G}(3^{4371} \cdot BM)$. Then $\widehat{G}/K$

splits over W/K since $\widetilde{G}/K$ is a complement. Similar observations can be made for the extensions of Co_2 and Aut Mat_{22}.

7.7 Some further s-coverings

It has been shown in [Wie97] and [BIP98] that the 2-coverings

$$\mathcal{G}(3^{23} \cdot Co_2) \to \mathcal{G}(Co_2) \quad \text{and} \quad \mathcal{G}(3^{4371} \cdot BM) \to \mathcal{G}(BM)$$

induce 1- and 2-coverings of certain subgeometries in $\mathcal{G}(Co_2)$ and $\mathcal{G}(BM)$.

We have seen in Section 4.13 that Co_2 contains the McLaughlin group McL as a subgroup and that the latter acts flag-transitively on a geometry $\mathcal{G}(McL)$ with the diagram

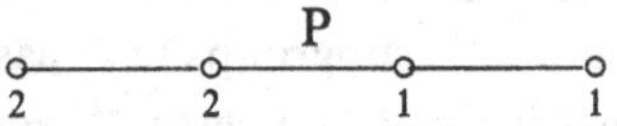

P

2 2 1 1

It can be seen that the parabolic subgroups associated with the action of McL on $\mathcal{G}(McL)$ are contained in the parabolic subgroups associated with the action of Co_2 on $\mathcal{G}(Co_2)$. Let ψ be the natural homomorphism of $3^{23} \cdot Co_2$ onto Co_2.

Proposition 7.7.1 [BIP98] *The full preimage of McL under ψ, which is a non-split extension of the form $3^{23} \cdot McL$, acts flag-transitively on a geometry $\mathcal{G}(3^{23} \cdot McL)$, and ψ induces a 2-covering $\mathcal{G}(3^{23} \cdot McL) \to \mathcal{G}(McL)$.* □

The residues in $\mathcal{G}(3^{23} \cdot McL)$ of elements of type 1 and 4 (isomorphic to $\mathcal{H}(3 \cdot Alt_7)$ and $\mathcal{G}(Mat_{22})$) are the universal (triple) covers of the corresponding residues in $\mathcal{G}(McL)$ (isomorphic to $\mathcal{H}(Alt_7)$ and $\mathcal{G}(Mat_{22})$, respectively). We do not know whether or not $\mathcal{G}(3^{23} \cdot McL)$ is simply connected.

Also in Section 4.13 we have indicated in $\mathcal{G}(McL)$ a subgeometry $\mathcal{G}(U_4(3))$ with the diagram

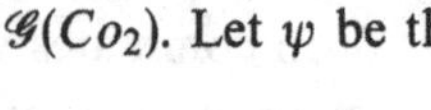

2 2 2

on which a subgroup of Co_2 isomorphic to $U_4(3).2^2$ induces a flag-transitive action.

Proposition 7.7.2 [BIP98] *The full preimage of $U_4(3).2^2$ under ψ, which is a non-split extension of the form $3^{23} \cdot U_4(3).2^2$, acts flag-transitively on a geometry $\mathcal{G}(3^{23} \cdot U_4(3))$ with the diagram*

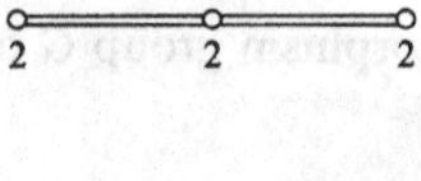

~ ~

2 2 2

There exists a covering of $\mathcal{G}(3^{23} \cdot U_4(3))$ *onto the residue of an element of type* 1 *in the geometry* $\mathcal{G}(3 \cdot M(24))$ *as in* (5.8.4). □

We have observed in Section 5.5 that BM contains a subgroup isomorphic to $F_4(2)$ which is the automorphism group of a Tits geometry $\mathcal{G}(F_4(2))$ with the diagram

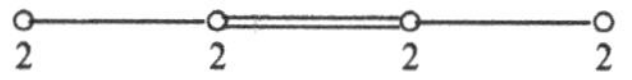

$$\underset{2}{\circ}\!\!-\!\!\!-\!\!\!-\!\!\underset{2}{\circ}\!\!=\!\!=\!\!=\!\!\underset{2}{\circ}\!\!-\!\!\!-\!\!\!-\!\!\underset{2}{\circ}$$

Let ψ denote the natural homomorphism of $\widetilde{G} = 3^{4371} \cdot BM$ onto BM.

Proposition 7.7.3 [Wie97] *Let* $\mathcal{A}$ *be the amalgam of minimal parabolics associated with the action of* $F_4(2)$ *on* $\mathcal{G}(F_4(2))$*. Then there is a subamalgam* $\widetilde{\mathcal{A}}$ *in* $\widetilde{G}$ *such that the restriction of* ψ *to* $\widetilde{\mathcal{A}}$ *is an isomorphism onto* $\mathcal{A}$*. The subgroup in* $\widetilde{G}$ *generated by* $\widetilde{A}$ *is a non-split extension of the form* $3^{833} \cdot F_4(2)$ *which acts flag-transitively on a geometry* $\mathcal{G}(3^{833} \cdot F_4(2))$ *with the diagram*

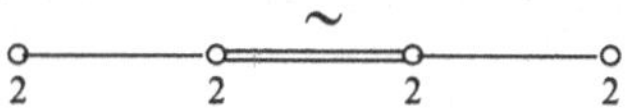

$$\underset{2}{\circ}\!\!-\!\!\!-\!\!\!-\!\!\underset{2}{\circ}\!\!\overset{\sim}{=\!=\!=}\!\!\underset{2}{\circ}\!\!-\!\!\!-\!\!\!-\!\!\underset{2}{\circ}$$

and ψ *induces a* 1*-covering* $\mathcal{G}(3^{833} \cdot F_4(2)) \to \mathcal{G}(F_4(2))$. □

It is not known up to now whether or not $\mathcal{G}(3^{833} \cdot F_4(2))$ is simply connected.

8

Y-groups

In early stages of studying the Monster group M in the 70's, B. Fischer had noticed that M can be generated by 15 involutions with pairwise products of order 2 or 3 corresponding to the following Coxeter diagram and found a few nice non-Coxeter relations satisfied by the involutions in the Monster.

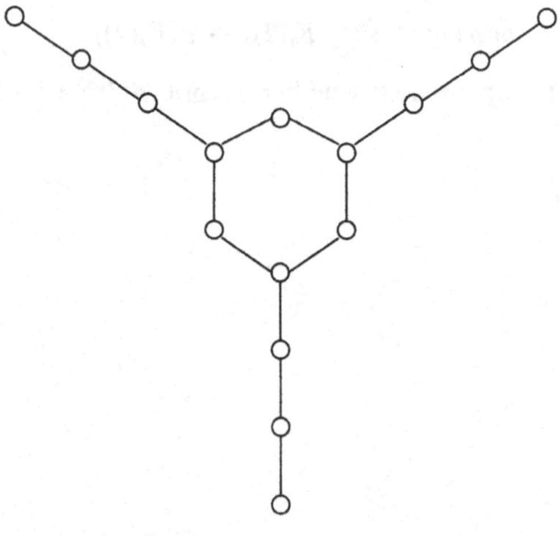

Around 1980 J.H. Conway conjectured that 16 involutions satisfying the Coxeter relations of the Y_{555} diagram given on the next page together with the so-called "spider" relation

$$(ab_1c_1ab_2c_2ab_3c_3)^{10} = 1$$

constitute a presentation for a group called the *Bimonster*, which is the wreath product $M \wr 2$ of the Monster and a group of order 2. Many people contributed to the proof of Conway's conjecture which has been completed in 1990 (1.13.5).

A crucial rôle in the proof of the Y_{555} theorem was played by the simple connectedness result for the tilde geometry of the Monster. In this chapter we review the original proof of the Y_{555} theorem and present an alternative proof based on an inductive approach to Y-groups.

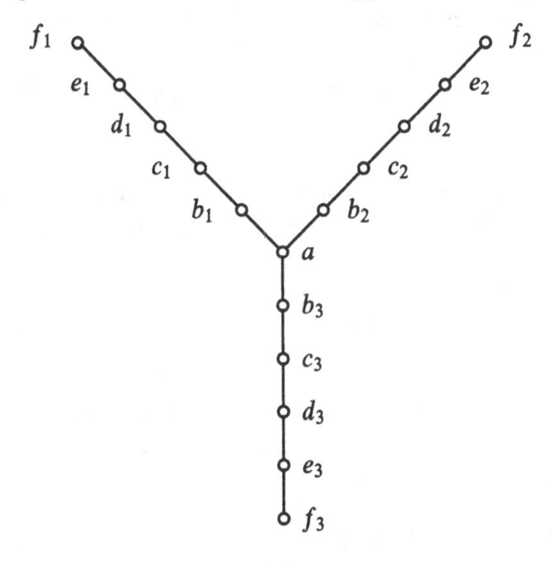

8.1 Some history

We start with the above Coxeter diagram known as the Y_{555} diagram and the following relation known as the "spider" relation:

$$(ab_1c_1ab_2c_2ab_3c_3)^{10} = 1.$$

For $2 \leq p, q, r \leq 5$ define Y_{pqr} be the quotient over the spider relation of the Coxeter group, whose generators are a and

first p terms from b_1, c_1, d_1, e_1, f_1,
first q terms from b_2, c_2, d_2, e_2, f_2,
first r terms from b_3, c_3, d_3, e_3, f_3,

and whose Coxeter relations correspond to the subdiagram of the Y_{555} diagram induced by the generators. A homomorphic image of the group Y_{pqr} will be called a Y_{pqr}-group and the Coxeter generators of Y_{pqr} will usually be identified with their images in a Y_{pqr}-group. If Z is a Y_{pqr}-group and $x, y, ..., z$ are some Coxeter generators of Z (or rather of Y_{pqr}) then $Z\lfloor x, y, ..., z\rfloor$ denotes the subgroup in Z generated by all its Coxeter generators except for $x, y, ..., z$. In these terms if x is the terminal node of the left arm of the Coxeter diagram of Y_{pqr} and $p \geq 3$ then $Y_{pqr}\lfloor x\rfloor$ is a $Y_{(p-1)qr}$-group.

If $\min\{p,q,r\} < 2$ then we define Y_{pqr} as $Y_{p_1q_1r_1}\lfloor x, ..., z\rfloor$ where $p_1 = \min\{2,p\}$, $q_1 = \min\{2,q\}$, $r_1 = \min\{2,r\}$ and $x, ..., z$ are the nodes in the Coxeter diagram of $Y_{p_1q_1r_1}$ whose removal gives the Coxeter diagram of Y_{pqr}. Suppose that $p-1, q, r \geq 2$ and that x is the terminal node of the left arm of the Coxeter diagram of Y_{pqr}. Then a Y_{pqr}-group Z is said to be *strong* if $Z\lfloor x\rfloor \cong Y_{(p-1)qr}$.

If $p, q, r \geq 2$ then every defining relation of Y_{pqr} has even length which implies the following.

Lemma 8.1.1 *Suppose that* $2 \leq p, q, r \leq 5$, *that* Z *is a* Y_{pqr}*-group and that* $O^2(Z) = Z$. *Then the direct product of* Z *and a group of order* 2 *is also a* Y_{pqr}*-group.* □

The structure of the groups Y_{pqr} is given in the following table.

pqr	Y_{pqr}	$[Y_{pqr} : Y_{(p-1)qr}]$
321	$2 \times Sp_6(2)$	56
421	$2 \cdot \Omega_8^+(2) : 2$	240
331	$2^7.(2 \times Sp_6(2))$	128
431	$2 \times Sp_8(2)$	255
441	$\Omega_{10}^-(2) : 2$	528
222	$3^5 : \Omega_5(3) : 2$	243
322	$2 \times \Omega_7(3)$	728
422	$2 \cdot \Omega_8^+(3) : 2$	2160
332	$2 \times 2 \cdot M(22)$	28 160
432	$2 \times M(23)$	31 671
442	$3 \cdot M(24)$	920 808
333	$2 \times 2^2 . ^2E_6(2)$	2 370 830 336
433	$2 \times 2 \cdot BM$	27 143 910 000
443	$2 \times M$	97 239 461 142 009 186 000
444	$M \wr 2$	$\|M\| \sim 10^{54}$

The groups above Y_{442} have been identified by means of coset enumeration on a computer in [CNS88], the group Y_{442} has been identified by D.Ž Djokovič also by coset enumeration on a computer and a computer-free identification, achieved by studying a certain hyperbolic reflection group, can be found in [CP92]. The group Y_{333} has been identified using double coset enumeration performed by S.A. Linton ([Lin89] and [Soi91]). The isomorphism type of Y_{443} was proved by combining the results in [Nor90], [Nor92] and in [Iv91a], [Iv92a] (see also [Con92]). The group Y_{433} has been identified in [Iv94]. It has been proved in [Soi89] that the isomorphism $Y_{443} \cong 2 \times M$ implies the isomorphism $Y_{444} \cong M \wr 2$. An independent characterization of Fischer groups as Y-groups can be found in [Vi97]. The groups Y_{p22}, $p \geq 5$, were identified in [Pr89] with certain orthogonal groups over $GF(3)$ (we do not present these results here). If $q \geq 3$, $r \geq 2$ then $Y_{5qr} = Y_{4qr}$, while Y_{632} and higher Y-groups collapse to a group of order 2 ((8.5.4) and (8.5.5)). It is worth mentioning that Y_{pqr} maps isomorphically onto its natural image in Y_{555} except for the groups Y_{421} and Y_{422} which lose their centres of order 2.

8.2 The 26-node theorem

In this section we discuss the 26-node theorem proved in [CNS88] and related results.

Theorem 8.2.1 *The group Y_{555} contains a set of* 26 *involutions including the set of* 15 *generators from the Y_{555} diagram, which satisfy the Coxeter relations given by the incidence graph Σ of the projective plane of order* 3. *The subgroup in Y_{555} which conjugates the vertex set of Σ onto itself induces the full automorphism group of Σ isomorphic to $L_3(3) : 2$.* □

Notice that the generators of M discovered by B. Fischer correspond to a subgraph of Σ.

By the 26-node theorem (8.2.1) Y_{555} is a quotient of the Coxeter group, whose diagram is the projective plane of order 3. L.H. Soicher [Soi91] found a very simple *hexagonal relation* which characterizes Y_{555} as such a quotient.

Theorem 8.2.2 *Let Σ be the incidence graph of the projective plane of order* 3 *and let (u, v, w, x, y, z) be a cycle of length* 6 *in Σ. Let C be the Coxeter* *group of* Σ *subject to the single additional relation*

$$(uxvywz)^4 = 1.$$

Then $C \cong Y_{555}$. □

By (1.13.5) Y_{555} is isomorphic to the Bimonster group $B := M \wr 2$ whose commutator subgroup D is isomorphic to the direct product of two copies of the Monster M. This shows that if p is a vertex of Σ then p is not contained in D and $C_B(p) \cong \langle p \rangle \times M$. Let Θ denote the set of involutions in Σ non-adjacent to p. Then every involution from Θ commutes with p and the subgraph of Σ induced by Θ is the incidence graph of the affine plane of order 3. It turns out that the subgroup in $C_B(p)$ generated by the involutions from Θ is isomorphic to the Monster. In [Mi95] the involutions from Θ are described explicitly in terms of the action of M on the Moonshine module so that the Coxeter relations can be checked, although it is not clear how difficult it would be to check a non-Coxeter relation, say the hexagonal relation.

The following very elegant characterization of Y_{555} proved in [CP92] was used originally for showing that the Bimonster is a quotient of Y_{555}.

Theorem 8.2.3 *Let* G *be a group (finite or infinite) which contains a subgroup* $A \cong Sym_5$ *such that* $C := C_G(A) \cong Sym_{12}$, *and if* B *is a 7-point stabilizer in the natural permutational action of* C *then* $B = A^g$ *for some* $g \in G$. *Suppose that no proper subgroup of* G *possesses the same property. Then either* $G \cong Sym_{17}$ *or* G *is a quotient of* Y_{555}. □

In [Nor90] S.P. Norton, using the 26-node theorem in a crucial way, has determined subgroups of Y_{555} of the shapes

$$2^{1+26}(2^{24} : Co_1) \quad \text{and} \quad (2^{10+16} \times 2^{10+16}).\Omega_{10}^+(2)$$

which correspond to some maximal parabolic subgroups associated with the action of M on its maximal 2-local parabolic geometry $\mathcal{H}(M)$.

It was realized during the Durham Symposium on Groups and Combinatorics in July 1990 that the simple connectedness result for the 2-local tilde geometry $\mathcal{G}(M)$ of the Monster [Iv92a] can be used to identify Y_{555} with the Bimonster. Using his earlier results on Y-groups Simon Norton proved in the course of the symposium that the derived subgroup of Y_{443} is generated by a subamalgam $\{C, N, L\}$ with

$$C \sim 2_+^{1+24}.Co_1, \quad N \sim 2^{2+11+22}.(Sym_3 \times Mat_{24}),$$
$$L \sim 2^{3+6+12+16}.(L_3(2) \times 3 \cdot Sym_6)$$

and $[N : N \cap C] = 3$, $[L : L \cap C] = [L : L \cap N] = 7$. As shown in [Iv92a], up to isomorphism there is only one such amalgam which consists of parabolic subgroups of the action of M on $\mathcal{G}(M)$ and M is the only completion of this amalgam. That is how the proof of the Y_{555} theorem was achieved.

After that, Y_{433} remained the only unknown case in the identification problem for Y-groups. The original proof of the Y_{555} theorem reduces the problem (via the 26-node theorem and the simple connectedness of $\mathcal{G}(M)$) to the triangulation problem for the second Monster graph. It turns out that there is a more direct way to associate a graph with a Y-group and to reduce the identification problem to the triangulation problem for that graph. This approach was realized in [Iv94] for Y_{433}. The identification problem was reduced to the triangulation problem of the graph $\widetilde{\Omega}$ as in Section 5.12. The triangulability of $\widetilde{\Omega}$ was proved in [Iv92c] within the simple connectedness proof for the Petersen type geometry of the Baby Monster. After that it became possible to apply an inductive approach to identify all the Y-groups. This approach, which gives an alternative proof of the Y_{555} theorem, is discussed in the remainder of the chapter.

8.3 From Y-groups to Y-graphs

We start this section with a definition. Let Δ be a graph and G be a vertex- and edge-transitive automorphism group of Δ. Let Ξ be another graph and H be an automorphism group of Ξ which is also assumed to be vertex- and edge-transitive. As usual, for a vertex $\alpha \in \Xi$ by $\Xi_i(\alpha)$ we denote the set of vertices at distance i from α and write $\Xi(\alpha)$ instead of $\Xi_1(\alpha)$, while $H(\alpha)$ denotes the stabilizer of α in H. Then (Ξ, H) is said to be *weakly locally* (Δ, G) if for every $\alpha \in \Xi$ there is an isomorphism

$$\varphi_\alpha : (\Delta, G) \to (\Xi(\alpha), H(\alpha))$$

of permutation groups such that whenever $\{x, y\}$ is an edge of Δ, $\{\varphi_\alpha(x), \varphi_\alpha(y)\}$ is an edge of Ξ. Notice that if (Ξ, H) is weakly locally (Δ, G) then H is a transitive extension of G ([Su86], p. 545). Identifying Δ and $\Xi(\alpha)$ via φ_α we can say that the subgraph in Ξ induced by $\Xi(\alpha)$ is a union of some orbitals of the action of G on Δ and this union contains the orbital formed by the edges of Δ. When H and G are clear from the context we simply say that Ξ is weakly locally Δ.

Suppose that Z is a Y_{pqr}-group, where $p \geq 2$, x is the terminal node of the left arm of the Coxeter diagram of Y_{pqr} and y is the node adjacent to x. We are dealing with the left arm just in order to simplify the notation.

Define a Y-graph $\Gamma = \Gamma(Z,x)$ to be a graph on the set of right cosets in Z of the subgroup $Z\lfloor x \rfloor$ in which two cosets $Z\lfloor x \rfloor g_1$, $Z\lfloor x \rfloor g_2$ are adjacent if there is an element h_1 in the former coset and an element h_2 in the latter coset such that $h_2 = xh_1$. In other terms the edges of Γ are the images under the natural action of Z of the pair $e := \{Z\lfloor x \rfloor, Z\lfloor x \rfloor x\}$. If $Z(e)$ is the elementwise stabilizer of the edge e then

$$Z(e) = Z\lfloor x \rfloor \cap Z\lfloor x \rfloor^x.$$

It is obvious that the latter group contains $Z\lfloor x, y \rfloor$ and the Y-graph Γ will be called *correct* if $Z(e) = Z\lfloor x, y \rfloor$.

Let $\alpha = Z\lfloor x \rfloor$, $\beta = Z\lfloor x \rfloor x$, $\gamma = Z\lfloor x \rfloor xy$, $H = \langle x, y \rangle \cong Sym_3$ and suppose that Γ is correct. Then $Z\lfloor x \rfloor = Z(\alpha)$ acts on $\Gamma(\alpha)$ as it acts on the cosets of $Z\lfloor x, y \rfloor$. Furthermore, since $(xy)^3 = 1$ and $y \in Z\lfloor x \rfloor$ we have

$$\gamma^x = Z\lfloor x \rfloor xyx = Z\lfloor x \rfloor yxy = Z\lfloor x \rfloor xy = \gamma,$$

which shows that $T := \{\alpha, \beta, \gamma\}$ is a triangle in Γ on which H induces the natural action. The images of T under Z will be called *Y-triangles.* Thus the action of $Z(\alpha)$ on $\Gamma(\alpha)$ is similar to its action on the vertex set of $\Delta := \Gamma(Z\lfloor x \rfloor, y)$ and two vertices in $\Gamma(\alpha)$ are adjacent whenever the corresponding vertices in Δ are adjacent. This shows that $\Gamma(Z,x)$ is weakly locally $\Gamma(Z\lfloor x \rfloor, y)$ (notice that $Z\lfloor x \rfloor$ is a $Y_{(p-1)qr}$-group). We summarize the most important case of this observation in the following.

Lemma 8.3.1 *Suppose that Z is a strong Y_{pqr}-group where $p-1, q, r \geq 2$ and that $\Gamma(Z,x)$ is correct. Then $\Gamma(Z,x)$ is weakly locally $\Gamma(Y_{(p-1)qr}, y)$.* □

Suppose that both $\Gamma(Z,x)$ and $\Gamma(Y_{pqr}, x)$ are correct. This is the case, for instance, when $\Gamma(Z,x)$ is correct and $Y_{(p-1)qr}\lfloor y \rfloor$ is a maximal subgroup of $Y_{(p-1)qr}$. Then the natural homomorphism

$$\varphi : Y_{pqr} \to Z$$

induces a covering

$$\psi : \Gamma(Y_{pqr}, x) \to \Gamma(Z,x)$$

of graphs such that the Y-triangles are contractible with respect to ψ. This gives the following

Lemma 8.3.2 *Suppose that Z is a strong Y_{pqr}-group and that both $\Gamma(Z,x)$ and $\Gamma(Y_{pqr}, x)$ are correct. Suppose further that the Y-triangles in $\Gamma(Z,x)$ generate the fundamental group of $\Gamma(Z,x)$. Then $Z \cong Y_{pqr}$.* □

The next lemma shows that in some cases examples of Y-groups can be constructed via their Y-graphs.

Lemma 8.3.3 *Let y be the terminal node of the left arm of the $Y_{(p-1)qr}$ diagram, where $p-1, q, r \geq 2$ and z be the node adjacent to y. Let Ξ be a graph and Z be a vertex- and edge-transitive automorphism group of Ξ and suppose that the following conditions hold for α being a vertex of Ξ.*

(i) $\Gamma(Y_{(p-1)qr}, y)$ *is correct;*

(ii) (Ξ, Z) *is weakly locally* $(\Gamma(Y_{(p-1)qr}, y), Y_{(p-1)qr})$ *and* φ_α *is the corresponding isomorphism;*

(iii) *if* $\beta = \varphi_\alpha(Y_{(p-1)qr}\lfloor y \rfloor)$ *then the setwise stabilizer in* Z *of* $\{\alpha, \beta\}$ *is the direct product of* $Z(\alpha) \cap Z(\beta)$ *and a group of order 2 generated by an element* x*;*

(iv) *the setwise stabilizer in* $Y_{(p-1)qr}$ *of* $\{Y_{(p-1)qr}\lfloor y \rfloor, Y_{(p-1)qr}\lfloor y \rfloor y\}$ *is the direct product* $\langle y \rangle \times Y_{(p-1)qr}\lfloor y, z \rfloor$ *and* $\langle y \rangle$ *is the centre of this stabilizer.*

Then Z is a strong Y_{pqr}-group.

Proof. The Coxeter generators of Z are x and the set K of (the images under φ_α of) the Coxeter generators of $Y_{(p-1)qr}$. By (ii) the generators in K satisfy the Coxeter relations and the spider relation. By (iii) x commutes with all the generators in K except for y. The product xy induces an action of order 3 on the triangle $T = \{\alpha, \beta, \gamma\}$ where $\gamma = \varphi_\alpha(Y_{(p-1)qr}\lfloor y \rfloor y)$. Hence $\sigma := (xy)^3$ is in the elementwise stabilizer L of this triangle. By (i) $L \cong Y_{(p-1)qr}\lfloor y, z \rfloor$, by (iii) and (iv) σ is in the centre of L and this centre is trivial by (iv). Hence $(xy)^3 = 1$ and the result follows. □

Our inductive approach to Y_{pqr} is the following. We consider a group Z acting vertex- and edge-transitively on a graph Ξ and we show eventually that Ξ is $\Gamma(Y_{pqr}, x)$ where x is the terminal node of the left arm of the Y_{pqr} diagram. First we show that Ξ is weakly locally $\Gamma(Y_{(p-1)qr}, y)$ where y is the node adjacent to x. Then we check the conditions in (8.3.3) and conclude that Z is a strong Y_{pqr}-group. Finally we show that the Y-triangles generate the fundamental group of Ξ and conclude from (8.3.2) that $Z \cong Y_{pqr}$. On the last step we often use (1.14.1). In some cases we will be able to show that the covering of Ξ under consideration induces another covering of graphs which is known to be an isomorphism. For this we use the strategy introduced in [Iv94].

8.4 Some orthogonal groups

In this section we identify Y_{222}, Y_{322} and Y_{422}. The diagram of Y_{222} is affine of extended E_6-type and hence the generators, when subject to the Coxeter relations only, produce a group isomorphic to $\mathbb{Z}^6 : \Omega_5(3).2$. By carrying out explicit calculations in the latter group we will find out the effect of adjoining the spider relation.

Let $H = Cox(Y_{222})$ be the Coxeter group and let $V = \mathbb{R}^6$ be a 6-dimensional real vector space with the natural inner product $(\ ,\ )$. Then H can be realized as a group of affine transformations of V in the following way.

Let $\mathscr{Y}$ be the set of Coxeter generators of H. It follows from the general theory of (affine) Coxeter groups ([Ebe94], [Hum90]) that there exists an (essentially unique) system $\{r(x) \mid x \in \mathscr{Y}\}$ of roots in V such that the angle between $r(x)$ and $r(y)$ is 120 or 90 degrees whenever x and y are adjacent or non-adjacent nodes, respectively. To wit, if $\{i, j, k\} = \{1, 2, 3\}$ then we choose $\{r(x) \mid x \in \mathscr{Y} \setminus \{c_i\}\}$ to be a fundamental system of roots in an E_6-lattice Γ and

$$r(c_i) = r(c_j) + r(c_k) + 2r(b_i) + 2r(b_j) + 2r(b_k) + 3r(a)$$

is the longest root in Γ with respect to this fundamental system.

For a root $r \in V$ let $l(r)$ be the reflexion with respect to the hyperplane orthogonal to r. Let L denote the group of orthogonal transformations generated by the reflexions $l(r(x))$ for all $x \in \mathscr{Y}$. Then L is the Coxeter group $Cox(E_6)$, which is known to be isomorphic to members of a number of series of classical groups over fields of characteristic 2 and 3:

$$L \cong \Omega_6^-(2).2 \cong U_4(2).2 \cong Sp_4(3).2 \cong \Omega_5(3).2.$$

In this chapter we will mainly use the isomorphism $L \cong \Omega_5(3).2$. The roots $r(x)$ generate the E_6-lattice Γ. Consider the semidirect product $\Gamma : L$ with respect to the natural action. Then every element $g \in \Gamma : L$ can be uniquely represented by a pair $(l(g), t(g))$ where $l(g)$ is a linear transformation of V and $t(g)$ is a translation. In this case the action of g on V is given by the following:

$$g : v \mapsto l(g) \cdot v + t(g)$$

where $v \in V$ and $l(g) \cdot v$ is the image of v under $l(g)$.

The Coxeter generators of H can be chosen in the following way: for every $x \in \mathscr{Y}$ we take $l(x) = l(r(x))$. Furthermore $t(x) = 0$ if $x \neq c_3$ and $t(c_3) = r(c_3)$. Then all the Coxeter relations are satisfied, the group generated is $\Gamma : L$ and we obtain the required realization of H.

Now let us analyse the effect of adjoining the spider relation. This relation can be written in the form $\varphi^{10} = 1$ where $\varphi = ab_1c_1ab_2c_2ab_3c_3$. We have chosen our generators so that $t(\varphi) = r(c_3)$. Then $l(\varphi^{10}) = l(\varphi)^{10}$ and

$$t(\varphi^{10}) = l(\varphi)^9 \cdot r(c_3) + \ldots + l(\varphi) \cdot r(c_3) + r(c_3).$$

One can easily calculate the roots $l(\varphi)^i \cdot r(c_3)$ as given below in the basis $\{r(x) \mid x \in \mathcal{Y} \setminus \{c_1\}\}$, which is another fundamental system of roots in Γ:

i	$l(\varphi)^i \cdot r(c_3)$
0	$r(c_3)$
1	$r(b_3)$
2	$-r(b_1) - r(a) - r(b_2) - r(c_2)$
3	$r(b_1) + 2r(a) + 2r(b_2) + r(c_2) + 2r(b_3) + r(c_3)$
4	$-2r(b_1) - 3r(a) - 2r(b_2) - r(c_2) - 2r(b_3) - r(c_3)$
5	$r(b_1) + r(a) + r(b_2) + r(c_2) + r(b_3) + r(c_3)$
6	$-r(b_1) - r(a) - r(b_2) - r(c_2)$
7	$r(c_2)$
8	$r(a) + r(b_1) + r(b_3) + r(c_3)$
9	$-r(b_1) - 2r(a) - r(b_2) - r(c_2) - r(c_3)$
10	$r(c_3)$

The above table shows that the orbit of $r(c_3)$ under the subgroup generated by $l(\varphi)$ is of length 10 and it is easy to see that the vectors from this orbit generate the lattice Γ. This gives

Lemma 8.4.1 *$l(\varphi)^{10}$ is the identity element of L.* □

The expression for $t(\varphi^{10})$ shows that this element is equal to the sum of all the rows in the above table which gives

$$t(\varphi^{10}) = -3r(b_1) - 3r(a) - r(b_2) + r(c_2) + r(b_3) + 2r(c_3).$$

Let K be the kernel of the homomorphism $H \to Y_{222}$. By (8.4.1) the image of K in $H/\Gamma \cong L$ is trivial and hence $K \leq \Gamma$. Since Γ is abelian, K is generated by the images of $t(\varphi^{10})$ under conjugation by elements of L.

Lemma 8.4.2 $K = \langle t(\varphi^{10}), 3\Gamma \rangle$.

Proof. By the definition K contains $t(\varphi^{10})$. Let us show that K also contains 3Γ. In fact, a direct calculation shows that

$$t(\varphi^{10}) - l(c_2) \cdot t(\varphi^{10}) = 3r(c_2).$$

Of course, the images of $3r(c_2)$ under L generate 3Γ. To show the reverse inclusion it is sufficient to show that $l(x) \cdot t(\varphi^{10}) \in \langle t(\varphi^{10}),\ 3\Gamma \rangle$ for every $x \in \mathcal{Y} \setminus \{c_1\}$. Since $l(x)$ is the reflexion associated with $r(x)$, we have the following:

$$l(x) \cdot t(\varphi^{10}) = t(\varphi^{10}) - (t(\varphi^{10}), r(x))r(x).$$

One can check that for $x \in \mathcal{Y} \setminus \{c_1\}$ the inner product $(t(\varphi^{10}), r(x))$ is divisible by 3 and hence the result follows. □

Since $t(\varphi)$ does not belong to 3Γ, we have the following.

Proposition 8.4.3 *$\Gamma/K \cong 3^5$ and $Y_{222} \cong 3^5 : \Omega_5(3).2$.* □

There is an orthogonal form on $O_3(Y_{222})$ and $Y_{222}/O_3(Y_{222})$ is the full automorphism group of this form. Then $\Gamma(Y_{222}, c_1)$ is a graph on the set of all vectors in a 5-dimensional $GF(3)$-space W with a non-singular orthogonal form, such that $v, w \in W$ are adjacent if $(v + w)$ is a plus vector, which means that the orthogonal complement $(v+w)^\perp$ contains a 2-dimensional totally singular subspace. It is straightforward to calculate that the suborbit diagram of $\Gamma(Y_{222}, c_1)$ is the following

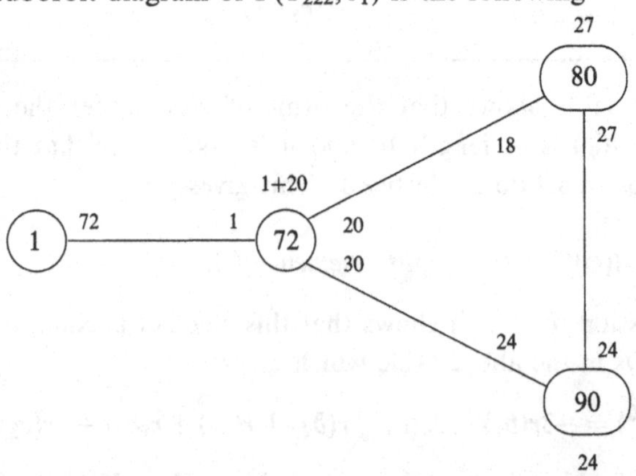

Let us turn to Y_{322}. Let W be a 7-dimensional $GF(3)$-space with a non-singular quadratic form and $Z \cong 2 \times \Omega_7(3)$ be the full automorphism group of this form. Let Ξ be a graph on the set of non-zero isotropic vectors in W in which two such vectors are adjacent if their inner product is 1. Direct calculations show that the suborbit diagram is the following

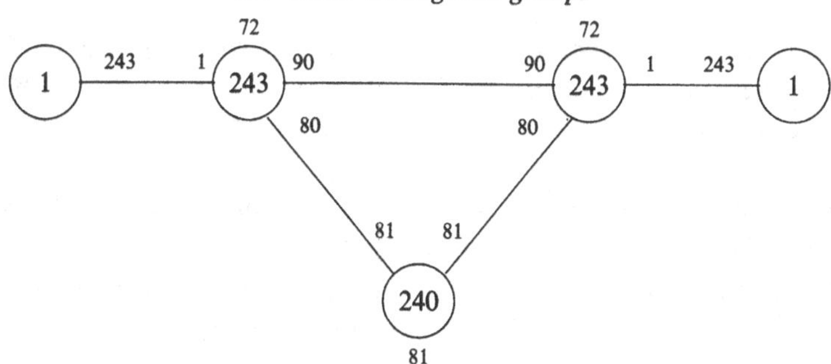

If $\alpha \in \Xi$ then $Z(\alpha) \cong 3^5 : \Omega_5(3) : 2 \cong Y_{222}$ and $O_3(Z(\alpha))$ acts regularly on $\Xi(\alpha)$ which immediately shows that Ξ is weakly locally $\Gamma(Y_{222}, c_1)$. It is easy to check that the remaining conditions in (8.3.3) are also satisfied and hence Z is a strong Y_{322}-group. From the above diagram we see that all triangles in Ξ are Y-triangles. Direct calculations in the orthogonal module W enable one to check the conditions in (1.14.1). Thus Ξ is triangulable, hence $Z \cong Y_{322}$ and $\Xi \cong \Gamma(Y_{322}, d_1)$ by (8.3.2).

For $\{i, j, k\} = \{1, 2, 3\}$ the nodes $a, b_i, c_i, d_i, b_j, c_j, b_k$ on the Y_{555} diagram induce a spherical E_7-diagram, so that the corresponding Coxeter group is isomorphic to $Sp_6(2) \times 2$ and its centre is generated by the following element [CNS88]:

$$f_{ijk} := (ab_ic_id_ib_jc_jb_k)^9.$$

For $i = 2$ and 3 put $X_i = Y_{322}\lfloor c_i \rfloor$ and let Σ_i be the subgraph in $\Gamma(Y_{322}, d_1)$ induced by the images of $Y_{322}\lfloor d_1 \rfloor$ under X_i. The Coxeter diagram of X_i is spherical of type E_7 and since all the Coxeter generators in Y_{322} are pairwise different, either $X_i \cong Sp_6(2) \times 2$ or $X_i \cong Sp_6(2)$. In the latter case $|\Sigma_i| = 28$ and X_i acts on Σ_i doubly transitively. By observing that $\Gamma(Y_{322}, d_1)$ does not contain cliques of size 28, or otherwise one concludes that $X_i \cong Sp_6(2) \times 2$ and the suborbit diagram of Σ_i with respect to the action of X_i is the following:

1 —27— 1 (27, 16) 10 —— 10 (27, 16) 1 —27— 1

Comparing the above diagram with the diagram of $\Gamma(Y_{322}, d_1)$, we immediately deduce that the centres of X_2, X_3 and Y_{322} coincide, and in terms of the above paragraph $f_{123} = f_{132}$.

Lemma 8.4.4 *If $q, r \geq 2$ then the element $f_{123} = f_{132}$ is in the centre of Y_{3qr}.*

Proof. A Coxeter generator of Y_{322} commutes with f_{1ij} since the latter element generates the centre of Y_{322}. On the other hand d_2 and higher terms clearly commute with f_{132} and the result follows. □

Permuting the indices p, q, r we obtain obvious analogues of (8.4.4) (compare the centres of Y-groups in the table).

By the diagram of $\Gamma(Y_{322}, c_2)$ given below and the list of maximal subgroups in $\Omega_7(3)$ [CCNPW], we have the following.

Lemma 8.4.5 *The graph $\Gamma(Y_{322}, c_2)$ is the unique orbital graph of valency 288 of $\Omega_7(3)$ acting on the cosets of $Sp_6(2)$ and every subgroup in $\Omega_7(3)$ of index 3159 is isomorphic to $Sp_6(2)$.* □

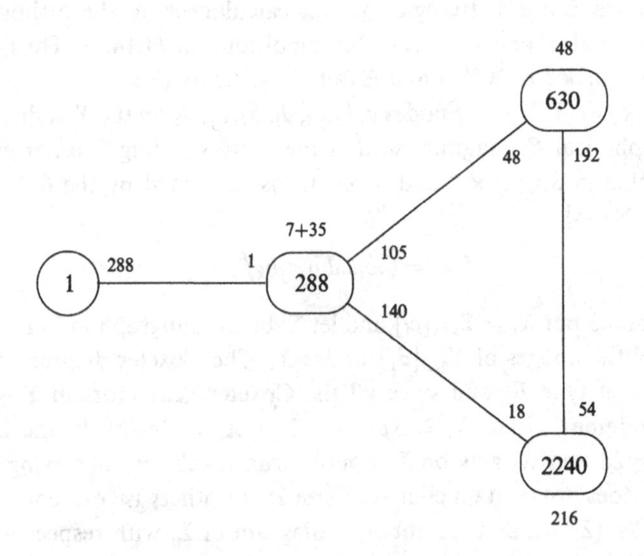

The last group to be considered in this section is Y_{422}. Let W be an 8-dimensional $GF(3)$-space with a non-singular quadratic form of plus type. The automorphism group of this form is $2 \cdot \Omega_8^+(3) : 2^2$ [CCNPW]. Let Z be a subgroup of index 2 in the automorphism group which contains a subgroup $H \cong 2 \times \Omega_7(3)$ trivially intersecting the centre. Then $Z \sim 2 \cdot \Omega_8^+(3).2_2$ in the atlas notation. Let O be the orbit of Z on the set of non-isotropic vectors in W such that H stabilizes a vector from O and let Ξ be a graph on O in which two vectors are adjacent if their inner product is 1. Then Ξ has the following suborbit diagram.

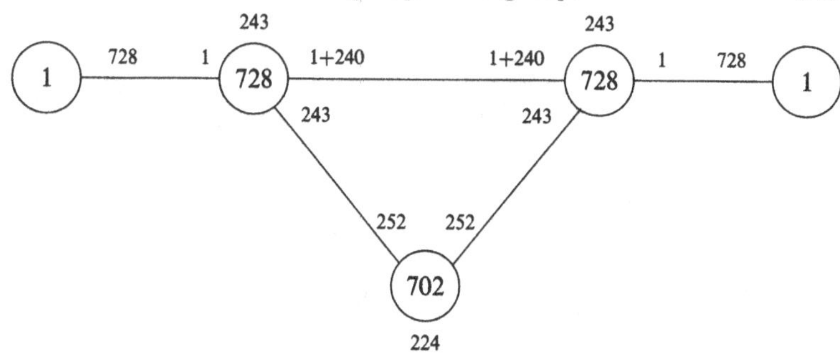

Performing some easy calculations in the orthogonal module W, we check that a triangle of Ξ is contained in 72 complete subgraphs on 4 vertices. In view of the suborbit diagram of $\Gamma(Y_{322}, c_1)$ this shows that Ξ is weakly locally $\Gamma(Y_{322}, c_1)$. It is straightforward to check the conditions in (8.3.3) and to conclude that Z is a strong Y_{422}-group. Finally the conditions in (1.14.1) hold, which show the isomorphism between Z and Y_{422}.

8.5 Fischer groups as Y-groups

In this section we identify Y_{332}, Y_{432} and Y_{442} with the groups $2 \cdot M(22)$, $M(23)$ and $3 \cdot M(24)$, respectively, and also discuss the group Y_{333}.

In this and next sections by $\Delta(M(23))$ and $\Delta(3 \cdot M(24))$ we denote, respectively the transposition graph of the Fischer group $M(23)$ and the triple cover of the transposition graph of the Fischer group $M(24)$. These graphs were introduced in Section 5.8 under the names Π_5 and Π_6, respectively. The suborbit diagrams of $\Delta(M(23)$ and $\Delta(3 \cdot M(24)$ with respect to the actions of $M(23)$ and $3 \cdot M(24)$ can also be found in Section 5.8.

The following result was proved in [Ron81a].

Proposition 8.5.1 *The graph $\Delta(M(23))$ and the graph $\Delta(3 \cdot M(24))$ are triangulable.* □

Consider the group Y_{332}. By (8.4.4) $\langle f_{213} \rangle$ is central in both Y_{332} and $Y_{332}\lfloor d_1 \rfloor$ and hence it is in the kernel of the action of Y_{332} on $\Gamma(Y_{332}, d_1)$. Consider the action of $Z := 2 \cdot M(22)$ (the non-split extension) on the cosets of a subgroup isomorphic to $\Omega_7(3)$. One of the orbital graphs (we denote it by Ξ) with respect to this action has the following suborbit diagram

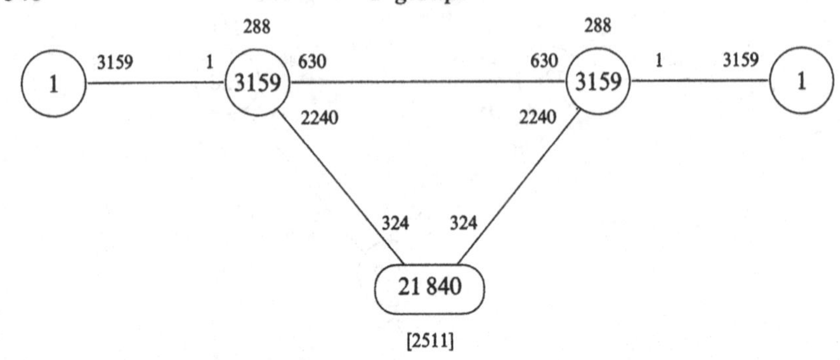

In view of the above diagram and (8.4.5) we conclude that Ξ is weakly locally $\Gamma(Y_{322}, c_2)$. Now it is easy to check the conditions in (8.3.3) and to conclude that Z is a strong Y_{332}-group.

One can see from the suborbit diagram of $\Delta(M(23))$ that the stabilizer in $M(23)$ of a vertex $\alpha \in \Delta = \Delta(M(23))$ is isomorphic to Z and its actions on the vertex-set of Ξ and on $\Delta_2(\alpha)$ are similar. Furthermore the subgraph in Δ induced by $\Delta_2(\alpha)$ is also an orbital of valency 3159. Noticing that the stabilizer in Z of a triangle in Ξ is isomorphic to Sym_7 while the stabilizer in $M(23)$ of a triangle in $\Delta(M(23))$ is of the form $[2^{11}].U_4(2)$ (in particular it does not involve Sym_7), we have the following

Lemma 8.5.2 *The subgraph in Δ induced by $\Delta_2(\alpha)$ and the graph Ξ with the above suborbit diagram are two different orbitals of valency* 3159 *of the action of* $2 \cdot M(22)$ *on the cosets of* $\Omega_7(3)$. □

Using (8.5.2) and calculating in the graph $\Delta(M(23))$ it is not difficult to check that the conditions in (1.14.1) are satisfied for Ξ, which gives the isomorphism $Y_{332}/\langle f_{213}\rangle \cong 2 \cdot M(22)$. Finally (8.1.1) completes the identification of Y_{332}.

Noticing that the Coxeter diagram of Y_{331} is affine of type E_7, it is not difficult to identify $Y_{332}\lfloor c_3 \rfloor/\langle f_{123}, f_{213}\rangle$ with a maximal subgroup in $M(22)$ of the form $2^6 : Sp_6(2)$. The subdegrees of $M(22)$ acting on the cosets of $2^6 : Sp_6(2)$, as calculated in [ILLSS], are the following:

$$1,\ 135,\ 1260,\ 2304,\ 8640,\ 10\,080,\ 45\,360,\ 143\,360,\ 241\,920^2.$$

Since $Y_{332}\lfloor b_3, c_3 \rfloor$ has index 2304 in $2^6 : Sp_6(2)$ the above subdegrees show that $\Gamma(Y_{332}, c_3)$ is correct and that it is isomorphic to the unique orbital graph of valency 2304 of the action of $M(22)$ on the cosets of $2^6 : Sp_6(2)$.

Let us turn to Y_{432}. Put $Z = M(23)$ and let Ξ be the complement of $\Delta(M(23))$. Then the vertex stabilizer $Z(\alpha)$ of the action of Z on Ξ is isomorphic to $2 \cdot M(22)$ which is the index 2 commutator subgroup of Y_{332}. The suborbit diagram of Ξ is

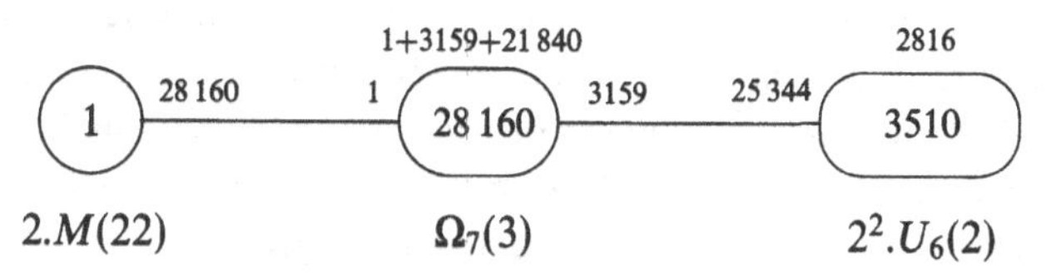

and by (8.5.2) Ξ is weakly locally $\Gamma(Y_{332}, d_1)$. Checking the conditions in (8.3.3), we conclude that Z is a Y_{432}-group.

The natural homomorphism $\varphi : \bar{Y}_{432} := Y_{432}/\langle f_{213}\rangle \to Z$ induces a covering

$$\psi : \Gamma(Y_{432}, e_1) \to \Xi$$

of graphs with respect to which the Y-triangles are contractible. Our nearest goal is to show that ψ induces a covering of $\Delta = \Delta(M(23))$.

Let $\widetilde{P} = (\tilde{s}_1, \tilde{s}_2, \tilde{s}_3)$ be a 2-path in $\Gamma(Y_{432}, e_1)$, $P = (s_1, s_2, s_3)$ be its image in Ξ and suppose that s_1 and s_3 are adjacent in Δ. Since ψ is a covering of graphs the stabilizer of $\widetilde{P}$ in $\bar{Y}_{432}$ maps isomorphically onto the stabilizer H_1 of P in Z. On the other hand the suborbit diagram of Ξ and (8.4.5) show that $H_1 \cong Sp_6(2)$. Without loss of generality we assume that $\{\tilde{s}_1, \tilde{s}_2\} = \{Y_{432}\lfloor e_1 \rfloor, Y_{432}\lfloor e_1 \rfloor e_1\}$, so that $\{s_1, s_2\} = \{Z\lfloor e_1 \rfloor, Z\lfloor e_1 \rfloor e_1\}$. Let $\widetilde{\Sigma}$ be the set of images of $\tilde{s}_1$ under $\bar{Y}_{432}\lfloor d_2 \rfloor$ and Σ be the set of images of s_1 under $Z\lfloor d_2 \rfloor$. Comparing the isomorphism $Y_{422} \cong 2 \cdot \Omega_7^+(3) : 2$ and the list of maximal subgroups in $M(23)$ or otherwise one concludes that $Z\lfloor d_2 \rfloor \cong \Omega_8^+(3) : 2$ and hence $|\Sigma| = 1080$. Thus the restriction of ψ to $\widetilde{\Sigma}$ is either a bijection, or has fibers of size 2. In either of the cases we can assume without loss of generality that $P \subset \Sigma$, $\widetilde{P} \subset \widetilde{\Sigma}$ and by the above sentence the stabilizer of $\{\tilde{s}_1, \tilde{s}_3\}$ in $\bar{Y}_{432}\lfloor d_2 \rfloor$ has index at most 2 in the stabilizer H_2 of $\{s_1, s_3\}$ in $Z\lfloor d_2 \rfloor$. From the suborbit diagram of $\Gamma(Y_{422}, e_1)$ we see that $H_2 \cong 2 \times 2 \cdot U_4(3) : 2$. Thus the stabilizer of $\{\tilde{s}_1, \tilde{s}_3\}$ in $\bar{Y}_{432}$ contains a subgroup isomorphic to $Sp_6(2)$ and a subgroup isomorphic to $2 \cdot U_4(3)$. On the other hand the stabilizer in $M(23)$ of an edge in Δ, isomorphic to $2^2 \cdot U_6(2)$ (a non-split extension), is generated by any two of its subgroups isomorphic to $Sp_6(2)$ and $2 \cdot U_4(3)$. Hence the stabilizer of $\{\tilde{s}_1, \tilde{s}_3\}$ in $\bar{Y}_{432}$ maps isomorphically onto the stabilizer of $\{s_1, s_3\}$ in Z which shows that ψ induces a covering

$$\chi : \widetilde{\Delta} \to \Delta$$

of graphs. Here the vertex set of $\tilde{\Delta}$ is that of $\Gamma(Y_{432}, e_1)$ and the edges are the images of $\{\tilde{s}_1, \tilde{s}_3\}$ under Y_{432}. Since χ is a covering of graphs, the restriction of ψ to $\tilde{\Sigma}$ must be a bijection and hence χ induces an isomorphism of the subgraph in $\tilde{\Delta}$ induced by $\tilde{\Sigma}$ onto the subgraph in Δ induced by Σ. The latter graph is the antipodal folding of $\Gamma(Y_{422}, e_1)$ and it is of rank 3. Since this subgraph obviously contains triangles and $M(23)$ acts transitively on the set of triangles in Δ, we conclude that all the triangles in Δ are contractible with respect to χ. But then χ and ψ must be isomorphisms since by (8.5.1) the triangles generate the fundamental group of Δ. Application of (8.1.1) completes the identification of Y_{432}.

Analysing the list of maximal subgroups of $M(23)$ [CCNPW] it is not difficult to identify $Y_{432}\lfloor c_3 \rfloor$ with $2 \times Sp_8(2)$. Consider the action of Y_{432} on $\Gamma(Y_{432}, c_3)$. By (8.4.4) f_{231} is in the kernel of the action. The stabilizer in $\bar{Y}_{432}$ of the edge $e := \{Y_{432}\lfloor c_3 \rfloor, Y_{432}\lfloor c_3 \rfloor c_3\}$ obviously contains $\bar{Y}_{432}\lfloor b_3, c_3 \rfloor \cong Sym_9$. On the other hand the subdegrees of the action of $M(23)$ on the cosets of $Sp_8(2)$ were calculated in [ILLSS]. The only non-trivial subdegree which divides the index 130 560 of Sym_9 in $Sp_8(2)$ is 13 056 and the corresponding 2-point stabilizer is isomorphic to Sym_{10}. Thus $\Gamma(Y_{432}, c_3)$ is not correct but in fact there is a way to "correct" the situation by adjoining an additional generator. Let $H \cong Sym_{10}$ be the stabilizer in $\bar{Y}_{432}$ of the edge e. Then the subdiagram of the Coxeter diagram of Y_{432} which is the Coxeter diagram of $\bar{Y}_{432}\lfloor b_3, c_3 \rfloor$ can be extended to that of H by adjoining a node adjacent to e_1 or to d_2. Since f_{213} is in the centre of Y_{432} the extra node (denote it by f_1) must be adjacent to e_1. Since H has no outer automorphisms, f_1 commutes with c_3. We claim that f_1 also commutes with b_3. This claim can be checked by noticing that every edge of $\Gamma(Y_{432}, c_3)$ is contained in $210 = [Sym_{10} : Sym_6 \times Sym_4]$ triangles [ILLSS] and that b_3 is involved in the expression for the element f_{213} and the latter commutes with f_1. Thus Y_{432} is a Y_{532}-group. Furthermore, f_1 commutes with $Y_{432}\lfloor e_1 \rfloor \cong 2 \times 2 \cdot M(22)$, the latter subgroup is self-centralized in Y_{432} and by (8.4.4) its centre is $\langle f_{123}, f_{213} \rangle$. Since e_1 has product of order 3 with both f_1 and f_{123} we conclude that the latter two elements are equal.

Lemma 8.5.3 $Y_{532} \cong Y_{432}$.

Proof. Suppose that $Y_{532}\lfloor f_1 \rfloor$ is a proper subgroup in Y_{532} and consider the action of $\bar{Y}_{532} := Y_{532}/\langle f_{231} \rangle$ on $\Gamma(Y_{532}, f_1)$. Then the structure of Y_{432}, Y_{332} and Y_{232} show that the elementwise stabilizers of a vertex, an edge and a triangle in $\bar{Y}_{543}$ are isomorphic to $M(23)$, $2 \cdot M(22)$ and $\Omega_7(3)$, respectively. Hence $\Gamma(Y_{532}, f_1)$ is weakly locally the complement Ξ of $\Delta(M(23))$ with the suborbit diagram given in this section. If the diameter of $\Gamma(Y_{532}, f_1)$ is 1 then the action of Y_{532} on the vertex set of the graph is doubly transitive and it is easy to show that this is not possible. On the other hand from the suborbit diagram of Ξ we see that the number of vertices at distance 2 from a given vertex is at most

$$31\,671 \cdot 3510/25\,345 < 4500.$$

Comparing this estimate with the indices of maximal subgroups in $M(23)$ we conclude that there is only one vertex at distance two. Since the action of $M(23)$ on Ξ is primitive, this gives a contradiction. □

By the above lemma and the paragraph before it, we obtain the following

Corollary 8.5.4

If $q \geq 3$ and $r \geq 2$ then $Y_{5qr} = Y_{4qr}$ and $f_1 = f_{123} = f_{132}$. □

Consider Y_{632} with the obvious meaning. Then the generator corresponding to the terminal node of the left arm of the Coxeter diagram commutes with f_{123} and the order of its product with f_1 divides 3. By Corollary 8.5.4 this gives the following.

Corollary 8.5.5

Y_{632} and higher Y-groups collapse to a group of order 2. □

In order to identify Y_{442} consider the action of $Z := 3 \cdot M(24)$ on $\Delta := \Delta(3 \cdot M(24))$. If $\alpha \in \Delta$ then $Z(\alpha) \cong 2 \times M(23) \cong Y_{432}$ and α can be identified with the unique non-trivial element in the centre of $Z(\alpha)$ (this element is an involution which maps onto a 3-transposition in $M(24)$). In these terms if $\beta \in \Delta_i(\alpha)$ then the product $\alpha\beta$ is of order 2, 3, 6 and 3 for $i = 1$, 2, 3 and 4, respectively.

Let $\beta \in \Delta_2(\alpha)$. Then $Z(\alpha) \cap Z(\beta) \cong \Omega_8^+(3) : 2 \cong Y_{432}\lfloor d_2 \rfloor$. Since the commutator subgroup Z' of Z acts distance-transitively on Δ, we conclude that $Z(\alpha) \cap Z(\beta)$ is not contained in the direct factor $M(23)$ of $Z(\alpha)$. Since all subgroups in $M(23)$ isomorphic to $\Omega_8^+(3) : 2$ are conjugate, this specifies the action of $Z(\alpha)$ on $\Delta_2(\alpha)$ and in particular shows that this action is similar to the action of Y_{432} on the vertex set of $\Gamma(Y_{432}, d_2)$. Since $Y_{432}\lfloor c_2, d_2 \rfloor \cong \Omega_8^+(2) : 2$ is a maximal subgroup of index 28 431 in $Y_{432}\lfloor d_2 \rfloor \cong \Omega_8^+(3) : 2$, we conclude that $\Gamma(Y_{432}, d_2)$ is correct of valency 28 431. The suborbit diagram of Δ shows that the subgraph in Δ induced by $\Delta_2(\alpha)$ is also an orbital of valency 28 431. We claim that they are different orbitals. Indeed, by (8.5.4) the stabilizer in $Y_{432} = Y_{532}$ of a triangle in $\Gamma(Y_{432}, d_2)$ contains $Y_{532}\lfloor b_2, c_2, d_2 \rfloor \cong Sym_9$ while the stabilizer in Z of a triangle in Δ is of the form $2^3.U_6(2)$ and does not involve Sym_9. Hence the claim follows.

Notice that the set

$$\Theta(\beta) = \{\gamma^\alpha \mid \gamma \in \Delta_2(\alpha) \cap \Delta(\beta)\} = \Delta_2(\alpha) \cap \Delta(\beta^\alpha)$$

is an orbit of length 28 431 of $Z(\alpha) \cap Z(\beta)$ on $\Delta_2(\alpha)$ containing vertices which are at distance 2 from β in Δ. It follows from [PS97] that the action of $M(23)$ on the cosets of $\Omega_8^+(3) : 2$ has subdegree 28 431 with multiplicity one, which means that if

$$\varphi : (\Gamma(Y_{432}, d_2), Y_{432}) \to (\Delta_2(\alpha)), Z(\alpha))$$

is an isomorphism of permutation groups which sends $Y_{432}\lfloor d_2 \rfloor$ onto β, then $\Theta(\beta)$ is the image under φ of the set of vertices adjacent to $Y_{432}\lfloor d_2 \rfloor$ in $\Gamma(Y_{432}, d_2)$. Thus we have the following.

Lemma 8.5.6 *Let Ξ be a graph on the set of vertices of $\Delta = \Delta(3 \cdot M(24))$ in which two vertices are adjacent if they are at distance 2 in Δ. Then Ξ is weakly locally $\Gamma(Y_{432}, d_2)$.* □

Now it is easy to see that the conditions in (8.3.3) are satisfied and hence $Z \cong 3 \cdot M(24)$ is a Y_{442}-group (by (8.5.4) it is also a Y_{552}-group).

Our next goal is to show that the natural homomorphism

$$\varphi : Y_{442} \to Z$$

induces a covering of Δ. Let $\widetilde{P} = (\tilde{s}_1, \tilde{s}_2, \tilde{s}_3)$ be a 2-path in $\Gamma(Y_{442}, \lfloor e_1 \rfloor)$, $P = (s_1, s_2, s_3)$ be its image in Ξ and suppose that s_1 and s_3 are adjacent in Δ. Since φ induces a covering of $\Gamma(Y_{442}, e_1)$ onto Ξ, the stabilizer of $\widetilde{P}$ in Y_{442} is isomorphic to $\Omega_8^+(2) : 2$ which is the stabilizer of P in Z. Let $\widetilde{\Sigma}$ be the set of images of $Y_{442}\lfloor e_1 \rfloor$ (considered as a vertex of $\Gamma(Y_{442}, e_1)$) under $Y_{442}\lfloor e_2 \rfloor$ and let Σ be the set of images of $Z\lfloor e_1 \rfloor$ under $Z\lfloor e_2 \rfloor$. Since $Y_{442}\lfloor e_2 \rfloor \cong Z\lfloor e_2 \rfloor \cong Y_{432} \cong 2 \times M(23)$, $\widetilde{\Sigma}$ maps bijectively onto Σ. Furthermore $Y_{442}\lfloor e_2 \rfloor$ acts on $\widetilde{\Sigma}$ with kernel of order 2 and the induced action is isomorphic to that of $M(23)$ on the vertex set of $\Delta(M(23))$. Without loss of generality we assume that $\widetilde{P} \subset \widetilde{\Sigma}$ in which case it follows from the suborbit digram of $\Delta(M(23))$ that the stabilizer of $\{\tilde{s}_1, \tilde{s}_3\}$ in $Y_{442}\lfloor e_2 \rfloor$ is of the form $2^3.U_6(2)$. Since the stabilizer of $\{s_1, s_3\}$ in Z, isomorphic to $2 \times 2 \cdot M(22)$ is generated by its subgroups isomorphic to

$\Omega_8^+(2):2$ and $2^3.U_6(2)$, we conclude that the stabilizer of $\{\tilde{s}_1, \tilde{s}_3\}$ in Y_{442} maps isomorphically onto the stabilizer of $\{s_1, s_3\}$ in Z which implies that φ induces a covering

$$\chi : \widetilde{\Delta} \to \Delta$$

of graphs. The subgraph in $\widetilde{\Delta}$ induced by $\widetilde{\Sigma}$ maps isomorphically onto the subgraph in Δ induces by Σ and both these subgraphs are isomorphic to $\Delta(M(23))$. Since the latter graph contains triangles and Z acts transitively on the set of triangles in Δ, we conclude that the triangles are contractible with respect to χ. Since Δ is triangulable by (8.5.1), both χ and ψ are isomorphisms and hence $Y_{442} \cong 3 \cdot M(24)$. Now analysing the maximal subgroups in $M(24)$ or otherwise one can check that $Y_{442}\lfloor c_3 \rfloor \cong \Omega_{10}^-(2):2$.

Finally, let us discuss the group Y_{333}. By (8.4.4) $\langle f_{123}, f_{213}, f_{312}\rangle$ is central in Y_{333} and $\langle f_{213}, f_{312}\rangle$ is contained in $Y_{333}\lfloor d_1 \rfloor$. Consider the action of $Z := {}^2E_6(2)$ on the cosets of $M(22) \cong Y_{233}/\langle f_{213}, f_{312}\rangle$. The intersection numbers of the centralizer algebra of this action have been calculated in [ISa96]. These calculations show in particular that there is an orbital graph Ξ of valency 694 980 with edge stabilizer isomorphic to a maximal subgroup of $M(22)$ isomorphic to $2^6 : Sp_6(2)$. Furthermore, every edge of Ξ is in exactly 13 644 triangles. Since

$$13\,644 = 1260 + 2304 + 10\,080$$

is the only decomposition of the number of triangles on an edge into the lengths of suborbits of $M(22)$ on the cosets of $2^6 : Sp_6(2)$, given above, we conclude that Ξ is weakly locally $\Gamma(Y_{332}, c_3)$. It is easy to check the conditions in (8.3.3). Hence Z is a Y_{333}-group. A possible way to identify Z with $Y_{333}/\langle f_{123}, f_{213}, f_{312}\rangle$ would be to show that the fundamental group of Ξ is generated by the Y-triangles. But this seems to be far too difficult, since the structure of Ξ is rather complicated and there are many classes of cycles in this graph. By this reason we refer to the original identification of Y_{333} which follow from the double coset enumeration performed by S.A. Linton ([Lin89], [Soi91]).

8.6 The monsters

In this section we identify Y_{433}, Y_{443} and Y_{444} with $2 \times 2 \cdot BM$, $2 \times M$ and $M \wr 2$, respectively.

Let $Z \cong 2 \cdot BM$ be the non-split extension of the Baby Monster BM by a centre of order 2, introduced in the first paragraph of Section 5.12 under the name $\widetilde{B}$. Let $\widetilde{\Omega}$ be the graph introduced in the paragraph before

(5.12.5). Then the vertex set of $\widetilde{\Omega}$ is the conjugacy class of involutions in Z with centralizers of the form $2^{2\cdot 2}E_6(2)$ [CCNPW]. The group Z acts on $\widetilde{\Omega}$ by conjugation, the centre of Z is the kernel and the induced action is similar to that of BM on the cosets of $2{\cdot}^2E_6(2)$. The suborbit diagram of this action is given in Section 5.12. The vertex α' antipodal to α is the product of α and the involution in the centre of Z. By (5.12.7) the product $\alpha\beta$ has order 2, 2, 3, 4, 6 and 4 if β is contained in $\widetilde{\Omega}(\alpha)$, $\widetilde{\Omega}(\alpha')$, $\widetilde{\Omega}_2^3(\alpha)$, $\widetilde{\Omega}_2^4(\alpha)$, $\widetilde{\Omega}_2^3(\alpha')$ and $\widetilde{\Omega}_3^4(\alpha)$, respectively. Notice that by joining in $\widetilde{\Omega}$ the antipodal vertices we obtain the graph isomorphic to the subgraph of the second Monster graph induced by the vertices adjacent to a given vertex.

The group $Z(\alpha) \cong 2^{2\cdot 2}E_6(2)$ (the commutator subgroup of Y_{333}) acts on $\widetilde{\Omega}_2^3(\alpha)$ as it acts on the cosets of its subgroup $2\cdot M(22)$ (the commutator subgroup of Y_{332}). Hence this action is similar to the action induced by Y_{333} on $\Gamma(Y_{333}, d_1)$. The graph $\Gamma(Y_{333}, d_1)$ has valency 694 980 and one can see from the diagram of $\widetilde{\Omega}$ that this is also the valency of the subgraph of $\widetilde{\Omega}$ induced by $\widetilde{\Omega}_2^3(\alpha)$. By [ISa96] the subdegree 694 980 appears with multiplicity one in the action of ${}^2E_6(2)$ on the cosets of $M(22)$. This means that for $\beta \in \widetilde{\Omega}_2^3(\alpha)$ the subgroup $Z(\alpha) \cap Z(\beta)$ has exactly two orbits of length 694 980 on $\widetilde{\Omega}_2^3(\alpha)$, namely $\widetilde{\Omega}_2^3(\alpha) \cap \widetilde{\Omega}(\beta)$ and $\Theta(\beta) := \widetilde{\Omega}_2^3(\alpha) \cap \widetilde{\Omega}(\beta^\alpha)$. If $\gamma \in \Theta(\beta)$ then $\beta\gamma$ is of order 3 and hence $\gamma \in \widetilde{\Omega}_2^3(\beta)$. We claim that $\Gamma(Y_{333}, d_1)$ is isomorphic to the graph on $\widetilde{\Omega}_2^3(\alpha)$ in which β is adjacent to $\Theta(\beta)$. In fact, the stabilizer in $Y_{333}/\langle f_{213}, f_{312}\rangle$ of a triangle in $\Gamma(Y_{333}, d_1)$ contains Sym_8 while if H is the stabilizer in BM of a triangle in $\widetilde{\Omega}$ then $H/O_2(H) \cong U_4(2)$ which does not involve Sym_8 and the claim follows.

Let Ξ be a graph on the set of vertices of $\widetilde{\Omega}$ in which α and β are adjacent if $\beta \in \widetilde{\Omega}_2^3(\alpha)$. By the above paragraph Ξ is weakly locally $\Gamma(Y_{333}, d_1)$ and checking the remaining conditions in (8.3.3) we conclude that $Z \cong 2 \cdot BM$ is a Y_{433}-group. Notice that we have realized the Coxeter generators of Y_{433} by involutions inside $2 \cdot BM$. By (8.1.1) the direct product $2 \times 2 \cdot BM$ is also a Y_{433}-group.

We claim that the homomorphism

$$\varphi : \bar{Y}_{433} := Y_{433}/\langle f_{213}, f_{312}\rangle \to BM$$

induces a covering χ of $\widetilde{\Omega}$. Let $\widetilde{P} = (\bar{s}_1, \bar{s}_2, \bar{s}_3)$ be a 2-path in $\Gamma(Y_{433}, e_1)$ which maps onto a 2-path $P = (s_1, s_2, s_3)$ in Ξ such that s_1 and s_3 are adjacent in $\widetilde{\Omega}$. Since φ induces a covering of $\Gamma(Y_{433}, e_1)$ onto Ξ, the stabilizer of $\widetilde{P}$ in $\bar{Y}_{433}$ maps isomorphically onto the stabilizer of P in BM and the latter is the edge stabilizer of the subgraph in $\widetilde{\Omega}$ induced by

$\widetilde{\Omega}_2^3(\alpha)$, isomorphic to $2^6 : Sp_6(2)$. We assume that $\tilde{s}_1 = \bar{Y}_{433}\lfloor e_1 \rfloor$, so that $s_1 = BM\lfloor e_1 \rfloor$. Let $\widetilde{\Sigma}$ be the set of images of $\tilde{s}_1$ under $\bar{Y}_{433}\lfloor d_2 \rfloor$ and let Σ be the set of images of s_1 under $BM\lfloor d_2 \rfloor$. Since

$$\bar{Y}_{433}\lfloor d_2 \rfloor \cong BM\lfloor d_2 \rfloor \cong Y_{432}/\langle f_{213} \rangle \cong M(23),$$

$\widetilde{\Sigma}$ maps bijectively onto Σ. Furthermore the action of $\bar{Y}_{433}\lfloor d_2 \rfloor$ on $\widetilde{\Sigma}$ is similar to that of $M(23)$ on $\Delta(M(23))$. Assuming without loss of generality that $\widetilde{P} \subset \widetilde{\Sigma}$ we conclude from the suborbit diagram of $\Delta(M(23))$ that the stabilizer of $\{\tilde{s}_1, \tilde{s}_3\}$ in $\bar{Y}_{433}\lfloor d_2 \rfloor$ is of the form $2^2 \cdot U_6(2)$. Finally the stabilizer of $\{s_1, s_3\}$ in BM, isomorphic to $2^{2+20}.U_6(2)$ is generated by its subgroups $2^6 : Sp_6(2)$ and $2^2 \cdot U_6(2)$ which implies that φ induces a covering $\chi : \widehat{\Omega} \to \widetilde{\Omega}$ of graphs. The subgraph in $\widehat{\Omega}$ induced by $\widetilde{\Sigma}$ maps isomorphically onto the subgraph in $\widetilde{\Omega}$ induced by Σ (both these subgraphs are isomorphic to $\Delta(M(23))$). Thus the triangles in $\widetilde{\Omega}$ are contractible with respect to χ. It has been proved in [Iv92b] and [Iv94] that $\widetilde{\Omega}$ is triangulable. Hence both χ and φ are isomorphisms and $\bar{Y}_{433} \cong BM$.

Now let Z be the Monster group M and Γ be the second Monster graph as in Section 5.12. Then Γ is a graph on the conjugacy class of $2a$- (Baby Monster) involutions in the Monster with two involutions being adjacent if their product is again a $2a$-involution. If $\alpha \in \Gamma$ then $Z(\alpha) \cong 2 \cdot BM$ (the commutator subgroup of Y_{433}) and the subgraph in Γ induced by $\Gamma(\alpha)$ is the graph $\widetilde{\Omega}$ as above together with a matching which joins pairs of antipodal vertices. This shows that Z has two orbits on the triangles in Γ. Every edge is contained in a unique triangle from one of the orbits (we call them short triangles) and in 3 968 055 triangles from another orbit (we call then long triangles). The suborbit diagram of Γ has been calculated in [Nor85] and we will use the following result from that paper.

Lemma 8.6.1 *Let $\alpha \in \Gamma$ and $\beta \in \Gamma^{3a}(\alpha)$. Then $Z(\alpha) \cap Z(\beta)$ acts transitively on the set of vertices $\gamma \in \Gamma^{3a}(\alpha) \cap \Gamma^{2a}(\beta)$ with stabilizer isomorphic to $Sp_8(2)$.* □

Let Ξ be a graph on the vertex set of Γ in which α and β are adjacent if $\beta \in \Gamma^{3a}(\alpha)$. By (5.12.10 (iii)) the isomorphism

$$\sigma : \bar{Y}_{433} := Y_{433}/\langle f_{213} \rangle \to Z(\alpha)$$

induces an isomorphism of the permutation group $(\Gamma(Y_{433}, d_3), \bar{Y}_{433})$ onto the permutation group $(\Xi(\alpha), Z(\alpha))$. We denote the latter isomorphism by the same letter σ. We claim that whenever u and v are adjacent vertices in $\Gamma(Y_{433}, d_3)$, $\sigma(u)$ and $\sigma(v)$ are adjacent vertices in Ξ. First the stabilizer of $\{u, v\}$ in $\bar{Y}_{433}$ is isomorphic to $Sp_8(2)$ and by (5.12.10) we have $\sigma(u) \in \Gamma^C(\sigma(v))$ where C is $2a$, $3a$ or $4b$. Let us first exclude the latter possibility. Without loss of generality we assume that $u = Y_{433}\lfloor d_3 \rfloor$ and $v = Y_{433}\lfloor d_3 \rfloor d_3$. Then the isomorphism σ sends the Coxeter generators of $Y_{433}\lfloor d_3 \rfloor$ into $Z(\sigma(u))$ and by (5.12.10 (iii)) the images of the generators are contained in $\Gamma(\alpha) \cap \Gamma(\sigma(u))$. The image under σ of d_3 maps $\sigma(u)$ onto $\sigma(v)$ and commutes with the images of the Coxeter generators of $Y_{433}\lfloor c_3, d_3 \rfloor$ which shows that $\sigma(u)$ and $\sigma(v)$ have at least 9 common neighbours in Γ and by (5.12.10 (v)) $\sigma(v) \notin \Gamma^{4b}(\sigma(u))$. Suppose that $\sigma(v) \in \Gamma^{2a}(\sigma(u))$. Then by (8.6.1 (iii)) $\Gamma(Y_{433}, d_3)$ maps isomorphically onto the subgraph in Γ induced by $\Xi(\alpha)$. We know from the previous section that the stabilizer in $\bar{Y}_{433}$ of a triangle in $\Gamma(Y_{433}, d_3)$ is isomorphic to Sym_{10}. Clearly this triangle is a long triangle in Γ, but if H is the stabilizer in Z of a long triangle then $H/O_2(H) \cong U_6(2)$ and the latter group does not involve Sym_{10}. This contradiction shows that $\sigma(u)$ and $\sigma(v)$ are adjacent in Ξ. Hence Ξ is weakly locally $\Gamma(Y_{433}, d_3)$. It is easy to check the conditions in (8.3.3) and to conclude that $Z \cong M$ is a Y_{443}-group.

We claim that the homomorphism

$$\varphi : \bar{Y}_{443} := Y_{443}/\langle f_{312} \rangle \to M$$

induces a covering $\chi : \widetilde{\Gamma} \to \Gamma$ of graphs with respect to which all long triangles are contractible. Consider a 2-path $\widetilde{P} = (\tilde{s}_1, \tilde{s}_2, \tilde{s}_3)$ in $\Gamma(Y_{443}, e_1)$ which maps onto a 2-path $P = (s_1, s_2, s_3)$ in Ξ such that s_1 and s_3 are adjacent in Γ. Then by (8.6.1) and since φ induces a covering of $\Gamma(Y_{443}, e_1)$ onto Ξ, the stabilizer of $\widetilde{P}$ in $\bar{Y}_{443}$ is isomorphic to $Sp_8(2)$. Assume without loss of generality that $\tilde{s}_1 = \bar{Y}_{443}\lfloor e_1 \rfloor$ and that $\tilde{s}_3$ is contained in the orbit of $\tilde{s}_1$ under $\bar{Y}_{443}\lfloor e_2 \rfloor \cong 2 \cdot BM$. Then the suborbit diagram of $\widetilde{\Omega}$ given in Section 5.12 shows that the stabilizer of $\{\tilde{s}_1, \tilde{s}_3\}$ in $Y_{443}\lfloor e_2 \rfloor$ is isomorphic to $2^{3+20}.U_6(2)$. Finally since the stabilizer of $\{s_1, s_3\}$ in Z, isomorphic to $2^{2.2}E_6(2)$ is generated by its subgroups isomorphic to $2^{3+20}.U_6(2)$ and $Sp_8(2)$ we conclude that φ indeed induces a covering $\chi : \widetilde{\Gamma} \to \Gamma$ of graphs. The subgraph in $\widetilde{\Gamma}$ induced by the images of $\tilde{s}_1$ under $\bar{Y}_{443}\lfloor e_2 \rfloor$ is isomorphic either to $\widetilde{\Omega}$ or to the subgraph in Γ induced by $\Gamma(\alpha)$ and in any case the long triangles are contractible with respect to χ. By (5.14.2) Γ is triangulable. In [Iv94] using this result it was shown that the long triangle already generate the fundamental group of Ξ. Hence $\bar{Y}_{443} \cong M$ and in view of (8.1.1) we have $Y_{443} \cong 2 \times M$.

Let us turn to Y_{444}. As above, let M be the Monster group, let D be the direct product of two copies of M:

$$D = \{(g,h) \mid g,h \in M\}; \quad (g_1,h_1)\cdot(g_2,h_2) = (g_1g_2, h_2h_1)$$

and define an action of D on M by $(g,h) : m \mapsto gmh$ for every $m \in M$ and $(g,h) \in D$. In this way we realize D as the group generated by the left and right regular representations of M. Let τ be the permutation on M acting by $\tau : m \mapsto m^{-1}$. Then τ can be considered as a permutation of D via $(g,h)^\tau = (h^{-1}, g^{-1})$, in particular τ normalizes D and permutes its direct factors. Let $Z = \langle D, \tau\rangle$ be the Bimonster. If $Z(1)$ is the stabilizer in Z of the identity element of M then $Z(1) = \langle\tau\rangle \times M'$ where $M' = \{(g,h) \in D \mid g = h^{-1}\}$ and hence every orbit of $Z(1)$ on M is of the form $C \cup C'$ where C is a conjugacy class of M and $C' = \{g^{-1} \mid g \in C\}$.

Let Ξ be a graph on M in which m_1 and m_2 are adjacent if and only if $m_1m_2^{-1}$ is an element of type $3a$ in M. Since the class of $3a$-elements is closed under taking inverses, Z acts on Ξ vertex- and edge-transitively. If t is of type $3a$ then the stabilizer in Z of the triple $T = \{1, t, t^{-1}\}$ is

$$\langle\tau\rangle \times N_{M'}(\langle t\rangle) \cong 2 \times 3 \cdot M(24),$$

which shows that the elementwise stabilizer in Z of the edge $\{1,t\}$ is isomorphic to $3 \cdot M(24)$ while the setwise stabilizer is of the form $3 \cdot M(24) \times 2$.

Let $A \cong Alt_4$ be a subgroup in M with the normalizer of the form $(Alt_4 \times \Omega_{10}^-(2)) : 2$ [Nor98]. Then all the elements of order 3 in A are of type $3a$ and hence by choosing t_1 and t_2 to be suitable such elements, we obtain a triangle $T_1 = \{1, t_1, t_2\}$ in Ξ whose elementwise stabilizer is isomorphic to $\Omega_{10}^-(2) : 2$ (notice that T_1 is fixed by the product of an element in the normalizer of A which inverts both t_1 and t_2 and the element τ). Hence Ξ is weakly locally $\Gamma(Y_{443}, d_3)$ and by checking the conditions in (8.3.3) we conclude that Z is a Y_{444}-group.

Let $\psi : \Gamma(Y_{444}, e_1) \to \Xi$ be the covering of graphs induced by the homomorphism of $Y_{444} \to Z$ and let Θ be a graph on M in which m_1 and m_2 are adjacent if $m_1m_2^{-1}$ is an element of type $2b$ in M. We are going to show that ψ induces a covering χ of Θ and that certain triangles in Θ are contractible with respect to χ. Notice that the elementwise stabilizer in Z of an edge of Θ is isomorphic to $2 \times 2_+^{1+24}.Co_1$. Consider in M the stabilizer P of an element of type 2 in $\mathcal{G}(M)$, so that

$$P \cong 2^{2+11+22}.(Sym_3 \times Mat_{24})$$

and let s be an element of order 3 in $O_{2,3}(P)$. Then $C_P(s) \cong 2^{11}.(3 \times Mat_{24})$,

s is of type $3a$ and the centre S of $O_2(P)$ is $2b$-pure of order 4. Hence for $s_1 \in sS \setminus \{s\}$ we obtain a triple $T_2 = \{1, s, s_1\}$ such that $\{s, 1, s_1\}$ is a 2-arc in Ξ and $\{s, s_1\}$ is an edge in Θ. The elementwise stabilizer of T_2 in Z contains Mat_{24}. Since ψ is a covering, there is a pair of vertices $\{\tilde{s}, \tilde{s}_1\}$ in $\Gamma(Y_{444}, e_1)$ which maps onto an edge of Θ and whose stabilizer in Y_{444} contains Mat_{24}.

Put $H = Z\lfloor e_3, \rfloor$ and let Σ be the set of images under H of the identity element of M. Then by the previous subsection $H \cong Y_{444}\lfloor e_3 \rfloor \cong 2 \times M$. Furthermore if $\tilde{\Sigma}$ is the vertex-set of a connected component of the subgraph in $\Gamma(Y_{444}, e_1)$ induced by $\psi^{-1}(\Sigma)$ then the subgraph in $\Gamma(Y_{444}, e_3)$ induced by $\tilde{\Sigma}$ and the subgraph in Ξ induced by Σ are isomorphic to $\Gamma(Y_{443}, e_1)$ and H acts on Σ with kernel of order 2. Now without loss of generality we can assume (in terms of the previous paragraph) that $s, s_1 \in \Sigma$ and $\tilde{s}, \tilde{s}_1 \in \tilde{\Sigma}$. Then the setwise stabilizer of $\{s, s_1\}$ in H (isomorphic to the stabilizer of $\{\tilde{s}, \tilde{s}_1\}$ in Y_{444}) is of the form $2 \times 2^{2+22}.Co_2$. Since the stabilizer in Z of an edge in Θ is generated by any two of its subgroups isomorphic to Mat_{24} and $2 \times 2^{2+22}.Co_2$, we conclude that the stabilizer in Y_{444} of $\{\tilde{s}, \tilde{s}_1\}$ maps bijectively onto the stabilizer in Z of $\{s, s_1\}$ and hence ψ induced a covering $\chi : \tilde{\Theta} \to \Theta$ of graphs. Notice that the vertex-set of $\tilde{\Theta}$ is that of $\Gamma(Y_{444}, e_3)$ and the edges are the images under Y_{444} of the pair $\{\tilde{s}, \tilde{s}_1\}$.

It is clear that the covering χ induces an isomorphism of the subgraph in $\tilde{\Theta}$ induced by $\tilde{\Sigma}$ onto the subgraph in Θ induced by Σ. This means that every triangle in Σ is contractible with respect to χ. Such a triangle is formed for instance by the non-identity elements from C. Thus we conclude that whenever z_1, z_2, z_3 are elements of type $2b$ in M such that $z_1 z_2 z_3 = 1$ and $z_i \in O_2(C_M(z_j))$ for $1 \leq i, j \leq 3$, then the triangle in Θ induced by $\{1, z_1, z_2\}$ is contractible with respect to χ.

In order to show that χ is an isomorphism we apply the result from [IPS96] that M is the universal representation group of its tilde geometry $\mathcal{G}(M)$. A direct factor M of D acts regularly on Θ and hence Θ can be considered as a Cayley graph of M so that the corresponding generators are the $2b$-involutions. Let

$$\delta : \hat{\Theta} \to \Theta$$

the covering of Θ with respect to the subgroup in its fundamental group generated by the images under M of the triangles $\{1, z_1, z_2\}$ such that $z_1, z_2, z_3 := z_1 z_2$ are $2b$-involutions and $z_i \in O_2(C_M(z_j))$ for $1 \leq i, j \leq 3$. Let $\widehat{M}$ be the group of all liftings of elements of M to automorphisms of $\hat{\Theta}$. It is clear that the subgroup of deck transformations acts regularly on each fiber and hence $\widehat{M}$ acts regularly on $\widehat{\Theta}$. This means that $\widehat{\Theta}$ is a Cayley graph of $\widehat{M}$ with respect to generators $t(z)$, one for every $2b$-involution z in M. Since $\widehat{\Theta}$ is undirected the generators are involutions and since the triangle $\{1, z_1, z_2\}$ as above is contractible with respect to δ, the corresponding generators satisfy the equality $t(z_1)t(z_2)t(z_3) = 1$. The following result was proved in [IPS96].

Lemma 8.6.2 *Let $\widehat{M}$ be a group generated by involutions $t(z)$, one for every $2b$-involution z in the Monster M such that $t(z_1)t(z_2)t(z_3) = 1$ whenever z_1, z_2, z_3 are $2b$-involutions in M such that $z_i \in O_2(C_M(z_j))$ for $1 \leq i, j \leq 3$ and $z_1z_2z_3 = 1$. Then $\widehat{M} \cong M$.* □

By (8.6.2) and the paragraph before it δ is an isomorphism. Hence χ is an isomorphism as well and $Y_{444} \cong M \wr 2$.

9
Locally projective graphs

In this chapter we study locally projective graphs. Let Γ be a graph and G be a vertex-transitive automorphism group of Γ. Then Γ is said to be a locally projective graph with respect to G if for every $x \in \Gamma$ the subconstituent $G(x)^{\Gamma(x)}$ is a projective linear group in its natural permutation representation. Incidence graphs of certain truncations of classical geometries are locally projective graphs with respect to their full automorphism groups. These examples can be characterized in the class of all locally projective graphs by the property that their girth is a small even number. We present a proof of this characterization based on the classification of Tits geometries and observe how a class of sporadic Petersen geometries naturally appear in this context via locally projective graphs of girth 5. In Section 9.1 we review some basic results on 2-arc-transitive actions of groups on graphs. In Section 9.2 we discuss examples of locally projective graphs coming from classical geometries. Locally projective lines and their characterizations are discussed in Section 9.3. In Section 9.4 we analyse the possibilities for the action of the vertex stabilizer $G(x)$ on the set of vertices at distance 2 from x. These possibilities determine the main types of locally projective graphs. In a locally projective graph there are virtual projective space structures defined on neighbourhoods of vertices. These virtual structures lead to the notion of geometrical subgraphs introduced in Section 9.5. Analysis of geometrical subgraphs enables us to specify further the structure of vertex stabilizers in Section 9.7. In Section 9.8 we show that if Γ contains a complete family of geometrical subgraphs then a flag-transitive geometry with a nice diagram is associated with the graph. In case Γ does not contain a complete family of geometrical subgraphs a procedure described in Section 9.6 enables us to associate with Γ a locally projective graph of smaller valency with a complete family of geometrical subgraphs and the same abstract group of automorphisms. In the remaining sections of the chapter we consider locally projective graphs of small girth g. In the case $g = 4$ we obtain parabolic geometries of orthogonal groups, in the case $g = 6$ the projective geometries and in the case $g = 5$ the Petersen geometries.

9.1 Groups acting on graphs

In this chapter we consider pairs (Γ, G) satisfying the following hypothesis.

Hypothesis LP. *Γ is a graph and G is a 2-arc-transitive automorphism group of Γ. There are an integer $n \geq 2$, a prime power $q = p^m$ and a group H, satisfying*

$$SL(V) \trianglelefteq H \leq \Gamma L(V),$$

where V is an n-dimensional $GF(q)$-space, such that for every $x \in \Gamma$ the action of $G(x)$ on $\Gamma(x)$ is similar to the action of H on the set of 1-dimensional subspaces of V. In other terms the subconstituent $G(x)^{\Gamma(x)}$ is a projective linear group of V in its natural permutation representation, in particular

$$L_n(q) \trianglelefteq G(x)^{\Gamma(x)} \leq P\Gamma L_n(q).$$

If Γ satisfies the above hypothesis for a subgroup G in its automorphism group, it is said to be a *locally projective graph of type* (n, q) (with respect to G).

In this section we prove a few standard results concerning actions of groups on graphs. We start with the following elementary lemma.

Lemma 9.1.1 *Let Γ be a connected graph and $e = \{x, y\}$ be an edge of Γ. Let K_1 and K_2 be subgroups of the automorphism group of Γ such that K_1 stabilizes x and acts transitively on $\Gamma(x)$ while K_2 stabilizes y and acts transitively on $\Gamma(y)$. Then the action of $K = \langle K_1, K_2 \rangle$ is edge-transitive and it is vertex-transitive if and only if Γ is not bipartite.*

Proof. Let Ω be the orbit of K on the edge set of Γ which contains e. We will prove that every edge is contained in Ω by induction on the distance from e. Every edge at distance 0 from e is either in the K_1- or in the K_2-orbit containing e and hence it is in Ω. Let $f = \{u, v\}$ be an edge at distance $s > 0$ from e in Γ. Without loss of generality we assume that $(x_0 = x, x_1, ..., x_s = u)$ is the shortest among the arcs joining a vertex from e with a vertex from f, which means in particular that $x_1 \neq y$. Let h be an element from K_1 which maps x_1 onto y. Then f^h is at distance $s-1$ from e and it is contained in Ω by the induction hypothesis. Hence f is also in Ω. If Γ is bipartite then both K_1 and K_2 preserve each part as a whole and hence K cannot be vertex-transitive. Let z be a vertex of Γ. If there is an arc of even length joining x and z then it is easy to see that x and z are in the same orbit of K. Suppose that Γ is not bipartite. Then it contains a cycle of odd length. Since K is edge-transitive, there is such a cycle which contains e, say $x_0 = x, x_1 = y, ..., x_{2t+1} = x$. In this case $(x_1, x_2, ..., x_{2t+1})$ is an arc of even length joining y and x. By the above observation this implies the vertex-transitivity of the action of K on Γ.□

The following elementary result is quite important.

Lemma 9.1.2 *Let Γ be a graph and $G \leq \mathrm{Aut}\,\Gamma$. Suppose that every vertex of Γ has valency at least 2. Then the following two conditions are equivalent:*

(i) *G acts 2-arc-transitively on Γ;*
(ii) *G is vertex-transitive and for every $x \in \Gamma$ the subconstituent $G(x)^{\Gamma(x)}$ is a doubly-transitive permutation group.*

Proof. (i) → (ii) Let $a, x \in \Gamma$. Since the valencies of both a and x are at least 2, there are 2-arcs $p = (b, a, c)$ and $q = (y, x, z)$. Let $g \in G$ be such that $p^g = q$. Then $a^g = x$ and we have vertex-transitivity. Let (y_1, z_1) be an ordered pair of vertices in $\Gamma(x)$ and $q_1 = (y_1, x, z_1)$. Then an element of G which maps q_1 onto q stabilizes x and maps (y_1, z_1) onto (y, z). This implies the double transitivity of $G(x)^{\Gamma(x)}$.

(ii) → (i) Let $p = (b, a, c)$ and $q = (y, x, z)$ be 2-arcs in Γ. By the vertex-transitivity there is an element $g \in G$ such that $a^g = x$. Then (b^g, c^g) is a pair of distinct vertices from $\Gamma(x) = \Gamma(a^g)$. Since $G(x)^{\Gamma(x)}$ is doubly transitive, there is $h \in G(x)$ which maps (b^g, c^g) onto (y, z). Then $p^{gh} = q$ and the action of G is 2-arc-transitive. □

The next result gives a necessary condition for $(s+1)$-arc-transitivity of an action which is known to be s-arc-transitive.

Lemma 9.1.3 *Let Γ be a graph, G be an automorphism group of Γ which acts s-arc-transitively for $s \geq 0$. Suppose that every vertex of Γ has valency at least 2. Then the following two conditions are equivalent:*

(i) *for an s-arc $p_s = (x_0, x_1, ..., x_s)$ its elementwise stabilizer $G(p_s)$ in G acts transitively on $\Gamma(x_s) \setminus \{x_{s-1}\}$;*
(ii) *G acts $(s+1)$-arc-transitively on Γ.*

Proof. (i) → (ii) Let x_{s+1} be a vertex in $\Gamma(x_s) \setminus \{x_{s-1}\}$, let $p_{s+1} = (x_0, ..., x_s, x_{s+1})$ and let $q = (y_0, y_1, ..., y_{s+1})$ be an arbitrary $(s+1)$-arc in Γ. Since G is s-arc-transitive, there exists $g \in G$ such that $y_i^g = x_i$ for $0 \leq i \leq s$. Since $G(p_s)$ is transitive on $\Gamma(x_s) \setminus \{x_{s-1}\}$, there exists $h \in G(p_s)$ which maps y_{s+1}^g onto x_{s+1}. Then gh maps q onto p_{s+1} and $(s+1)$-arc-transitivity follows.

(ii) → (i) Since the valency of every vertex in Γ is at least 2, every s-arc is contained in an $(s+1)$-arc and the s-arc-transitivity is implied by the $(s+1)$-arc-transitivity. Let $x'_{s+1} \in \Gamma(x_s) \setminus \{x_{s-1}\}$ and $p'_{s+1} = (x_0, ..., x_s, x'_{s+1})$. Then an element in G which maps p'_{s+1} onto p_{s+1} is contained in $G(p_s)$ and maps x'_{s+1} onto x_{s+1}. Hence $G(p_s)$ acts transitively on $\Gamma(x_s) \setminus \{x_{s-1}\}$. □

Lemma 9.1.4 *Let G act vertex-transitively on a connected graph Γ and suppose that for an integer $i \geq 0$ and a vertex $x \in \Gamma$ the subgroup $G_i(x)$ acts trivially on $\Gamma_{i+1}(x)$, that is $G_i(x) = G_{i+1}(x)$. Then $G_i(x) = 1$.*

Proof. Since G acts vertex-transitively on Γ, the hypothesis implies that $G_i(z) = G_{i+1}(z)$ for every $z \in \Gamma$. Let y be an arbitrary vertex from $\Gamma(x)$. Since $G_i(x) = G_{i+1}(x)$, $G_i(x)$ fixes every vertex at distance at most i from y which means that $G_i(x) \leq G_i(y)$. By the connectivity of Γ we obtain that $G_i(x) \leq G_i(z)$ for every $z \in \Gamma$, which means that $G_i(x) = 1$. □

Lemma 9.1.5 *Let G act 1-arc-transitively on a connected graph Γ and $\{x, y\} \in E(\Gamma)$. Suppose that $G(x)$ is finite. Then*

(i) *if $G_1(x)^{\Gamma(y)}$ is a p-group for a prime number p, then $G_1(x)$ is a p-group,*
(ii) *every composition factor of $G(x)$ is isomorphic either to a composition factor of $G(x)/G_1(x)$ or to a composition factor of $G_1(x)/G_1(x, y)$.*

Proof. By (9.1.4) if $G_i = G_{i+1}$ then $G_i(x) = 1$. Hence for some n the following is a normal series of $G(x)$:

$$1 = G_n(x) \lhd ... \lhd G_1(x) \lhd G(x).$$

In order to prove (i) it is sufficient to show that $G_i(x)/G_{i+1}(x)$ is a p-group for $1 \leq i \leq n-1$. By the definition $G_i(x)/G_{i+1}(x)$ is the action induced by $G_i(x)$ on $\Gamma_{i+1}(x)$. For every $u \in \Gamma_{i+1}(x)$ there is a 2-arc (w, v, u) such that $w \in \Gamma_{i-1}(x)$ and $v \in \Gamma_i(x)$. Then $G_i(x) \leq G_1(w)$ and by the hypothesis $G_i(x)^{\Gamma(u)} \leq G_1(w)^{\Gamma(u)}$ is a p-group. Hence $G_i(x)^{\Gamma_{i+1}(x)}$ is a p-group as well and (i) follows. The assertion (ii) can be proved in a similar way. □

Lemma 9.1.6 *Let G act distance-transitively on Γ and suppose that the action of G is strictly s-arc-transitive. Then the girth of Γ is at most $2(s+1)$.*

Proof. Suppose that the girth of Γ is greater than $2(s+1)$. Then for every $i \leq s+1$ there is a natural bijection between the pairs of vertices at distance i and i-arcs in Γ. Hence in this case distance-transitivity would imply the $(s+1)$-arc-transitivity. □

The following lemma shows that regular generalized polygons appear as extremal cases in s-arc-transitive actions along with the Moore graphs. Recall that a *Moore graph* is a regular graph of valency $k \geq 3$ and girth $2d+1$ where d is the diameter of the graph.

Lemma 9.1.7 *Let G act s-arc-transitively on a graph Γ of girth g and suppose that every vertex of Γ has valency at least 3. Then*

(i) $g \geq 2s-2$,

(ii) *if $g = 2s-2$ then Γ is a generalized $(s-1)$-gon and the action of G on Γ is distance-transitive,*

(iii) *if $g = 2s-1$ then Γ is a Moore graph of diameter $s-1$ and the action of G on Γ is distance-transitive.*

Proof. Let $(x_0, x_1, ..., x_{g-1}, x_g = x_0)$ be a shortest cycle in Γ. For $0 \leq t \leq g$ let $G(p_t)$ denote the elementwise stabilizer in G of the arc $p_t = (x_0, x_1, ..., x_t)$.

(i) Let t be the least integer greater than or equal to $(g+1)/2$. Since the girth of Γ is g there is a unique arc of length $g-t$ joining x_t and x_g. Hence $G(p_t)$ stabilizes x_{t+1}. By (9.1.3) this means that G cannot act $(t+1)$-arc-transitively on Γ.

(ii) Let $t = g/2 = s-1$. By (9.1.3) $G(p_t)$ acts transitively on the vertices in $\Gamma(x_t) \setminus \{x_{t-1}\}$ and one of these vertices, namely x_{t+1}, is in $\Gamma_{t-1}(x_0)$. Hence all the vertices adjacent to x_t are in $\Gamma_{t-1}(x_0)$. This means that Γ does not contain cycles of odd length (*i.e.* Γ is bipartite) and the diameter of Γ is $t = s-1$. So the result follows.

(iii) Let $t = (g-1)/2$. Then $G(p_t)$ acts transitively on the vertices in $\Gamma(x_t) \setminus \{x_{t-1}\}$ and one of these vertices, namely x_{t+1}, is in $\Gamma_t(x_0)$. Hence every vertex other than x_{t-1} which is adjacent to x_t is in $\Gamma_t(x_0)$. This means that Γ is a distance-transitive Moore graph. □

9.2 Classical examples

In this section we describe a few infinite families of locally projective graphs associated with classical geometries and present some motivations for the general interest in locally projective graphs of small girth.

$\mathscr{A}_n^1(q)$. Consider an $(n+1)$-dimensional $GF(q)$-vector–space V and form a bipartite graph $\mathscr{A}_n^1(q)$ whose vertices are 1- and n-dimensional subspaces of V with the adjacency relation defined via inclusion. Then $\mathscr{A}_n^1(q)$ is the incidence graph of the symmetric 2-design having the 1-dimensional subspaces of V as elements and the n-dimensional subspaces as blocks. The parameters of this design are $v = \begin{bmatrix} n+1 \\ 1 \end{bmatrix}_q$, $k = \begin{bmatrix} n \\ 1 \end{bmatrix}_q$, $\lambda = \begin{bmatrix} n-1 \\ 1 \end{bmatrix}_q$. If G is an extension of $PGL_{n+1}(q)$ by a contragredient automorphism then G acts 2-arc-transitively on $\mathscr{A}_n^1(q)$, $G(x)^{\Gamma(x)} \cong PGL_n(q)$ and $|G_1(x,y)| = q$. From the description of $\mathscr{A}_n^1(q)$ as the incidence graph of a symmetric 2-design it is easy to see that it is distance-transitive of diameter 3 with the following intersection array:

$$\left\{ \begin{bmatrix} n \\ 1 \end{bmatrix}_q, \begin{bmatrix} n \\ 1 \end{bmatrix}_q - 1, \begin{bmatrix} n \\ 1 \end{bmatrix}_q - \begin{bmatrix} n-1 \\ 1 \end{bmatrix}_q ; 1, \begin{bmatrix} n-1 \\ 1 \end{bmatrix}_q, \begin{bmatrix} n \\ 1 \end{bmatrix}_q \right\}.$$

In particular the girth of $\mathscr{A}_n^1(q)$ is 4.

$\mathscr{A}_{2n-2}^2(q)$. Now let V be a $(2n-1)$-dimensional $GF(q)$-space. The vertices of $\mathscr{A}_{2n-2}^2(q)$ are all $(n-1)$- and n-dimensional subspaces of V with the adjacency relation defined via inclusion. This graph is known as the *q-analogue of the double cover of the odd graph*. The extension G of $PGL_{2n-1}(q)$ by a contragredient automorphism acts 3-arc-transitively on $\mathscr{A}_{2n-2}^2(q)$ with $G(x)^{\Gamma(x)} \cong PGL_n(q)$ and with $G_1(x)$ containing a section isomorphic to $SL_{n-1}(q)$. The action is distance-transitive and the intersection numbers are the following:

$$c_{2i-1} = c_{2i} = \begin{bmatrix} i \\ 1 \end{bmatrix}_q ; \; b_{2i-1} = b_{2i} = \begin{bmatrix} n-i \\ 1 \end{bmatrix}_q \cdot q^i.$$

In particular the girth of $\mathscr{A}_{2n-2}^2(q)$ is 6.

$\mathscr{D}_n(q)$. In this case V is a $2n$-dimensional $GF(q)$-space equipped with a non-singular quadratic form f of Witt index n. The maximal totally singular subspaces of V have dimension n and they are partitioned into two classes in such a way that whenever two such subspaces intersect in an $(n-1)$-dimensional subspace, they are from different classes. Every $(n-1)$-dimensional totally singular subspace is contained in exactly two maximal ones (from different classes). The vertices of $\mathscr{D}_n(q)$ are the maximal totally singular subspaces of V, with two subspaces adjacent if their intersection is of dimension $(n-1)$. In view of the above this means that the graph is bipartite. The extension G of the Lie type group $D_n(q)$ by a diagram automorphism (that is, the group of all linear transformations of V preserving f) acts 2-transitively on $\mathcal{D}_n(q)$ with $G(x)^{\Gamma(x)} \cong PGL_n(q)$ and $O_p(G_1(x))$, as a $GF(q)$-module for $G(x)/O_p(G_1(x))$, is isomorphic to $\bigwedge^2 V_n(q)$. The graph is distance-transitive of diameter n with parameters

$$b_i = \begin{bmatrix} n \\ 1 \end{bmatrix}_q - \begin{bmatrix} i \\ 1 \end{bmatrix}_q, \quad c_i = \begin{bmatrix} i \\ 1 \end{bmatrix}_q, \quad \text{for } 0 \le i \le n.$$

The above three families of graphs possess a uniform description in terms of truncations of the corresponding classical geometries. To wit, the vertices of the graphs are the elements corresponding to the black nodes in the respective diagrams with the adjacency relation induced by the incidence relation in the geometry.

$\mathcal{A}_n^1(q)$: 1 2 3 ... $n-2$ $n-1$ n (nodes labelled q)

$\mathcal{A}_{2n-2}^2(q)$: 1 2 ... $n-1$ n ... $2n-3$ $2n-2$ (nodes labelled q)

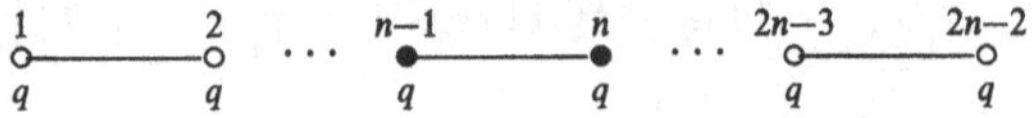

$\mathcal{D}_n(q)$: 1 2 3 ... $n-3$ $n-2$ $n-1$ (nodes labelled q)

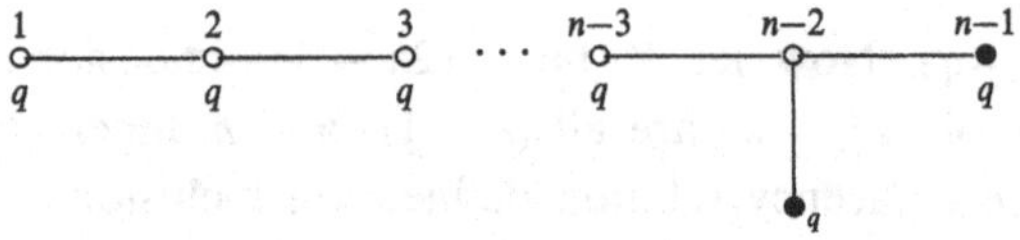

The graphs $\mathcal{A}_2^1(q)$ and $\mathcal{A}_2^2(q)$ are both isomorphic to the incidence graph of the projective plane over $GF(q)$ (*i.e.* to the corresponding generalized triangle), and we will denote it by $\mathcal{A}_2(q)$. Its full automorphism group isomorphic to $\mathrm{Aut}\, L_3(q)$ acts strictly 4-arc-transitively and distance-transitively.

There are two more series of rank 2 Lie type geometries possessing diagram automorphisms. The corresponding graphs are locally projective lines.

$\mathcal{B}_2(q)$: q q

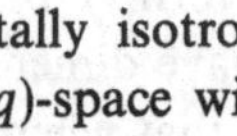

The vertices of $\mathcal{B}_2(q)$ are the totally isotropic 1- and 2-dimensional subspaces of a 4-dimensional $GF(q)$-space with respect to a fixed non-singular symplectic form. So $\mathcal{B}_2(q)$ is the generalized quadrangle of symplectic type. The diagram automorphism exists if and only if q is even (that is, a power of 2). The graph $\mathcal{B}_2(2^m)$ is strictly 5-arc-transitive and distance-transitive.

$\mathcal{G}_2(q)$: 6 q q

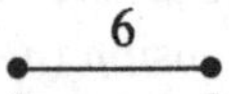

The graph $\mathcal{G}_2(q)$ is the generalized hexagon associated with the Lie type group $G_2(q)$. The diagram automorphism exists if and only if q is a power of 3, and $G_2(3^m)$ is strictly 7-arc-transitive and distance-transitive.

Another series of locally projective graphs is associated with the Lie type groups of type $F_4(q)$ and in the above terms can be described by the following diagram:

$$\mathscr{F}_4(q): \quad \overset{1}{\underset{q}{\circ}} \text{———} \overset{2}{\underset{q}{\bullet}} \text{===} \overset{3}{\underset{q}{\bullet}} \text{———} \overset{4}{\underset{q}{\circ}}$$

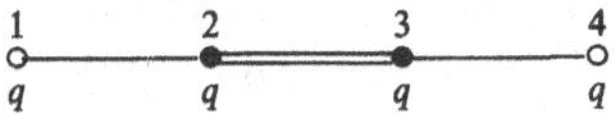

In order for $\mathscr{F}_4(q)$ to be vertex transitive we need a diagram automorphism. Such an automorphism exists if and only if q is even (that is a power of 2). The graph $\mathscr{F}_4(q)$ is not distance-transitive for any q and its girth is 8.

The following very elegant result characterizing the graphs $\mathscr{A}_n^1(q)$ and $\mathscr{D}_n(q)$ was proved in [CPr82].

Theorem 9.2.1 *Let (Γ, G) be a pair satisfying Hypothesis LP with $G_1(x) \neq 1$ and suppose that the girth of Γ is 4. Then one of the following holds:*

(i) $\Gamma \cong \mathscr{A}_n^1(q)$ *and* $L_{n+1}(q).\langle\tau\rangle \leq G \leq \operatorname{Aut} L_{n+1}(q)$, *where τ is a diagram automorphism;*

(ii) $\Gamma \cong \mathscr{D}_n(q)$ *and* $\Omega_{2n}^+(q).\langle\tau\rangle \leq G \leq \operatorname{Aut}\Omega_{2n}^+(q)$, *where τ is a diagram automorphism;*

(iii) $\Gamma \cong K_{m,m}$ *is the complete bipartite graph,* $m = \left[{n \atop 1}\right]_q$ *and* $L_n(q) \times L_n(q) < G \leq \operatorname{Aut}(L_n(q) \times L_n(q))$. □

Later in this chapter we will extend the above characterization to graphs of girth 5 and 6. In the girth 6 case this will provide us with a characterization of the graphs $\mathscr{A}_{2n-2}^2(q)$. In the girth 5 case a class of sporadic Petersen type geometries arises. Before proceeding to this let us discuss another motivation for the interest in locally projective graphs, especially in those of small girth.

The motivation comes from a general problem of bounding the order of the vertex stabilizer $G(x)$ for a group G acting 2-arc-transitively on a graph Γ in terms of the subconstituent $G(x)^{\Gamma(x)}$. For an arbitrary doubly transitive permutation group H it is possible to produce a 2-arc-transitive action with $G(x)^{\Gamma(x)} \cong H$ and $G_1(x) \neq 1$. For this we take Γ to be the complete bipartite graph $K_{m,m}$ (where m is the degree of H) and G to be the wreath product $H \wr 2$. However, in all these cases $G_1(x, y) = 1$. The situation when $G_1(x, y) \neq 1$ turns out to be much more specific as is shown by the following result known as the Thompson–Wielandt theorem (see [Wei79a] for the proof of an improved version of it).

Theorem 9.2.2 *Suppose G acts 2-arc-transitively on Γ and $G_1(x,y) \neq 1$ for $\{x,y\} \in E(\Gamma)$. Then $G_1(x,y)$ is a p-group for a prime number p.* □

A non-trivial element from $G_1(x,y)$ is called an *elation*. It is easy to see from (9.2.2) and (9.1.4) that if G acts 2-arc-transitively on Γ and contains elations then the point stabilizer $G(x,y)^{\Gamma(x)}$ in the subconstituent $G(x)^{\Gamma(x)}$ is p-local (*i.e.* it has a non-trivial normal p-subgroup). Using a case-by-case analysis of the known doubly transitive permutation groups with p-local point stabilizers, the following result was proved (the survey [Wei81a]).

Theorem 9.2.3 *Let G act 2-arc-transitively on Γ and suppose that for an edge $\{x,y\}$ of Γ, $G_1(x,y) \neq 1$. Then Γ is locally projective with respect to the action of G.* □

Thus the problem of bounding the order of $G(x)$ for a 2-arc-transitive action was reduced to the case of projective subconstituents. A possible way to solve this problem is to find a constant c such that $G_c(x) = 1$ for every 2-arc-transitive action of G on Γ with a projective subconstituent. (Although there is no *a priori* reason at all for such a constant to exist.) The solution along these lines was announced in [Tro91] and some particular cases are now published in [Tro92] and [Tro94]. The result says that the constant c exists and $c = 6$ works.

The next problem of great interest is to describe all the possibilities for the point stabilizer $G(x)$ coming from 2-arc-transitive actions with projective subconstituents. For this type of problem it is a standard strategy to consider actions on trees.

The formalism is the following. Let G act 1-arc-transitively on a graph Γ. Then there always exist a tree $\tilde{\Gamma}$ and an automorphism group $\tilde{G}$ of $\tilde{\Gamma}$ such that the actions of G on Γ and of $\tilde{G}$ on $\tilde{\Gamma}$ are "locally isomorphic". We can define $\tilde{\Gamma}$ and $\tilde{G}$ in the following manner. Let $x \in \Gamma$, $\{x,y\} \in E(\Gamma)$, $G(x)$ be the stabilizer of x in G and $G[x,y]$ be the setwise stabilizer of $\{x,y\}$. Because of 1-arc-transitivity, $G[x,y] \cap G(x) = G(x,y)$ is of index 2 in $G[x,y]$. We define $\tilde{G}$ to be the universal completion of the amalgam $\mathscr{A} = \{G(x), G[x,y]\}$, that is, the free amalgamated product of $G(x)$ and $G[x,y]$ over the common subgroup $G(x,y)$. Now define the vertices of $\tilde{\Gamma}$ to be the right cosets of $G(x)$ in $\tilde{G}$ and declare two vertices adjacent if and only if there exists a right coset of $G[x,y]$ in $\tilde{G}$ which intersects them both. Since Γ is connected, $G(x)$ and $G[x,y]$ generate G. This means that there are a covering $\varphi : \tilde{\Gamma} \to \Gamma$ of graphs and a homomorphism $\psi : \tilde{G} \to G$ of groups such that the fibres of φ are the orbits on $\tilde{\Gamma}$ of the kernel of ψ and the kernel consists of all the elements from $\tilde{G}$ which are deck transformations with respect to φ. In other words the vertices of $\tilde{\Gamma}$ can be identified with the arcs in Γ originating at x with two such arcs adjacent in $\tilde{\Gamma}$ if and only if one of them can be obtained from the other one by deleting the terminal vertex. This shows the meaning of the local isomorphism between the actions of G on Γ and of $\tilde{G}$ on $\tilde{\Gamma}$. In particular G acts s-arc-transitively on Γ if and only if $\tilde{G}$ does so on $\tilde{\Gamma}$.

In view of the above, as long as we are only concerned about local properties of a 2-arc-transitive action such as the structure of vertex stabilizers, the underlying graph can be assumed to be a tree. Eventually we are interested in actions of finite groups on finite graphs, so only locally finite actions (with finite vertex stabilizers) have to be considered. In certain circumstances it is much more convenient to work with actions on trees, basically because there are no cycles to cause problems. On the other hand it is known [Ser77] that a locally finite 1-arc-transitive action is always locally isomorphic to an action on a finite graph. Although a construction procedure exists for producing a finite graph and an action on it which is locally isomorphic to a given action, the resulting graph is usually rather large. However, in practice it often happens that important actions can be realized on surprisingly small graphs. For instance it turns out (see (9.3.2) in the next section) that for $s \geq 4$ every action which is strictly s-arc-transitive is locally isomorphic to such an action on a graph of girth $2(s-1)$. At the same time a graph of valency at least 3 is never s-arc-transitive if its girth is less than $2(s-1)$ and every s-arc-transitive graph of girth $2(s-1)$ is a generalized $(s-1)$-gon (9.1.7). The fact that every s-arc-transitive action for $s \geq 4$ is locally isomorphic to an action on a generalized $(s-1)$-gon has played an important rôle in the classification of such actions, particularly in specification of the corresponding vertex stabilizers ([Wei79b], [DGS85]).

It is believed [Iv93a] that every 2-arc-transitive action with a projective subconstituent can be realized on a graph with small girth, say up to 8. We consider this to be a motivation for the particular interest in (9.2.1) and in its generalizations. The fact that sporadic groups and their geometries appear in this generalization is also quite remarkable.

9.3 Locally projective lines

In this section we present a brief survey of what is known about locally projective graphs of type $(2, q)$, also known as *locally projective lines*. Thus we consider pairs (Γ, G) where Γ is a graph, G is a 2-arc-transitive automorphism group of Γ and

$$L_2(q) \trianglelefteq G(x)^{\Gamma(x)} \leq P\Gamma L_2(q)$$

for every vertex $x \in \Gamma$. In view of the discussions in the previous section, when considering local properties of the action of G on Γ we can always assume that the latter is a tree (of valency $q+1$).

The graphs which are locally projective lines are of particular importance because of the following result proved in [Wei81b] using the classification of finite doubly transitive permutation groups.

Theorem 9.3.1 *Let G act s-arc-transitively on a graph Γ and $s \geq 4$. Then $s \leq 7$ and Γ is locally a projective line with respect to the action of G.* □

It turns out that every 4-arc-transitive action is locally isomorphic to an action on a classical generalized polygon ([Wei79b], [DGS85]).

Theorem 9.3.2 *Let G act strictly s-arc-transitively for $4 \leq s \leq 7$ on a graph Γ. Then $s \neq 6$ and the action is locally isomorphic either to an s-arc-transitive action on the generalized $(s-1)$-gon $\mathcal{A}_2(q)$, $\mathcal{B}_2(2^m)$, $\mathcal{G}_2(3^m)$ for $s = 4, 5, 7$, respectively, or to a 4-arc-transitive action of $PGL_2(9)$ on $\mathcal{B}_2(2)$.* □

Notice that a group which acts s-arc-transitively on a classical generalized $(s-1)$-gon contains the corresponding simple group of Lie type of rank 2 (1.6.5).

The actions on vertex-transitive classical generalized polygons were characterized in [Wei85] in the context of distance-transitive graphs.

Theorem 9.3.3 *Let G act on a graph Γ distance-transitively and s-arc-transitively for $s \geq 4$. Then Γ is isomorphic to one of the following:*

(i) *the generalized polygon $\mathcal{A}_2(q)$, $\mathcal{B}_2(2^m)$ or $\mathcal{G}_2(3^m)$;*
(ii) *the incidence graph of the rank 2 tilde geometry $\mathcal{G}(3 \cdot Sp_4(2))$;*
(iii) *a cubic distance-transitive graph on 102 vertices with the automorphism group isomorphic to $L_2(17)$.* □

Some particular cases of the above theorem were known long before [Wei85]. Specifically it was proved in [Glea56] that every distance-transitive generalized triangle is isomorphic to $\mathcal{A}_2(q)$, in [Hig64] that every distance-transitive generalized quadrangle is isomorphic to $\mathcal{B}_2(2^m)$ and in [Yan76] that every distance-transitive generalized hexagon is isomorphic to $\mathcal{G}_2(3^m)$.

By (9.3.2) all local properties of an s-arc-transitive action for $s \geq 4$ can be checked for suitable classical generalized polygon. For $s = 4$ and 5 all calculations are elementary and for $s = 7$ one can use a model of $\mathcal{G}_2(q)$ given in [Kan86].

We will make use of the following result of this type. Let $V_2(q)$ denote the natural 2-dimensional $GF(q)$-module for the group $SL_2(q)$ and let $V_1(q)$ denote the trivial 1-dimensional $GF(q)$-module for this group.

Lemma 9.3.4 *Let G act on a graph Γ strictly s-arc-transitively for $s \geq 4$, so that Γ is locally a projective line over $GF(q)$. Let p be the characteristic of the field and let $W_i = O_p(G_i(x))/G_{i+1}(x)$ for $i \geq 1$. Then $G(x)/O_p(G(x))$ contains a characteristic subgroup K isomorphic to $SL_2(q)$. Moreover, if W_i is considered as a $GF(p)K$-module then the following hold:*

(i) *if $s = 4$ then $W_1 \cong V_2(q)$ and $W_2 = 1$;*
(ii) *if $s = 5$ then $W_1 \cong V_2(q)$, $W_2 \cong V_1(q)$ and $W_3 = 1$;*
(iii) *if $s = 7$ then $W_1 \cong W_2 \cong V_2(q)$, $W_3 \cong V_1(q)$ and $W_4 = 1$.*

In particular $|G_1(x, y)| = q^{s-3}$ and if $y, z \in \Gamma(x)$ then $G_1(x, y)$ induces on $\Gamma(z) \setminus \{x\}$ a regular action whose kernel coincides with $G_2(x)$. □

For the sake of completeness we present the following result whose proof can be achieved by completely elementary methods.

Lemma 9.3.5 *Let G act strictly 3-arc-transitively on a graph which is locally a projective line. Then $G_1(x, y) = 1$.* □

The following well-known characterization of distance-transitive Moore graphs (Section 6.7 in [BCN89]) is in a certain sense analogous to (9.3.3).

Lemma 9.3.6 *Let Γ be a Moore graph of valency $k \geq 3$ and diameter d. Then $d = 2$ and $k \in \{3, 7, 57\}$. If in addition $G = \mathrm{Aut}\,\Gamma$ acts on Γ distance-transitively then one of the following holds:*

(i) *$k = 3$, Γ is the Petersen graph, $G \cong Sym_5$ acts 3-arc-transitively on Γ, $G(x)^{\Gamma(x)} \cong Sym_3 \cong L_2(2)$ and $G_1(x) \cong 2$;*
(ii) *$k = 7$, Γ is the Hoffman–Singleton graph, $G \cong P\Sigma L_3(5)$ acts 3-arc-transitively on Γ, $G(x)^{\Gamma(x)} \cong Sym_7$ and $G_1(x) = 1$.* □

It is clear that a 3-arc-transitive graph of girth 4 must be a complete bipartite graph. Thus by (9.1.7), (9.3.3), (9.3.6) we have the following result.

Lemma 9.3.7 *Let G act s-arc-transitively on Γ for $s \geq 3$ and suppose that the girth g of Γ is less than or equal to $2s-1$. Then one of the following holds:*

(i) $g = 4$, $s = 3$ *and Γ is complete bipartite;*
(ii) $g = 2s-2$, $s \geq 4$ *and Γ is a classical generalized $(s-1)$-gon $\mathcal{A}_2(q)$, $\mathcal{B}_2(2^m)$ or $\mathcal{G}_2(3^m)$ for $s = 4$, 5 or 7, respectively;*
(iii) $g = 5$, $s = 3$ *and Γ is either the Petersen graph or the Hoffman–Singleton graph.* □

In the above lemma distance-transitive Moore graphs together with vertex-transitive classical generalized polygons appear as extremal cases of s-arc-transitive graphs of small girth. Notice that the Petersen graph is locally projective while the Hoffman–Singleton graph is not.

Later in this chapter we will observe that the way in which sporadic Petersen geometries are built from the Petersen graph is similar to the way classical geometries are built from generalized polygons. The important property of the Petersen graph which allows us to build complicated geometries on its base is the non-triviality of the kernel at a vertex in the full automorphism group (9.3.6).

9.4 Main types

In this section we consider some basic properties of locally projective graphs of type (n, q) for $n \geq 3$. By considering the action of $G(x)$ on $\Gamma_2(x)$ for a vertex x of the graph we will distinguish the main types of such graphs.

Let Γ be a graph which is locally projective of type (n, q), $n \geq 3$, $q = p^m$, with respect to a group G of its automorphisms. The action of $G(x)$ on $\Gamma(x)$ induces on the latter the structure of a projective geometry of rank $(n-1)$ which we denote by π_x. To wit, a subset in $\Gamma(x)$ of size $\begin{bmatrix} i \\ 1 \end{bmatrix}_q$ is an element of type i in π_x if and only if its setwise stabilizer in $G(x)$ contains a Sylow p-subgroup of $G(x)$. The incidence relation is via inclusion. In particular the points of π_x are the vertices of $\Gamma(x)$. Let L_x and H_x denote the sets of lines and hyperplanes of π_x, respectively (these two sets coincide when $n = 3$).

For a vertex $y \in \Gamma(x)$ let $\pi_x(y)$ denote the set of subspaces of π_x containing y (where y is considered as a point of π_x). Let $L_x(y)$ and $H_x(y)$ denote the sets of lines and hyperplanes of π_x which contain y. For $y, z \in \Gamma(x)$ let $l_x(y, z)$ denote the unique line of π_x which contains both y and z. The set $L_x(y)$ can be naturally treated as the point set of a projective geometry of rank $(n-2)$. This geometry (which is the residue of y in π_x) will also be denoted by $\pi_x(y)$. In what follows we sometimes identify projective geometries with their point sets.

Let P denote the action of $O_p(G(x))$ on $\Gamma_2(x)$ which is abstractly isomorphic to $O_p(G(x))/G_2(x)$. We will see below that P is non-trivial whenever $G_1(x) \neq 1$. Let Σ denote the set of orbits of P on $\Gamma_2(x)$. Let $\overline{G(x)}$ denote the permutation action induced by $G(x)$ on Σ and use similar notation for subgroups in $G(x)$, so that $\overline{P}$ is the identity group. Our nearest goal is to determine the possibilities for $\overline{G(x)}$ (compare [Wei78]).

Proposition 9.4.1 *Let (Γ, G) satisfy Hypothesis LP for some $n \geq 3$ and $q = p^m$. Suppose that $G_1(x) \neq 1$. Let $P = O_p(G(x))/G_2(x)$ and let Σ denote the set of orbits of P on $\Gamma_2(x)$. Then for $S \in \Sigma$ we have $|S| = q$ and P induces on S an elementary abelian group (of order q).*

Let s be the integer such that G acts strictly s-arc-transitively on Γ and let g denote the girth of Γ. Then one of the following cases (1) and (2) holds.

(1) *$s = 3$, $L_n(q) \times L_{n-1}(q) \trianglelefteq \overline{G(x)} \leq P\Gamma L_n(q) \times P\Gamma L_{n-1}(q)$, a projective geometry δ_x of rank $(n-2)$ over $GF(q)$ is associated with x so that $\overline{G(x)}$ acts flag-transitively on $\pi_x \times \delta_x$, there is a mapping φ_x of Σ onto the point set of δ_x which commutes with the action of $G(x)$ and for $y \in \Gamma(x)$ the restriction of φ_x to $\pi_y(x)$ is a collineation. Moreover, either*

(1.1) *$g = 4$, Γ is the complete bipartite graph $K_{m,m}$ where $m = \left[{n \atop 1}\right]_q$, or*

(1.2) *$g > 4$, $\Sigma = \{S(u, \alpha) \mid u \in \pi_x,\ \alpha \in \delta_x\}$, and $S(u, \alpha) \subset \Gamma(y)$ exactly when $u = y$.*

(2) *$s = 2$, $L_n(q) \trianglelefteq \overline{G(x)} \leq P\Gamma L_n(q)$, for every vertex $y \in \Gamma(x)$ there is an isomorphism $\psi = \psi_{xy}$ of $\pi_x(y)$ onto $\pi_y(x)$ which commutes with the action on $G(x, y)$ and for all edges one of the following subcases occurs:*

(2.1) *ψ is a collineation and either*

(2.1.i) *$g = 4$, $\Sigma = \{S(l) \mid l \in L_x\}$, and $S(l) \subset \Gamma(y)$ exactly when $l \in L_x(y)$, or*

(2.1.ii) *$g > 4$, $\Sigma = \{S(u, l) \mid u \in \pi_x, l \in L_x(u)\}$ and $S(u, l) \subset \Gamma(y)$ exactly when $u = y$;*

(2.2) *ψ is a correlation $(n \geq 4)$ and either*

(2.2.i) $g = 4$, $\Sigma = \{S(h) \mid h \in H_x\}$, *and* $S(h) \subset \Gamma(y)$ *exactly when* $h \in H_x(y)$, *or*

(2.2.ii) $g > 4$, $\Sigma = \{S(u,h) \mid u \in \pi_x, h \in H_x(u)\}$, *and* $S(u,h) \subset \Gamma(y)$ *exactly when* $u = y$.

Proof. Let $H = G(x,y)^{\Gamma(y)}$. Since $G(x)^{\Gamma(x)}$ contains as a normal subgroup the group $L_n(q)$ in its natural doubly transitive action, by (2.4.3 (i)) $H \cong A : B$, where $A = O_p(H)$ is an elementary abelian group of order q^{n-1} and $SL_{n-1}(q) \trianglelefteq B \leq \Gamma L_{n-1}(q)$ with B acting faithfully and irreducibly on A. This implies in particular that every nontrivial normal subgroup of H contains A. By (2.4.4) if S is an orbit of A on $\Gamma(y) \setminus \{x\}$ then $S = l \setminus \{x\}$ for a line $l \in L_y(x)$ and the action induced by A on S is of order q.

Let $M = G_1(x)^{\Gamma(y)}$. It is standard that M is non-trivial. Indeed, if $M = 1$ then since $G_1(x) \trianglelefteq G(x)$ and $G(x)^{\Gamma_2(x)}$ is transitive, we obtain that $G_1(x)^{\Gamma_2(x)} = 1$. But the latter implies $G_1(x) = 1$ by (9.1.4), a contradiction. Hence M is a nontrivial normal subgroup in H and $M \geq A$ by the above paragraph. Let $z \in \Gamma(x) \setminus \{y\}$ and suppose that $N := G_1(x,z)^{\Gamma(y)} \neq 1$. Then by (9.2.2) N is of a prime power order. Since N is normal in M and B acts faithfully on A, N must intersect A properly and hence the primes in (9.2.2) and in the present proposition are the same. Since $O_p(G(x)/G_1(x)) = 1$ this implies that $P^{\Gamma(y)} = A$ and the first paragraph of the proposition is proved.

Furthermore, since $L_{n-1}(q) \trianglelefteq H^{L_y(x)} \leq P\Gamma L_{n-1}(q)$, and $M \trianglelefteq H$, the action of M on $L_y(x)$ is either (1) transitive or (2) trivial. Consider possibility (1). In this case $G_1(x)$ is transitive on $\Gamma(y) \setminus \{x\}$ and applying (9.1.3) one can see that $s \geq 3$. Second, by (2.4.1) the group $L_{n-1}(q)$ is simple except for the cases $(n,q) = (3,2)$ and $(3,3)$. Hence we conclude that either (a) $M^{L_y(x)} \geq L_{n-1}(q)$ or (b) $(n,q) = (3,2)$ and $M^{L_y(x)} = 3$ or (c) $(n,q) = (3,3)$ and $M^{L_y(x)} = 2^2$.

We show that the last two cases cannot be realized. With this end in mind, let us consider the centralizer $\overline{C(x)}$ of the group $\overline{G_1(x)}$ in $\overline{G(x)}$ and its complete preimage $C(x)$ in $G(x)$. We claim that $\overline{G_1(y)} \leq \overline{C(x)}$. Indeed, $G_1(x) \trianglelefteq G(x,y)$, $G_1(y) \trianglelefteq G(x,y)$ and $G_1(x) \cap G_1(y) = G_1(x,y) \leq P$; therefore $\overline{G_1(x)} \cap \overline{G_1(y)} = 1$, *i.e.* $\overline{G(x,y)} \geq \overline{G_1(x)} \times \overline{G_1(y)}$. Since $G_1(y)^{\Gamma(x)}$ is non-trivial and $C(x) \trianglelefteq G(x)$, we conclude that $L_n(q) \trianglelefteq C(x)^{\Gamma(x)}$. The factor group $G(x,y)/(C(x) \cap G(x,y))$ is isomorphic to 2 in case (b) and to 3 in case (c). At the same time, by what we said above and since $\overline{G_1(x)}$ is abelian in both the cases under consideration, we have $G(x) = C(x)$ in case (b) and $[G(x) : C(x)] \leq 2$ in case (c), a contradiction.

By this we see that $\overline{G_1(x)}$ contains $L_{n-1}(q)$, and since the centre of the latter group is trivial, $\overline{G(x)} \geq \overline{G_1(x)} \times \overline{C(x)}$. From this it is straightforward to conclude that

$$L_n(q) \times L_{n-1}(q) \trianglelefteq \overline{G(x)} \leq P\Gamma L_n(q) \times P\Gamma L_{n-1}(q).$$

Therefore $L_{n-1}(q) \trianglelefteq \overline{G_1(x)} \leq P\Gamma L_{n-1}(q)$ and for $y \in \Gamma(x)$ the group $\overline{G_1(x)}$ induces on $L_y(x)$ the natural doubly transitive action. For another vertex $z \in \Gamma(x)$ the action is either (a) isomorphic or (b) dual and there is an isomorphism ξ^x_{yz} between $\pi_y(x)$ and $\pi_z(x)$ which commutes with the action of $G_1(x)$. In case (a) it is a collineation and in case (b) it is a correlation. Since $G(x)^{\Gamma(x)}$ is doubly transitive, the form of ξ^x_{yz} is independent of the particular choice of the pair (y, z). For a third vertex $u \in \Gamma(x)$ the isomorphism ξ^x_{yu} can be realized as a product of ξ^x_{yz} and ξ^x_{zu}. Since the product of two collineations as well as of two correlations is a collineation we see that ξ^x_{yz} must be a collineation for all $y, z \in \Gamma(x)$. So we can define a single projective geometry δ_x of rank $(n-2)$ and an action of $G(x)$ on this space so that for every $y \in \Gamma(x)$ there is a collineation χ^x_y of $\pi_y(x)$ onto δ_x which commutes with the action of $G(x, y)$. The points of δ_x are equivalence classes of orbits from Σ, where two orbits are equivalent if their setwise stabilizers in $G_1(x)$ coincide. So we have the desired mapping φ_x of Σ onto (the point set of) δ_x. Let (u, y, x, z, v_1) and (u, y, x, z, v_2) be two 4-arcs in Γ such that $l_y(x, u) \setminus \{x\}$ is equivalent to $l_z(x, v_1) \setminus \{x\}$ but not to $l_z(x, v_2) \setminus \{x\}$ with respect to the above defined equivalence relation. Then clearly these two 4-arcs are in different orbits of G and so G acts strictly 3-transitively on Γ. If the girth of Γ is 4 it must be complete bipartite by (9.1.7). Otherwise different pairs (u, α) with $u \in \Gamma(x)$ and $\alpha \in \delta_x$ determine different orbits $S(u, \alpha) \in \Sigma$. Thus, case (1) is completely settled.

Let us turn to case (2). Here $M^{L_y(x)} = 1$, and hence $\overline{G_1(x)} = 1$. From this it follows that $L_n(q) \trianglelefteq \overline{G(x)} \leq P\Gamma L_n(q)$. Furthermore, $L_{n-1}(q) \leq \overline{G(x, y)} \leq P\Gamma L_{n-1}(q)$ induces natural doubly transitive actions on $L_x(y)$ and on $L_y(x)$. Hence there is an isomorphism ψ_{xy} between $\pi_x(y)$ and $\pi_y(x)$ commuting with the action of $G(x, y)$. In this case ψ_{xy} can be a collineation as well as a correlation. Certainly the type of ψ_{xy} is independent of the choice of the edge $\{x, y\}$. If the girth of Γ is at least 5, then we arrive at situations (2.1.i) or (2.2.ii), respectively.

Suppose that Γ is of girth 4. We consider here the case of collineation (the correlation case can be treated quite analogously). Consider the set $Y = \{(u, l) \mid u \in \Gamma(x), l \in L_x(u)\}$. A pair $(u, l) \in Y$ determines a unique orbit from Σ, namely $\psi_{xu}(l) \setminus \{x\}$. Since the girth of Γ is 4 some pairs correspond to the same orbits from Σ and we have an equivalence relation on Y preserved by the action of $G(x)$. The group $G(x)$ induces on Y an action of $L_n(q) \trianglelefteq G(x)/G_1(x) \leq P\Gamma L_n(q)$ on the cosets of a premaximal parabolic subgroup with respect to the natural projective geometry. So it follows from (2.4.2 (v)) that a proper subgroup of $G(x)/G_1(x)$ which contains the stabilizer of $(u, l) \in Y$ must be either the stabilizer of u or the stabilizer of l. From this we easily see that the orbits from Σ are in a bijection with the lines of π_x, thus leading to subcase (2.1.i). □

By (9.2.1) if Γ corresponds to the subcase (1.1), (2.1.i) or (2.2.i) in the above proposition then it is isomorphic to $K_{m,m}$, $\mathscr{D}_n(q)$ or $\mathscr{A}^1_n(q)$, respectively. One may notice that the graphs $\mathscr{A}^2_{2n-2}(q)$ and $\mathscr{F}_4(q)$ correspond to the subcase (1.2).

We formulate explicitly the following results which are implicit in the proof of (9.4.1).

Corollary 9.4.2 *If $s = 2$ then for every edge $\{x, y\}$ of Γ there is a unique isomorphism ψ_{xy} of $\pi_x(y)$ onto $\pi_y(x)$ which commutes with the action of $G(x, y)$. The type of ψ_{xy} (i.e. whether it is collineation or correlation) is independent of the choice of $\{x, y\}$.* □

Corollary 9.4.3 *If $s = 3$ and $\{x, y\}$ is an edge of Γ then there is a collineation χ^x_y of $\pi_y(x)$ onto δ_x commuting with the action of $G(x, y)$. If $\{y, x, z\}$ is a 2-arc in Γ then $\xi^x_{yz} = (\chi^x_z)^{-1}\chi^x_y$ is a collineation of $\pi_y(x)$ onto $\pi_z(x)$ which commutes with the action of $G(x, y, z)$.* □

For the following corollary compare (2.4.3).

Corollary 9.4.4 *If $s = 2$ then $G_1(x)/O_p(G(x))$ is a cyclic group whose order divides $q - 1$; if $s = 3$ then $G_1(x)/O_p(G(x))$ contains a characteristic subgroup K isomorphic to $SL_{n-1}(q)$ and $O_p(G(x))/G_1(x, y)$ is a natural module for K.* □

9.5 Geometrical subgraphs

For the remainder of the chapter we assume that Γ is a locally projective graph of type (n, q), $n \geq 3$, $q = p^m$, with respect to a subgroup G in the automorphism group of Γ such that $G_1(x) \neq 1$ for $x \in \Gamma$. Let s be the integer such that the action of G on Γ is strictly s-arc-transitive. By (9.4.1) either $s = 3$ or $s = 2$ and Γ is of collineation (case (2.1)) or of correlation (case (2.2)) type.

Let $x \in \Gamma$. The action of $G(x)$ induces on $\Gamma(x)$ a projective geometry structure π_x. If $s = 3$ then the action of $G(x)$ induces on the set of orbits of $O_p(G(x))$ on $\Gamma_2(x)$ a structure of a direct product $\pi_x \times \delta_x$ of projective geometries. So far these structures are defined virtually in the sense that they were not attached to the combinatorial structure of Γ. On the other hand, if Γ is one of the classical examples $\mathscr{A}_n^1(q)$, $\mathscr{A}_n^2(q)$, $\mathscr{D}_n(q)$ and $\mathscr{F}_4(q)$, then with every element of the original geometry we can associate a subgraph in Γ induced by the vertices which are incident to this element in the geometry. In those cases the structures π_x and $\pi_x \times \delta_x$ are realized by the subgraphs of this type passing through x. We will attempt to find a similar system of subgraphs in an arbitrary locally projective graph. It turns out that such a system does not necessarily exist, but if it does, then it is unique.

Definition 9.5.1 *Let Γ be a locally projective graph of type (n, q) with respect to $G \leq \operatorname{Aut}\Gamma$. A connected subgraph Ξ of Γ will be called a* geometrical subgraph *if the following conditions hold:*

(G1) *for every vertex $x \in \Xi$ the intersection $\Xi \cap \Gamma(x)$ is a subspace in π_x;*

(G2) *if $s = 2$ then for every $x \in \Xi$ the subgroup $G(x) \cap G[\Xi \cap \Gamma(x)]$ stabilizes Ξ setwise;*

(G3) *if $s = 3$ then for every $x \in \Xi$ the subgroup $G(x) \cap G[\Xi \cap \Gamma_2(x)]$ stabilizes Ξ setwise.*

We start by discussing some properties of geometrical subgraphs in the case $s = 2$.

Lemma 9.5.2 *Let Ξ be a geometrical subgraph in Γ. If $s = 2$ then for every edge $\{x, y\}$ of Ξ we have $\Xi \cap \Gamma(y) = \psi_{xy}(\Xi \cap \Gamma(x))$.*

Proof. By (G1) $\Psi := \Xi \cap \Gamma(x)$ is a subspace in π_x containing y and $\Phi := \Xi \cap \Gamma(y)$ is a subspace in π_y containing x. It is easy to deduce from (G2) that Φ must be stable under $G(x, y) \cap G[\Psi]$ and that Ψ must be stable under $G(x, y) \cap G[\Phi]$. It follows from (9.4.1) and (2.4.2 (ii)) that this is possible only if $\Phi = \psi_{xy}(\Psi)$ and the result follows. □

Assuming that we are still in the case $s = 2$, let x be a vertex of Γ, Ψ be a subspace in π_x and suppose that there exists a geometrical subgraph Ξ in Γ containing x such that $\Xi \cap \Gamma(x) = \Psi$. Since Ξ is connected, for every $z \in \Xi$ there exists an arc $(x_0 = x, x_1, ..., x_t = z)$ in Ξ joining x with

z. Put $\Psi(x_i) = \Xi \cap \Gamma(x_i)$. Then we have the following (9.5.2):

$$(*)\quad \Psi(x_0) = \Psi, \quad x_{i-1}, x_{i+1} \in \Psi(x_i) \ \text{ for } \ 1 \le i \le t-1$$

and

$$\Psi(x_i) = \varphi_{x_{i-1}x_i}(\Psi(x_{i-1})) \ \text{ for } \ 1 \le i \le t.$$

Now let $X = (x_0 = x, x_1, ..., x_t)$ be an arbitrary arc in Γ, originating at x. Suppose $(*)$ holds for a family of subspaces Ψ_i in π_{x_i}, $0 \le i \le t$. Then we say that X *transfers* Ψ to $\Psi(x_t)$. Notice that the subspaces $\Psi(x_i)$ (if they exist) are uniquely determined by X and Ψ. The following result is a direct consequence of (9.5.2).

Lemma 9.5.3 *In the case $s = 2$ let x be a vertex of Γ, Ψ be a subspace in π_x and suppose that there exists a geometrical subgraph Ξ containing x such that $\Xi \cap \Gamma(x) = \Psi$. Let z be a vertex in Γ. Then z is contained in Ξ if and only if there exists an arc in Γ joining x and z which transfers Ψ to a subspace $\Psi(z)$ in π_z. If such an arc exists then $\Psi(z) = \Xi \cap \Gamma(z)$.* □

The following lemma gives necessary and sufficient conditions for existence of geometrical subgraphs in the case $s = 2$.

Lemma 9.5.4 *Let $s = 2$, $x \in \Gamma$ and Ψ be a subspace in π_x. Then a geometrical subgraph Ξ containing x such that $\Xi \cap \Gamma(x) = \Psi$ exists if and only if Γ does not contain cycles through x transferring Ψ to subspaces in π_x different from Ψ. Moreover, if Ξ exists then it is unique.*

Proof. Suppose that Γ contains a cycle $(x_0 = x, x_1, ..., x_t = x)$ which transfers Ψ to a subspace Ψ' in π_x and $\Psi' \neq \Psi$. Then a vertex $u \in \Psi' \setminus \Psi$ must be in Ξ by (9.5.3) and it must not be in Ξ since $\Xi \cap \Gamma(x) = \Psi$. Hence Ξ does not exist in this case. Suppose that Γ does not contain cycles passing through x which transfer Ψ to different subspaces in π_x. By vertex-transitivity this is true for every vertex of Γ. Let $\Xi(x, \Psi)$ be the set of vertices in Γ defined as follows: $z \in \Xi(x, \Psi)$ if and only if there exists an arc $X = (x_0 = x, x_1, ..., x_t = z)$ in Γ which joins x with z and transfers Ψ to a subspace $\Psi(X)$ in π_z. We claim that $\Psi(X)$ depends only on z but not on X. Let t be minimal with the property that there is another arc $Y = (y_0 = x, y_1, ..., y_s = z)$ with $\Psi(Y) \neq \Psi(X)$. In view of vertex-transitivity t is independent of x and $x_{t-1} \neq y_{s-1}$ by the minimality assumption. Let r be the largest index such that $x_r = y_q$ for some $0 \le q \le s$. Then $Z = (z = x_t, x_{t-1}, ..., x_r = y_q, y_{q+1}, ..., y_s = z)$ is a cycle which transfers $\Psi(X)$ to $\Psi(Y)$, a contradiction. Hence $\Psi(X)$ is independent of the choice of the arc and $\Psi(X) = \Xi(x, \Psi) \cap \Gamma(z)$. Since $\Xi(x, \Psi)$ is defined only in terms of x and Ψ, it is stable under $G(x) \cap G[\Psi]$. Finally if $z \in \Xi(x, \Psi)$ and $\Psi(z) = \Xi(x, \Psi) \cap \Gamma(z)$ the connectivity of $\Xi(x, \Psi)$ implies that $\Xi(x, \Psi) = \Xi(z, \Psi(z))$. This means that $\Xi(x, \Psi)$ is stable under $G(z) \cap G[\Psi(z)]$ and hence $\Xi(x, \Psi)$ is a geometrical subgraph with the required properties. The uniqueness claim is by the construction. □

In what follows, if $s = 2$, x is a vertex of Γ and Ψ is a subspace in π_x then $\Xi(x, \Psi)$ will denote the geometrical subgraph containing x such that $\Xi \cap \Gamma(x) = \Psi$. This subgraph may or may not exist.

Let us consider geometrical subgraphs in the case $s = 3$.

Lemma 9.5.5 *Suppose that $s = 3$ and let Ξ be a geometrical subgraph in Γ. Then for every 2-arc (y, x, z) in Ξ we have $\Xi \cap \Gamma(z) = \xi^x_{yz}(\Xi \cap \Gamma(y))$.*

Proof. Since Ξ is a geometrical subgraph containing (y, x, z), $\Phi := \Xi \cap \Gamma(y)$ is a subspace in π_y containing x and $\Lambda := \Xi \cap \Gamma(z)$ is a subspace in π_z containing x. The action of $G_1(x)$ on δ_x contains $L_{n-1}(q)$ and hence by (2.4.2 (ii)) $\chi^x_y(\Phi)$ and $\chi^x_z(\Lambda)$ are the only subspaces in δ_x stabilized by $G_1(x) \cap G[\Phi]$ and $G_1(x) \cap G[\Lambda]$, respectively. Since both of these subgroups are contained in $G[\Xi \cap \Gamma_2(x)]$, we have $\chi^x_y(\Phi) = \chi^x_z(\Lambda)$ and the result follows. □

Still assuming that $s = 3$, consider a geometrical subgraph Ξ in Γ. For an edge $\{x, y\}$ in Ξ put $\Psi = \Xi \cap \Gamma(x)$ and $\Phi = \Xi \cap \Gamma(y)$. Let $z \in \Xi$ and let $(x_0 = x, x_1, ..., x_t = z)$ be an arc joining x and z in Ξ (here x_1 may or may not be equal to y). Let $\Lambda(x_i) = \Xi \cap \Gamma(x_i)$. Then we have the following properties (9.5.5):

$$(**)\quad \Lambda(x_0) = \Psi, \quad x_{i-1}, x_{i+1} \in \Lambda(x_i) \ \text{ for } \ 1 \le i \le t-1, \quad \Lambda(x_1) = \xi^x_{yx_1}(\Phi)$$

$$\text{and } \ \Lambda(x_{i+1}) = \xi^{x_i}_{x_{i-1}x_{i+1}}(\Lambda(x_{i-1})) \ \text{ for } \ 1 \le i \le t-1.$$

In the case $x_1 = y$ the mapping $\xi^x_{yx_1}$ in (**) is assumed to be the identity. Similarly to the case $s = 2$ let $X = (x_0 = x, x_1, ..., x_t)$ be an arbitrary arc originating in x, but in this case we assume that t is even. Suppose that for every i, $0 \le i \le t$, there is a subspace $\Lambda(x_i)$ in π_{x_i} such that (**) holds. Notice that if the $\Lambda(x_i)$ exist, then they are uniquely determined by X, Ψ and Φ. In this case we will say that X *transfers* Ψ to $\Lambda(x_t)$ with respect to Φ.

Now it is easy to prove the following analogue of (9.5.3) and (9.5.4).

Lemma 9.5.6 *In the case $s = 3$ let $\{u_1, u_2\}$ be an edge in Γ. For $i = 1$ and 2 let Ψ_i be a subspace in π_{u_i} containing u_{3-i}. Then*

(i) *a geometrical subgraph Ξ such that $\Xi \cap \Gamma(u_i) = \Psi_i$ for $i = 1$ and 2 exists if and only if for $j = 1$ and 2 there is no arc in Γ transferring Ψ_j with respect to Ψ_{3-j} to a subspace in π_{u_1} different from Ψ_1 or to a subspace in π_{u_2} different from Ψ_2,*

(ii) *if Ξ as above exists then a vertex z of Γ is contained in Ξ if and only if for $i = 1$ or 2 there is an arc in Γ which transfers Ψ_i with respect to Ψ_{3-i} to a subspace Λ in π_z, and if such an arc exists then $\Lambda = \Xi \cap \Gamma(z)$.* □

In what follows, if $s = 3$, $\{x, y\}$ is an edge in Γ, Ψ is a subspace in π_x containing y and Φ is a subspace in π_y containing x, then $\Xi(\Psi, \Phi)$ denotes the geometrical subgraph containing $\{x, y\}$ such that $\Xi \cap \Gamma(x) = \Psi$ and $\Xi \cap \Gamma(y) = \Phi$. Similarly to the case $s = 2$ such a subgraph may or may not exist. Notice that Ψ can be taken to be the whole space π_x or just a point (similarly for Φ), in particular $\Xi(\pi_x, x) = \{x\} \cup \Gamma(x)$.

Definition 9.5.7 *We will say that Γ contains a complete family of geometrical subgraphs if either*

(i) *$s = 2$ and for every vertex x and every subspace Ψ in π_x the geometrical subgraph $\Xi(x, \Psi)$ exists, or*

(ii) *$s = 3$ and for every edge $\{x, y\}$, every subspace Ψ in π_x containing y and every subspace Φ in π_y containing x the geometrical subgraph $\Xi(\Psi, \Phi)$ exist.*

We formulate two direct consequences of (9.5.4) and (9.5.6).

Corollary 9.5.8 *Suppose that Γ is a tree which is locally projective with respect to the action of $G \leq \operatorname{Aut} \Gamma$. Then Γ contains a complete family of geometrical subgraphs.* □

Corollary 9.5.9

(i) *Let $s = 2$ and Γ be of collinearity type. Let Ψ and Ψ' be subspaces in π_x and suppose that the geometrical subgraphs $\Xi(x, \Psi)$ and $\Xi(x, \Psi')$ exist. Then $\Xi(x, \Psi \cap \Psi')$ exists and is equal to $\Xi(x, \Psi) \cap \Xi(x, \Psi')$. In particular $\Xi(x, \Psi)$ contains $\Xi(x, \Psi')$ whenever Ψ contains Ψ'.*

(ii) *Let* $s = 3$. *Let* $\{x, y\} \in E(\Gamma)$, Ψ *and* Ψ' *be subspaces in* π_x *and* Φ *and* Φ' *be subspaces in* π_y. *Suppose that* $\Xi(\Psi, \Phi)$ *and* $\Xi(\Psi', \Phi')$ *exist. Then* $\Xi(\Psi \cap \Psi', \Phi \cap \Phi')$ *exists and is equal to* $\Xi(\Psi, \Phi) \cap \Xi(\Psi', \Phi')$. *In particular* $\Xi(\Psi, \Phi)$ *contains* $\Xi(\Psi', \Phi')$ *whenever* Ψ *contains* Ψ' *and* Φ *contains* Φ'. □

9.6 Further properties of geometrical subgraphs

Let Ξ be a geometrical subgraph of Γ and let H be the setwise stabilizer of Ξ in G. If $x \in \Xi$ then the stabilizer $H(x)$ of x in H induces on the set $\Xi(x)$ of vertices adjacent to x in Ξ the natural permutation action of $L_n(q)$ of degree $\left[{n \atop 1}\right]_q = (q^n - 1)/(q - 1)$ (the case $n = 1$ is also included), possibly extended by outer automorphisms. In particular the action of H on Ξ is edge-transitive. By (9.1.1) this means that for an edge $e = \{x, y\}$ in Ξ the group $H^+ = \langle H(x), H(y) \rangle$ acts edge-transitively on Ξ and either $H^+ = H$, or $[H : H^+] = 2$ and Ξ is bipartite. In any case we can redefine Ξ as the subgraph induced on the set of images under H^+ of the vertices on $e = \{x, y\}$. This enables us to give a group-theoretical version of the necessary and sufficient conditions for existence of geometrical subgraphs.

Lemma 9.6.1

(i) *Let* $s = 2$, $\{x, y\}$ *be an edge of* Γ, Ψ *be a subspace in* π_x *containing* y *and* $\Phi = \psi_{xy}(\Psi)$. *Put*

$$H^+ = \langle G(x) \cap G[\Psi], G(y) \cap G[\Phi] \rangle.$$

Then the geometrical subgraph $\Xi(x, \Psi)$ *exists if and only if*

$$H^+ \cap G(x) = G(x) \cap G[\Psi].$$

(ii) *Let* $s = 3$, $\{x, y\}$ *be an edge of* Γ, Ψ *be a subspace in* π_x *containing* y *and* Φ *be a subspace in* π_y *containing* x. *Put*

$$H^+ = \langle G(x) \cap G[\Psi] \cap G[\chi_y^x(\Phi)], G(y) \cap G[\Phi] \cap G[\chi_x^y(\Psi)] \rangle.$$

Then the geometrical subgraph $\Xi(\Psi, \Phi)$ *exists if and only if*

$$H^+ \cap G(x) = G(x) \cap G[\Psi] \cap G[\chi_y^x(\Phi)].$$

Proof. We consider the case (i); the case (ii) can be proved analogously. Suppose that the geometrical subgraph $\Xi(x, \Psi)$ exists and let F be the setwise stabilizer of Ξ. Then H^+ defined as in the lemma is contained in F and $G(x) \cap G[\Psi]$ is the stabilizer of x in F. Hence $H^+ \cap G(x) =$

$G(x) \cap G[\Psi]$. If $\Xi(x, \Psi)$ does not exist then by (9.5.4) Γ contains a cycle $(x = x_0, x_1, ..., x_t = x)$ which transfers Ψ to a different subspace in π_x. This implies that there are subspaces $\Psi(x_i)$ in π_{x_i} for $0 \le i \le t$ for which (*) holds and $\Psi(x_t) \ne \Psi$. Then arguing as in (9.1.1) we can show that for every i with $0 \le i \le t$ the subgroup H^+ contains an element which maps x onto x_i and conjugates $G(x) \cap G[\Psi]$ to $G(x_i) \cap G[\Psi(x_i)]$. Hence H^+ contains $G(x) \cap G[\Psi(x_t)]$ for $\Psi(x_t) \ne \Psi$ and by (2.4.2 (i)) $H^+ \cap G(x) = G(x)$. □

A vertex-transitive geometrical subgraph is locally projective with respect to its setwise stabilizer. The following characterization of vertex-transitive geometrical subgraphs is quite straightforward.

Lemma 9.6.2 *Let Γ be a locally projective graph with respect to a group G. A geometrical subgraph Ξ is acted on vertex-transitively by its setwise stabilizer in G if and only if one of the following holds:*

(i) *$s = 2$ and Γ is of collineation type;*
(ii) *$s = 2$, Γ is of correlation type, $\Xi = \Xi(x, \Psi)$ and Ψ is of dimension $(n - 1)/2$;*
(iii) *$s = 3$ and $\Xi = \Xi(\Psi, \Phi)$ where Ψ and Φ have the same dimension.*□

If G acts s-arc-transitively on Γ and Ξ is a vertex-transitive geometrical subgraph in Γ, then it is easy to see that the action on Ξ of its setwise stabilizer is t-arc-transitive for $t \ge s$. In some cases geometrical subgraphs happen to have higher degrees of transitivity than the original graph. Some of those cases are described in the following lemma.

Lemma 9.6.3

(i) *Suppose that $s = 2$, Γ is of collineation type and $G_1(x) \ne 1$ for $x \in \Gamma$. Let Ψ be a line in π_x and suppose that $\Xi = \Xi(x, \Psi)$ exists. Then the action on Ξ of its setwise stabilizer is strictly t-arc-transitive for $t \ge 3$.*
(ii) *Suppose that $s = 3$ and $G_1(x, y) \ne 1$ for an edge $\{x, y\}$ of Γ. Let $\Psi \in L_x(y)$ and $\Phi \in L_y(x)$ and suppose that $\Xi = \Xi(\Psi, \Phi)$ exists. Then the action on Ξ of its setwise stabilizer is strictly t-arc-transitive for $t \ge 4$.*

Proof. (i) Let $y \in \Psi$ and $\Phi = \psi_{xy}(\Psi)$. By (9.4.1) $\Phi \setminus \{x\}$ is an orbit of $O_p(G_1(x))$ on $\Gamma_2(x)$ and $\Phi \setminus \{x\} = \Xi \cap \Gamma(y)$ by (9.5.2). So the result follows directly from (9.1.3).

(ii) Let $z \in \Psi \setminus \{y\}$ and $\Lambda = \xi_{yz}^{x}(\Phi)$. Then by (9.4.1) $\Lambda \setminus \{x\}$ is an orbit of $G_1(x,y)$ on $\Gamma_2(x)$ and $\Lambda \setminus \{x\} = \Xi \cap \Gamma(z)$ by (9.5.5). So again it is sufficient to apply (9.1.3). □

For the remainder of the section let us assume that either $s = 3$, or $s = 2$, and Γ is of collineation type. Let $\{x, y\}$ be an edge of Γ, Ψ be a d-dimensional subspace in π_x which contains y, $d \geq 2$. Let Φ be a d-dimensional subspace in π_y containing x if $s = 3$ and put $\Phi = \psi_{xy}(\Psi)$ if $s = 2$. Let $\Xi = \Xi(\Psi, \Phi)$ if $s = 3$ and $\Xi = \Xi(x, \Psi)$ if $s = 2$. Put $H_1 = G[x,y] \cap G[\Psi \cup \Phi]$ and $H_2 = G(x) \cap G[\Psi] \cap G[\chi_y^x(\Phi)]$ if $s = 3$ and $H_2 = G(x) \cap G[\Psi]$ if $s = 2$. If the geometrical subgraph Ξ exists then $H = \langle H_1, H_2 \rangle$ is the setwise stabilizer of Ξ in G with H_1 and H_2 being the stabilizers of $\{x, y\}$ and x, respectively. Independently of the existence of Ξ define $\Omega = \Omega(\Gamma, d)$ to be a graph whose vertices are the (right) cosets of H_2 in G with two such vertices adjacent if they intersect a common coset of H_1 in G. Clearly, in this case the edges of Ω are indexed by the cosets of H_1 in G. It is possible to describe $\Omega(\Gamma, d)$ in combinatorial terms as follows.

If $s = 2$, then the vertices of $\Omega(\Gamma, d)$ are all pairs (z, Λ) where z is a vertex of Γ and Λ is a d-dimensional subspace in π_z with (z, Λ) being adjacent to (z', Λ') if and only if $z' \in \Lambda$ and $\Lambda' = \psi_{zz'}(\Lambda)$.

If $s = 3$, then the vertices of $\Omega(\Gamma, d)$ are all triples (z, Λ, α) where z is a vertex of Γ, Λ is a d-dimensional subspace in π_z and α is a $(d-1)$-dimensional subspace in δ_x. This vertex is adjacent to a similar vertex (z', Λ', α') if and only if $z' \in \Lambda$, $z \in \Lambda'$, $\chi_z^{z'}(\Lambda) = \alpha'$ and $\chi_{z'}^{z}(\Lambda') = \alpha$.

It is clear that G acts 2-arc-transitively on Ω and with respect to this action Ω is a locally projective graph of type (d, q). Let Ω^c denote the connected component of Ω containing (x, Ψ) and let ω be the mapping of Ω onto Γ defined by

$$\omega : (z, \Lambda) \mapsto z \text{ if } s = 2,$$

$$\omega : (z, \Lambda, \alpha) \mapsto z \text{ if } s = 3.$$

Lemma 9.6.4 *In the above notation the following three conditions are equivalent:*

(i) *Ξ exists;*
(ii) *if H is defined to be $\langle H_1, H_2 \rangle$ then $H \cap G[x, y] = H_1$;*
(iii) *Ω is disconnected and the restriction of ω to Ω^c is an isomorphism onto Ξ.*

Proof. To simplify the notation we only consider the case $s = 2$. The equivalence of (i) and (ii) is just the vertex-transitive version of (9.6.1 (i)). The definition of Ω and (9.5.4) imply that Ξ exists if and only if (x, Ψ) and (x, Ψ') are in distinct connected components of Ω whenever $\Psi \neq \Psi'$, *i.e.* if and only if (iii) holds. □

Let us consider in more detail the situation when $s = 2$ and Γ is of collineation type. If Ξ does not exist then $H \cap G(x)$ contains H_2 as a proper subgroup. Since H_2 is maximal in $G(x)$ this means that H contains the whole of $G(x)$. Since H contains H_1 as well and Γ is connected, this means that $H = G$ and Ω is connected. For $x \in \Gamma$ the set $\omega^{-1}(x)$ is an imprimitivity system of the action of G on Ω and $|\omega^{-1}(x)| = \begin{bmatrix} n-1 \\ d \end{bmatrix}_q$ where $n - 1$ is the rank of π_x. The graph Γ can be reconstructed from Ω by factorizing over this imprimitivity system. Thus when Ξ does not exist we can study Ω instead of Γ since both these graphs are locally projective with respect to the same abstract group G. The next lemma shows that if Ω satisfies a certain minimality condition, then it contains a complete family of geometrical subgraphs.

Lemma 9.6.5 *Let $s = 2$, Γ be of collineation type and let d be the smallest number such that $\Xi(x, \Psi)$ does not exist for any d-dimensional subspace Ψ in π_x. Then $\Omega = \Omega(\Gamma, d)$ is connected and contains a complete family of geometrical subgraphs.*

Proof. If $f < d$ then the graph $\Theta = \Omega(\Gamma, f)$ possesses the following description in terms of Γ. The vertices of Θ are triples (z, Φ, Λ) where $z \in \Gamma$, Φ and Λ are f- and d-dimensional subspaces in π_z, respectively, with $\Phi \subset \Lambda$ and this vertex is adjacent to a similar vertex (z', Φ', Λ') if and only if $z' \in \Phi$, $\Phi' = \psi_{zz'}(\Phi)$ and $\Lambda' = \psi_{zz'}(\Lambda)$. Since $\Omega(\Gamma, f)$ is disconnected by the minimality assumption (9.6.4), it is clear that Θ is also disconnected. □

Consider the action of H_2 on the connected component Ω^c of Ω containing (x, Ψ). H_2 acts on the set of vertices adjacent to (x, Ψ) as it acts on the point set of Ψ. By (9.4.4) the elementwise stabilizer of these vertices induces on the set of vertices adjacent to (y, Φ) a p-group extended by a cyclic group whose order divides $q - 1$. Since the action of G on Ω is 2-arc-transitive, by (9.2.2) the elementwise stabilizer in H_2 of the vertices at distance 1 from the edge $\{(x, \Psi), (y, \Phi)\}$ induces a p-group on Ω^c. On the other hand, if m is the codimension of Ψ in π_x, then the elementwise stabilizer of Ψ in H_1 induces on the set of subspaces in π_x

containing Ψ an action which contains $L_m(q)$ as a section. This means that Ω^c cannot be the whole of Ω if m is at least 2, and we have the following result.

Lemma 9.6.6 *Suppose that $s = 2$ and that Γ is of collineation type. If for a vertex $x \in \Gamma$ and a subspace Ψ from π_x the geometrical subgraph $\Xi(x, \Psi)$ does not exist, then Ψ is a hyperplane.* □

In the next section we show that $\Xi(x, \Psi)$ always exists when Ψ is a line.

It should be mentioned that we know only one example of a locally projective graph which does not contain a complete family of geometrical subgraphs. This is a graph of valency 31 related to the fourth sporadic Janko simple group J_4, which is locally projective of type $(5, 2)$. This graph appeared in (1.13.2 (ii)).

9.7 The structure of P

In this section we restrict ourselves to the situation when either $s = 3$, or $s = 2$, and Γ is of collineation type. Here we study the action of $O_p(G(x))$ on $\Gamma_2(x)$. We will deduce important information about this action from analysis of vertex-transitive geometrical subgraphs which are locally projective lines (*i.e.* of valency $q + 1$).

We use the notation introduced before (9.4.1). In particular Σ stays for the set of orbits of $P = O_p(G(x))/G_2(x)$ on $\Gamma_2(x)$. For an orbit $S \in \Sigma$ let $P(S)$ be the elementwise stabilizer of S in P. By (9.4.1) $P/P(S)$ is elementary abelian of order q. This means that the whole of P is an elementary abelian p-group.

We consider first the case $s = 3$ and assume as usual that $n \geq 3$. Directly from (9.4.4) we have the following.

Lemma 9.7.1 *Let $s = 3$ and $G_1(x, y) = 1$. Then P is elementary abelian of order q^{n-1} and it is the natural $GF(q)$-module for the characteristic subgroup of $G_1(x)/P$ isomorphic to $SL_{n-1}(q)$.* □

Lemma 9.7.2 *Let $s = 3$ and $G_1(x, y) \neq 1$. Let $x \in \Gamma$, $\Psi \in L_x$, $u_1 = y, u_2, u_3$ be distinct points on Ψ and let α be a point of δ_x. Then*

$$P(S(u_1, \alpha)) \cap P(S(u_2, \alpha)) \leq P(S(u_3, \alpha)).$$

Proof. Let Φ be the line from $L_y(x)$ such that $\chi_y^x(\Phi) = \alpha$. We can and will assume that the geometrical subgraph $\Xi = \Xi(\Psi, \Phi)$ exists. In fact, if it does not exist, then instead of Γ we can consider its covering tree (which contains a complete family of geometrical subgraphs by (9.5.8)) and instead of G the free amalgamated product of $G(x)$ and $G[x,y]$. The local properties of the action of G on Γ including the one stated in the lemma are preserved when we switch to the covering tree. By (9.6.3 (ii)) the setwise stabilizer H of Ξ in G induces on Ξ a 4-arc-transitive action. Clearly P stabilizes Ξ as a whole and it stabilizes every vertex adjacent to x in Ξ. Now $S(u_i, \alpha)$ is exactly the set of vertices other than x adjacent to u_i in Ξ and the result follows by applying (9.3.4) to the action of P on Ξ. □

Let K be the characteristic subgroup in $G_1(x)/O_p(G(x))$ isomorphic to $SL_{n-1}(q)$ (compare (9.4.4)). We are going to specify P as a $GF(p)K$-module.

Lemma 9.7.3 *As a $GF(p)$-module for $K \cong SL_{n-1}(q)$ the group P is the direct sum of l copies of the natural $GF(q)$-module and $l \leq n$.*

Proof. We will identify subgroups in $O_p(G(x))$ with their images in P. Let $Z = \{z_1, ..., z_l\}$ be a non-empty subset of $\Gamma(x)$ and put $R(Z) = \bigcap_{i=1}^{l} G_1(x, z_i)$. Since P fixes $\Gamma(x)$ elementwise, $R(Z)$ is a K-submodule in P. Let $y, z \in \Gamma(x)$ with $y \neq z$ and Q be a submodule in P. Since the natural module of K is irreducible, either $G_1(x,y) \cap Q = G_1(x,z) \cap Q$ or $\langle G_1(x,y) \cap Q, G_1(x,z) \cap Q \rangle = Q$. Let Z as above be maximal with the property that $R(Z)$ is not contained in $G_1(x,y)$. Then $R(Z)$ is a complement to $G_1(x,y)$ in P. Hence P is the direct sum of natural $GF(q)K$-modules. Let $Z = \{z_1, z_2, ..., z_n\}$ be a maximal set of independent points in π_x, which means that Z is not contained in any hyperplane of π_x. Applying (9.7.2) it is easy to show that $R(Z) = G_2(x)$ and the result follows. □

Let us consider the subgroup of $G(x)$ which commutes with the action of K on P. Since $G_1(x)$ is normal in $G(x)$ and K is characteristic in $G_1(x)/O_p(G(x))$, the group $G(x)/O_p(G(x))$ acts on K. Let F be the kernel of this action, that is the centralizer of K in $G(x)/O_p(G(x))$. Since $G(x)^{\Gamma(x)}$ contains a normal subgroup isomorphic to $L_n(q)$ and $\mathrm{Aut}\, SL_{n-1}(q)$ does not have sections isomorphic to $L_n(q)$, we conclude that F contains a section isomorphic to $L_n(q)$. This means that the centralizer in $G(x)/O_p(G(x))$ of the action of K on P contains a section isomorphic to $L_n(q)$. On the other hand the centralizer of the action of K on its natural module is isomorphic to the multiplicative group of the field $GF(q)$ and the centralizer of its action on l copies of the natural module is isomorphic to $GL_l(q)$ by Schur's lemma. Since $SL_n(q)$ is not involved in $SL_l(q)$ for $l < n$, the equality holds in (9.7.3) and we have the following.

Lemma 9.7.4 *If $s = 3$ then P is isomorphic to the tensor product of the natural $GF(q)$-module for $SL_{n-1}(q)$ and the natural $GF(q)$-module for $SL_n(q)$. In particular $|P| = q^{n^2-n}$.* □

Let us turn to the situation when $s = 2$ and Γ is of collineation type.

Lemma 9.7.5 *Let $s = 2$, Γ be of collineation type and $G_1(x) \neq 1$. Let P^* denote the dual of P. Then for every incident point–line pair (y, Ψ) in π_x there is a subgroup $P^*(y, \Psi)$ of order q in P^* such that*

(i) *the subgroups $P^*(y, \Psi)$ taken for all incident point–line pairs in π_x generate P^*,*
(ii) *for a fixed $y_0 \in \Gamma(x)$, $P^*(y_0) := \langle P^*(y_0, \Psi) \mid y_0 \in \Psi \rangle$ is the natural $GF(q)$-module for the $SL_n(q)$-section of $G(x, y_0)$,*
(iii) *for a fixed $\Psi_0 \in L_x$, $P^*(\Psi_0) := \langle P^*(y, \Psi_0) \mid y \in \Psi_0 \rangle$ is either the natural $GF(q)$-module for the $SL_2(q)$-section of $G(x) \cap G[\Psi_0]$ or the trivial 1-dimensional $GF(q)$-module for this section.*

Proof. We define $P^*(y, \Psi)$ to be the dual of $P(S(y, \Psi))$. Then (i) and (ii) follow directly from (9.4.1). As in the case $s = 3$ we assume that Γ contains a complete family of geometrical subgraphs. Let $\Psi_0 \in L_x$, $\Xi = \Xi(x, \Psi_0)$ and H be the stabilizer of Ξ in G. By (9.6.3 (i)) the action of H on Ξ is strictly t-arc-transitive for $t \geq 3$. Let $y = u_1, u_2, u_3$ be distinct points on Ψ_0. Then $S(u_i, \Psi_0)$ contains all vertices other than x adjacent to u_i in Ξ. Clearly P is contained in H and it stabilizes every vertex adjacent to x in Ξ. If $t = 3$ then $P(S(u_i, \Psi)) = P(S(u_j, \Psi))$ for all i, j, $1 \leq i, j \leq 3$ by (9.3.5) and if $t \geq 4$ then $P(S(u_1, \Psi)) \cap P(S(u_2, \Psi)) \leq P(S(u_3, \Psi))$ by (9.3.4). This implies the result. □

By (9.7.5) and (2.4.6) we have the following.

Lemma 9.7.6 *In the notation of (9.7.5) if $P^*(\Psi_0)$ is 1-dimensional then P^* is the exterior square of the natural $SL_n(q)$-module, in particular $|P| = q^{(n^2-n)/2}$.* □

Lemma 9.7.7 *In the notation of (9.7.5) if the module $P^*(\Psi_0)$ is 2-dimensional then $|P| \geq q^{2(n-1)}$.*

Proof. Let y and z be distinct points on Ψ_0. We are going to show that the action of $G_1(x,y)$ on $\Gamma(z)$ coincides with the action of $G_1(x)$ on this set. This will immediately imply the result. Since $G_1(y)$ is non-trivial, it acts transitively on $\Psi_0 \setminus \{y\}$ and hence the action of $G(x,y,z)$ on $\Gamma(y)$ coincides with the action of $G(x,y) \cap G[\Psi_0]$ on this set. On the other hand $G(x,y) \cap G[\Psi_0]$ normalizes in $P/G_1(x,y)$ a unique proper subspace, namely the dual of $P(S(y,\Psi_0))$. Since P induces on $S(y,\Psi_0) \cup S(z,\Psi_0)$ an action of order q^2 we observe that $G_1(x,y)$ induces on $S(z,\Psi_0)$ an action of order q, in particular $G_1(x,y) \neq 1$. But applying the obvious symmetry between y and z to the above observation, we obtain that the dual of $\widetilde{R}(S(z,\Psi_0))$ is the only proper subspace in $P/G_1(\dot{x},z)$ normalized by $G(x,y,z)$, so the result follows. □

We conclude the section with the following.

Lemma 9.7.8 *Suppose that either $s = 3$, or $s = 2$, and Γ is of collineation type. Then Γ contains geometrical subgraphs which are locally projective lines.*

Proof. If the claim fails then the graph $\Omega = \Omega(\Gamma, 1)$ defined before (9.6.4) is connected. The action of G on Ω is strictly t-arc-transitive for $3 \leq t \leq 7$. By the construction the subgroup H_2 is the vertex stabilizer of the action of G on Ω. By (9.3.4) and since Ω is connected, we have $|O_p(H_2)| = q^{t-2}$. On the other hand, $O_p(H_2)$ contains $O_p(G(x))$ and by (9.7.4), (9.7.6) and (9.7.7) the latter has order at least $q^{2(n-1)}$. In addition $O_p(H_2)$ induces on $\Gamma(x)$ an action of order $q^{2(n-2)}$. Since $n \geq 2$ this is a contradiction. □

9.8 Complete families of geometrical subgraphs

If Γ is a classical locally projective graph, that is $\mathscr{A}_n^1(q)$, $\mathscr{A}_n^2(q)$, $\mathscr{D}_n(q)$ or $\mathscr{F}_4(q)$, then Γ contains a complete family of geometrical subgraphs. Moreover, every geometrical subgraph of Γ is induced by the vertices incident to a certain flag in the underlying classical geometry. Conversely, if Γ is an arbitrary locally projective space which contains a complete family of geometrical subgraphs, then some of these subgraphs can be considered as elements of a diagram geometry associated with Γ. In this section we specify the diagrams of geometries arising in this way.

Let us start with the case when $s = 2$ and Γ is of collineation type. Suppose that π_x is of rank $n-1$ and that for every subspace Ψ in π_x the geometrical subgraph $\Xi(x,\Psi)$ exists. We also include the degenerate cases. That is, if $\Psi = \{y\}$ is a point, then $\Xi(x, \Psi)$ is the edge $\{x, y\}$ and, if Ψ is empty, then $\Xi(x, \Psi)$ is just $\{x\}$. Define $\mathscr{H}(\Gamma)$ to be the geometry of rank n whose set of elements of type i consists of the geometrical subgraphs $\Xi(x, \Psi)$ for all vertices $x \in \Gamma$ and all subspaces Ψ in π_x of dimension $n - i$. Two elements are adjacent if one of them contains the other one considered as subgraphs in Γ. This means in particular that the elements of type $n - 1$ and n are the edges and the vertices of Γ.

Lemma 9.8.1 *Let Γ be a locally projective graph of type (n, q) with respect to the action of G. Suppose that $s = 2$ and that Γ is of collineation type. If Γ contains a complete family of geometrical subgraphs, then G acts flag-transitively on the above defined geometry $\mathscr{G}(\Gamma)$ and the geometry has the following diagram:*

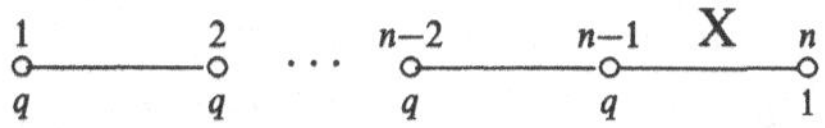

$$\overset{1}{\underset{q}{\circ}}\!\!-\!\!-\!\!-\!\!\overset{2}{\underset{q}{\circ}}\ \cdots\ \overset{n-2}{\underset{q}{\circ}}\!\!-\!\!-\!\!-\!\!\overset{n-1}{\underset{q}{\circ}}\!\!\overset{\mathrm{X}}{-\!\!-\!\!-}\!\!\overset{n}{\underset{1}{\circ}}$$

where X is the geometry of edges and vertices of a geometrical subgraph $\Xi(x, \Psi)$ where Ψ is a line in π_x.

Proof. Let $\Phi_0 = \{\Xi_{i_j} \mid 1 \le i_1 < ... < i_m \le n\}$ be a flag in $\mathscr{G}(\Gamma)$ where Ξ_{i_j} is of type i_j. Let x be a vertex in Ξ_{i_m}. Since the incidence relation is by inclusion, x is contained in Ξ_{i_j} for $1 \le j \le m$ and without loss of generality we can assume that $\Xi_{i_m} = \{x\}$. This means that the subspaces $\Psi_{i_j} = \Gamma(x) \cap \Xi_{i_j}$ for $1 \le j \le m - 1$ form a flag in π_x and hence there is a maximal flag $\{\Psi_k \mid 1 \le k \le n - 1\}$ containing it. Let $\Phi_1 = \{\Xi(x, \Psi_k) \mid 1 \le k \le n - 1\} \cup \{x\}$. Then by (9.5.9) Φ_1 is a maximal flag in $\mathscr{G}(\Gamma)$ containing Φ_0. It is clear from the above that $\mathscr{G}(\Gamma)$ belongs to a string diagram and that the residue of x is isomorphic to π_x. Let $\Lambda = \{\Xi_j \mid 1 \le j \le n - 2\}$ be a flag in $\mathscr{G}(\Gamma)$ where Ξ_j is of type j. Then $\Xi_{n-2} = \Xi(x, \Psi)$ where Ψ is a line in π_x and the residue of Λ is formed by the edges and vertices of Ξ_{n-2} with the natural incidence relation. Finally it is clear that the action of G on $\mathscr{G}(\Gamma)$ is flag-transitive. □

In order to get closer to the classical diagrams we need Γ to be bipartite. If Γ is not bipartite, then instead of Γ we can consider its standard double cover $2 \cdot \Gamma$. If Γ is not bipartite then $2 \cdot \Gamma$ is bipartite and connected, otherwise it is a disjoint union of two copies of Γ. The action of $2 \times G$ on $2 \cdot \Gamma$ is locally isomorphic to the action of G on Γ, in particular both actions are strictly s-arc-transitive for the same s. In fact $2 \cdot \Gamma$ shares more properties with Γ. For instance if the girth g of Γ is even then g is also the girth of $2 \cdot \Gamma$; $2 \times G$ acting on $2 \cdot \Gamma$ preserves an equivalence relation whose classes have size 2 and intersect both parts. We will use this property to recognize locally projective spaces which are standard double covers of other locally projective spaces.

Let Γ be as in (9.8.1) and assume in addition that Γ is bipartite with parts Γ^1 and Γ^2. Let us modify the geometry $\mathcal{G}(\Gamma)$ as follows. We exclude the edges from the element set of the geometry and consider the vertices from Γ^1 as elements of type $n-1$ and the vertices from Γ^2 as elements of type n. The incidence relation is as it used to be, except that two vertices from distinct parts are incident if and only if they are adjacent in Γ. Let us denote the geometry obtained in this way by $\mathcal{H}(\Gamma)$. The proof of the following result is similar to that of (9.8.1).

Lemma 9.8.2 *Let Γ be as in* (9.8.1) *and in addition assume that Γ is bipartite. Let G^+ be the subgroup of index* 2 *in G preserving the parts of Γ. Then G^+ acts flag-transitively on the above defined geometry $\mathcal{H}(\Gamma)$. The geometry $\mathcal{H}(\Gamma)$ is described by the following diagram:*

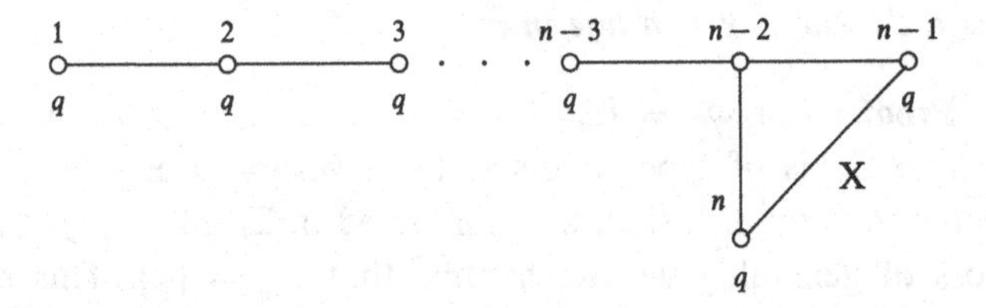

where X *stays for the rank* 2 *geometry realized by the vertices of the (bipartite) geometrical subgraph $\Xi(x,\Psi)$ where Ψ is a line in π_x. The elements from $G \setminus G^+$ perform diagram automorphisms of $\mathcal{H}(\Gamma)$ permuting the types $n-1$ and n.* □

Now let $s = 3$ and suppose that Γ is bipartite with parts Γ^1, Γ^2 and that Γ contains a complete family of geometrical subgraphs. Let us associate with Γ an incidence system $\mathcal{G}(\Gamma)$ of rank $2n-2$ by the following rule. For $1 \le i \le n-1$ the elements of type i in $\mathcal{G}$ are the subgraphs $\Xi(\pi_x, \Phi)$ where $x \in \Gamma^1$ and Φ is a subspace of dimension $n-1-i$ in δ_x; for $n \le i \le 2n-2$ the elements of type i are the geometrical subgraphs $\Xi(\pi_x, \Phi)$ where $x \in \Gamma^2$ and Φ is a subspace of dimension $i-n$ in δ_x. This means in particular that the elements of type $n-1$ and n are the subgraphs $\{x\} \cup \Gamma(x)$ for $x \in \Gamma^1$ and $x \in \Gamma^2$, respectively. Let $\Xi_1 = \Xi(\pi_x, \Phi_1)$ and $\Xi_2 = \Xi(\pi_y, \Phi_2)$. If x and y are in the same part then Ξ_1 and Ξ_2 are incident if and only if either $\Xi_1 \subseteq \Xi_2$ or $\Xi_2 \subseteq \Xi_1$. If x and y are in different parts then Ξ_1 and Ξ_2 are incident if and only if $\Xi_1 \cap \Xi_2 \neq \emptyset$. As above let G^+ be the index 2 subgroup in G which preserves the parts of Γ.

Lemma 9.8.3 *Suppose that $s = 3$, Γ is bipartite and contains a complete family of geometrical subgraphs. Then the above defined incidence system $\mathcal{G}(\Gamma)$ is a geometry on which G^+ acts flag-transitively; the diagram of $\mathcal{G}(\Gamma)$ is*

$$\overset{1}{\underset{q}{\circ}}\!\!-\!\!\!-\!\!\!-\overset{2}{\underset{q}{\circ}}\cdots\overset{n-2}{\underset{q}{\circ}}\!\!-\!\!\!-\!\!\!-\overset{n-1}{\underset{q}{\circ}}\!\!\overset{X}{-\!\!\!-\!\!\!-}\!\!\overset{n}{\underset{q}{\circ}}\!\!-\!\!\!-\!\!\!-\overset{n+1}{\underset{q}{\circ}}\cdots\overset{2n-3}{\underset{q}{\circ}}\!\!-\!\!\!-\!\!\!-\overset{2n-2}{\underset{q}{\circ}}$$

where X is the geometry of rank 2 whose incidence graph is $\Xi(\Psi,\Phi)$ where Ψ and Φ are lines. Every element from $G \setminus G^+$ performs a diagram automorphism of $\mathcal{G}(\Gamma)$ permuting the type i and $2n-1-i$ for $1 \leq i \leq n-1$.

Proof. It is an easy exercise to check that $\mathcal{G} = \mathcal{G}(\Gamma)$ is a geometry and G^+ is a flag-transitive automorphism group of $\mathcal{G}$. Let $\Xi = \{x\} \cup \Gamma(x)$ be an element of type $n-1$ in $\mathcal{G}$. If $\Xi_1 \in \operatorname{res}^-_{\mathcal{G}}(\Xi)$ and $\Xi_2 \in \operatorname{res}^+_{\mathcal{G}}(\Xi)$, then $\Xi_1 = \Xi(\pi_x, \Phi_1)$ for $\Phi_1 \in \pi_y$ and $\Xi_2 = \Xi(\pi_y, \Phi_2)$ for $y \in \Gamma(x)$ and $\Phi_2 \in \pi_x$. Since $x \in \Xi_1 \cap \Xi_2$, the elements Ξ_1 and Ξ_2 are incident and $\operatorname{res}_{\mathcal{G}}(\Xi)$ is the direct sum of the $\operatorname{res}^{\varepsilon}_{\mathcal{G}}(\Xi)$ for $\varepsilon \in \{+,-\}$. In addition it is easy to see from the above that $\operatorname{res}^+_{\mathcal{G}}(\Xi) \cong \delta_x$ and $\operatorname{res}^-_{\mathcal{G}}(\Xi) \cong \pi_x$. Since the elements of type n have similar residues, we conclude that $\mathcal{G}$ has a string diagram as above and all we have to do is to specify the residue of a flag of cotype $\{n-1, n\}$. Notice that an element $\Xi' = \{y\} \cup \Gamma(y)$ of type n is incident to Ξ if and only if $\{x, y\} \in E(\Gamma)$. Let Θ_1 and Θ_2 be incident elements of type $n-2$ and $n+1$, respectively. Then $\Theta_1 \cap \Theta_2$ contains an edge $\{x, y\}$, say, and $\Theta_1 = \Xi(\pi_x, \Psi)$ for $\Psi \in L_y$ and $\Theta_2 = \Xi(\pi_y, \Phi)$ for $\Phi \in L_x$. An element $\Theta = \{z\} \cup \Gamma(z)$ of type $n-1$ or n is incident to both Θ_1 and Θ_2 if and only if $z \in \Theta_1 \cap \Theta_2$. On the other hand by (9.5.9) the latter intersection is exactly $\Xi(\Psi, \Phi)$ and the result follows. □

9.9 Graphs of small girth

Let Γ be locally projective of type (n, q), $n \geq 3$, with respect to a group G with $G_1(x) \neq 1$ and $G_1(x, y) \neq 1$ in the case $s = 3$. In this section if $s = 2$ then Γ will be assumed to be of collineation type. Then by (9.7.8) Γ contains a family of geometrical subgraphs of valency $q + 1$ which are locally projective lines. Suppose that the girth g of Γ satisfies the inequality $g \leq 2s + 1$. By (9.1.6) this is always the case when the action of G on Γ is distance-transitive. We are going to show that under these circumstances a shortest cycle is contained in a geometrical subgraph of valency $q+1$.

Lemma 9.9.1 *Under the above assumptions if the girth g of Γ satisfies $g \leq 2s+1$ then a shortest cycle in Γ can be found inside a geometrical subgraph of valency $q+1$ which is a locally projective line.*

Proof. We consider different values of g separately. Let $k = \left[{n \atop 1}\right]_q$ be the valency of Γ.

$g=4$. If $s=3$ then by (9.3.7) Γ is complete bipartite and the claim is obvious. Let $s=2$ and (x,y,z) be a 2-arc (which is contained in a 4-cycle). Let $\Psi = l_y(x,z)$, $\Phi = \psi_{yx}(\Psi)$ and $\Theta = \Gamma(x) \cap \Gamma(z)$. We know that Θ contains y and at least one extra vertex and that Θ is a union of $G(x,y,z)$-orbits. There are exactly three orbits of $G(x,y) \cap G[\Psi]$ on $\Gamma(x)$, namely $\{y\}$, $\Phi \setminus \{y\}$ and $\Gamma(x) \setminus \Phi$. Since $G_1(x)$ acts transitively on $\Psi \setminus \{x\}$, the orbits of $G(x,y,z)$ on $\Gamma(x)$ are the same. If there are no 4-cycles in $\Xi(x,\Phi)$, then $\Theta = \{y\} \cup (\Gamma(x) \setminus \Phi)$ and

$$|\Gamma_2(x)| = \frac{|\Gamma(x)| \cdot (|\Gamma(x)|-1)}{|\Theta|} = \frac{k \cdot (k-1)}{k-q}$$

which is not an integer if $n \geq 3$ and $q \geq 2$.

$g=5$. Since $n \geq 3$ the case $s=3$ is impossible by (9.3.7) and we assume that $s=2$. As above let Σ be the set of orbits of $P = O_p(G(x))/G_2(x)$ on $\Gamma_2(x)$. Since Γ contains cycles of length 5, there are vertices $u, v \in \Gamma_2(x)$ which are adjacent. Let $S_1, S_2 \in \Sigma$ so that $u \in S_1$, $v \in S_2$. Since there are no triangles in Γ, we have $S_1 \neq S_2$. Moreover, if $S_1 \subset \Gamma(y)$ and $S_2 \subset \Gamma(z)$ for $y, z \in \Gamma(x)$, then $y \neq z$ since otherwise (y,u,v,z) would be a 3-cycle. Since P acts regularly on S_1 and S_2, we conclude that every vertex from S_1 is adjacent to exactly one vertex from S_2 and hence $P(S_1) = P(S_2)$. Thus we obtain a non-trivial equivalence relation on Σ, defined by

$$S_1 \sim S_2 \text{ if and only if } P(S_1) = P(S_2),$$

which is invariant under $G(x)$. Let $S = S(w,l) \in \Sigma$ for $l \in L_x(w)$ and T be the union of P-orbits equivalent to S. Then $G[S] = G(x) \cap G(w) \cap G[l]$ and $G[T]$ contains $G[S]$ as a proper subgroup. By (2.4.2 (v)) either $G[T] = G(x,w)$ or $G[T] = G(x) \cap G[l]$ and since clearly the former is impossible we conclude that $S(w,l) \sim S(t,m)$ if and only if $l = m$, in particular the cycle (x,y,u,v,z) is contained in the geometrical subgraph $\Xi(x,\Psi)$, where $\Psi = l_x(y,z)$.

$g = 6$. In this and the next cases we have $s = 3$. Let (u, y, x, z) be a 3-arc in Γ contained in a 6-cycle and let $\Theta = \Gamma(z) \cap \Gamma_2(u)$. Then Θ contains x and at least one extra vertex and Θ is a union of $G(u, y, z, x)$-orbits. Let $\Psi = l_y(u, x)$ and $\Psi' = \xi^x_{yz}(\Psi)$. Then $G(y, x, z) \cap G[\Psi]$ has three orbits on $\Gamma(z)$, namely $\{x\}$, $\Psi' \setminus \{x\}$ and $\Gamma(z) \setminus \Psi'$. Since $G_1(x, y)$ acts transitively on $\Psi \setminus \{x\}$, the orbits of $G(u, y, x, z)$ on $\Gamma(z)$ are the same. Suppose that there are no 6-cycles containing (u, y, x, z) and contained in $\Xi(\Psi, \Phi)$, where $\Phi = l_x(y, z)$. Then $\Theta = \{x\} \cup (\Gamma(z) \setminus \Psi')$ and similarly to the case $g = 4$ we achieve a contradiction with the integrality of

$$|\Gamma_3(x)| = \frac{k \cdot (k-1)^2}{|\Theta|}.$$

$g = 7$. Let $X = (x, y, u, z)$ be a 3-arc in Γ (contained in a 7-cycle) and let $\Xi = \Xi(l_y(x, u), l_u(y, z))$ be the unique geometrical subgraph of valency $q + 1$ which contains X. Then $G(X)$ acting on $\Gamma(z) \setminus \{u\}$ has exactly two orbits, Θ_1 and Θ_2, with length q and $k - q - 1$ such that $\Theta_1 \cup \{u\} = \Gamma(z) \cap \Xi$. If Ξ does not contain 7-cycles then the vertices in Θ_1 are all in $\Gamma_4(x)$. Let Ω be the connected component containing z of the subgraph in Γ induced on $\Gamma_3(x)$. Then the valency k_1 of Ω is $k - q - 1$ and its girth is at least the girth of Γ which is 7. Hence Ω contains at least

$$1 + k_1 + k_1(k_1 - 1) + k_1(k_1 - 1)^2 = k_1^3 - k_1^2 + k_1 + 1$$

vertices and without loss of generality we can assume that

$$|\Omega \cap \Gamma_2(y)| \geq \frac{k_1^3 - k_1^2 + k_1 + 1}{k}.$$

Let us produce an upper bound on $|\Omega \cap \Gamma_2(y)|$. We claim that an element $h \in G_1(x) \cap G(z)$ fixes Ω elementwise. In fact, suppose that $v, w \in \Gamma_3(x) \cap \Gamma(z)$ and $v^h = w$. Let (v, t, s, x) be the shortest arc joining v and x. Since $h \in G_1(x)$, $s^h = s$ and (z, v, t, s, t^h, w, z) is a 6-cycle, a contradiction to $g = 7$. Since Ω is connected, this implies that h must fix it elementwise. Let $F = G_1(x) \cap G_1(y) \cap G(z)$, S be the orbit of $O_p(G(y))$ on $\Gamma_2(y)$ containing z and $\Psi = l_y(u, x)$. Since by (9.7.4) $O_p(G(y))/G_2(y)$ is the tensor product of the natural module for $SL_n(q)$ and the natural module for $SL_{n-1}(q)$, it is easy to show that F fixes elementwise an orbit T of $O_p(G(y))$ on $\Gamma_2(y)$ if and only if $T \subset \Gamma(w)$ for $w \in \Psi$ and $T = \xi^y_{uw}(S)$. This shows that F fixes exactly q^2 vertices in $\Gamma_2(y) \cap \Gamma_3(x)$, which contradicts the above established lower bound on $|\Omega \cap \Gamma_2(y)|$. □

By (9.6.3), (9.3.7) and the above lemma we have the following.

Proposition 9.9.2 *Let Γ be a locally projective graph of type (n,q), $n \geq 3$, with respect to a group G. Suppose that $G_1(x,y) \neq 1$ in the case $s = 3$, $G_1(x) \neq 1$ and Γ is of collineation type in the case $s = 2$. Suppose also that $g \leq 2s+1$. Then a geometrical subgraph Ξ in Γ of valency $q+1$ is $(s+1)$-arc transitive of girth g, so that one of the following holds:*

(i) *$g = 4$ and $\Xi = K_{q+1,q+1}$;*
(ii) *$g = 5$, $s = 2$, $q = 2$ and Ξ is the Petersen graph;*
(iii) *$g = 6$, $s = 3$ and $\Xi \cong \mathscr{A}_2(q)$.* □

By (9.2.1) in case (i) of the above lemma Γ is either complete bipartite or isomorphic to $\mathscr{D}_n(q)$. We will show later in this chapter that in case (iii) Γ is isomorphic to $\mathscr{A}^2_{2n-2}(q)$ and in case (ii) Γ can be constructed from the derived graph of a P-geometry.

9.10 Projective geometries

In this section we classify the pairs (Γ, G) which satisfy (9.9.2 (iii)). If (Γ, G) satisfies the conditions and Γ is not bipartite then it corresponds to the pair $(2\cdot\Gamma, 2\times G)$ which satisfies the same conditions (since the girth of Γ is even) and the graph is bipartite. On the other hand the pair $(2\cdot\Gamma, 2\times G)$ is specified by the property that $2\times G$ preserves on $2\cdot\Gamma$ an imprimitivity system with classes of size 2. We will show that if (Γ, G) satisfies (9.9.2 (iii)) and Γ is bipartite then $\Gamma \cong \mathscr{A}^2_{2n-2}(q)$. By (1.6.5) and since $G_1(x,y) \neq 1$ this will imply that $L_{2n-1}(q) \trianglelefteq G^+$. Since $L_{2n-1}(q)$ does not preserve on $\mathscr{A}^2_{2n-2}(q)$ an equivalence system with classes of size 2, there are no non-bipartite examples at all.

Thus we assume that (Γ, G) satisfies (9.9.2 (iii)) and Γ is bipartite. By (9.8.3) and (1.6.3) to establish the isomorphism $\Gamma \cong \mathscr{A}^2_{2n-2}(q)$ it is sufficient to show that Γ contains a complete family of geometrical subgraphs.

Lemma 9.10.1 *Suppose that (Γ, G) satisfies* (9.9.2(iii)). *Then Γ contains a complete family of geometrical subgraphs.*

Proof. Let $\{x,y\} \in E(\Gamma)$. In view of (9.5.9) it is sufficient to show that for every subspace Φ in π_y the geometrical subgraph $\Xi(\pi_x, \Phi)$ exists. Certainly we may assume Φ to be a proper subspace. Let $\Lambda = \chi^x_y(\Phi)$ be the image of Φ in δ_x (compare (9.4.3)) and let $G_1(x,\Lambda)$ denote the largest subgroup in $G_1(x)$ which stabilizes elementwise every orbit $S \in \Sigma$ whose image $\varphi_x(S)$ in δ_x is contained in Λ. Let Θ be the connected component containing x of the subgraph in Γ induced on the vertices fixed by $G_1(x,\Lambda)$. We claim that $\Theta = \Xi(\pi_x, \Phi)$. Since $O_p(G(x))/G_2(x)$ is the tensor product on natural modules (9.7.4), it is easy to see the following: if $v \in \Gamma_2(x)$ and S is the orbit of $O_p(G(x))$ containing v, then v is fixed by $G_1(x,\Lambda)$ if and only if $\varphi_x(S) \in \Lambda$. Thus in order to establish the claim it is sufficient to show that whenever v is at an even distance from x in Θ, there is a subspace M in δ_v (whose dimension is equal to the dimension of Λ) such that $G_1(x,\Lambda) = G_1(v,M)$. Moreover, since Θ is connected, it is sufficient to consider the case when $v \in \Gamma_2(x) \cap \Theta$ and in view of the obvious symmetry it is sufficient to show that $G_1(x,\Lambda) \leq G_1(v,M)$. Thus let $v \in \Gamma_2(x) \cap \Theta$ and let $\{y\} = \Gamma(x) \cap \Gamma(v)$. We first show that $G_1(x,\Lambda) \leq G_1(v)$. Let $W = (x,y,v,u)$ be a 3-arc and let Ξ be the unique geometrical subgraph of valency $q+1$ which contains W. Then $\varphi_x(\Xi \cap \Gamma_2(x)) \in \Lambda$ and hence $G_1(x,\Lambda)$ fixes $\Xi \cap \Gamma_2(x)$. Since $\Xi \cong \mathscr{A}_2(q)$, by (9.3.4 (i)) $G_1(x,\Lambda) \leq G(\Xi) \leq G(u)$, which means that $G_1(x,\Lambda) \leq G_1(v)$.

As above let Φ be the largest subspace in π_y fixed elementwise by $G_1(x,\Lambda)$. Then $\Phi \setminus \{v\}$ is contained in $\Gamma_2(x)$ and the image M of Φ in δ_v is a subspace whose dimension is equal to the dimension of Λ. Let $z \in \Gamma_2(v)$, $z \notin \Gamma(y)$ and $\varphi_v(z) \in M$. Let us show that $G_1(x,\Lambda) \leq G(z)$. Towards this end consider a 3-arc $U = (z,w,v,y)$ where $\{w\} = \Gamma(z) \cap \Gamma(v)$ and let Ξ' be the unique geometrical subgraph of valency $q+1$ which contains U. Then $\varphi_v(\Xi'(y) \setminus \{v\}) = \varphi_v(z) \in M$ and hence $G_1(x,\Lambda)$ fixes every vertex adjacent to y and contained in Ξ'. Let us show that $G_1(x,\Lambda)$ fixes every vertex at distance at most 2 from y in Ξ'. We have noticed that every vertex t adjacent to y in Ξ' is fixed by $G_1(x,\Lambda)$. The vertex t is contained in $\Gamma_2(x) \cup \{x\}$ and it is fixed by $G_1(x,\Lambda)$. We have shown in the previous paragraph that in this case $G_1(x,\Lambda) \leq G_1(t)$ and hence $G_1(x,\Lambda)$ indeed fixes every vertex at distance at most 2 from y in Ξ'. Since $\Xi' \cong \mathscr{A}_2(q)$ this implies that $G_1(x,\Lambda) \leq G(\Xi')$, in particular $G_1(x,\Lambda) \leq G(z)$. Thus $G_1(x,\Lambda) \leq G_1(x,M)$ and the result follows. □

In view of the discussion at the beginning of the section we obtain the following.

Proposition 9.10.2 *Let Γ be a locally projective graph of type (n,q), $n \geq 3$, with respect to a group G, such that the action of G on Γ is 3-arc-transitive and $G_1(x,y) \neq 1$ for $\{x,y\} \in E(\Gamma)$. Suppose that the girth of Γ is 6. Then $\Gamma \cong \mathscr{A}^2_{2n-2}(q)$ and $L_{2n-1}(q) \trianglelefteq G^+$.* □

9.11 Petersen geometries

In this section we consider the pairs (Γ, G) corresponding to (9.9.2 (ii)) and show that there is a P-geometry $\mathscr{G}$ on which G acts as a flag-transitive automorphism group and Γ either is the derived graph $\Delta(\mathscr{G})$ or can be obtained from $\Delta(\mathscr{G})$ by factorizing over an imprimitivity system of G. The following result is immediate from (9.9.2 (ii)), (9.8.1) and the definition of the derived graph of a P-geometry.

Lemma 9.11.1 *Suppose that the pair* (Γ, G) *satisfies* (9.9.2(ii)) *and that* Γ *contains a complete family of geometrical subgraphs. Then* $\mathscr{H}(\Gamma)$ *defined before* (9.8.1) *is a P-geometry of rank n, G acts on* $\mathscr{H}(\Gamma)$ *as a flag-transitive automorphism group and* Γ *is the derived graph of* $\mathscr{H}(\Gamma)$. □

Suppose now that Γ fails to contain a complete family of geometrical subgraphs. By (9.6.6) if $x \in \Gamma$ and Ψ is a subspace in π_x then $\Xi(x, \Psi)$ exists unless Ψ is a hyperplane (*i.e.* unless the dimension of Ψ is $n-1$) and by (9.7.8) Ψ is not a line, which means that $n \geq 4$. Furthermore by (9.6.5) the graph $\Omega = \Omega(\Gamma, n-1)$ is connected, locally projective of type $(n-1, 2)$ with respect to the action of G, and Ω contains a complete family of geometrical subgraphs. Notice that Γ can be obtained from Ω by factorizing over an imprimitivity system of G with blocks of size $2^n - 1$. Thus to achieve our goal we have to show that the pair (Ω, G) also satisfies (9.9.2 (ii)) which means that the girth of Ω is 5.

Lemma 9.11.2 *Under the above assumptions the girth of* Ω *is* 5.

Proof. Let $\{x, y\} \in E(\Gamma)$, $K = \{\Psi_1 = \{y\}, \Psi_2, ..., \Psi_{n-1}\}$ be a maximal flag in π_x containing $\{y\}$ where Ψ_j is j-dimensional. Let $\Phi_j = \psi_{xy}(\Psi_j)$ for $1 \leq j \leq n-1$, so that $L = \{\Phi_1 = \{x\}, \Phi_2, ..., \Phi_{n-1}\}$ is a maximal flag in π_y containing $\{x\}$. Then $\alpha = (x, \Psi_{n-1})$ and $\beta = (y, \Phi_{n-1})$ are adjacent vertices in Ω and $\Xi = \Xi(x, \Psi_2) = \Xi(y, \Phi_2)$ is a geometrical Petersen subgraph in Γ. Since $q = 2$, it follows from (2.4.2 (iii)) that $G(\Xi)$ induces on $\mathrm{res}^+_{\pi_x}(\Psi_2)$ the full automorphism group of the latter residue isomorphic to $L_{n-2}(2)$. Let Q be the stabilizer in $G(\Xi)$ of K (equivalently the stabilizer in $G(\Xi)$ of L). Then Q is a Sylow 2-subgroup of $G(\Xi)$ and $N_{G(\Xi)}(Q) = Q$. Put $R = O_2(G(x, y))$. Since $q = 2$ and Γ is of collineation type, we have $G(x, y)/R \cong L_{n-1}(2)$, $G[x, y]/R \cong L_{n-1}(2) \times 2$ and hence there is an element τ in $G[x, y] \setminus G(x, y)$ such that $\tau^2 \in R$ and $K^\tau = L$. This means that τ stabilizes the edge $\{\alpha, \beta\}$ of Ω and normalizes Q. Let $C = (x, y, z, u, v, x)$ be a 5-cycle contained in Ξ. Since $O_2(G(x))$ is a non-trivial subgroup in R, the element τ can be chosen to induce on

C the permutation $(x,y)(z,v)(u)$. Let σ be an element in $G[\Xi]$ which induces on G the permutation $(x)(y,v)(z,u)$. Since $G(\Xi)$ induces the full automorphism group of $\mathrm{res}^+_{\pi_x}(\Psi_2)$, we can choose σ to stabilize K and hence to normalize Q. In this case σ stabilizes the vertex α of Ω. The element $(\tau\sigma)^5$ fixes C elementwise and since Ξ is the Petersen graph, it is easy to see that $(\tau\sigma)^5 \in G(\Xi)$. By the construction $(\tau\sigma)^5$ normalizes Q and since Q is self-normalized in $G(\Xi)$ we have $(\tau\sigma)^5 \in Q$. Hence $(\tau\sigma)^5$ stabilizes the vertex α and the subgraph in Ω induced on the set of images of α under $\langle \tau, \sigma \rangle$ is a 5-cycle. □

The following lemma handles the graphs of correlation type.

Lemma 9.11.3 *Let Γ be a locally projective graph of type (n,q), $n \geq 4$, with respect to a group G acting strictly 2-arc-transitively with $G_1(x) \neq 1$. If Γ is of correlation type then $g \neq 5$.*

Proof. Let u and v be adjacent vertices from $\Gamma_2(x)$ and let $u \in S(y,h)$, $v \in S(z,k)$ in the notation of (9.4.1). Then arguing as in the proof of (9.9.1) one can show that $h = k$. On the other hand $G(x,y,u)$ acts transitively on the vertices in h other than y and hence for every $z \in h \setminus \{y\}$ there is a vertex $v \in S(z,h)$ which is adjacent to u. Let $X = X(h)$ be the subgraph of Γ induced on the union of the $S(y,h)$ for all $y \in h$. Then X is of valency $\begin{bmatrix} n-1 \\ 1 \end{bmatrix}_q$ on $q \cdot \begin{bmatrix} n-1 \\ 1 \end{bmatrix}_q$ vertices and its girth is at least the girth of Γ which is 5. Since $n \geq 4$ this is impossible. □

Thus we have established the main result of the chapter.

Proposition 9.11.4 *Let Γ be a locally projective graph of type (n,q), $n \geq 3$, with respect to a group G such that $G_1(x) \neq 1$ and suppose that the girth of Γ is 5. Then $q = 2$ and there is a P-geometry $\mathscr{G}$ of rank m on which G acts as a flag-transitive automorphism group and either $m = n$ and Γ is the derived graph $\Delta(\mathscr{G})$, or $m = n-1$ and Γ can be obtained from $\Delta(\mathscr{G})$ by factorizing over an imprimitivity system of G with blocks of size $2^n - 1$.* □

The classification of flag-transitive P-geometries together with (9.11.4) provides us with the complete classification of locally projective graphs of type (n,q), $n \geq 3$, with $G_1(x) \neq 1$ and girth 5, as stated in (1.13.2).

It was shown in [Iv90] using the classification of P_3-geometries in [Sh85] that a distance-transitive graph which is either the derived graph $\Delta(\mathscr{G})$ of a P-geometry $\mathscr{G}$ or a quotient of $\Delta(\mathscr{G})$ must be the derived graph of a P_3-geometry. This together with (9.1.6), (9.2.1), (9.2.3), (9.3.3), (9.9.2), (9.11.1) and (9.11.3) gives the following.

Theorem 9.11.5 *Let G be a group acting 2-arc-transitively and distance-transitively on a graph Γ so that $G_1(x,y) \neq 1$ for $\{x,y\} \in E(\Gamma)$. Then Γ is locally projective of type (n,q) with respect to G and Γ is isomorphic to one of the following graphs:*

(i) *the point–hyperplane incidence graph $\mathscr{A}_n^1(q)$;*
(ii) *the q-analogue $\mathscr{A}_{2n-2}^2(q)$ of the double cover of the odd graph;*
(iii) *the orthogonal graph $\mathscr{D}_n(q)$;*
(iv) *the generalized polygon $\mathscr{B}_2(q)$, $q = 2^m$, or $\mathscr{G}_2(q)$, $q = 3^m$;*
(v) *the incidence graph of the rank 2 tilde geometry $\mathscr{G}(3 \cdot Sp_4(2))$;*
(vi) *a cubic distance-transitive graph on 102 vertices with the automorphism group isomorphic to $L_2(17)$;*
(vii) *the derived graph of the P_3-geometry $\mathscr{G}(Mat_{22})$ or $\mathscr{G}(3 \cdot Mat_{22})$.* □

In the above context it is natural to ask for the classification of locally projective graphs of girth 5 with $G_1(x) = 1$. In [IP98] the classification problem has been reduced to analysis of a family of Cayley graphs defined as follows.

For $n \geq 3$ let T be the group freely generated by the involutions from the set

$$D = \{t_i \mid 1 \leq i \leq \left[{n \atop 1}\right]_4\}.$$

Suppose that the structure Π of an $(n-1)$-dimensional projective $GF(4)$-space is defined on D, so that $\{t_1, ..., t_5\}$ is a line. Let $A \cong PGL_n(4)$ be a subgroup in the automorphism group of Π and G be the semidirect product of T and A with respect to the natural action. Let R be the normal closure in G of the element $t_1t_2t_3t_4t_5$. Let $W(n)$ be the Cayley graph of T/R with respect to the (bijective) image in T/R of the generating set D. Then it can be seen that $W(n)$ is a locally projective graph of type $(n,4)$. If T/R is abelian (equivalently if $t_1t_2t_3t_5t_4 \in R$) then the girth of $W(n)$ is 4, otherwise the girth is 5. The main result of [IP98] is the following.

Theorem 9.11.6 *Let Γ be a locally projective graph of type (n,q) with $n \geq 3$, and of girth 5, with respect to a subgroup G of automorphisms of Γ. Suppose that $G_1(x) = 1$. Then one of the following holds:*

(i) *$n = 4$, $q = 2$, Γ is the derived graph of the P-geometry $\mathscr{G}(Mat_{23})$;*
(ii) *$n \geq 3$, $q = 4$ and Γ is a quotient of the graph $W(n)$; moreover*

(a) *$W(3)$ is of girth 5 and has exactly 2^{20} vertices,*
(b) *the regular subgroup T of the automorphism group of $W(n)$ satisfies $[T,T,T]=1$, both $T/[T,T]$ and $[T,T]$ are elementary abelian 2-groups of rank less than $\left[{n \atop 1}\right]_4$ and $\left[{n \atop 2}\right]_4$, respectively, in particular $W(n)$ is finite.* □

We conjecture that $W(n)$ is of girth 5 for all $n \geq 3$. Notice that a geometrical subgraph of valency 5 in $W(3)$ is isomorphic to the so-called *Wells graph*. The Wells graph is a 2-fold antipodal cover of the folded 5-cube. The automorphism group $2_-^{1+4} : Alt_5$ of the Wells graph is isomorphic to involution centralizers in the sporadic simple Janko groups J_2 and J_3.

Bibliography

[Alp65] J.L. Alperin, On a theorem of Manning, *Math. Z.* **88** (1965), 434–435.

[AB95] J.L. Alperin and R.B. Bell, *Groups and Representations*, Springer, New York, 1995.

[A83] M. Aschbacher, Flag structures on Tits geometries, *Geom. Dedic.* **14** (1983), 21–32.

[A84] M. Aschbacher, Finite geometries of type C_3 with flag-transitive automorphism group, *Geom. Dedic.* **16** (1984), 195–200.

[A94] M. Aschbacher, *Sporadic Groups*, Cambridge Univ. Press, Cambridge, 1994.

[A97] M. Aschbacher, 3-*Transposition Groups*, Cambridge Univ. Press, Cambridge, 1997.

[ASeg91] M. Aschbacher and Y. Segev, The uniqueness of groups of type J_4, *Invent. Math.* **105** (1991), 589–607.

[ASeg92] M. Aschbacher and Y. Segev, Extending morphisms of groups and graphs, *Ann. Math.* **135** (1992), 297–324.

[ASei76] M. Aschbacher and G.M. Seitz, Involutions in Chevalley groups over fields of even order, *Nagoya Math. J.* **63** (1976), 1–91.

[BI84] E. Bannai and T. Ito, *Algebraic Combinatorics I, Association Schemes*, Benjamin-Cummings Lect. Notes, Benjamin, Menlo Park, Calif., 1984.

[BIP98] B. Baumeister, A.A. Ivanov and D.V. Pasechnik, A characterization of the Petersen-type geometry of the McLaughlin group, *Math. Proc. Camb. Phil. Soc.* **18** (2000).

[Ben80] D.J. Benson, The simple group J_4, Ph.D. Thesis, Cambridge Univ., 1980.

[BJL86] T. Beth, D. Jungnickel and H. Lenz, *Design Theory*, Cambridge Univ. Press, Cambridge, 1986.

[Beu86] A. Beutelspacher, $21 - 6 = 15$: A connection between two distinguished geometries, *Amer. Math. Monthly* **93** (1986), 29–41.

[Big75] N.L. Biggs, Designs, factors and codes in graphs, *Quart. J. Math. Oxford* (2) **26** (1975), 113–119.

[BW92a] J. van Bon and R. Weiss, A characterization of the groups Fi_{22}, Fi_{23} and Fi_{24}, *Forum Math.* **4** (1992), 425–432.

[BW92b] J. van Bon and R. Weiss, An existence lemma for groups generated by 3-transpositions, *Invent. Math.* **109** (1992), 519–534.

[BCN89] A.E. Brouwer, A.M. Cohen and A. Neumaier, *Distance-Regular Graphs*, Springer Verlag, Berlin, 1989.

[BvL84] A.E. Brouwer and J.H. van Lint, Strongly regular graphs and partial geometries, in: *Enumeration and Design*, D.M. Jackson and S.A. Vanstone eds., Academic Press, Toronto, 1984, pp. 85–122.

[Bro82] K.S. Brown, *Cohomology of Groups*, Springer Verlag, Berlin, 1982.

[Bue79] F. Buekenhout, Diagrams for geometries and groups, *J. Combin. Theory*, **A27** (1979), 121–151.

[Bue81] F. Buekenhout, The basic diagram of a geometry, in: *Geometries and Groups*, Lect. Notes Math. **893**, Springer Verlag, Berlin, 1981, pp. 1–29.

[Bue85] F. Buekenhout, Diagram geometries for sporadic groups, In: *Finite Groups – Coming of Age*, Proc. 1982 Montreal Conf., J. McKay ed., Contemp. Math. **45**, AMS, Providence, R.I., 1985, pp. 1–32.

[BF83] F. Buekenhout and B. Fischer, A locally dual polar space for the Monster, unpublished manuscript around 1983.

[BH77] F. Buekenhout and X. Hubaut, Locally polar spaces and related rank 3 groups, *J. Algebra*, **45** (1977), 391–434.

[BP95] F. Buekenhout and A. Pasini, Finite diagram geometries extending buildings, in: *Handbook of Incidence Geometry*, F. Buekenhout ed., Elsevier, Amsterdam, 1995, pp. 1143–1254.

[Chi94] K. Ching, On 3-transitive graphs of girth 6, *J. Combin. Theory* (B) **62** (1994), 316–322.

[CPr82] P.J. Cameron and C.E. Praeger, Graphs and permutation groups with projective subconstituents, *J. London Math. Soc.* **25** (1982), 62–74.

[Con69] J.H. Conway, A characterization of Leech's lattice, *Invent. Math.* **7** (1969) 137–142.

[Con71] J.H. Conway, Three lectures on exceptional groups. in: *Finite Simple Groups*, M.B. Powell and G. Higman eds., Acad. Press, New York, 1971, pp. 215–247.

[Con85] J.H. Conway, A simple construction of the Fischer–Griess monster group, *Invent. Math.* **79** (1985), 513–540.

[Con92] J.H. Conway, Y_{555} and all that, in: *Groups, Combinatorics and Geometry*, Durham 1990, M. Liebeck and J. Saxl eds., London Math. Soc. Lect. Notes **165**, Cambridge Univ. Press, Cambridge, 1992, pp. 22–23.

[CCNPW] J.H. Conway, R.T. Curtis, S.P. Norton, R.A. Parker and R.A. Wilson, *Atlas of Finite Groups*, Clarendon Press, Oxford, 1985.

[CNS88] J.H. Conway, S.P. Norton and L.H. Soicher, The bimonster, the group Y_{555} and the projective plane of order 3, in: *Computers in Algebra*, M.C. Tangora ed., Marcel Dekker, 1988, pp. 27–50.

[CP92] J.H. Conway and A.D. Pritchard, Hyperbolic reflexions for the bimonster and $3Fi_{24}$, in: *Groups, Combinatorics and Geometry*, Durham 1990, M. Liebeck and J. Saxl eds., London Math. Soc. Lect. Notes **165**, Cambridge Univ. Press, Cambridge, 1992, pp. 23–45.

[CS88] J.H. Conway and N.J.A. Sloane, *Sphere Packing, Lattices and Groups*, Grundlehren Math. Wiss., 290, Springer, Berlin, 1988.

[DGMP] A. Del Fra, D. Ghinelli, T. Meixner and A. Pasini, Flag-transitive extensions of C_n geometries, *Geom. Dedic.* **37** (1991), 253–273.

[DGS85] A. Delgado, D.M. Goldschmidt and B. Stellmacher, *Groups and Graphs: New Results and Methods*, Birkhäuser Verlag, Basel, 1985.

[Ebe94] W. Ebeling, *Lattices and Codes*, Vieweg, Wiesbaden, 1994.

[Edge65] W. L. Edge, Some implications of the geometry of the 21-point plane, *Math. Z.* **87** (1965), 348–362.

[FII86] I.A. Faradjev, A.A. Ivanov and A.V. Ivanov, Distance-transitive graphs of valency 5, 6 and 7, *Europ. J. Combin.* **7** (1986), 303–319.

[FH64] W. Feit and D. Higman, The nonexistence of certain generalized polygons, *J. Algebra* **1** (1964), 114–131.

[Glea56] A.M. Gleason, Finite Fano planes, *Amer. J. Math.* **78** (1956), 797–807.

[GS75] J.M. Goethals and J.J. Seidel, The regular two-graph on 276 vertices, *Discrete Math.*, **12** (1975), 143–158.

[GM93] G. Grams and T. Meixner, Some results about flag-transitive diagram geometries using coset enumeration, *Ars Combinatoria* **36** (1993), 129–146.

[Gri73] R.L. Griess, Automorphisms of extra special groups and nonvanishing degree 2 cohomology, *Pacific J. Math.* **28** (1973), 355–421.

[Gri76] R.L. Griess, The structure of the "Monster" simple group, in: *Proc. of the Conference on Finite Groups*, W.R. Scott and R. Gross eds, Academic Press, New York, 1976, pp. 113–118.

[Gri82] R.L. Griess, The Friendly Giant, *Invent. Math.* **69** (1982), 1–102.

[GMS89] R.L. Griess, U. Meierfrankenfeld and Y. Segev, A uniqueness proof for the Monster, *Ann. Math.* **130** (1989), 567–602.

[Hei91] St. Heiss, On a parabolic system of type M_{24}, *J. Algebra* **142** (1991), 188–200.

[Hig64] D.G. Higman, Finite permutation groups of rank 3, *Math. Z.* **86** (1964), 145–156.

[Hig76] D.G. Higman, A monomial character of Fischer's Baby Monster, in: *Proc. of the Conference on Finite Groups*, W.R. Scott and R. Gross eds, Academic Press, New York, 1976, pp. 277–283.

[HP85] D.R. Hughes and F.C. Piper, *Design Theory*, Cambridge Univ. Press, Cambridge, 1985.

[Hum90] J.E. Humphreys, *Reflection Groups and Coxeter Groups*, Cambridge Univ. Press, Cambridge, 1990.

[Ito82] T. Ito, Bipartite distance-regular graphs of valency three, *Linear Algebra and Appl.* **46** (1982), 195–213.

[Iv87] A.A. Ivanov, On 2-transitive graphs of girth 5, *Europ. J. Combin.* **8** (1987), 393–420.

[Iv88] A.A. Ivanov, Graphs of girth 5 and diagram geometries related to the Petersen graphs, *Soviet Math. Dokl.* **36** (1988), 83–87.

[Iv90] A.A. Ivanov, The distance-transitive graphs admitting elations, *Math. USSR Izvestiya*, **35** (1990), 307–335.

[Iv91a] A.A. Ivanov, Geometric presentation of groups with an application to the Monster, In: *Proc. ICM-90*, Kyoto, Japan, August 1990, Springer Verlag, Berlin, 1991 pp. 385–395.

[Iv91b] A.A. Ivanov, A geometric approach to the question of uniqueness for sporadic simple groups, *Soviet Math. Dokl.* **43** (1991), 226–229.

[Iv92a] A.A. Ivanov, A geometric characterization of the Monster, in: *Groups, Combinatorics and Geometry*, Durham 1990, M. Liebeck and J. Saxl eds., London Math. Soc. Lect. Notes **165**, Cambridge Univ. Press, Cambridge, 1992, pp. 46–62.

[Iv92b] A.A. Ivanov, A presentation for J_4, *Proc. London Math. Soc.* **64** (1992), 369–396.

[Iv92c] A.A. Ivanov, A geometric characterization of Fischer's Baby Monster, *J. Algebraic Combin.* **1** (1992), 43–65.

[Iv92d] A.A. Ivanov, The minimal parabolic geometry of the Conway group Co_1 is simply connected, in: *Combinatorics 90: Recent Trends and Applications*, North-Holland, Amsterdam, 1992, pp. 259–273.

[Iv93a] A.A. Ivanov, Graphs with projective subconstituents which contain short cycles, in: *Surveys in Combinatorics* 1993, K. Walker ed., London Math.

Soc. Lect. Notes **187**, pp. 173–190, Cambridge Univ. Press, Cambridge, 1993, pp. 173–190.

[Iv93b] A.A. Ivanov, Constructing the Monster via its Y-presentation, In: *Combinatorics, Paul Erdös is Eighty*, vol. 1, Bolyai Soc. Math. Studies, Budapest, 1993, pp. 253–270.

[Iv94] A.A. Ivanov, Presenting the Baby Monster, *J. Algebra* **163** (1994), 88–108.

[Iv95] A.A. Ivanov, On geometries of the Fischer groups, *Europ. J. Combin.*, **16** (1995), 163–183.

[Iv96] A.A. Ivanov, On the Buekenhout–Fischer geometry of the Monster, in: *Moonshine, the Monster and Related Topics*, C. Dong and G. Mason eds., *Contemp. Math.* **193** AMS, 1996, pp. 149–158.

[Iv97] A.A. Ivanov, Exceptional extended dual polar spaces, *Europ. J. Combin.*, **18** (1997), 859–886.

[Iv98a] A.A. Ivanov, Affine extended dual polar spaces, In: *Trends in Mathematics*, A. Pasini ed., Birkhäuser Verlag, Basel, 1998, pp. 107–121.

[Iv98b] A.A. Ivanov, Flag-transitive extensions of dual polar spaces, *Doklady Math.* **58** (1998), 180–182.

[ILLSS] A.A. Ivanov, S.A. Linton, K. Lux, J. Saxl and L.H. Soicher, Distance-transitive representations of the sporadic groups, *Comm. Algebra* **23** (1995), 3379–3427.

[IMe93] A.A. Ivanov and U. Meierfrankenfeld, A computer free construction of J_4, preprint, 1993, submitted to *J. Algebra*

[IMe97] A.A. Ivanov and U. Meierfrankenfeld, Simple connectedness of the 3-local geometry of the Monster, *J. Algebra,* **194** (1997), 383–407.

[IPS96] A.A. Ivanov, D.V. Pasechnik and S.V. Shpectorov, Non-abelian representations of some sporadic geometries, *J. Algebra* **181** (1996), 523–557.

[IP98] A.A. Ivanov and C.E. Praeger, On locally projective graphs of girth 5, *J. Algebraic Combin.* **7** (1998), 259–283.

[ISa96] A.A. Ivanov and J. Saxl, The character table of ${}^2E_6(2)$ acting on the cosets of Fi_{22}, in: *Advanced Studies in Pure Math.* **24** (1996), 165–196.

[ISh88] A.A. Ivanov and S.V. Shpectorov, Geometries for sporadic groups related to the Petersen graph. I, *Comm. Algebra* **16** (1988), 925–954.

[ISh89a] A.A. Ivanov and S.V. Shpectorov, Geometries for sporadic groups related to the Petersen graph. II, *Europ. J. Combin.* **10** (1989), 347–362.

[ISh89b] A.A. Ivanov and S.V. Shpectorov, On geometries with the diagram $\circ\overset{\sim}{=\!=\!=}\circ\!-\!\!-\!\!-\circ\ \cdots\ \circ\!-\!\!-\!\!-\circ$, preprint, 1989.

[ISh90a] A.A. Ivanov and S.V. Shpectorov, The P-geometry for M_{23} has no nontrivial 2-coverings, *Europ. J. Combin.* **11** (1990), 373–379.

[ISh90b] A.A. Ivanov and S.V. Shpectorov, P-geometries of J_4-type have no natural representations, *Bull. Soc. Math. Belgique* (A) **42** (1990), 547–560.

[ISh93a] A.A. Ivanov and S.V. Shpectorov, An infinite family of simply connected flag-transitive tilde geometries, *Geom. Dedic.* **45** (1993), 1–23.

[ISh93b] A.A. Ivanov and S.V. Shpectorov, The last flag-transitive P-geometry. *Israel J. Math.* **82** (1993), 341–362.

[ISh94a] A.A. Ivanov and S.V. Shpectorov, Natural representations of the P-geometries of Co_2-type, *J. Algebra* **164** (1994), 718–749.

[ISh94b] A.A. Ivanov and S.V. Shpectorov, Flag-transitive tilde and Petersen type geometries are all known, *Bull. Amer. Math. Soc.* **31** (1994), 173–184.

[ISh94c] A.A. Ivanov and S.V. Shpectorov, Application of group amalgams to algebraic graph theory, in: *Investigations in Algebraic Theory of*

Combinatorial Objects, Kluwer Acad. Publ., Dordrecht, NL, 1994, pp. 417–441.

[ISh97] A.A. Ivanov and S.V. Shpectorov, The universal non-abelian representation of the Petersen type geometry related to J_4, *J. Algebra*, **191** (1997), 541–567.

[ISh98] A.A. Ivanov and S.V. Shpectorov, A new cover of the 3-local geometry of Co_1, *Geom. Dedic.* **73** (1998), 237–244.

[ISt96] A.A. Ivanov and G. Stroth, A characterization of 3-local geometry of $M(24)$, *Geom. Dedic.* **63** (1996), 227–246.

[J76] Z. Janko, A new finite simple group of order 86,775,571,046,077,562,880 which possesses M_{24} and the full covering group of M_{22} as subgroups, *J. Algebra* **42** (1976), 564–596.

[JLPW] Ch. Jansen, K. Lux, R. Parker and R. Wilson, *An Atlas of Brauer Characters*, Clarendon Press, Oxford, 1995.

[JP76] W. Jones and B. Parshall, On the 1-cohomology of finite groups of Lie type, in: *Proc. of the Conference on Finite Groups*, W.R. Scott and R. Gross eds, Academic Press, New York, 1976, pp. 313–327.

[Kan81] W.M. Kantor, Some geometries that are almost buildings, *Europ. J. Combin.* **2** (1981), 239–247.

[Kan86] W.M. Kantor, Generalized polygons, SCABs and GABs, *Lect. Notes Math.* **1181**, Springer Verlag, Berlin, 1986, pp. 79–158.

[Kit84] M. Kitazume, A 2-local geometry for the Fischer group Fi_{24}, *J. Fac. Sci. Univ. Tokyo, Sec. 1A, Math.* **31** (1984), 59–79.

[KKM91] M. Kitazume, T. Kondo and I. Miyamoto, Even lattices and doubly even codes, *J. Math. Soc. Japan* **43** (1991), 67–87.

[LSi77] J.S. Leon and C.C. Sims, The existence and uniqueness of a simple group generated by 3,4-transpositions, *Bull. Amer. Math. Soc.* **83** (1977), 1039–1040.

[Lin89] S. Linton, The maximal subgroups of the sporadic groups Th, Fi_{24} and Fi'_{24} and other topics, Ph.D. Thesis, Cambridge Univ. 1989.

[Lün69] H. Lüneburg, *Transitive Erweiterungen endlicher Permutationgruppen*, Lect. Notes Math. **89**, Springer Verlag, Berlin, 1969.

[MS77] F.J. MacWilliams and N.J.A. Sloane, *The Theory of Error-Correcting Codes*, North-Holland, Amsterdam, 1977.

[MSm82] G. Mason and S.D. Smith, Minimal 2-local geometries for the Held and Rudvalis sporadic groups, *J. Algebra*, **79** (1982), 286–306.

[Maz79] P. Mazet, Sur le multiplicateur du groupe de Mathieu M_{22}, *C.R. Acad. Sci. Paris* **289** (1979), 659–661.

[Mei91] T. Meixner, Some polar towers, *Europ. J. Combin.* **12** (1991), 397–417.

[Mi95] M. Miyamoto, 21 Involutions acting on the Moonshine module, *J. Algebra* **175** (1995), 941–965.

[Neu84] A. Neumaier, Some sporadic geometries related to $PG(3,2)$, *Arch. Math.* **42** (1984), 89–96.

[Nor80] S.P. Norton, The construction of J_4, in: *Proc. Symp. Pure Math.* No. 37, B. Cooperstein and G. Mason eds., AMS, Providence, R.I., 1980, pp. 271–278.

[Nor85] S.P. Norton, The uniqueness of the Fischer–Griess Monster, in: *Finite Groups – Coming of Age*, Proc. 1982 Montreal Conf., J. McKay ed., Contemp. Math. **45**, AMS, Providence, R.I., 1985, pp. 271–285.

[Nor90] S.P. Norton, Presenting the Monster? *Bull. Soc. Math. Belgique* (A) **42** (1990), 595–605.

[Nor92] S.P. Norton, Constructing the Monster, in: *Groups, Combinatorics and Geometry*, Durham 1990, M. Liebeck and J. Saxl eds., London Math. Soc. Lect. Notes **165**, Cambridge Univ. Press, Cambridge, 1992, pp. 63–76.

[Nor98] S.P. Norton, Anatomy of the Monster: I, in: *The Atlas of Finite Groups: Ten Years on*, R. Curtis and R. Wilson eds., London Math. Soc. Lect. Notes **249**, Cambringe Univ. Press, Cambridge 1998, pp. 198–214.

[Par92] C. Parker, Groups containing a subdiagram $\circ\!\!-\!\!-\!\!-\!\!\circ\!\overset{\sim}{=\!=\!=}\!\circ$, *Proc. London. Math. Soc.* **65** (1992), 85–120.

[Pase94] D.V. Pasechnik, Geometric characterization of the sporadic groups Fi_{22}, Fi_{23} and Fi_{24}, *J. Combin. Theory* (A) **68** (1994), 100–114.

[Pasi85] A. Pasini, Some remarks on covers and apartments, in: *Finite Geometries*, C.A. Baker and L.M. Batten eds., Marcel Dekker, New York, 1985, pp. 233–250.

[Pasi94] A. Pasini, *Diagram Geometries*, Clarendon Press, Oxford, 1994.

[PT84] S.E. Payne and J.A.Thas, *Finite Generalized Quadrangles*, Pitman Publ. Ltd, London, 1984.

[PS97] C.E. Praeger and L.H. Soicher, *Low Rank Representations and Graphs for Sporadic Groups*, Cambridge Univ. Press, Cambridge, 1997.

[Pr89] A.D. Pritchard, Some reflection groups and their quotients, Ph. D. Thesis, Cambridge, Univ. 1989.

[Ron80] M.A. Ronan, Coverings and automorphisms of chamber systems, *Europ. J. Combin.* **1** (1980), 259–269.

[Ron81a] M.A. Ronan, Coverings of certain finite geometries, in: *Finite Geometries and Designs*, Cambridge Univ. Press, Cambridge, 1981, pp. 316–331.

[Ron81b] M.A. Ronan, On the second homotopy group of certain simplicial complexes and some combinatorial applications, *Quart. J. Math. (2)* **32** (1981) 225-233.

[Ron82] M.A. Ronan, Locally truncated buildings and M_{24}, *Math. Z.* **180** (1982), 489–501.

[Ron87] M.A. Ronan, Embeddings and hyperplanes of discrete geometries, *Europ. J. Combin.* **8** (1987), 179–185.

[RSm80] M.A. Ronan and S. Smith, 2-Local geometries for some sporadic groups, in: *Proc. Symp. Pure Math.* No. 37, B. Cooperstein and G. Mason eds., AMS, Providence, R.I., 1980, pp. 283–289.

[RSm86] M.A. Ronan and S.D. Smith, Universal presheaves on group geometries, and modular representations, *J. Algebra* **102** (1986), 135–154.

[RSm89] M.A. Ronan and S.D. Smith, Computation of 2-modular sheaves and representations for $L_4(2)$, A_7, $3S_6$ and M_{24}, *Comm. Algebra* **17** (1989), 1199–1237.

[RSt84] M.A. Ronan and G. Stroth, Minimal parabolic geometries for the sporadic groups, *Europ. J. Combin.* **5** (1984), 59–91.

[Row89] P. Rowley, On the minimal parabolic system related to M_{24}, *J. London Math. Soc.* **40** (1989), 40–56.

[Row91] P. Rowley, Minimal parabolic systems with diagram $\circ\!\!-\!\!-\!\!-\!\!\circ\!\!-\!\!-\!\!-\!\!\circ\!\overset{\sim}{=\!=\!=}\!\circ$, *J. Algebra* **141** (1991), 204–251.

[Row92] P. Rowley, Pushing down minimal parabolic systems, in: *Groups, Combinatorics and Geometry*, Durham 1990, M. Liebeck and J. Saxl eds., London Math. Soc. Lect. Notes **165**, Cambridge Univ. Press, Cambridge, 1992, pp. 144–150.

[Row94] P. Rowley, Sporadic group geometries and the action of involutions, *J. Austral. Math. Soc.* (A) **57** (1994), 35–48.

[GAP] M. Schönert *et al.*, *GAP: Groups, Algorithms and Programming*, Lehrstuhl D für Mathematik, RWTH, Aachen, 1995.

[Seg88] Y. Segev, On the uniqueness of the Co_1 2-local geometry, *Geom. Dedic.* **25** (1988), 159–219.

[Seg91] Y. Segev, On the uniqueness of Fischer's Baby Monster, *Proc. London Math. Soc.* (3) **62** (1991), 509–536.

[Sei73] G. Seitz, Flag-transitive subgroups of Chevalley groups, *Ann. Math.* **97** (1973), 27–56.

[Ser73] J.-P. Serre, *A Course in Arithmetic*, Springer, New York, 1973.

[Ser77] J.-P. Serre, *Arbres, amalgams,* SL_2, Astérisque **46**, Soc. Math. de France, Paris, 1977.

[Sh85] S.V. Shpectorov, A geometric characterization of the group M_{22}, in: *Investigations in Algebraic Theory of Combinatorial Objects*, VNIISI, Moscow, 1985, pp. 112–123 [In Russian, English translation by Kluwer Acad. Publ., Dordrecht, NL, 1994]

[Sh88] S.V. Shpectorov, On geometries with diagram P^n, preprint, 1988. [In Russian]

[Sh92] S.V. Shpectorov, The universal 2-cover of the P-geometry $\mathcal{G}(Co_2)$, *Europ. J. Combin.* **13** (1992), 291–312.

[Sh93] S.V. Shpectorov, Natural representations of some tilde and Petersen type geometries, *Geom. Dedic.* **54** (1995), 87–102.

[ShSt94] S.V. Shpectorov and G. Stroth, Classification of certain types of tilde geometries, *Geom. Dedic.* **49** (1994), 155–172.

[ShY80] E. Shult and A. Yanushka, Near *n*-gons and line systems, *Geom. Dedic.* **9** (1980) 1–72.

[Sm92] S.D. Smith, Universality of the 24-dimensional embedding of the .1 2-local geometry, *Comm. Algebra*, **22(13)** (1994), 5159–5166.

[Soi89] L.H. Soicher, From the Monster to the Bimonster, *J. Algebra* **121** (1989), 275–280.

[Soi91] L.H. Soicher, More on the group Y_{555} and the projective plane of order 3, *J. Algebra* **136** (1991), 168–174.

[Sp66] E.H. Spanier, *Algebraic Topology*, McGraw-Hill, New York, 1966.

[Str84] G. Stroth, Parabolics in finite groups. in: *Proc. Rutgers Group Theory Year*, 1983 – 1984, Cambridge Univ. Press, Cambridge 1984, pp. 211–224.

[Str96] G. Stroth, Some sporadic geometries, in: *Groups, Difference Sets and the Monster*, Walter de Gruyter, Berlin, 1996, pp. 99–116.

[StW88] G. Stroth and R. Weiss, Modified Steinberg relations for the group J_4, *Geom. Dedic.* **25** (1988), 513–525.

[Su86] M. Suzuku, *Group Theory II*. Springer Verlag, 1986.

[Tay92] D.E. Taylor, *The Geometry of the Classical Groups*, Heldermann Verlag, Berlin, 1992.

[Th79] J.G. Thompson, Uniqueness of the Fischer–Griess Monster, *Bull. London Math. Soc.* **11** (1979), 340–346.

[Tim84] F.G. Timmesfeld, Tits geometries and revisionism of the classification of finite simple groups of Char. 2-type. in: *Proc. Rutgers Group Theory Year*, 1983 – 1984, Cambridge Univ. Press, Cambridge 1984, pp. 229–242.

[Tim89] F.G. Timmesfeld, Classical locally finite Tits chamber systems of rank 3, *J. Algebra*, **124** (1989), 9–59.

[Ti74] J. Tits, *Buildings of Spherical Type and Finite BN-pairs*, Lect. Notes Math. **386**, Springer-Verlag, Berlin 1974.

[Ti82] J. Tits, A local approach to buildings, in: *The Geometric Vein* (*Coxeter-Festschrift*), Springer Verlag, Berlin, 1982, pp. 519–547.

[Ti85] J. Tits, Le Monstre, in: *Seminar Bourbaki* exposè no 620, 1983/84, Astérisque **121–122** (1985), 105–122.

[Ti86] J. Tits, Ensembles ordonnés, immeubles et sommes amalgameés, *Bull. Soc. Math. Belg.* **A38** (1986) 367–387.

[Tro91] V.I. Trofimov, Stabilizers of the vertices of graphs with projective suborbits, *Soviet Math. Dokl.* **42** (1991), 825–827.

[Tro92] V.I. Trofimov, Graphs with projective suborbits, *Math. USSR Izvestiya* **39** (1992) 869–894.

[Tro94] V.I. Trofimov, Graphs with projective suborbits. The case of small characteristics, I and II, *Math. Izvestiya Russian Acad. Sci.* **58** (1994), No 5, 124–173 and No 6, 137–156.

[Vi97] M.-M. Virotte Ducharme, Some Y-groups, *Geom. Dedic* **65** (1997), 1–30.

[Wei78] R. Weiss, Symmetric graphs with projective subconstituents, *Proc. Amer. Math. Soc.* **72** (1978), 213–217.

[Wei79a] R. Weiss, Elations of graphs, *Acta Math. Acad. Sci. Hungar.*, **34** (1979), 101–103.

[Wei79b] R. Weiss, Groups with a (B, N)-pair and locally transitive graphs, *Nagoya Math. J.*, **74** (1979), 1–21.

[Wei81a] R. Weiss, s-Transitive graphs, in: *Algebraic Methods in Graph Theory*, North-Holland, Amsterdam, 1981, pp. 827–847.

[Wei81b] R. Weiss, The nonexistence of 8-transitive graphs, *Combinatorica*, **1** (1981), 309–311.

[Wei85] R. Weiss, Distance-transitive graphs and generalized polygons, *Arch. Math.* **45** (1985), 186–192.

[WY90] R. Weiss and S. Yoshiara, A geometric characterization of the groups Suz and HS, *J. Algebra* **133** (1990), 182–196.

[Wie97] C. Wiedorn, A tilde geometry for $F_4(2)$, *Cont. Algebra and Geom.* **38** (1997), 337–342.

[Wil89] R.A. Wilson, Vector stabilizers and subgroups of Leech lattice groups, *J. Algebra* **127** (1989), 387–408.

[Wil92] R.A. Wilson, The action of a maximal parabolic subgroup on the transpositions of the Baby Monster, *Proc. Edinburgh Math. Soc.* **37** (1992), 185–189.

[Wil93] R.A. Wilson, A new construction of the baby Monster and its applications, *Bull. London Math. Soc.* **25** (1993), 431–437.

[Wit38] E. Witt, Über Steinersche Systeme, *Abh. Math. Sem. Univ. Hamburg* **12** (1938), 265–275.

[Yan76] A. Yanushka, Generalized hexagons of order (t, t), *Israel J. Math.* **23** (1976), 309–324.

[Yos91] S. Yoshiara, A classification of flag-transitive classical $c.C_2$-geometries by means of generators and relations, *Europ. J. Combin.* **12** (1991), 159–181.

[Yos92] S. Yoshiara, Embeddings of flag-transitive classical locally polar geometries of rank 3, *Geom. Dedic.* **43** (1992), 121–165.

[Yos94] S. Yoshiara, On some extended dual polar spaces I, *Europ. J. Combin.* **15** (1994), 73–86.

[Yos96] S. Yoshiara, The flag-transitive C_3-geometries of finite order, *J. Algebraic Combin.* **5** (1996), 251–284.

Index